Frontiers on Sustainable Food Packaging

Frontiers on Sustainable Food Packaging

Editors

Theodoros Varzakas
Rui M.S. Cruz

MDPI • Basel • Beijing • Wuhan • Barcelona • Belgrade • Manchester • Tokyo • Cluj • Tianjin

Editors
Theodoros Varzakas
Food Science and Technology
University of the Peloponnese
Kalamata
Greece

Rui M.S. Cruz
Department of Food
Engineering & Institute of
Engineering
Universidade do Algarve
Campus da Penha
Faro
Portugal

Editorial Office
MDPI
St. Alban-Anlage 66
4052 Basel, Switzerland

This is a reprint of articles from the Special Issue published online in the open access journal *Foods* (ISSN 2304-8158) (available at: www.mdpi.com/journal/foods/special_issues/frontiers_sustainable_food_packaging).

For citation purposes, cite each article independently as indicated on the article page online and as indicated below:

LastName, A.A.; LastName, B.B.; LastName, C.C. Article Title. *Journal Name* **Year**, *Volume Number*, Page Range.

ISBN 978-3-0365-6567-5 (Hbk)
ISBN 978-3-0365-6566-8 (PDF)

Contents

About the Editors

Theodoros Varzakas

Theodoros Varzakas has a Bachelor's (Honours) in Microbiology and Biochemistry (1992), a Ph.D. in Food Science and Technology, and an MBA in Food and Agricultural Management from Reading University, UK (1998). He has also worked as postdoctoral research staff at the same university. He worked in large pharmaceutical and multinational food companies in Greece for 5 years and has at least 22 years of experience in the public sector. Since 2005, he has served as Assistant, Associate, and Full Professors at the Department of Food Science and Technology, University of Peloponnese, ex Technological Educational Institute of Peloponnese, Greece, specializing in issues of food technology, food processing/engineering, and food quality and safety. He is also a Section Editor-in-Chief (Food Security and Sustainability) for the journal *Foods* (2020–), was an ex Editor-in-Chief for *Current Research in Nutrition and Food Science* (2015–2019), and is a reviewer and member of the editorial board in many international journals. He has written more than 250 research papers and chapters in books and has presented more than 160 papers and posters at national and international conferences. He has written and edited 15 books in Greek and 11 in English on sweeteners, biosensors, food engineering, food processing, chemometrics, and authenticity published by CRC. He has participated in many European and national research programs as a coordinator or scientific member. His work has been cited in over 4000 citations and has an h-index of 31. He is also an ex EFSA advisor/scientific expert on a panel about biological hazards; a visiting professor at Ghent University global campus, South Korea; and a Research Fellow at the University Technology Malaysia. According to an update on data from September 2022 for "Updated science-wide author databases of standardized citation indicators", Varzakas Theodoros is in the top 2% of citations worldwide.

Rui M.S. Cruz

Rui M. S. Cruz has a Bachelor's in Food Engineering (2001) and a Ph.D. in Biotechnology, with expertise in Food Science and Engineering, from the Portuguese Catholic University (2009). He is currently an Adjunct Professor at the Department of Food Engineering, ISE, at the University of Algarve and a researcher at MED (Mediterranean Institute for Agriculture, Environment and Development) and CHANGE (Global Change and Sustainability Institute). His area of interest is the application of emerging and sustainable technologies in food preservation, particularly in the development of new biodegradable packaging systems and the application of new technologies in food processing. He has collaborated with several food companies in the development of new products and the application of processes, as well as in the characterization and determination of the shelf-lives of different food products. He has evaluated and participated in several research projects, written more than 30 research papers and chapters in books, and presented several posters and oral presentations at national and international conferences. He has also edited five international books in food processing, food analysis, and food packaging. He has reviewed more than 2500 research papers in the area of Food Science and Technology. He is also a member of the editorial board in international journals such as *Heliyon, Journal of Food Processing and Preservation, All Life, Applied Food Research, Frontiers in Food Science and Technology, Open Agriculture, International Journal of Food Studies,* and *LWT—Food Science and Technology.*

Preface to "Frontiers on Sustainable Food Packaging"

The implementation of sustainable food-packaging solutions within future circular food supply chains is essential today to protect customers and to ensure food quality, safety, and optimal shelf-life. This will be improved by new innovative packaging materials and a reduction in food waste. In this vein, it is important to employ lifecycle assessments (LCAs) to define the impacts of the food supply chain, taking into consideration food waste, the global food industry's environmental impacts, and shipping distances, with the final target being to achieve consumer satisfaction.

The purpose of this Special Issue was to attract papers in the new era of sustainability and food packaging that address all major sectors of the food industry from cereals and confectionery to fruits and vegetables, meat, and dairy products.

It is important to share data on (i) the consequences of specific food product–package interactions, (ii) the utilization of novel packaging biomaterials, and (iii) overall consumer behavior and satisfaction as a critical focus.

Theodoros Varzakas and Rui M.S. Cruz
Editors

Editorial

Frontiers on Sustainable Food Packaging

Rui M. S. Cruz [1,2,]* and **Theodoros Varzakas** [3,]*

1. Department of Food Engineering, Institute of Engineering, Campus da Penha, Universidade do Algarve, 8005-139 Faro, Portugal
2. MED—Mediterranean Institute for Agriculture, Environment and Development and CHANGE—Global Change and Sustainability Institute, Faculty of Sciences and Technology, Campus de Gambelas, Universidade do Algarve, 8005-139 Faro, Portugal
3. Department of Food Science and Technology, University of Peloponnese, 24100 Kalamata, Greece
* Correspondence: rcruz@ualg.pt (R.M.S.C.); t.varzakas@uop.gr (T.V.)

Citation: Cruz, R.M.S.; Varzakas, T. Frontiers on Sustainable Food Packaging. *Foods* **2023**, *12*, 349. https://doi.org/10.3390/foods12020349

Received: 30 December 2022
Accepted: 7 January 2023
Published: 11 January 2023

The implementation of sustainable food packaging solutions within future circular food supply chains is essential to protect customers and ensure food quality, safety, and optimal shelf-life. This will be improved by new innovative packaging materials and will contribute to reducing food waste. In this direction, it is important to employ lifecycle assessment (LCA) to define food supply chain impacts, taking into consideration food waste, global food industry environmental impacts, and shipping distances, with the aim of achieving consumer satisfaction. It is important to share data on (i) the consequences of specific food product–package interactions, (ii) the consideration of the utilization of novel packaging biomaterials, and (iii) overall consumer behavior and satisfaction as a critical focus. The aim of this Special Issue was to bring the most updated information in the new era of sustainability and food packaging.

Dörnyei et al. [1] proposed a literature-based attribute-cue matrix as a tool for analyzing packaging solutions. Using a 2021 snapshot of the wafer market in nine European countries, the study demonstrated the tool's utility by analyzing the cues found that signal environmentally friendly packaging attributes. Although the literature suggests that environmentally friendly packaging is increasingly used by manufacturers, the analysis of 164 wafer packages showed that communication is very limited except for information related to recyclability and disposal.

The work of Wang et al. [2] presented a supply chain traceability system framework based on blockchain and radio frequency identification (RFID) technology. The system consisted of a decentralized blockchain-enabled data storage platform for data management and an RFID system at the packaging level for data collection and storage. The new traceability system has the potential to simplify the tracking of products and can be scaled for industrial use.

The study of Shin et al. [3] showed the effects of chitosan and duck fat-based emulsion coatings on the quality characteristics and microbial stability of chicken meat during refrigerated storage. The results suggested that chitosan/duck fat-based edible coatings can be used to maintain the quality of raw chicken meat during refrigeration.

Pleva et al. [4] investigated biofilm formation on selected biodegradable polymer films involving selected bacterial strains isolated from dairy products. The antibacterial properties of the films were enhanced with thymol and eugenol. The results showed that these films can be used to prepare novel active food packaging for the dairy industry to prevent biofilm formation and enhance food quality and safety in the future.

Chen et al. [5] developed an edible starch-based film for packaging seasonings in instant noodles. The results showed that the developed starch-based film meets the general requirements of the flavor bag packaging used in instant noodles. Thus, the developed edible film can quickly dissolve into hot water so that the seasoning bag can mix into the soup of instant noodles during preparation.

López-Gálvez et al. [6] assessed the potential cross-contamination of fresh cauliflowers with *Salmonella enterica* via different contact materials (polypropylene from reusable plastic crates (RPCs), corrugated cardboard, and medium-density fiberboard (MDF) from wooden boxes). The survival of the pathogenic microorganism was studied in cauliflowers and the contact materials during storage. The LCA approach was used to evaluate the environmental impact of produce-handling containers fabricated from the different food-contact materials tested. The results showed a higher risk of cross-contamination via polypropylene compared with cardboard and MDF. Another outcome of the study was the potential of *Salmonella* surviving both in cross-contaminated produce and in contact materials under supply chain conditions. Regarding environmental sustainability, RPCs showed a lower environmental impact than single-use containers (cardboard and wooden boxes).

Cruz et al. [7] presented the environmental impact, trends, and regulatory aspects of bioplastics for food packaging. This review showed that further research is needed to improve the production of bioplastics and their potential applications according to different properties, mechanisms of biodegradation, environmental impacts, markets, and how consumers perceive bioplastics.

In the article of Miller et al. [8] various physical, chemical, and biochemical modifications of potato constituents were identified, and the resulting structural and property changes were presented. The review provided an up-to-date and comprehensive overview of the possibilities and implications of modifying potato components for potential further valorization, particularly in bio-based food packaging.

The review from Krauter et al. [9] contextualized packaging, sustainability, and related LCA methods. They displayed and discussed how and to what extent food packaging is included in existing LCAs in the cereal and confectionary sector, pointed out the environmental impact of cereal and confectionary packaging in relation to food products with a special focus on GHG emissions, and highlighted improvement strategies to optimize (cereal and confectionary) packaging systems, as well as an LCA of the same. The results revealed that only a few studies sufficiently include (primary, secondary, and tertiary) packaging in LCAs, and when they do, the focus is mainly on their direct (e.g., the material used) rather than indirect environmental impacts (e.g., food losses and waste).

Bauer et al.'s [10] study aimed at building a comprehensive basis for future sustainable packaging development activities in the area of cereal and confectionary by presenting relevant information on the functions and properties of packaging materials. They detailed product group-specific decay mechanisms and frequently used packaging solutions and highlighted packaging-related shelf-life extension technologies.

In another study, Bauer et al. [11] presented the benefits of multilayer flexible food packaging and showed its negative recyclability trade-offs, especially for food technologists. The review showed that the substitution of non-recyclable flexible barrier packaging is challenging because only a limited number of barriers are available. In the worst case, the restriction on material choice can result in a higher environmental burden through shortened food shelf-life and increased packaging weights.

Junior et al. [12] presented the latest trends in sustainable polymeric food packaging films. This review showed development and advances in bio-based and functional food packaging produced by conventional methodologies and by 3D printing, as well as advances in bio-based alternative feedstock for 3D printing with potential applications in the food packaging area.

Funding: This research received no external funding.

Conflicts of Interest: The authors declare no conflict of interest.

References

1. Dörnyei, K.R.; Bauer, A.-S.; Krauter, V.; Herbes, C. (Not) Communicating the Environmental Friendliness of Food Packaging to Consumers—An Attribute- and Cue-Based Concept and Its Application. *Foods* **2022**, *11*, 1371. [CrossRef] [PubMed]
2. Wang, L.; He, Y.; Wu, Z. Design of a Blockchain-Enabled Traceability System Framework for Food Supply Chains. *Foods* **2022**, *11*, 744. [CrossRef] [PubMed]
3. Shin, D.-M.; Kim, Y.-J.; Yune, J.-H.; Kim, D.-H.; Kwon, H.-C.; Sohn, H.; Han, S.-G.; Han, J.-H.; Lim, S.-J.; Han, S.-G. Effects of Chitosan and Duck Fat-Based Emulsion Coatings on the Quality Characteristics of Chicken Meat during Storage. *Foods* **2022**, *11*, 245. [CrossRef] [PubMed]
4. Pleva, P.; Bartošová, L.; Máčalová, D.; Zálešáková, L.; Sedlaříková, J.; Janalíková, M. Biofilm Formation Reduction by Eugenol and Thymol on Biodegradable Food Packaging Material. *Foods* **2022**, *11*, 2. [CrossRef] [PubMed]
5. Chen, H.; Alee, M.; Chen, Y.; Zhou, Y.; Yang, M.; Ali, A.; Liu, H.; Chen, L.; Yu, L. Developing Edible Starch Film Used for Packaging Seasonings in Instant Noodles. *Foods* **2021**, *10*, 3105. [CrossRef] [PubMed]
6. López-Gálvez, F.; Rasines, L.; Conesa, E.; Gómez, P.A.; Artés-Hernández, F.; Aguayo, E. Reusable Plastic Crates (RPCs) for Fresh Produce (Case Study on Cauliflowers): Sustainable Packaging but Potential *Salmonella* Survival and Risk of Cross-Contamination. *Foods* **2021**, *10*, 1254. [CrossRef] [PubMed]
7. Cruz, R.M.S.; Krauter, V.; Krauter, S.; Agriopoulou, S.; Weinrich, R.; Herbes, C.; Scholten, P.B.V.; Uysal-Unalan, I.; Sogut, E.; Kopacic, S.; et al. Bioplastics for Food Packaging: Environmental Impact, Trends and Regulatory Aspects. *Foods* **2022**, *11*, 3087. [CrossRef]
8. Miller, K.; Reichert, C.L.; Schmid, M.; Loeffler, M. Physical, Chemical and Biochemical Modification Approaches of Potato (Peel) Constituents for Bio-Based Food Packaging Concepts: A Review. *Foods* **2022**, *11*, 2927. [CrossRef] [PubMed]
9. Krauter, V.; Bauer, A.-S.; Milousi, M.; Dörnyei, K.R.; Ganczewski, G.; Leppik, K.; Krepil, J.; Varzakas, T. Cereal and Confectionary Packaging: Assessment of Sustainability and Environmental Impact with a Special Focus on Greenhouse Gas Emissions. *Foods* **2022**, *11*, 1347. [CrossRef]
10. Bauer, A.-S.; Leppik, K.; Galić, K.; Anestopoulos, I.; Panayiotidis, M.I.; Agriopoulou, S.; Milousi, M.; Uysal-Unalan, I.; Varzakas, T.; Krauter, V. Cereal and Confectionary Packaging: Background, Application and Shelf-Life Extension. *Foods* **2022**, *11*, 697. [CrossRef] [PubMed]
11. Bauer, A.-S.; Tacker, M.; Uysal-Unalan, I.; Cruz, R.M.S.; Varzakas, T.; Krauter, V. Recyclability and Redesign Challenges in Multilayer Flexible Food Packaging—A Review. *Foods* **2021**, *10*, 2702. [CrossRef] [PubMed]
12. Junior, E.G.S.S.; Cardoso, S.; Bettencourt, A.F.; Ribeiro, I.A.C. Latest Trends in Sustainable Polymeric Food Packaging Films. *Foods* **2023**, *12*, 168. [CrossRef]

Review

Latest Trends in Sustainable Polymeric Food Packaging Films

Edilson G. S. Silva, Sara Cardoso, Ana F. Bettencourt and Isabel A. C. Ribeiro *

Research Institute for Medicines (iMed.ULisboa), Faculty of Pharmacy, Universidade de Lisboa, Avenida Prof. Gama Pinto, 1649-003 Lisboa, Portugal
* Correspondence: iribeiro@ff.ulisboa.pt; Tel.: +351-217946400; Fax: +351-217946470

Abstract: Food packaging is the best way to protect food while it moves along the entire supply chain to the consumer. However, conventional food packaging poses some problems related to food wastage and excessive plastic production. Considering this, the aim of this work was to examine recent findings related to bio-based alternative food packaging films by means of conventional methodologies and additive manufacturing technologies, such as 3D printing (3D-P), with potential to replace conventional petroleum-based food packaging. Based on the findings, progress in the development of bio-based packaging films, biopolymer-based feedstocks for 3D-P, and innovative food packaging materials produced by this technology was identified. However, the lack of studies suggests that 3D-P has not been well-explored in this field. Nonetheless, it is probable that in the future this technology will be more widely employed in the food packaging field, which could lead to a reduction in plastic production as well as safer food consumption.

Keywords: food packaging; 3D printing; plastic; biopolymer; films

Citation: Silva, E.G.S.; Cardoso, S.; Bettencourt, A.F.; Ribeiro, I.A.C. Latest Trends in Sustainable Polymeric Food Packaging Films. *Foods* **2023**, *12*, 168. https://doi.org/10.3390/foods12010168

Academic Editors: Theodoros Varzakas and Rui M. S. Cruz

Received: 25 November 2022
Revised: 19 December 2022
Accepted: 20 December 2022
Published: 29 December 2022

1. Introduction

Nowadays, food packaging is fundamental to ensuring food distribution and protection around the world, especially when considering the solid growth of the population. Without packaging, food would easily spoil and the distribution of enormous quantities of food, raw and processed, to different areas around the globe would hardly be possible [1,2]. Among its functions, food packaging protects food from contamination and physical damage, maintains its freshness, improves its shelf-life, and gives relevant information about its contents [1,2]. Despite its effectiveness, conventional food packaging poses some concerns, such as food spoilage, since plastic itself has no effect on microorganism contamination, as well as excessive production of fossil-based plastic—this sector being one of those that employ this type of material most heavily, food packaging representing more than 40% of total plastic production [3]. Since fossil-based plastic is inherently non-renewable and non-biodegradable and its production has been massively increasing in the last seven decades (from 2 million tons in 1950 to 367 million tons in 2020) [3,4], new alternative materials for the manufacture of food packaging have been sought.

One emerging set of alternatives that have been studied as potential solutions to the above-mentioned problems are bio-based food packaging films functionalized with compounds of natural origin, since they are characterized by both biodegradability and renewability, in addition to active and/or intelligent functions [5–7]. In the production of these bio-based films, different methodologies may be employed depending on the purpose in question, the most common methods being solvent casting, layer-by-layer assembly, and extrusion. Moreover, additive manufacturing technologies, also known as "3D printing" have been, although scarcely, used in the production of bio-based films and other types of bio-based packaging. Considering the importance of food packaging, its current drawbacks, and the potentialities of 3D printing, the objective of this review is to explore progress in the production of bio-based films by conventional methodologies and additive manufacturing and to investigate how this technology can contribute for the development of bio-based sustainable primary food packaging. For this, the literature was examined to find recent research on developments and advances in bio-based and functional food packaging produced by conventional methodologies and by 3D printing, as well as advances in bio-based alternative feedstocks for 3D printing with potential application in the food packaging area. Furthermore, perspectives on and limitations of additive manufacturing applied in the production and development of bio-based primary food packaging are presented in order to try to understand why this innovative and powerful technology has been barely explored in this field.

2. Novel Materials for Food Packaging Films: Biopolymers and Additives

As alternatives to conventional plastic-based packaging, bio-based polymer films have been explored as potential candidates for the development of food packaging. Regarding the advantages over conventional petroleum-based food packaging, these bio-based films can decrease carbon dioxide levels, do not release dangerous substances into the environment, can be degraded by naturally occurring bacteria, can reduce the amount of waste generated, and are non-toxic [8,9]. Furthermore, the main materials used in the preparation of these films, that is, the biopolymers, are generally abundant in nature and can be derived from plenty of sources, namely, microorganisms, plants, animals, and food/agricultural wastes. Examples of biopolymers used in food packaging applications are summarized in Figure 1.

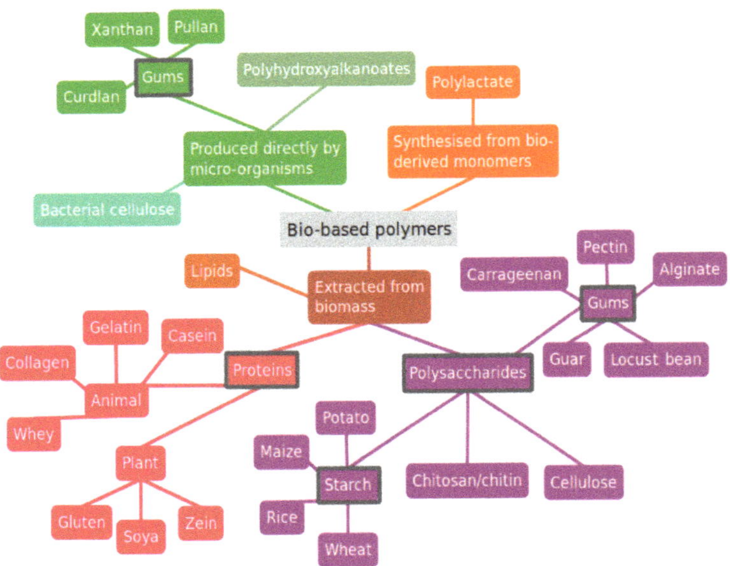

Figure 1. Bio-based polymers used in food packaging applications.

In addition to biopolymers for bio-based films, there is an increasing interest in compounds of natural origin, such as extracts and essential oils, as additives in the production of food packaging. When these natural compounds are incorporated into polymeric films, they can provide active properties, such as antioxidant, antimicrobial, and scavenging properties, that are crucial for food packaging, due to food deterioration and microbial contamination, which can produce off-flavors, lead to food spoilage, and cause food-borne diseases [7,10,11]. The active substances present in these compounds differ in composition and quantity and can be divided according to their structures and modes of action: phenols and phenolic acids, quinones, flavones and flavonoids, tannins, coumarins, terpenoids, alkaloids, and lectins and polypeptides [12]. In terms of antimicrobial activity, the modes of action of these substances can vary and include leakage from the bacterial cell, cell-shape damage, the destruction of cell walls, and alterations to membrane composition, among other mechanisms [12,13]. In addition to active properties, these compounds have also been used to provide extra and/or intelligent properties in films, such as sensitivity to pH changes; improved mechanical, thermal, optical, and barrier properties; and sensing abilities, among others [7]. In sustainable food packaging, the incorporation of these compounds can lead to the development of active materials that, in addition to being biodegradable/sustainable, can also increase food shelf life, reduce microbial contamination, and give information on food freshness [10]. Many natural compounds, such as natural extracts and essential oils, have been studied for the above-mentioned properties. Some examples of natural products that have been employed as sources of antioxidant and/or antibacterial compounds in bio-based food packaging include extracts of cranberries, cabbage, amaranth leaves, rosemary, cinnamon, broccoli, kale, and others, as presented [14–19].

3. Main Methods Used in the Production of Bio-Based Films

In the production of bio-based films, the methodology employed depends on the film application and on the objectives.

In brief, the fundamental step in processing any biopolymer film involves solubilizing and/or melting a biopolymer mixture, which is followed by the implementation of the desired technique [20]. Some of the most common methods for the production and application of bio-based films include the solvent casting method, layer-by-layer assembly, and coating and extrusion methods, which are briefly described in Table 1 [21,22]. Some of these methods are limited to lab scale, while others can be scaled up for industrial settings.

Table 1. Main methods for preparation of biopolymer-based films (Kumar et al., 2020; Wang et al., 2018).

Method	Main Characteristics
Solution casting	The film-forming solution is cast on a surface (e.g., a Petri dish), appropriately dried, and the formed film is peeled off; It is the simplest method for film preparation; The conditions used are relatively mild; It is time-consuming; It is limited to lab scale;
Coating	The film-forming solution is directly applied onto the food by means of dipping, spraying, or brushing and dried afterwards in appropriate conditions; Often applied on fresh food; Materials must be of food grade if the coating is meant to be eaten; Some applications in the food industry (mostly wax coatings);
Layer-by-layer assembly	Based on the deposition of alternating layers; Deposition can be achieved either by submersion in or spraying the film-forming solutions on the food; Potential for industrial applications, though currently it is mostly limited to lab scale;
Extrusion	The mixture containing the biopolymer is poured into an extruder system, which produces a uniform film at the end of the process; Faster and less energy-demanding than the solution casting method; Produces films with superior mechanical and thermal properties; Conditions may be aggressive for biopolymers; Can be scaled up for industrial settings;

Solution casting is the simplest and most reported method in the literature on the production of bio-based films. The method consists of preparing a film-forming solution with an appropriate polymer–solvent concentration and casting it on a surface (e.g., a Petri dish, a glass plate) according to the desired thickness and uniformity of the films [22]. The drying conditions can vary and can merely involve drying the solution at room temperature or in an oven at a high temperature, with or without an auxiliary air system [23]. When the solvent evaporates, the film can be peeled off from the surface. Due to its simplicity and the mild conditions involved, solution casting is the method of choice in laboratory and scale-up experiments, but it is not practical at an industrial scale.

As an alternative, a film-forming solution can also be applied directly onto a food surface. This methodology, known as coating, can vary according to the nature of the food, the coating objective, and the film-solution viscosity [24,25].

Among the existing methods for coating are dipping, spraying, and brushing. The dipping method is based on the immersion of food in a film-forming solution and is suitable for viscous solutions. Spraying is based on the diffusion of film-forming-solution droplets through a spraying tool and is most suitable for less viscous solutions [24]. In brushing, the solution is applied on food using a brush or a similar tool. As in the other methods, the food is properly dried after the coating [25]. As it is a requirement that it be consumable, the composition of the coating must be of food grade; thus, the food can be eaten along with the coating.

The layer-by-layer assembly (LBL) approach is based on the deposition of alternating layers of oppositely charged compounds. This method allows for working with a variety of molecules, such as polysaccharides, nucleic acids, proteins, carbohydrates, synthetic polymers, and others. The formation of the multilayers can vary and be achieved by different methods, such as casting different layers of a solution on a surface, spray-coating layers directly onto the substrate's surface, and dipping or immersing the substrate in different solutions [22,26]. Among the mentioned methods, dipping is advantageous in that it is not subject to geometrical restrictions, and, for that reason, it is one of the most frequently employed methods.

In extrusion methods, polymers are mixed and extruded at high pressures and temperatures. In brief, a model of an extruder can be described as being composed of a hopper, which is where the raw materials enter the system; a barrel containing one or two rotating screws; and a die, where the polymer mixture leaves the equipment with the desired shape [27]. The rotating screws mix and transport the raw materials to the barrel, where the temperature is high and the polymer is melted, allowing for the incorporation of additives or other substances. An initial extrusion is conducted to produce pellets with the desired formulation, and these then pass through a second extrusion process to produce the films [22,28]. Figure 2 summarizes all four described methods.

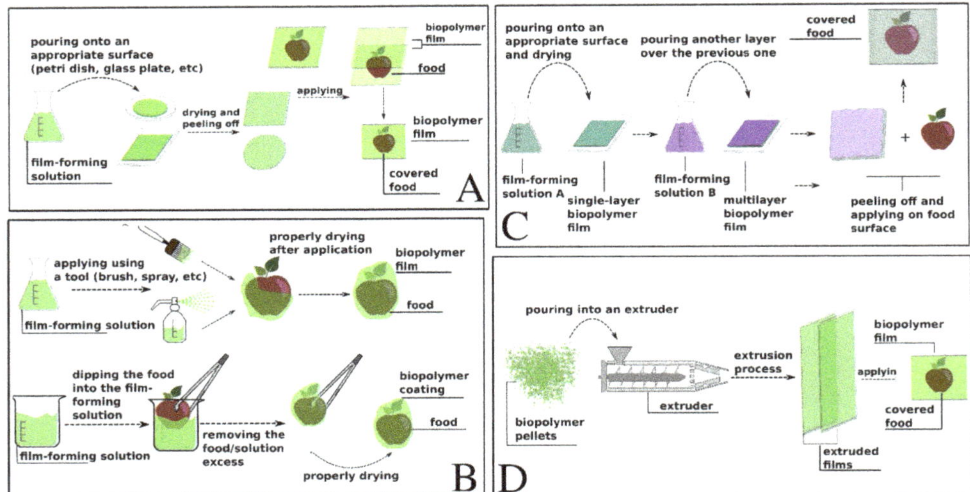

Figure 2. Conventional methods for the production of bio-based films. (**A**) Solution casting. (**B**) Coating. (**C**) Layer-by-layer assembly. (**D**) Extrusion.

Presently, there are many studies focusing on the use of different biopolymers functionalized with active compounds that can be obtained by different processes, as exemplified in Table 2. Nevertheless, there is still room for innovation. For example, an emerging methodology in the field of food packaging is additive manufacturing, which will be discussed in the following section.

Table 2. Advances in the development of bio-based food packaging films.

Methodology	Aim	Used Biopolymers	Other Components	Properties	Ref.
Casting	Effect of chitosan molecular weight on film performance	Chitosan	-	Improved preservation abilities; Improved performance with high chitosan weights/contents	[29]
	Influence of chitosan molecular weight on the film properties	Chitosan + bacterial cellulose	Curcumin	Improved performance with high chitosan weights/contents	[30]
	Develop a multifunctional food packaging film	Chitosan	Alizarin	pH-responsive film (4–10 range); Improved thermal stability, hydrophobicity, and UV-blocking properties	[31]
	Effect of a starch source on the performance of edible starch-based films	Starch (tapioca, rice, potato, and wheat)	-	Tapioca, potato, and rice starch had better mechanical strength and less color difference	[32]
	Production of biodegradable cellulose/alginate films	Cellulose, alginate, and carrageenan	-	Films showed weight loss of up to 50% after 60 days buried in soil; Activity against *E. coli*, *Pseudomonas syringae*, and *S. aureus* strains	[33]
	Develop UV absorbent films	Polylactic acid	Grape syrup	High UV absorption property	[34]
	Develop functional bio-hydrogel films for food packaging	Alginate, agar, and collagen	Grapefruit seed extract Silver NPs	Improved mechanical properties; High UV screening; Strong antimicrobial activity: prevents greening of fresh potatoes	[35]
	Study the influence of nano-SiO$_2$ concentration on the properties of the films	Agar + alginate	Silicon oxide NPs	Improved mechanical properties	[36]
	Evaluated the effect of fatty acid chain length on the properties of edible films	Basil seed gum-based	Caprylic, lauric, and palmitic acids	Improved barrier properties; Improved mechanical properties (lauric and caprylic acids)	[37]
	Develop intelligent films for food packaging	Gellan gum and soy protein	*Clitoria ternatea* extract	pH-responsive (3–11 range); Bacteriostatic activity	[38]
	Evaluate the influence of the concentration of the extract on the properties of films	Starch	Red cabbage extract	Antioxidant activity; Significantly increased food shelf life (meat)	[17]
	Develop a composite film and evaluate its activity as primary food packaging for fresh poultry	Chitosan	Rosemary essential oil and montmo-rillonite	Antioxidant and antimicrobial activity; Improved barrier properties; Films were able to retard lipid peroxidation (poultry)	[19]
	Develop a film with mesoporous silica NPs loaded with clove essential oil	Polylactic acid	Clove essential oil and silica NPs	Antimicrobial activity; Controlled the release of the active compound	[16]

Table 2. Cont.

Methodology	Aim	Used Biopolymers	Other Components	Properties	Ref.
Coating and solvent casting	Develop a film-forming formulation and compare the effects of different applications (coating, wrapping, and direct application of active compounds) on food	Alginate and cellulose	*Ziziphora* essential oil, apple peel extract, and zinc oxide nanoparticles	Coating showed the lowest bacterial population and best sensory attributes among the studied methodologies	[39]
	Investigate the potential of cranberry extract as an antibiofilm additive for a chitosan-based film	Chitosan	Cranberry extract	Antioxidant and antimicrobial activities	[18]
Coating	Develop a superhydrophobic food-grade coating	-	Candelilla and rice bran waxes	Highly hydrophobic coating; Excellent coating resistance to physical damage	[40]
LBL and coating	Compare the effects of coatings by means of LBL and standard coating	Chitosan; Cellulose	-	Single-layer and LBL coatings had positive effects on strawberry conservation; LBL coating showed better performance at reducing firmness and volatile compound loss	[41]
LBL	Development of a bilayered chitosan/FucoPo film	Chitosan and FucoPol	-	Improved gas barrier towards O_2 and CO_2 in comparison with monolayer film	[42]
	Prepare and characterize an antibacterial film	Chitosan and modified polyethylene	Hyaluronic acid	Excellent antibacterial activity; Improved degradability	[43]
LBL and solution casting	Evaluate the effects of preparation methods on the properties of the films	Chitosan and alginate	Ferulic acid	Crosslinked LBL films showed better results with improved mechanical, thermal, optical, and barrier properties	[44]
Extrusion	Study the effect of nanofillers on extruded films	Chitosan and starch	Nanoclay and bamboo fibers	Improved mechanical, thermal, and barrier properties	[45]
	Investigate the effects of nanoclay contents and pH levels on the properties of the films	Soy protein	Nanoclay	Nanoclay addition improved mechanical and rheological properties; pH changes demonstrated to have positive effects on film properties	[46]
	Evaluated the potential of the extrusion process and wax source on edible film properties	Rennet casein	Potassium sorbate; bee, candelilla, and carnauba waxes;	Beeswax had the best performance in terms of improving mechanical properties and hydrophobicity; Wax incorporation allowed a controlled release of potassium sorbate	[47]
	Develop a composite film and evaluate its activity as primary food packaging for fresh minced meat	Starch	Sappan and cinnamon herbal extracts	Improved barrier properties; Reduced microbial counts; Preservation of redness of packaged meat	[15]

4. Three-Dimensional Printing of Food Packaging and Films

Three-dimensional printing or additive manufacturing (AM) is a relatively new technology that has been revolutionizing a range of industries, research, and the overall manufacturing of new products because of its advantages, such as the reduction of manufacturing times, the possibility of producing complex shapes and parts, and the potential for innovation, and it has also been, although scarcely, used as a means to develop bio-based packaging materials. With this set of technologies, solid models are fabricated through the layer-by-layer deposition of raw materials, followed by their solidification, and it is possible to work with powder-based, liquid-, and solid-state feedstocks, depending on the chosen technology [48–51].

In general, the 3D printing process can be described as a sequence of steps, the first one being the generation of a computer-aided design of the desired object, followed by its conversion into a 3D object file, which will be read by the slicing software and built on the platform afterwards [49]. The principle of operation and type of 3D printing can vary depending on the application. In total, there are seven standardized processes (or techniques) that 3D printing is based on: binder jetting, directed energy deposition, material extrusion, material jetting, powder bed fusion, sheet lamination, and vat photopolymerization [50,52]. These processes differ in terms of the type and the state of the raw material used, in the degree of detail of the printed object, and in the fundamentals behind the printing process. According to Zhou et al., among the cited methods, material jetting, powder bed fusion, vat photopolymerization, and material extrusion are the most suitable techniques for printing 3D objects made of soft materials, such as polymers, and they will be briefly summarized below (see also Figure 2) [52].

4.1. Vat Photopolymerization

In vat photopolymerization, the 3D object is created by the solidification of a photopolymer resin when it is hit by light. In this process, a liquid photo-reactive polymer, which is contained in a vessel, is selectively cured by a UV light coming from a light source, forming a thin-layer cured polymer as a result [50]. The main 3D printing techniques that are based on this principle are stereolithography (SLA) and digital light processing (DLP), mainly differing in the sources of light [52]. The main advantages of vat photopolymerization are the high degree of accuracy and the smooth surface of the produced 3D object. Drawbacks include the need to use supports during the printing process and the inherent natures of the photopolymers employed as raw materials, such as their physical fragility and susceptibility to sunlight, which limit the range of applications of these products and make them less durable.

4.2. Material Jetting

Similar to vat photopolymerization, in material jetting, the object is formed by the solidification of a photo-sensible resin, but unlike the previous technique, this method is based on the deposition of tiny droplets of the photopolymer resin on the build platform, followed by their solidification by ultraviolet light [50]. This technique is regarded as the most accurate 3D printing technique and can produce objects with smooth surfaces and high degrees of detail. Analogous to vat photopolymerization, the main drawbacks of material jetting are related to the intrinsic properties of the raw materials, including the poor mechanical properties and the susceptibility to sunlight of the produced objects [49].

4.3. Powder Bed Fusion

The powder bed fusion process is based on the fusion of a powder-based material by a laser or an electron beam [50]. In this mode of operation, a thin layer of powder (e.g., a metal, ceramic, polymer, or composite) is distributed on the build platform and a laser automatically fuses layers of the material. This technology includes three printing techniques: electron beam melting (EBM), selective laser sintering (SLS), and selective heat sintering (SHS). The EBM and SHS techniques are mainly employed with metals, whereas SLS is employed for polymer materials (Redwood et al., 2017). The resolution of SLS is inversely proportional to the particle size, and it is preferable to use low-thermal-conductivity polymers as raw materials due to their stability in the fusing step [52]. Among its advantages, SLS produces objects with isotropic natures, making them stronger and more resistant than other printing technologies, such as FDM. In addition, SLS has a high degree of accuracy and, unlike vat photopolymerization, does not require extra supports to build objects [49].

4.4. Material Extrusion

Material extrusion is one of the most widely used 3D printing processes. The principles of this technique can be divided into two main groups based on whether the raw material is melted or not [53]. The technology based on the melting of a material is known as fused filament deposition modeling (FDM) and uses thermoplastics in the form of thin filaments as raw materials [50,52]. Another technology is direct ink writing (DIW), which is based on the extrusion of viscoelastic materials by means of pneumatic (air or pressure) or mechanical (screw- and piston-based) action, followed by the curing of the extruded material using photopolymerization or thermal processes [52,53]. Among the cited techniques, FDM is the most common method used in 3D printing. In FDM, a solid filament is extruded through a heated nozzle, melted, and selectively deposited on the build platform where it solidifies, forming a layer of the object. The advantages of this technique include the low costs of the materials and machines, the easy mode of operation, and a broad range of workable materials [48].

The main limitation of extrusion-based 3D printing is related to the anisotropic nature of the produced objects, that is, the fragility of objects in one of their directions. The rheological and thermal properties of the material employed are also critical and depend on the nature of the extrusion process. Additionally, as is often the case with other AM technologies, it is likely that the final object will require some post-treatment to remove undesirable layer lines, the formation of which is inherent to the layer-by-layer building process [48,49]. Representations of the four described additive manufacturing techniques are presented in Figure 3.

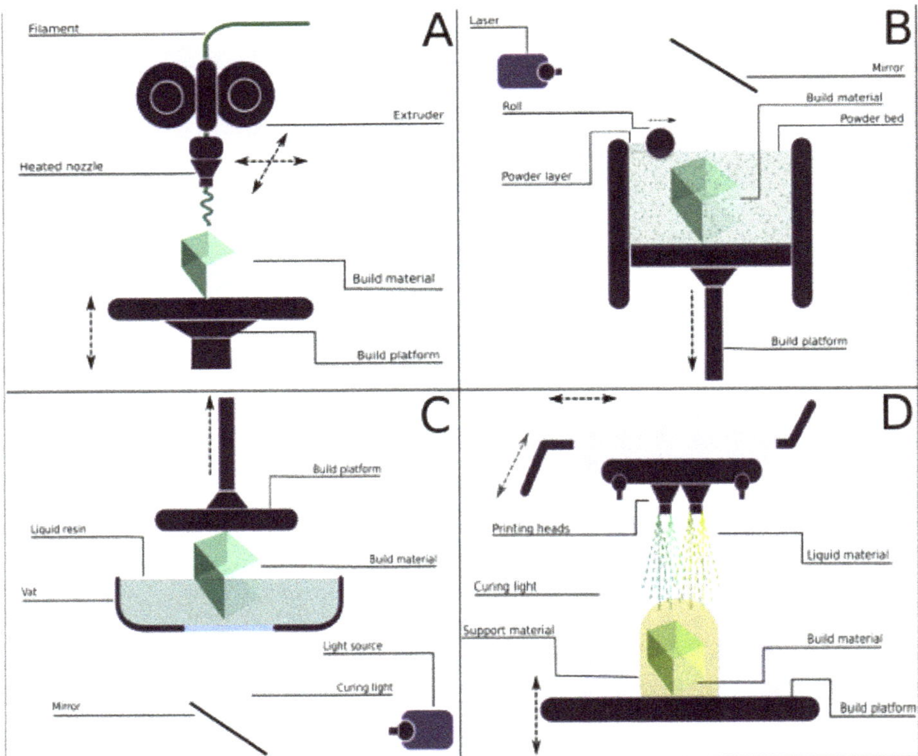

Figure 3. Three-dimensional printing techniques suitable for working with polymers. (**A**) Material extrusion. (**B**) Powder bed fusion. (**C**) Vat photopolymerization. (**D**) Material jetting.

5. Perspectives on AM in the Production of Bio-Based Films

As previously discussed, there are different additive manufacturing technologies that are compatible with polymers. However, not all of these technologies seem appropriate for applications in the food industry, especially for food processing and the production of primary food packaging. In the following section, research on the development of bio-based feedstocks for these technologies, as well as applications and/or potentialities in the food area regarding packaging production, will be addressed. Some of the studies presented may not be directly concerned with food packaging or related fields; however, most of the materials employed may be or have already been used in the production of films intended for food packaging.

5.1. Vat Photopolymerization and Material Jetting

One of the main driving factors of the research on bio-based photopolymers for 3D printing is the concern with sustainability issues, since most of the resins used in photo-based 3D printing technologies are derived from fossil resources [54]. Among sustainable alternatives to fossil-based materials are vegetable oils, lignin, chitosan, starch, and many others, which, after functionalization with photo-sensitive groups, such as acrylic or epoxy groups, can form solid shapes when cured by UV light [54].

Among the studies in this area is the work of Ding and coworkers [55], where they produced a high-biorenewable-content blend composed of natural phenolic acrylates, which was further evaluated as a photo-curable resin for SLA 3D printing. The acrylate compounds were synthesized from guaiacol, vanillyl alcohol, and eugenol and printed by a vat-photopolymerization-based 3D printer. The blends were then evaluated by real-time infrared and SEM, tensile strength, and thermal analyses. Based on the results, the researchers found that the produced blend had a high curing rate and a high glass-transition temperature, while the produced prototype showed good thermal and mechanical properties, although a few defects were observed on the printed surface.

Another interesting work in this area was conducted by Kim et al. [56], in which they produced a modified silk fibroin as a bioink for digital light processing intended for bioengineering applications. In their work, silk fibroin, a natural protein produced by silkworms, was functionalized with methacrylate groups, and its printability was evaluated by a DLP 3D printer. The mechanical, rheological, and water-uptake properties of the produced silk fibroin-based (Sil-MA) hydrogel were assessed, and, as a result, the research group found that the mechanical properties, such as compressive strain and compressive stress, increased as the concentration of Sil-MA increased, up to a 30% content of Sil-MA, at which the hydrogel prototype was able to support a 7 kg weight without being deformed after the weight's removal.

Despite the advances in developing bio-based feedstocks for these AM technologies, it is unlikely that they will find application in food packaging fields. Firstly, resin-based AM technologies are reported to produce brittle and UV-sensitive objects, both characteristics inappropriate for food packaging films. In addition, many compounds used to prepare photo-sensitive resins, which are employed in these techniques, are considered toxic to some degree; therefore, due to safety and legal issues, it is unlikely that the produced objects will be suitable for contact with food.

5.2. Powder Bed Fusion

Unlike the above-mentioned techniques, powder bed fusion technology produces objects by means of the fusion of a powder material; therefore, a functionalization step with photo-sensitive groups is not required. Since this technique can employ less chemically modified materials as feedstocks, it is likely that this technology will find more applications with biopolymers in food-related fields than the previous two techniques. By contrast, the thermal properties and particle sizes of powder materials are of great relevance to this methodology. With respect to research on bio-based feedstocks for this technique, most of the published works consulted are concerned with regenerative medicine and similar fields.

In one of these studies, Dechet et al. [57] reported the production of spherical poly(L-lactide) particles for powder bed fusion using a sustainable method. The method, known as liquid–liquid phase separation, involves preparing a polymer solution with a poor solvent at a high temperature and subsequently cooling the solution so that the polymer precipitates and forms microspheres. In this work, triacetin, a green solvent derived from glycerol, was employed to solubilize the polymer. After producing the particles, by SEM analysis, the researchers found that, with increasing polymer concentration, the efficiency of the process increased, producing as a result more spherical particles with greater flowability. The specimens produced by poly(L-lactide) particles via powder bed fusion 3D printing showed good layer adhesion and good mechanical properties, comparable to those produced by the FDM process.

In another recent work, Gayer et al. [58] produced a solvent-free biodegradable PLA/calcium carbonate composite intended for bone-tissue engineering applications. The powder was prepared by processing a mixture of the two compounds in an impact mill, followed by a sieving step to obtain a narrow range of particle sizes. At the end of these processes, four powder mixtures, with calcium carbonate contents ranging from 22% to 27%, were obtained and characterized. The printability of the composite powders was assessed using an SLS 3D printer, and the obtained specimens were evaluated by mechanical strength, cell viability, and porosity assays. The results showed that the composite powder with 23% calcium carbonate content had the best processability, good mechanical strength, low melt viscosity, and small particle size, in addition to good cell compatibility.

In another interesting work using biopolymers and powder bed fusion, Diermann and coworkers [59] produced and evaluated scaffolds made of poly(3-hydroxybutyrate-co-3hydroxyvalerate) (PHBV) and Åkermanite, a sorosilicate mineral, as a filler, in vitro. The scaffold was intended for tissue engineering, taking advantage of PHBV properties, such as slow degradation and compatibility with the components of human blood [60]. For the preparation of the composite powder, PHBV powder was sieved to obtain a narrow particle size distribution and some of the as-received Åkermanite powder was ball-milled to obtain particles at micro- and nanoscales. Both powders were obtained commercially. After these steps, the powders were blended in a mixer for 8 h and sintered in an SLS machine to produce four scaffolds with different PHBV/Åkermanite ratios and different particle sizes. As demonstrated by the authors, the Åkermanite particles were well dispersed throughout the PHBV matrix, and the scaffold with microparticles had the best mechanical performance over the Åkermanite nanoparticles. Additionally, the incorporation of Åkermanite into the blend improved the water uptake of the scaffold—an important property for the intended application [59].

Although powder bed fusion seems more promising for working with bio-based polymers without further chemical modification, in contrast to vat photopolymerization and material jetting, this technique only works with solid-state materials, limiting its versatility in the production of films. In fact, no studies on the production of bio-based films using powder bed fusion were found.

5.3. Material Extrusion

Among the AM technologies studied, material extrusion seems to be the most appropriate for developing bio-based films and other materials for food packaging applications using either filaments or gels. Despite the fact that powder bed fusion technologies use polymers as feedstocks, no studies on the production of films using these materials were found. Other technologies discussed herein, such as vat photopolymerization and material jetting, seem not to be suitable for the production of bio-based food packaging due to the use of resins as their main materials, which are often non-compatible with food safety. Additionally, the objects produced with these technologies are known for having characteristics undesirable in films, such as brittleness and sensitivity to UV light. Extrusion-based technologies have a broader range of workable materials in comparison with resin-based AM technologies. Additionally, unlike powder bed fusion, extrusion-based 3D printers allow for working with biopolymers in solid (FDM) and gel–liquid (DIW) states, making them more versatile tools for working with bio-based polymers than the other AM technologies discussed herein. Considering this, some advances in the production of feedstocks for extrusion-based technologies using biopolymers with potential for application in food packaging, as well as advances in the production of films and food packaging using these materials, will be presented below. Among the biopolymers that can be used to produce packaging films by material extrusion techniques are lactic acid-based polymers, lignin, alginate, chitosan, starch, gums, cellulose and its derivatives, whey, and many others, some of which will be discussed below.

In one of these studies, Domínguez-Robles et al. developed a lignin/lactic acid-based filament with antioxidant properties intended for fused filament deposition modeling (FDM) [61]. The produced filament was extruded by a 3D printer, showing good mechanical properties and stability, keeping its integrity even after being immersed in phosphate-buffered saline solution for 30 days. The researchers were also able to successfully incorporate an antibiotic into the filament using a hot-melt extrusion process, demonstrating the possibility of incorporating multiple active compounds into the filament's composition by the methodology employed.

Another bio-based filament for 3D printing was developed by Umerah and coworkers [62]. The filament was produced using a blend of coconut shell powder, polylactic acid, and a starch-based bioplastic. To produce the filament, coconut shell powder was immersed and subsequently precipitated in a solution containing the polymers. After being filtered, the precipitate was turned into a powder and extruded in the form of a filament. The produced filament was shown to have improved thermal and mechanical properties compared to the bioplastic per se, which was attributed to the coconut shell powder addition. The eco-friendly aspect of the composite, along with its non-toxicity, makes it a potential raw material suitable for food packaging applications.

In another interesting study with biopolymers, Hafezi et al. produced several chitosan-based films incorporating genipin—a fruit-derived compound with antibacterial properties [63]. For the film production, an appropriate gel using low-molecular chitosan and genipin as a crosslinker was prepared and further extruded by a 3D printer. After being extruded, the films were thermally cured in an oven and properly characterized. The researchers were also able to incorporate an organic compound into the films' composition as a model drug. The films with the model drug incorporated into them were further evaluated in a drug-release assay, showing appropriate release rates for the intended application (wound healing) and demonstrating the possibility of incorporating active substances into the films' matrices.

In one of the few studies found concerning the use of 3D printing and bio-based polymers in food packaging applications, Li et al. developed a double-composite intelligent film intended for monitoring and extending meat shelf life [64]. The film, which was chitosan-based, consisted of two layers, one prepared with lemongrass essential oils, the other with mulberry anthocyanin in its composition, both encapsulated by a starch-based film. For the production of the films, a chitosan solution was prepared with the active components and extruded by a 3D printer, followed by its curing at a controlled temperature. Starch films were also prepared and heat-sealed onto the active films. The final films had antioxidant and antibacterial properties due to the lemongrass essential oil presence and the ability to change color according to the pH of the medium in which they were placed due the pH-responsiveness of anthocyanin. The latter was further evaluated in the monitoring of fresh-meat spoilage, where the film successfully responded to changes in pH, changing in color from a reddish tone (at pH 2–6) to a blueish one (at pH 7–12). By an antibacterial assay, the researchers found that the addition of anthocyanin to the films had a bacteriostatic effect toward *E. coli*, in addition to the antibacterial effect provided by the lemongrass essential oil, which was effective in inhibiting both *E. coli* and *S. aureus*. By a release-rate assay, the researchers also found that the release rates of the active compounds supplied by the essential oil increased with increasing pH, suggesting that the active properties could be even more effective in increasing food shelf life. Overall, the produced films showed great promise as innovative primary food packaging materials.

In the work of Wang et al., a chitosan-based active film was produced by the solvent casting method and 3D-printed after the appropriate formulation was found [65]. In the film production, the researchers used chitosan as the film-forming substance, tea polyphenols as a source of active compounds, and nanotubes of halloysite—a naturally occurring aluminosilicate—as fillers to improve some properties of the films and control the release of active compounds. After evaluating the films made by the solution casting method, the researchers found the best formulation to produce the bio-based ink for an extrusion-based 3D printer. The ink was successfully extruded, producing thin smooth films with both good antioxidant and antibacterial activities against a variety of bacteria, including *E. coli* and *S. aureus*. Furthermore, the halloysite addition improved the films' mechanical properties, with no further reduction in printability. In a further work, the researchers employed a similar formulation to produce a bio-based food packaging container by means of 3D printing [66]. The container was evaluated with respect to the preservation of fresh blueberries and was able to maintain fruit freshness for a longer period in comparison with a blank control and a pure chitosan container, showing less loss of weight, firmness, and ascorbic acid contents [66].

In the work of Biswas and coworkers, another active food packaging film was formulated and 3D-printed. For this, they synthesized and incorporated silica–carbon–silver nanoparticles into a biodegradable polymer known by its brand name "Ecoflex" [67]. The objective of using nanoparticles in this work was to add antibacterial properties to the films as well as to improve the films' mechanical and thermal properties. The nanoparticles were synthesized using rice husks, an agro-industrial waste, and silver nitrate by means of thermal treatment and a ball-milling process. After being synthesized, the nanoparticles were incorporated into a film-forming solution containing the polymer, and the resultant solution was printed by an extrusion-based 3D printer. The researchers evaluated the antibacterial activity of the films against *Salmonella enteritidis* and found that the films possessed a bacteriostatic effect which was able to effectively inhibit the studied bacteria by contact. In order to evaluate the release of the films' nanocomponents, the team conducted a silver-release test, in which the films were immersed in water for one week. No trace of silver was found in the studied period, suggesting that the produced films have the potential to be used as food packaging materials [67].

Other work worth mentioning in the food packaging field was performed by Ahmed et al., in which they developed a composite gelatin-based film with zinc oxide and clove essential oil [68]. In the film's formulation, zinc oxide (considered a Generally Recognized as Safe (GRAS) substance by the FDA) was employed to improve the film's properties and add inhibitory activity, and clove essential oil was used to add antibacterial and antioxidant properties. According to the authors, the presence of both active compounds would have a symbiotic effect on the film's properties: while the addition of clove essential oil would negatively affect some mechanical properties, the zinc oxide, which does not possess the same efficiency in terms of active properties, would act as a filler and improve the film's overall properties. After finding an appropriate film formulation by the solvent casting method, the researchers produced a semi-solid paste by hot-melt extrusion which was further extruded by a 3D printer to produce the bio-based films. The produced films showed improved mechanical properties in comparison with the control (pure gelatin), besides complete antibacterial activity towards both *L. monocytogenes* and *Salmonella typhimuriums*. Additionally, as suggested by the authors, the use of hot-melt extrusion in conjunction with 3D printing has the potential to optimize film production by means of this technology, which is beneficial, since 3D printing technologies are generally considered slow methods of production.

Another interesting work involving 3D printing in the food packaging field was conducted by Zhou et al., in which a bio-based active food packaging container was produced [69]. The container was produced by means of coaxial 3D printing, where a core–shell structure made of cellulose nanofibers incorporated with blueberry anthocyanin was loaded with chitosan and 1-methylcyclopropene (1-MCP)—a compound used for slowing the ripening of fruit. The idea behind the coaxial structure was to effectively control the release of the active components. For this, a cellulose-based ink was prepared using anthocyanin and both sodium alginate and K-carrageenan gums in order to improve the ink viscoelastic properties. This ink was subsequently printed, along with chitosan and 1-MCP in its core, and the resultant object was appropriately cured. By a pH evaluation, the researchers confirmed the pH sensibility of the container, and the release behavior of 1-MCP was evaluated by gas chromatography. In a further assay, the labels, as the authors refer to the printed containers, were evaluated for the monitoring and extension of the freshness of litchis and were found to successfully prolong fruit shelf life for six days, in addition to visually indicating changes in the litchis' freshness.

Besides the production of new 3D printing feedstocks using bio-based polymers, in the literature there are also reports concerning the reuse or recycling of materials with similar purposes. One interesting work on the recycling of materials for 3D printing was conducted by Cisneros-López et al., in which they evaluated the production of biocomposites for material-extrusion-based 3D printers based on recycled polylactic acid [70]. The blends that the researchers produced were made with 30% recycled polylactic acid in a matrix of virgin polylactic acid, along with microcrystalline cellulose and an epoxy-based chain extender. The blend was extruded by a twin-screw extrusion process to produce the filaments, and the latter were printed using a FDM 3D printer. The researchers compared the performance of the 3D-printed objects with an injection-molding process utilizing the same blend and found that the 3D-printed objects had lower viscosities compared to the ones produced by the injection-molding process. Furthermore, the addition of micro-crystalline cellulose and the epoxy-based chain had a positive effect on the blend, improving both the mechanical and thermal properties of the produced filament [70]. A summary of research on 3D printing with biopolymers relevant to the food industry can be found in Table 3.

Table 3. Studies on 3D printing with biopolymers of relevance to the food industry.

AM Technology	Polymer/Active Compounds and Fillers	Proposed Application	Properties	Ref.
Vat photo-polymeriza-tion	Guaiacol, vanillyl alcohol, and eugenol (acrylates)	Sustainable 3D-printing feedstock formulation	Good thermal and mechanical properties	[55]
	Silk fibroin (acrylate)	3D bioprinting in tissue engineering applications	Improved mechanical properties	[56]
Powder bed fusion	Polylactide	Sustainable 3D-printing feedstock formulation	Good layer adhesion and good mechanical properties	[57]
	Polylactic acid/calcium carbonate	Tissue engineering	Good processability, mechanical properties, low melt viscosity, and small particle size	[58]
	Hard keratin	Sustainable 3D-printing feedstock formulation	Weaker mechanical properties; Successful keratin incorporation/proce-ssing	[71]
	Polyhydroxyalkanoate/akermanite	Tissue engineering	Improved water-uptake properties	[59]

Table 3. *Cont.*

AM Technology	Polymer/Active Compounds and Fillers	Proposed Application	Properties	Ref.
Material extrusion	Lignin and polylactic acid	Wound healing	Good mechanical properties and stability; Successful incorporation of an antibiotic	[61]
	Polylactic acid and starch/coconut shell	Sustainable 3D-printing feedstock formulation	Improved thermal and mechanical properties	[62]
	Chitosan/genipin	Wound healing	Good release rate of the active compound	[63]
	Chitosan and starch/lemongrass essential oil and mulberry anthocyanin	Food packaging	Color-changing properties; Antibacterial effect	[64]
	Chitosan/tea polyphenols and halloysite nanotubes	Food packaging	Good antioxidant and antibacterial activity; Improved mechanical properties	[65]
	Bio-based plastic "Ecoflex"/silica–carbon–silver nanoparticles	Food packaging	Bacteriostatic effect	[67]
	Gelatin/zinc oxide and clove essential oil	Food packaging	Improved mechanical properties and antibacterial activity	[68]
	Chitosan and cellulose/blueberry anthocyanin and methylcyclopropene	Food packaging	Color changing properties and preservation ability	[69]
	Polylactic acid (virgin and recycled)	Sustainable 3D-printing feedstock formulation	Improved both mechanical and thermal properties	[70]

6. Limitations of 3D Printing in the Production of Films

Undoubtedly, additive manufacturing technology has great potential in the food packaging field; however, research in this area is still very limited. Most studies on AM technology and biopolymers are concerned with medical, textile, and pharmaceutical applications and the "tailor-made" characteristics of 3D printing, along with the biodegradability, abundance, low cost, and biocompatibility of the biopolymers used which make them suitable for the fabrication of biodegradable scaffolds, tissue and organ engineering, drug delivery systems, and innovative textile products [72–75]. With regard to the food industry, most research on AM and biopolymers aims at the production of customized food, as discussed in the previous section. In the few studies found on the development of films or materials for food packaging, AM technology proved to be very useful, allowing for the production of innovative and functional bio-based packages with controlled release of active substances.

Considering the lack of research on foodstuff packaging and the fact that additive manufacturing is a relatively new technology that has been on the market for no more than a couple of decades, it is obvious that more studies on AM focused on the development of food packaging are needed. Additionally, in order to explore the potentiality of 3D printing in the food packaging area, some challenges must be overcome. To begin with, one must bear in mind that, given the current state of AM technology, its uses are confined to the development and research of bio-based packaging films rather than their industrial-scale production. This is due to the fact that, despite being faster than conventional methods for producing complex objects, AM is still considered a slow process and can take from hours to days to produce an object, depending on the object's complexity [76]. In addition to the above, depending on the printer specifications and the final purpose, AM technology can be very costly and can include the costs for 3D printer machines, materials, and post-processing [51]. Adding these two shortcomings together, it is unlikely that large-scale production of 3D-printed objects will be possible without further modifications or improvements.

Another challenge in the AM technology field is presented by the physicochemical properties of the biopolymers used, such as the minimum requirements for the biopolymers to be processed by 3D printing technologies, as well as the properties that are desired in the final products after processing. For instance, in FDM, which is by far the most intensively explored AM technology, one requirement is that the biopolymer should be thermally stable and melt-processable, this being a challenge for most biopolymers, since they generally have lower thermal stability, heat-flowability, and a narrower range of workable temperatures in comparison with their petroleum-based counterparts [77]. In the preparation of feedstocks for AM, solubility is another key property. Some biopolymers, such as cellulose, have inherently low solubilities in common solvents, making it difficult for them to be processed by AM technologies. In the case of cellulose, strategies to properly dissolve and regenerate it have been employed using ionic liquids and other non-standard solvents, but it still poses a challenge for AM processing [78,79]. In contrast, the highly hydrophilic natures of some biopolymers may compromise their final applications, especially if they are to be used in packaging films, where good barrier properties are essential to the packages' providing effective protection. These and other drawbacks, such as thermal instability, brittleness, stiffness, low barrier properties, and vulnerability to degradation, need to be improved in order for these alternative materials to be successfully used in food packaging applications [8]. Regarding the production of intelligent and active films, another interesting issue is the evaluation of as-produced films in order to identify possible alterations to the films' active properties after processing by AM technologies.

Some strategies to overcome these challenges include the study of appropriate formulations and/or functionalization of the biopolymers aiming at improving their properties for better AM processability. Adaptations of AM technology may also be necessary to improve efficiency and performance in the packaging field by means of bio-based polymers, including greater compatibility with alternative feedstocks, better processing speed, and general optimizations of the overall technology to reduce costs. Nonetheless, the precision, automation, and versatility of AM technology can clearly contribute to significant advances in the production and development of bio-based packaging films.

7. Conclusions and Future Perspectives

The aims of this work were to explore the progress in developing bio-based alternatives to conventional plastic packaging as well as to examine how additive manufacturing technologies can contribute to the development of bio-based food packaging films. To attain these goals, the literature on the production and development of biopolymer-based films and primary food packaging by conventional methods and by means of AM, as well as alternative feedstocks for AM relevant to food packaging development, was reviewed and discussed. Based on the information extracted from the studies, bio-based films and food packages developed by means of AM technologies, as well as promising feedstocks for these technologies, were identified. Among the employed biopolymers, we highlight chitosan, polylactic acid, cellulose and its derivatives, starch, gums, and polyhydroxyalkanoates—all of which can be used, individually or in blends, in the production of sustainable films. Additionally, the use of active substances of natural origin was also found in the development of active bio-based packaging. Along with the biopolymers, these compounds allow for the development of packaging formulations that are not only biodegradable and sustainable, but also possess active and intelligent properties, such as antibacterial activity, antioxidant activity, sensitivity to pH changes, and resistance to ultraviolet radiation. From the findings, it was concluded that, despite the promising works directly related to the development of bio-based food packaging by AM, this technology has not been well explored in this field. Most of the research concerning the development of bio-based feedstocks for AM is aimed at biomedical, pharmaceutical, and textile fields, where the precision, automation, and the ability to build complex shapes and tailor-made objects, along with the biodegradability, biocompatibility, and the abundance of biopolymers in AM, promote advances in the development of tissues, organs, scaffolds, drug delivery systems, and smart and innovative textile products, among other tailor-made objects in these areas. At present, in the food industry, AM applications are mainly directed at the production and development of customized food. A brief overall SWOT analysis of the potential of 3D printing as a tool in the production of biopolymer-based films for food packaging applications is presented in Figure 4.

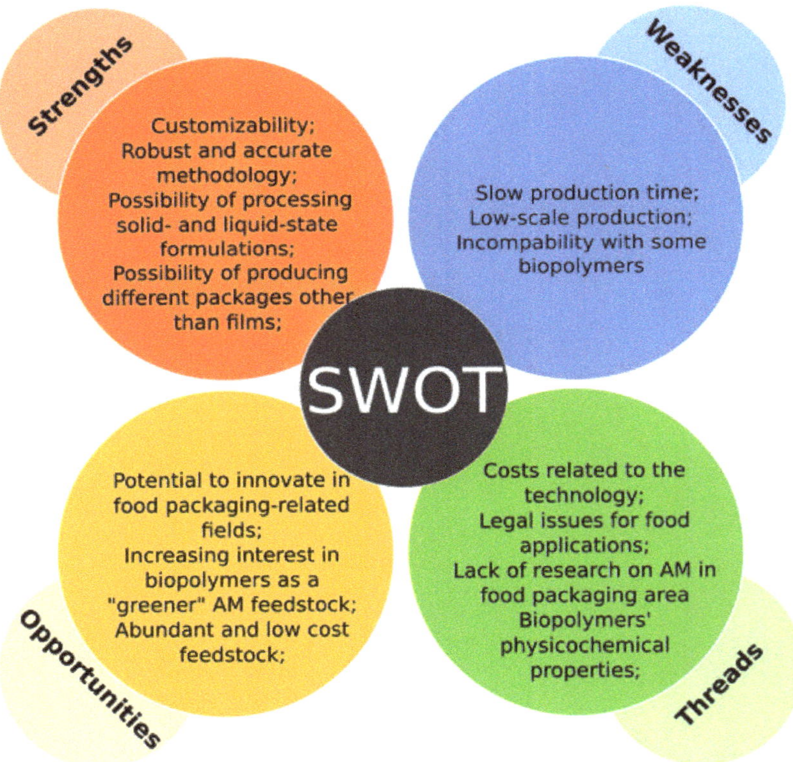

Figure 4. SWOT analysis of 3D printing as a tool for the production of bio-based films for food packaging applications.

The factors that contribute to the lack of research on food packaging films produced by AM might include the high costs associated with AM technologies, the incompatibility of biopolymers with 3D printing, the relatively slow production methods, the scaling-up difficulties, and the need to develop biopolymer blends/formulations with not only good printability but also the properties that meet the necessary criteria for food packaging materials.

Despite the lack of studies on the production of bio-based materials for food packaging applications by AM, this technology still seems very promising in this field. Furthermore, it is very likely that this area will benefit from the advances related to AM and biopolymers in other fields. As the feedstocks and the technology employed are the same, adaptations in terms of better compatibility/processability in AM regarding biopolymers of relevance to these various fields would probably benefit the production of bio-based food packaging by means of this technology as well. In this respect, it is very likely that, as AM is gradually better adapted for the processing of biopolymers and these materials are increasingly explored in relation to this technology, the potential for 3D printing as a more effective and less limited tool in the production and development of biopolymer-based primary food packaging will increase.

Author Contributions: E.G.S.S.J.: writing—original draft preparation. S.C.: writing—original draft preparation. A.F.B.: writing—review and editing. I.A.C.R.: supervision, writing—review and editing. All authors have read and agreed to the published version of the manuscript.

Funding: Fundação para a Ciência e Tecnologia (FCT), national funds under the project UIDB/04138/2020 and UIDP/04138/2020.

Acknowledgments: The authors would like to thank the Portuguese government, Fundação para a Ciência e Tecnologia (FCT), for financial support through national funds under the projects UIDB/04138/2020 and UIDP/04138/2020. Minder Software, version: 1.13.1 was used to generate Figure 1, and Inkscape software, version 1.2.2 was used to generate the graphical abstract, Figures 2–4.

Conflicts of Interest: The authors declare no conflict of interest.

References

1. Piergiovanni, L.; Limbo, S. Introduction to Food Packaging Materials. In *Food Packaging Materials*; Springer: Cham, Switzerland, 2016; pp. 1–3. [CrossRef]
2. Tajeddin, B.; Arabkhedri, M. Polymers and food packaging. In *Polymer Science and Innovative Applications*; Elsevier: Amsterdam, The Netherlands, 2020; pp. 525–543. [CrossRef]
3. Geyer, R.; Jambeck, J.R.; Law, K.L. Production, use, and fate of all plastics ever made. *Sci. Adv.* **2017**, *3*, e1700782. [CrossRef] [PubMed]
4. Statista. Annual Production of Plastics Worldwide from 1950 to 2020. Available online: https://www.statista.com/statistics/2827 32/global-production-of-plastics-since-1950/ (accessed on 24 November 2022).
5. Vartiainen, J.; Vähä-Nissi, M.; Harlin, A. Biopolymer Films and Coatings in Packaging Applications—A Review of Recent Developments. *Mater. Sci. Appl.* **2014**, *05*, 708–718. [CrossRef]
6. Varghese, S.A.; Siengchin, S.; Parameswaranpillai, J. Essential oils as antimicrobial agents in biopolymer-based food packaging –A comprehensive review. *Food Biosci.* **2020**, *38*, 100785. [CrossRef]
7. Rodrigues, C.; Souza, V.; Coelhoso, I.; Fernando, A. Bio-Based Sensors for Smart Food Packaging—Current Applications and Future Trends. *Sensors* **2021**, *21*, 2148. [CrossRef] [PubMed]
8. Jabeen, N.; Majid, I.; Nayik, G.A.; Yildiz, F. Bioplastics and food packaging: A review. *Cogent Food Agric.* **2015**, *1*, 1117749. [CrossRef]
9. Jariyasakoolroj, P.; Leelaphiwat, P.; Harnkarnsujarit, N. Advances in research and development of bioplastic for food packaging. *J. Sci. Food Agric.* **2020**, *100*, 5032–5045. [CrossRef]
10. Sullivan, D.J.; Azlin-Hasim, S.; Cruz-Romero, M.; Cummins, E.; Kerry, J.P.; Morris, M.A. Natural Antimicrobial Materials for Use in Food Packaging. In *Handbook of Antimicrobial Coatings*; Elsevier: Amsterdam, The Netherlands, 2018; pp. 181–233. [CrossRef]
11. Suppakul, P. Natural extracts in plastic food packaging. In *Multifunctional and Nanoreinforced Polymers for Food Packaging*; Woodhead Publishing Limited: Cambridge, UK, 2011; pp. 421–459. [CrossRef]
12. Beoletto, V.; Oliva, M.d.L.M.; Marioli, J.; Carezzano, M.; Demo, M. Antimicrobial Natural Products Against Bacterial Biofilms. In *Antibiotic Resistance Mechanisms and New Antimicrobial Approaches*; Academic Press: Cambridge, MA, USA, 2016; pp. 291–307. [CrossRef]
13. Jafri, H.; Ansari, F.A.; Ahmad, I. Prospects of Essential Oils in Controlling Pathogenic Biofilm. In *New Look to Phytomedicine: Advancements in Herbal Products as Novel Drug Leads*; Elsevier Inc.: Amsterdam, The Netherlands, 2018; pp. 203–236. [CrossRef]
14. Hu, W.S.; Nam, D.M.; Choi, J.Y.; Kim, J.S.; Koo, O.K. Anti-attachment, anti-biofilm, and antioxidant properties of Brassicaceae extracts on Escherichia coli O157:H7. *Food Sci. Biotechnol.* **2019**, *28*, 1881–1890. [CrossRef]
15. Khumkomgool, A.; Saneluksana, T.; Harnkarnsujarit, N. Active meat packaging from thermoplastic cassava starch containing sappan and cinnamon herbal extracts via LLDPE blown-film extrusion. *Food Packag. Shelf Life* **2020**, *26*, 100557. [CrossRef]
16. Lu, W.; Cui, R.; Zhu, B.; Qin, Y.; Cheng, G.; Li, L.; Yuan, M. Influence of clove essential oil immobilized in mesoporous silica nanoparticles on the functional properties of poly(lactic acid) biocomposite food packaging film. *J. Mater. Res. Technol.* **2021**, *11*, 1152–1161. [CrossRef]

17. Sanches, M.A.R.; Camelo-Silva, C.; Carvalho, C.D.S.; de Mello, J.R.; Barroso, N.G.; Barros, E.L.D.S.; Silva, P.P.; Pertuzatti, P.B. Active packaging with starch, red cabbage extract and sweet whey: Characterization and application in meat. *LWT Food Sci. Technol.* **2021**, *135*. [CrossRef]
18. Severo, C.; Anjos, I.; Souza, V.G.; Canejo, J.P.; Bronze, M.; Fernando, A.L.; Coelhoso, I.; Bettencourt, A.F.; Ribeiro, I.A. Development of cranberry extract films for the enhancement of food packaging antimicrobial properties. *Food Packag. Shelf Life* **2021**, *28*, 100646. [CrossRef]
19. Souza, V.G.L.; Pires, J.R.; Vieira, T.; Coelhoso, I.M.; Duarte, M.P.; Fernando, A.L. Activity of chitosan-montmorillonite bionanocomposites incorporated with rosemary essential oil: From in vitro assays to application in fresh poultry meat. *Food Hydrocoll.* **2019**, *89*, 241–252. [CrossRef]
20. Priyadarshi, R.; Rhim, J.-W. Chitosan-based biodegradable functional films for food packaging applications. *Innov. Food Sci. Emerg. Technol.* **2020**, *62*, 102346. [CrossRef]
21. Kumar, S.; Mukherjee, A.; Dutta, J. Chitosan based nanocomposite films and coatings: Emerging antimicrobial food packaging alternatives. *Trends Food Sci. Technol.* **2020**, *97*, 196–209. [CrossRef]
22. Wang, H.; Qian, J.; Ding, F. Emerging Chitosan-Based Films for Food Packaging Applications. *J. Agric. Food Chem.* **2018**, *66*, 395–413. [CrossRef]
23. da Rocha, M.; de Souza, M.M.; Prentice, C. Biodegradable Films: An Alternative Food Packaging. In *Food Packaging and Preservation*; Elsevier Inc.: Amsterdam, The Netherlands, 2018; pp. 307–342. [CrossRef]
24. Menezes, J.; Athmaselvi, K. Report on Edible Films and Coatings. In *Food Packaging and Preservation*; Elsevier Inc.: Amsterdam, The Netherlands, 2018; pp. 177–212. [CrossRef]
25. Suhag, R.; Kumar, N.; Petkoska, A.T.; Upadhyay, A. Film formation and deposition methods of edible coating on food products: A review. *Food Res. Int.* **2020**, *136*, 109582. [CrossRef]
26. Taketa, T.B.; Bataglioli, R.A.; Neto, J.B.M.R.; de Carvalho, B.G.; de la Torre, L.G.; Beppu, M.M. Fundamentals and biomedical applications of biopolymer-based layer-by-layer films. In *Biopolymer Membranes and Films*; Elsevier: Amsterdam, The Netherlands, 2020; pp. 219–242. [CrossRef]
27. Ashter, S.A. Extrusion of biopoymers. In *Introduction to Bioplastics Engineering*; Elsevier: Oxford, UK, 2016; pp. 211–225.
28. McKeen, L.W. Production of Films. In *Film Properties of Plastics and Elastomers*, 3rd ed.; McKeen, L.W., Ed.; Elsevier: Oxford, UK, 2017; pp. 65–79. [CrossRef]
29. Liu, Y.; Yuan, Y.; Duan, S.; Li, C.; Hu, B.; Liu, A.; Wu, D.; Cui, H.; Lin, L.; He, J.; et al. Preparation and characterization of chitosan films with three kinds of molecular weight for food packaging. *Int. J. Biol. Macromol.* **2020**, *155*, 249–259. [CrossRef]
30. Xu, Y.; Liu, X.; Jiang, Q.; Yu, D.; Xu, Y.; Wang, B.; Xia, W. Development and properties of bacterial cellulose, curcumin, and chitosan composite biodegradable films for active packaging materials. *Carbohydr. Polym.* **2021**, *260*, 117778. [CrossRef]
31. Ezati, P.; Rhim, J.-W. pH-responsive chitosan-based film incorporated with alizarin for intelligent packaging applications. *Food Hydrocoll.* **2020**, *102*, 105629. [CrossRef]
32. Wilfer, P.B.; Giridaran, G.; Jeevahan, J.J.; Joseph, G.B.; Kumar, G.S.; Thykattuserry, N.J. Effect of starch type on the film properties of native starch based edible films. *Mater. Today Proc.* **2021**, *44*, 3903–3907. [CrossRef]
33. Rajeswari, A.; Christy, E.J.S.; Swathi, E.; Pius, A. Fabrication of improved cellulose acetate-based biodegradable films for food packaging applications. *Environ. Chem. Ecotoxicol.* **2020**, *2*, 107–114. [CrossRef]
34. Nikvarz, N.; Khayati, G.R.; Sharafi, S. Preparation of UV absorbent films using polylactic acid and grape syrup for food packaging application. *Mater. Lett.* **2020**, *276*, 128187. [CrossRef]
35. Wang, L.-F.; Rhim, J.-W. Preparation and application of agar/alginate/collagen ternary blend functional food packaging films. *Int. J. Biol. Macromol.* **2015**, *80*, 460–468. [CrossRef]
36. Hou, X.; Xue, Z.; Xia, Y.; Qin, Y.; Zhang, G.; Liu, H.; Li, K. Effect of SiO_2 nanoparticle on the physical and chemical properties of eco-friendly agar/sodium alginate nanocomposite film. *Int. J. Biol. Macromol.* **2019**, *125*, 1289–1298. [CrossRef] [PubMed]
37. Gahruie, H.H.; Mostaghimi, M.; Ghiasi, F.; Tavakoli, S.; Naseri, M.; Hosseini, S.M.H. The effects of fatty acids chain length on the techno-functional properties of basil seed gum-based edible films. *Int. J. Biol. Macromol.* **2020**, *160*, 245–251. [CrossRef]
38. Wu, L.-T.; Tsai, I.-L.; Ho, Y.-C.; Hang, Y.-H.; Lin, C.; Tsai, M.-L.; Mi, F.-L. Active and intelligent gellan gum-based packaging films for controlling anthocyanins release and monitoring food freshness. *Carbohydr. Polym.* **2021**, *254*, 117410. [CrossRef]
39. Rezaei, F.; Shahbazi, Y. Shelf-life extension and quality attributes of sauced silver carp fillet: A comparison among direct addition, edible coating and biodegradable film. *LWT Food Sci. Technol.* **2018**, *87*, 122–133. [CrossRef]
40. Liu, B.-Y.; Xue, C.-H.; An, Q.-F.; Jia, S.-T.; Xu, M.-M. Fabrication of superhydrophobic coatings with edible materials for super-repelling non-Newtonian liquid foods. *Chem. Eng. J.* **2019**, *371*, 833–841. [CrossRef]
41. Yan, J.; Luo, Z.; Ban, Z.; Lu, H.; Li, D.; Yang, D.; Aghdam, M.S.; Li, L. The effect of the layer-by-layer (LBL) edible coating on strawberry quality and metabolites during storage. *Postharvest Biol. Technol.* **2019**, *147*, 29–38. [CrossRef]
42. Ferreira, A.R.V.; Torres, C.A.V.; Freitas, F.; Sevrin, C.; Grandfils, C.; Reis, M.A.M.; Alves, V.D.; Coelhoso, I.M. Development and characterization of bilayer films of FucoPol and chitosan. *Carbohydr. Polym.* **2016**, *147*, 8–15. [CrossRef]
43. del Hoyo-Gallego, S.; Pérez-Álvarez, L.; Gómez-Galván, F.; Lizundia, E.; Kuritka, I.; Sedlarik, V.; Laza, J.M.; Vila-Vilela, J.L. Construction of antibacterial poly(ethylene terephthalate) films via layer by layer assembly of chitosan and hyaluronic acid. *Carbohydr. Polym.* **2016**, *143*, 35–43. [CrossRef] [PubMed]

44. Li, K.; Zhu, J.; Guan, G.; Wu, H. Preparation of chitosan-sodium alginate films through layer-by-layer assembly and ferulic acid crosslinking: Film properties, characterization, and formation mechanism. *Int. J. Biol. Macromol.* **2019**, *122*, 485–492. [CrossRef] [PubMed]

45. Llanos, J.H.R.; Tadini, C.C.; Gastaldi, E. New strategies to fabricate starch/chitosan-based composites by extrusion. *J. Food Eng.* **2021**, *290*, 110224. [CrossRef]

46. Felix, M.; Martinez, I.; Romero, A.; Partal, P.; Guerrero, A. Effect of pH and nanoclay content on the morphology and physic-ochemical properties of soy protein/montmorillonite nanocomposite obtained by extrusion. *Compos. Part B Eng.* **2018**, *140*, 197–203. [CrossRef]

47. Chevalier, E.; Chaabani, A.; Assezat, G.; Prochazka, F.; Oulahal, N. Casein/wax blend extrusion for production of edible films as carriers of potassium sorbate—A comparative study of waxes and potassium sorbate effect. *Food Packag. Shelf Life* **2018**, *16*, 41–50. [CrossRef]

48. Kumar, J.L.; Pandey, P.M.; Wimpenny, D.I. *3D Printing and Additive Manufacturing Technologies*; Springer: Singapore, 2019. [CrossRef]

49. Redwood, B.; Schöffer, F.; Garret, B. *The 3D-Printing Handbook: Technologies, Design and Applications*; 3D Hubs: Amsterdam, The Netherlands, 2017.

50. *ISO/ASTM 52900*; Additive Manufacturing—General Principles and Terminology. International Organization for Standardization: Geneva, Switzerland, 2015. Available online: https://www.iso.org/obp/ui/#iso:std:69669:en%0Ahttps://www.iso.org/standard/69669.html%0Ahttps://www.astm.org/Standards/ISOASTM52900.htm (accessed on 24 November 2022).

51. Noorani, R. *3D-Printing, Technology, Applications, and Selection*; CRC Press: Boca Raton, FL, USA, 2017. [CrossRef]

52. Zhou, L.; Fu, J.; He, Y. A Review of 3D Printing Technologies for Soft Polymer Materials. *Adv. Funct. Mater.* **2020**, *30*. [CrossRef]

53. Rajabi, M.; McConnell, M.; Cabral, J.; Ali, M.A. Chitosan hydrogels in 3D printing for biomedical applications. *Carbohydr. Polym.* **2021**, *260*, 117768. [CrossRef]

54. Voet, V.S.D.; Guit, J.; Loos, K. Sustainable Photopolymers in 3D Printing: A Review on Biobased, Biodegradable, and Recyclable Alternatives. *Macromol. Rapid Commun.* **2021**, *42*, e2000475. [CrossRef]

55. Ding, R.; Du, Y.; Goncalves, R.; Francis, L.F.; Reineke, T.M. Sustainable near UV-curable acrylates based on natural phenolics for stereolithography 3D printing. *Polym. Chem.* **2019**, *10*, 1067–1077. [CrossRef]

56. Kim, S.H.; Yeon, Y.K.; Lee, J.M.; Chao, J.R.; Lee, Y.J.; Seo, Y.B.; Sultan, T.; Lee, O.J.; Lee, J.S.; Yoon, S.-I.; et al. Precisely printable and biocompatible silk fibroin bioink for digital light processing 3D printing. *Nat. Commun.* **2018**, *9*, 1620. [CrossRef]

57. Dechet, M.A.; Demina, A.; Römling, L.; Gómez Bonilla, J.S.; Lanyi, F.J.; Schubert, D.W.; Bück, A.; Peukert, W.; Schmidt, J. Development of poly(L-lactide) (PLLA) microspheres precipitated from triacetin for application in powder bed fusion of polymers. *Addit. Manuf.* **2020**, *32*, 100966. [CrossRef]

58. Gayer, C.; Ritter, J.; Bullemer, M.; Grom, S.; Jauer, L.; Meiners, W.; Pfister, A.; Reinauer, F.; Vučak, M.; Wissenbach, K.; et al. Development of a solvent-free polylactide/calcium carbonate composite for selective laser sintering of bone tissue engineering scaffolds. *Mater. Sci. Eng. C* **2019**, *101*, 660–673. [CrossRef] [PubMed]

59. Diermann, S.H.; Lu, M.; Dargusch, M.; Grøndahl, L.; Huang, H. Akermanite reinforced PHBV scaffolds manufactured using selective laser sintering. *J. Biomed. Mater. Res. Part B Appl. Biomater.* **2019**, *107*, 2596–2610. [CrossRef] [PubMed]

60. Sultana, N.; Khan, T.H. In Vitro Degradation of PHBV Scaffolds and nHA/PHBV Composite Scaffolds Containing Hydroxyapatite Nanoparticles for Bone Tissue Engineering. *J. Nanomater.* **2012**, *2012*, 190950. [CrossRef]

61. Domínguez-Robles, J.; Martin, N.K.; Fong, M.L.; Stewart, S.A.; Irwin, N.J.; Rial-Hermida, M.I.; Donnelly, R.F.; Larrañeta, E. Antioxidant PLA Composites Containing Lignin for 3D Printing Applications: A Potential Material for Healthcare Applications. *Pharmaceutics* **2019**, *11*, 165. [CrossRef]

62. Umerah, C.O.; Kodali, D.; Head, S.; Jeelani, S.; Rangari, V.K. Synthesis of carbon from waste coconutshell and their application as filler in bioplast polymer filaments for 3D printing. *Compos. Part B Eng.* **2020**, *202*, 108428. [CrossRef]

63. Hafezi, F.; Scoutaris, N.; Douroumis, D.; Boateng, J. 3D printed chitosan dressing crosslinked with genipin for potential healing of chronic wounds. *Int. J. Pharm.* **2019**, *560*, 406–415. [CrossRef]

64. Li, S.; Jiang, Y.; Zhou, Y.; Li, R.; Jiang, Y.; Hossen, A.; Dai, J.; Qin, W.; Liu, Y. Facile fabrication of sandwich-like anthocyanin / chitosan / lemongrass films via 3D printing for intelligent evaluation of pork freshness. *Food Chem.* **2021**, *370*, 131082. [CrossRef]

65. Wang, Y.; Yi, S.; Lu, R.; Sameen, D.E.; Ahmed, S.; Dai, J.; Qin, W.; Li, S.; Liu, Y. Preparation, characterization, and 3D printing verification of chitosan/halloysite nanotubes/tea polyphenol nanocomposite films. *Int. J. Biol. Macromol.* **2021**, *166*, 32–44. [CrossRef]

66. Liu, Y.; Yi, S.; Sameen, D.E.; Hossen, M.A.; Dai, J.; Li, S.; Qin, W.; Lee, K. Designing and utilizing 3D printed chitosan/halloysite nanotubes/tea polyphenol composites to maintain the quality of fresh blueberries. *Innov. Food Sci. Emerg. Technol.* **2021**, *74*, 102808. [CrossRef]

67. Biswas, M.C.; Tiimob, B.J.; Abdela, W.; Jeelani, S.; Rangari, V.K. Nano silica-carbon-silver ternary hybrid induced antimicrobial composite films for food packaging application. *Food Packag. Shelf Life* **2019**, *19*, 104–113. [CrossRef]

68. Ahmed, J.; Mulla, M.; Joseph, A.; Ejaz, M.; Maniruzzaman, M. Zinc oxide/clove essential oil incorporated type B gelatin nanocomposite formulations: A proof-of-concept study for 3D printing applications. *Food Hydrocoll.* **2020**, *98*, 105256. [CrossRef]

69. Zhou, W.; Wu, Z.; Xie, F.; Tang, S.; Fang, J.; Wang, X. 3D printed nanocellulose-based label for fruit freshness keeping and visual monitoring. *Carbohydr. Polym.* **2021**, *273*, 118545. [CrossRef] [PubMed]

70. Cisneros-López, E.O.; Pal, A.K.; Rodriguez, A.U.; Wu, F.; Misra, M.; Mielewski, D.K.; Kiziltas, A.; Mohanty, A.K. Recycled poly(lactic acid)–based 3D printed sustainable biocomposites: A comparative study with injection molding. *Mater. Today Sustain.* **2020**, *7–8*, 100027. [CrossRef]

71. Singamneni, S.; Velu, R.; Behera, M.P.; Scott, S.; Brorens, P.; Harland, D.; Gerrard, J. Selective laser sintering responses of keratin-based bio-polymer composites. *Mater. Des.* **2019**, *183*. [CrossRef]

72. Biazar, E.; Najafi, S.M.; Heidari, K.S.; Yazdankhah, M.; Rafiei, A.; Biazar, D. 3D bio-printing technology for body tissues and organs regeneration. *J. Med. Eng. Technol.* **2018**, *42*, 187–202. [CrossRef]

73. Bose, S.; Ke, D.; Sahasrabudhe, H.; Bandyopadhyay, A. Additive manufacturing of biomaterials. *Prog. Mater. Sci.* **2018**, *93*, 45–111. [CrossRef]

74. Goel, A.; Meher, M.K.; Gulati, K.; Poluri, K.M. Fabrication of Biopolymer-Based Organs and Tissues Using 3D Bioprinting. In *3D Printing Technology in Nanomedicine*; Elsevier: St. Louis, MO, USA, 2019; pp. 43–62. [CrossRef]

75. Mobaraki, M.; Ghaffari, M.; Yazdanpanah, A.; Luo, Y.; Mills, D. Bioinks and bioprinting: A focused review. *Bioprinting* **2020**, *18*, e00080. [CrossRef]

76. Attaran, M. The rise of 3-D printing: The advantages of additive manufacturing over traditional manufacturing. *Bus. Horiz.* **2017**, *60*, 677–688. [CrossRef]

77. Chaunier, L.; Guessasma, S.; Belhabib, S.; Della Valle, G.; Lourdin, D.; Leroy, E. Material extrusion of plant biopolymers: Opportunities & challenges for 3D printing. *Addit. Manuf.* **2018**, *21*, 220–233. [CrossRef]

78. Gauss, C.; Pickering, K.L.; Muthe, L.P. The use of cellulose in bio-derived formulations for 3D/4D printing: A review. *Compos. Part C Open Access* **2021**, *4*, 100113. [CrossRef]

79. Le-Bail, A.; Maniglia, B.C.; Le-Bail, P. Recent advances and future perspective in additive manufacturing of foods based on 3D printing. *Curr. Opin. Food Sci.* **2020**, *35*, 54–64. [CrossRef]

foods

Review

Bioplastics for Food Packaging: Environmental Impact, Trends and Regulatory Aspects

Rui M. S. Cruz [1,2,*], Victoria Krauter [3,*], Simon Krauter [3], Sofia Agriopoulou [4], Ramona Weinrich [5], Carsten Herbes [6], Philip B. V. Scholten [7], Ilke Uysal-Unalan [8,9], Ece Sogut [8,10], Samir Kopacic [11], Johanna Lahti [12], Ramune Rutkaite [13] and Theodoros Varzakas [4]

1   Department of Food Engineering, Institute of Engineering, Campus da Penha, Universidade do Algarve, 8005-139 Faro, Portugal
2   MED-Mediterranean Institute for Agriculture, Environment and Development and CHANGE-Global Change and Sustainability Institute, Faculty of Sciences and Technology, Campus de Gambelas, Universidade do Algarve, 8005-139 Faro, Portugal
3   Packaging and Resource Management, Department Applied Life Sciences, FH Campus Wien, University of Applied Sciences, 1100 Vienna, Austria
4   Department of Food Science and Technology, University of Peloponnese, 24100 Kalamata, Greece
5   Department of Consumer Behaviour in the Bioeconomy, University of Hohenheim, Wollgrasweg 49, 70599 Stuttgart, Germany
6   Institute for International Research on Sustainable Management and Renewable Energy, Nuertingen Geislingen University, Neckarsteige 6-10, 72622 Nuertingen, Germany
7   Bloom Biorenewables, Route de l'Ancienne Papeterie 106, 1723 Marly, Switzerland
8   Department of Food Science, Aarhus University, Agro Food Park 48, 8200 Aarhus, Denmark
9   CiFOOD—Center for Innovative Food Research, Aarhus University, Agro Food Park 48, 8200 Aarhus, Denmark
10 Department of Food Engineering, Suleyman Demirel University, 32200 Isparta, Turkey
11 Institute for Bioproducts and Paper Technology, Graz University of Technology, Inffeldgasse 23, 8010 Graz, Austria
12 Sustainable Products and Materials, VTT Technical Research Centre of Finland, Visiokatu 4, 33720 Tampere, Finland
13 Department of Polymer Chemistry and Technology, Kaunas University of Technology, Radvilenu Rd 19, 50254 Kaunas, Lithuania
*   Correspondence: rcruz@ualg.pt (R.M.S.C.); victoria.krauter@fh-campuswien.ac.at (V.K.)

Citation: Cruz, R.M.S.; Krauter, V.; Krauter, S.; Agriopoulou, S.; Weinrich, R.; Herbes, C.; Scholten, P.B.V.; Uysal-Unalan, I.; Sogut, E.; Kopacic, S.; et al. Bioplastics for Food Packaging: Environmental Impact, Trends and Regulatory Aspects. *Foods* **2022**, 11, 3087. https://doi.org/10.3390/foods11193087

Academic Editors: Rafael Gavara and Long Yu

Received: 4 July 2022
Accepted: 29 September 2022
Published: 5 October 2022

Publisher's Note: MDPI stays neutral with regard to jurisdictional claims in published maps and institutional affiliations.

Abstract: The demand to develop and produce eco-friendly alternatives for food packaging is increasing. The huge negative impact that the disposal of so-called "single-use plastics" has on the environment is propelling the market to search for new solutions, and requires initiatives to drive faster responses from the scientific community, the industry, and governmental bodies for the adoption and implementation of new materials. Bioplastics are an alternative group of materials that are partly or entirely produced from renewable sources. Some bioplastics are biodegradable or even compostable under the right conditions. This review presents the different properties of these materials, mechanisms of biodegradation, and their environmental impact, but also presents a holistic overview of the most important bioplastics available in the market and their potential application for food packaging, consumer perception of the bioplastics, regulatory aspects, and future challenges.

Keywords: food packaging; bioplastics; environmentally-friendly; consumer perception; biodegradation; sustainability

1. Introduction

Packaging is an integral part and enabler of modern food systems. As a result, there is hardly any food item today that is not packaged at least once on its way from farm to fork [1,2]. The background to this is the underlying and essential service functions that it performs. Even the most trivial function, namely containment, is what makes liquid foodstuffs, for example, manageable and transportable in the first place—a key function for our modern economy. Moreover, and most importantly, it provides protection to the food, thus, enabling high levels of food quality, safety, and security to be achieved. This is rounded off by the functions of communication (e.g., information about the product) and convenience (e.g., easy-to-open) [3].

The needs of a food product are strongly dependent on the type of packaging (e.g., design, type of construction) and packaging material chosen (e.g., paper, glass, metal, and corrugated or non-corrugated cardboard, plastic, and composite materials with more than one material, such as plastic-coated cardboard). Hence, careful consideration of the material's properties is a key step in designing packaging that is fit for its purpose and, thus, effective. Properties include features, such as a barrier against gasses (e.g., oxygen, carbon dioxide, water vapor), physical and mechanical strength, aroma, fat, lightness, and migration, as well as hygiene and, as a result, are strongly dependent on the nature of the material itself [3–5].

Taking a closer look at plastic materials, it quickly becomes clear this material group comprises a wide range of different materials, including polyolefins, such as polyethylene (PE), polypropylene (PP), and polyethylene terephthalate (PET), each with very different properties. Accordingly, they also offer a wide range of advantages and disadvantages. In terms of suitability for packaging applications, it can be said that plastics are often preferred because of their lightness, formability, low cost, versatile and controllable properties (e.g., mechanical, physical, and chemical properties, barrier, color, temperature stability, and sealability), convenience (e.g., transportability and resistance to breakage) and usability in the preparation of multilayer materials [3–5]. Despite the, per se, very good suitability, it is above all the environmental aspect and the careless handling of raw materials and packaging waste, such as (marine) litter, microplastics, limited recyclability and (bio)degradability, and the use of fossil resources, that pose a major disadvantage and have been the focus of public and political debate in recent decades [6–9].

Among different sectors, the packaging sector is the main user of plastics (around 40%). For example, plastic packaging in the European Union (EU) makes up around 60% of post-consumer plastic waste [10]. Most of the packaging is used only once, and the lack of reuse associated with failures in the recycling systems contribute to generating huge amounts of solid wastes that are discarded, contributing to a negative impact on land and marine environments [11]. On average, the amount of plastic packaging waste generated per capita increased from 27 kg to 35 kg between 2009 and 2019 [12].

The EU is trying to solve these problems with approaches, such as circular economy and bioeconomy, to promote innovation and research for guaranteeing resource utilization efficiency. The circular economy highlights the 4R concept (reduce, reuse, recycle, and recover), and stresses that sustainable production and consumption of resources should be developed and used where the evidence clearly shows that they are more sustainable compared to conventional petrochemical plastic production. The bio-economy is related to the renewable part of managing agricultural waste [8,13]. Furthermore, the United Nations 2030 Agenda for Sustainable Development aims, among other goals, to substantially reduce waste generation through prevention, reduction, recycling, and reuse (Goal 12.5) and to prevent and significantly reduce marine pollution of all kinds, in particular from land-based activities, including marine debris and nutrient pollution (Goal 14.1) [14].

Thus, one of the challenges to our society is to decrease the amount of durable and non-biodegradable packaging materials, such as glass, metal, and mainly plastic, and to find new solutions. The search for viable alternatives with suitable packaging properties is continuously under study, and the reduction in these wastes can be achieved with the development of new environmentally friendly packaging systems [11].

New packaging systems with bioplastics have been developed in the last two decades. This packaging includes materials derived from renewable resources and/or biodegradable polymers, and ranges from flexible films to rigid materials that have a high potential to produce sustainable packaging. These bio-materials are usually blended to control and achieve desirable mechanical, physical, and barrier properties [15]. Although cultural, economic, and even culinary factors from different geographic areas may contribute in different manners to shaping and selecting these different environmental friendly materials [16], the main objectives of this work are to present a literature review of the different properties of these materials, regulations, and mechanisms of biodegradation, to create a holistic overview of the most important bioplastics available in the market at an international level and their potential application for food packaging, environmental impact, a systematic review how consumers perceive bioplastics, and future trends.

2. Definitions and Regulations

According to the European Bioplastics Organization (EBO), the term 'bioplastics' refers to both the bio-based origin of plastic and/or its biodegradable character (Figure 1) [17]. Those derived from plant-based materials (also known as biomass) are bio-based plastics according to the European Standard EN 16575 from 2014 [18,19]. However, it is not only bio-based plastics that are biodegradable, and not all types of bio-based plastics are biodegradable [18], as will be discussed in Section 3.

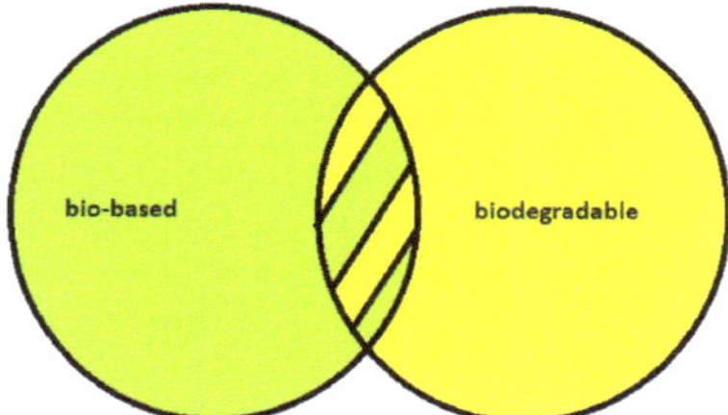

Figure 1. Bioplastics are bio-based, biodegradable, or both (adapted from European Bioplastics [17]).

Hence, we should appropriately define the vocabulary surrounding bio-plastics. From a chemical point of view, and in contrast to the most frequently used types of plastics worldwide (polyolefins are by far the most abundant [20]), the vast majority of substances among biopolymers are linked via heteroatom bonds. This is due to the fact that selective linkage of C–C bonds is chemically very challenging, and regioselective cleavage of non-polarized bonds even more so [21,22]. In nature, reversibility and energetically favorable activatability are essential in the enzymatically catalyzed biosynthesis of structural and storage polymers (in fact the utilization of artificial enzymes for chemical synthesis is an increasingly studied field, with the potential to shift synthetic chemistry toward more environmentally friendly and less energy-intense methods) [23]. This is usually based on nucleophilic substitution of carbon centers (mostly carbonyl or acetals/ketals) positively polarized by doubly- or singly-bound oxygen, with the linkage of C–O or C–N hetero bonds. In addition, the monomer building blocks must be capable of aqueous solvation to enable polymerization and are activated with suitable leaving groups to provide the energy needed for biosynthesis (typically nucleotide activation of building blocks, such as carbohydrates or amino acids). The substitution reactions are catalyzed by selective enzymes, such as peptidyl transferases [24], glycosyl transferases [25], or polyester synthases [26], while enzymatic polymerization, as well as artificial enzymes, are also important objects of research [27,28]. Less polar monomers, such as lignin precursor molecules, are typically conjugated with polar compounds, such as carbohydrates, to enable transport in the cytosol, which is mandatory for the further biosynthesis of wood [29]. This results in the classes of substances available as biopolymers, most of which are derived from functionalized carbonyl groups. These include carboxylic acid derivatives, such as proteins or polyesters, and acetals/ketals, such as carbohydrates. Due to the aforementioned requirements for monomers and enzymatic reactions, namely water solubility and the possibility of forming hetero-bonds, an increased functionalization with polar groups, such as alcohols, amines, or carboxylic acids, is found and, thus, a tendency towards the polar character is identified.

This results in significant physicochemical properties of the material due to the increased intra- and intermolecular interactions, which influence processability, barrier properties, and several other factors. These properties include higher crystallinity and melting or glass transition temperatures, whereby the intermolecular interactions outweigh the intramolecular interactions due to strong hydrogen bonds (strong interactions lead to higher heat resistance but also a higher tendency of water absorption) in the extreme case of carbohydrates. This results in decomposition instead of melting, and the number of hydrogen bonds must, therefore, be reduced either by additives, e.g., when obtaining thermoplastic starch (TPS), or by chemical modifications to enable thermoplastic processability [30]. An important exception to this is lignin, which contains a mixture of phenolic ethers and radically linked carbons, i.e., it is comparable to phenolic resins, such as Bakelite, can be used as a basis for similar materials, and, thus, has a much more apolar character, as well as poor water solubility [31–33]. The typical thermoset networks are, therefore, particularly stable and also require organisms capable of degrading lignin to expend more energy than other biopolymers. This, and the inhomogeneity of the material, also depending on the starting material, have led to the fact that lignin has hardly been used for packaging so far, despite its abundance and inexpensive availability. Nevertheless, it has a lot of potentials to be utilized for water vapor barrier functionality [34].

In relation to the bio-based origin, there is no general agreement on a specific reference limit; however, threshold values of renewable content that mark the bio-based nature of a material can be found in national regulations [35]. For example, the United States Department of Agriculture (USDA) BioPreferred Program depended on product category factors to determine a wide range of minimum acceptable bio-based content of between 7–95% [36]. However, certifiers, such as the certification organization of TÜV Rheinland, the German Technical Inspection Association, and DIN, the German Institute for Standardization (DIN CERTCO), and the Technical Inspection Association (TÜV) AUSTRIA Belgium, provide standardized labels that indicate the biomass content of bio-based materials [35,37].

According to the European Commission (EC) policy recommendation, waste-to-energy (WtE) processes respect the waste hierarchy, making co-combustion processes energy-efficient techniques. This leads to the maximization of the circular economy's contribution to decarbonization [38].

The EU has addressed the problem of plastic food packaging in its plastic strategy and Circular Economy Action Plan [39]. The transition towards a circular economy is offered as a comprehensive solution for the plastic crisis. This requires various collaborations and the engagement of different societal actors, such as citizens and consumers, authorities, policymakers, and non-governmental organizations (NGOs), whose aim is the creation of novel producing methodologies for packaging materials and the manufacturing of sustainable foods.

The negative environmental impacts have raised increasing concerns, both in public media forums and in the cabinets of policymakers [8]. Several policies and regulation measures include the reduction or ban of single-use plastics [40]. Voluntary measures, such as collaborative commitments [41] and pacts [42] to foster the circular economy of plastics, have been proposed by public and private bodies to address the problems caused by plastic food packaging.

Since the establishment of the United Nations' 17 sustainable development goals (SDGs) [43], many companies have advocated sustainable practices. These goals aim to make use of renewable sources without causing impacts on human health (SDG3), climate change (SDG13), to preserve life below water in oceans, seas, and marine resources for sustainable development (SDG14), and to protect life on land (SDG15). Circularity is one of these goals which aims to tackle SDG11 (sustainable cities and communities) and SDG12 (responsible production and consumption). However, the transition toward environmentally-friendly plastics following the adoption of the SDGs is still slow and requires country-specific policies.

This is due to the many choices and approaches followed by producers, consumers, and policy-makers. A shift towards the circularity and sustainability of plastics is required.

Policy measures are essential for the management of plastic waste and mitigation of its generation. They should be enforced at all stages of collection, storage, transportation, and final disposal or recycling. Of course, these policies should be financially sustainable, and technically sustainable, and should incorporate social, legal, and environmental aspects [44].

These measures will include prevention strategies for the reduction in waste and control of types of waste and materials through bans, restrictions, and control strategies by the adoption of standards and protocols, and practices on the ground. Allocation of different roles and responsibilities for each party among stakeholders is also essential [45–47].

The Chinese waste import ban of 2017 showed the highest impact on the reduction in plastic waste. This pushed several countries to find other solutions for their plastic waste.

Table 1 shows the percentage of imports and exports of plastic waste referring to some European and non-European countries, while Table 2 shows the countries with regulations about types of banned plastic materials.

Table 1. Percentage of imports and exports of plastic waste (adapted from Plastic Atlas [48]; Filiciotto and Rothenberg [49]).

	Malaysia	Thailand	Vietnam	USA	Japan	Germany
Imports	11%	6%	5%			
Exports				16%	15%	13%

The EU-28 represents the largest exporter of plastic waste, accounting for around one-third of all exports of plastic waste from 1988 to 2016 [50]. Most of this waste has now been halved and re-routed to Vietnam, Thailand, and Malaysia [51].

Table 2. Countries with regulations about types of banned plastic materials.

Countries	Level	Types of Banned Plastic Materials	References
Canada, Costa Rica, Taiwan, Belize, India, and the USA (California and Florida)	National bans	Single-use plastics (SUPs), including plastic bags, straws, and cutlery	[52]
The Netherlands, Tanzania, Australia, Italy, South Korea, New Zealand, the UK, the USA, and Canada	National bans	Microbead plastics	
25 African countries	National bans	Plastic bags	[53]
Australia	National bans	Lightweight plastic bags	
Papua New Guinea	National bans	Nonbiodegradable plastic bags	

Retailers have taken voluntary actions to reduce plastic bag consumption within the European Union. For instance, many supermarkets have voluntarily abolished the provision of (free) plastic bags (such as in Austria and Lithuania) and others have introduced a value of around EUR 0.05–0.10 per single-use plastic bag (Belgium, Estonia, France, Germany, Hungary, Latvia, the Netherlands, Portugal, Sweden, Slovakia, and the UK) or have substituted them with biodegradable plastic bags (Austria, France, and Sweden) or used alternative bags made of cotton, hessian, or linen. Plastic pollution of the environment can be reduced by interventions, such as 'Operation Clean Sweep', organized by non-governmental organizations (NGOs) to clean beaches and drains [54]. Reusable bags are produced by NGOs who sell them to finance their activities in part. Raising awareness through media campaigns or billboards to remind customers to reuse their bags is another strategy adopted by the UK. Finally, paying customers a small amount of money (around EUR 0.10) if they do not take any plastic bags is supported in the UK [44].

Extended producer responsibility (EPR) is another policy mechanism that aims to mitigate the risks associated with waste management. With EPR, the mitigation of the environmental impacts of products throughout their lifecycle stages is accomplished by producers who are legally and financially responsible. Indeed, EPR can help in plastic pollution prevention and mitigation by limiting the health, safety, environmental, and social impacts of plastic products [55]. However, difficulties with enforcement have been reported.

Hence, the implementation of recycling processes and the development of biodegradable plastics are some of these strategies. Europe halved its monthly plastic waste export with these restrictions (from 300 to 150 kton) [49] and, in 2019, the Basel Convention called for more domestic solutions in dealing with (hazardous) waste [56]. This is signed by 187 countries worldwide (excluding the US, among others).

At the European level, the new EU Green Deal 2020 is targeting (illegal) waste exports to third countries. At the same time, a regulatory framework for biodegradable and bio-based plastics is set to be implemented aiming at the local improvement of waste management techniques and leading to the push of recycling processes forward, hence, reducing the need for biodegradable plastics. The development of both circular and bio-economies will be implemented by the amelioration of rural areas with a new financial plan [57].

Financially speaking, setting clear criteria for the assessment of green investment funds is one of the goals of the 2018 EU regulation facilitating sustainable investment in this direction [58,59]. Europe imposes fees to discourage plastic production under the extended producer responsibility (EPR) concept [60]. Moreover, the European Chemical Agency (ECHA) has recently discussed intentionally-added microplastics (e.g., microbeads in cosmetics) by the provision of a socio-economic assessment [61].

Substances of very high concern (SVHCs, i.e., carcinogenic, mutagenic, or toxic for reproduction, or CMR, and persistent and/or bio-accumulative substances) are being banned by the REACH regulation at the EU level [62,63] due to the cumulative and detrimental effects of (micro) plastics. In the future, the EU Green Deal [6], as well as the REACH registration of polymers, might aid in the classification and management of hazardous substances in (new) waste streams.

Currently, California law wants to phase out plastics that cannot be compostable or recyclable, but even this legislation faces bureaucratic resistance [64]. Other countries, such as China, support research on biodegradable plastics via funding, but also have limited policies [65].

California developed policy concepts in 2013 to make the producers of selected products responsible not only for recycling but also for litter prevention and mitigation. This new policy required a reduction in their products' total volume in the environment by 95% in 11 years [55]. Bureaucracy might be a major obstacle in achieving these goals. Moreover, it might work well for some products but not for others. Difficulties with enforcement might also occur, and the problem of data scarcity has been reported in developing countries [66].

Finally, political will might be lacking due to countries having other priorities. Some ways to promote the political will are to make this the priority of the country analyzing the impact of environmental changes on health and society. Governments should employ tools that allow all consumers to enhance their awareness of the management of plastic and plastic waste. Consumers should change habits and lifestyles that require plastic usage, e.g., by means of a reduction in the reliance on single-use plastics or through source preparation and social awareness, and public education programs should also be included [44].

3. The Common Misconception in the Definition of Biodegradable and Compostable Polymers

Degradable polymers are polymers that disintegrate by different mechanisms, including physical, chemical, and/or biological processes, resulting in a loss of some properties that may vary as measured by standard test methods appropriate to the plastic. A biodegradable polymer is defined as a polymer that undergoes degradation due to the action of various microorganisms within a specific period and environment. A compostable polymer is a polymer that is degraded by biological actions during composting to yield carbon dioxide (CO_2), water (H_2O), and inorganic compounds. However, the terms "biodegradable" and "compostable" may lead to confusion among consumers and other stakeholders. The simplified distinction between the two terms is accepted that all compostable plastics are biodegradable but not all biodegradable plastics are compostable, so the two terms are not to be used interchangeably. In addition to these two main terms, there are some other complex definitions, such as home compostable, industrial compostable, and marine-degradable, regarding biodegradable polymers. *Industrial compostable polymers* are composted under a controlled process (very strictly controlled oxygen, water, and heat input) in industrial composting plants to be used in agricultural applications, while *home compostable polymers* are defined as polymers that can fully decompose in the soil [67]. On the other hand, *marine-degradable* plastics are plastics that can be degraded into CO_2 and H_2O_2 in marine environments, including coastal and ocean waters, lakes, lake-connecting waters, subsoils, submerged lands, and sea and coastal habitats, under light, heat, or microbial effect. However, a harmonized EN standard for only industrially compostable packaging exists, whereas no general standard for marine biodegradation is implemented. Currently, no detailed EU law is present for bio-based, biodegradable, and compostable plastics. The EU Commission announced a policy framework where resources of bio-based feedstock and the environmental benefits of using biodegradable and compostable plastics will be evaluated, as well as the conditions for these uses [68].

Brief Overview of Degradation Pathways for Polymers

At present, the complexity of biodegradation is accepted, as it includes several steps, such as biodeterioration, depolymerization, assimilation, and mineralization [69]. The biodegradation steps and mechanisms behind this process have been exclusively addressed elsewhere [69–73]. In this part, a very brief overview of degradation pathways is provided, which is then to be associated with the environmental impact of bioplastics.

Biodegradation is a process that degrades materials into CO_2, H_2O, biomass, and CH_4 with the help of living microorganisms under various environmental conditions, such as compost, soil, marine conditions, or other mediums [74]. Abiotic degradation, such as oxidative or hydrolytic degradation, may initiate or enhance biodegradation by increasing the surface area of the organism–polymer interface [69,75,76]. In general, enzyme-catalyzed or biotic reactions are efficient methods for the biodegradation of polymers. Furthermore, after the abiotic and/or biotic degradation of polymers, the final products are bio-assimilated by microorganisms to be used as growth factors or in cellular respiration (Figure 2) [70].

Polymer biodegradation results in various products depending on whether it occurs under aerobic or anaerobic conditions. As mentioned earlier, in aerobic degradation, oxygen is utilized as the final electron acceptor while, in anaerobic degradation, CO_2, nitrates, or sulfates are used as the electron acceptors by microorganisms to produce the energy needed to maintain cell functions [77,78]. However, most of the biodegradable polymers biodegrade under both aerobic and anaerobic conditions [79] and, in enzymatically degradable polymers, such as PLA (polylactic acid), temperature plays an important role in how polymer scission occurs [71].

Aerobic biodegradation is the conversion of organic carbon into CO_2 and water as a result of microbial metabolism in the presence of oxygen. In anaerobic biodegradation, methane is produced, while some CO_2 can be obtained depending on the residual oxygen or the type of degraded material. Soil biodegradation, composting, and marine biodegradation are the main areas of aerobic biodegradation standards, whereas sewage sludge biodegradation, anaerobic digestion biodegradation, and (accelerated) landfill biodegradation are the main areas of anaerobic biodegradation standards [70]. Landfills may result in the uncontrolled biodegradation of plastic materials with methane release to the environment, while biogas facilities are a part of anaerobic digestion systems, capturing the released methane for energy conversion [80]. Inappropriate applications in the biodegradation of polymers may result in methane release in the environment due to the switching from anaerobic to aerobic conditions [70].

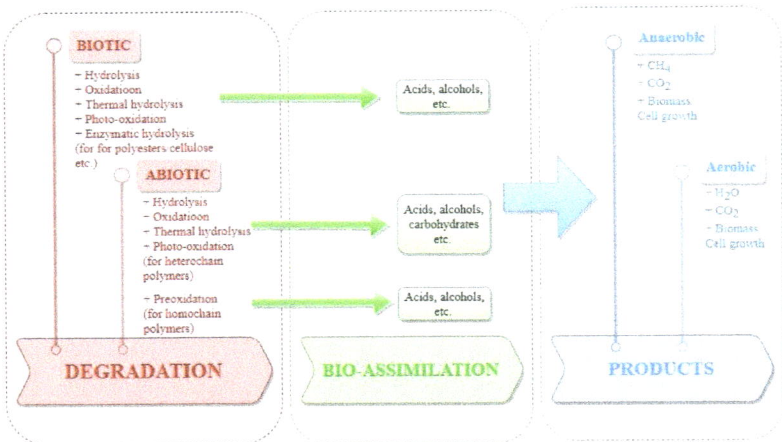

Figure 2. Theoretical biodegradation pathway for polymers (adapted from Meereboer et al. [70]).

4. Research on Bioplastics

The lifespan of plastics produced from petrochemicals has been proven to be several decades, and the need to replace them with bioplastics is more urgent than ever. For example, packaging materials made of PET (such as beverage bottles) have a proven lifespan of more than 90 years [81].

The production of biopolymers is based on living organisms and takes advantage of various properties, such as strength, stability, and flexibility. Plants, crops, animals, and microorganisms are the basic raw materials that can be used to produce biopolymers [82]. Producing innovative bioplastics using biological raw materials is expected to lead to significant benefits in certain areas, such as the environment and the economy [83]. The classification of biopolymers into different categories can be carried out in different ways, since the number of resources from which they arise is extremely large [84]. A classification system concerns the division into categories based on how biodegradable they are and according to their biomass content. Based on these criteria, there are (i) bio-based and non-biodegradable, (ii) biodegradable and bio-based, and (iii) biodegradable and fossil-based alternatives [85]. Another classification can be made according to the origin of the resources, which means that it is possible to have biopolymers derived exclusively from renewable resources and polymers which are mixtures of biopolymers and commercial polyesters [84]. Bio-based and biodegradable biopolymers can also be categorized into synthetic biopolymers (synthesized from bio-derived monomers), microbial biopolymers (produced by microorganisms), and natural biopolymers (extracted from biomass) [86]. Polysaccharide-based films, protein-based films, or a combination of both are the biopolymers with the greatest potential in film making. In food packaging, important pathogens, such as *Listeria monocytogenes*, *Salmonella*, *Campylobacter*, *Bacillus cereus*, *Saccharomyces cerevisiae*, *Staphylococcus aureus*, *Aspergillus niger*, and *Clostridium perfringens*, may survive and develop depending both on the conditions inside the packaging but also on the conditions of the external environment of the packaging. Much biodegradable green packaging has significant antimicrobial functions due to the bioactive compounds contained in plant by-products [87].

4.1. Protein-Based Bioplastics

Protein-based bioplastics can be derived from raw materials of both plant and animal origin. Common sources of plant origin are wheat gluten, soy, pea, corn zein, and cottonseed proteins. On the other hand, whey, casein, collagen, gelatin, and keratin are some proteins of animal origin [88]. Because proteins consist of different types of amino acids, the strong intermolecular binding of proteins affects the functional properties of protein-based bioplastics, giving them superior characteristics in comparison with carbohydrates and lipids [89]. Protein-based films are extremely popular, as they are abundant, inexpensive, non-ecotoxic, biodegradable, and have very good film-forming properties [90].

4.2. Polysaccharide-Based Bioplastics

Polysaccharides have also been proposed as a biopolymer source for bioplastics [91]. Alginate, cellulose, pectin, and starch are derived from plants, while glycogen and chitin are of animal origin [92].

4.2.1. Cellulose-Based Bioplastics

Cellulose is the most abundant biopolymer available on the planet, gaining an important role in the production of new materials. Cellulose is renewable, widely available, non-toxic, low-cost, environmentally friendly, biocompatible, biodegradable, thermally and chemically stable, and derivable [93,94]. Fruit and vegetable waste is very rich in this valuable biopolymer. Cellulose esters and cellulose ethers are the main cellulose derivatives that are used in industrial applications, as the production of pure cellulose bioplastics still remains quite difficult, due to the structural complexity and difficulty in melting and dissolving it through standard processes [95]. Mechanical properties, thermal stability, and water absorption are some properties of bioplastics that could be improved with the addition of cellulose [96].

4.2.2. Starch-Based Bioplastics

Potato is the main source of starch for the production of bioplastics. Cereals and legumes, such as wheat, rice, barley, oat, corn, beans, and soy, are also significant sources [97]. Starch must be incorporated with many plasticizers, as the main problem with starch in the food packaging industry is its low plasticity [98].

4.3. Synthetic Bioplastics

The main synthetic bioplastics are PBS (polybutylene succinate), PLA (polylactic acid), PVOH (polyvinyl alcohol), PGA (polyglycolic acid), and PCL (polycaprolactone), [97].

Indeed, PLA is one of the most commonly used bioplastics and, in the year 2021, had the largest market share for the production capacity of biodegradable bioplastics worldwide [99].

On the other hand, PCL is easily processable, belongs to semi-crystalline polymers, and is fully biodegradable. As a result, 11% of the total market of biodegradable polyesters is held by PCL. It is a bioplastic with excellent compatibility with other polymers and additives, which makes it very promising in food packaging in the future. The PGA bioplastic has a similar chemical structure to PLA, but it is characterized by improved degradability, mechanical properties, and gas barrier properties that make this a beneficial supplement to PLA. Indeed, PBS is extremely flexible, elastic, and biodegradable, with a low glass transition temperature. Another bioplastic, PVOH, is widely used for food packaging due to its good film-forming ability, biodegradability, non-toxicity, water processability, and low cost [100].

The following tables (Tables 3 and 4) present studies on bioplastic materials for food packaging and their properties developed using fruit and vegetable by-products during 2017–2021. European countries, the USA, China, and India are among the countries that contributed to the development of these bioplastic materials.

Table 3. Studies on bioplastic materials for food packaging developed from fruit by-products during 2017–2021.

Fruit By-Products	Type of Bioplastic Materials	Target Microorganisms	Physical and Mechanical Properties	References
Apricot kernel essential oil	Chitosan films	Reduction in fungal growth on packaged bread slices	Improved water resistance, increased tensile strength	[101]
Grapefruit seed extract	Coating of alginate and chitosan films	Reduced bacteria count by 2 log CFU	Increased barrier properties	[102]
Grapefruit seed extract	Carrageenan films	Large inhibitory zone against *Listeria monocytogenes*, *Escherichia coli*, and *Bacillus cereus*	Increased water vapor permeability and surface hydrophilicity	[103]
Coconut husk extract	Nanocomposite films or gelatin films	-	Improved water sensitivity	[87]
Mango peel flour and extracts of mango seed kernel	Biodegradable coatings and films	-	Good barrier and antioxidant activity	[104]
Mango kernel extract	Soy protein isolate and fish gelatin films	-	Thicker and more translucent films, increased tensile strength, decreased the water solubility, and increased antioxidant activity	[105]
Apple peel polyphenols	Chitosan films	-	Increased thickness, density, solubility, opacity, and swelling ratio, and antioxidant and antimicrobial activities	[106]
Apple skin extract	Carboxymethylcellulose films	*Listeria monocytogenes*, *Staphylococcus aureus*, *Salmonella enterica*, and *Shigella flexneri*	Enhanced mechanical, water barrier, solubility, and antioxidant and antimicrobial activities	[107]
Banana peel extract	Chitosan films	-	Reduced hydrophilicity and excellent antioxidant activity	[108]
Pomegranate peel extract	Chitosan–pullulan composite edible coatings	-	Resistance to water loss and gas transpiration	[109]
Pomegranate peel powder	Gelatin films	*Staphylococcus aureus*, *Listeria monocytogenes*, and *Escherichia coli*	Increased antioxidant and antimicrobial activities	[110]
Pomegranate peel extract	Zein films	*Staphylococcus aureus*, *Escherichia coli*, *Pseudomonas perfringens*, *Micrococcus luteus*, *Enterococci faecalis*, *Proteus vulgaris*, and *Salmonella typhii*	Increased tensile strength and antioxidant Activity, and decreased film solubility and water vapor transmission rate	[111]
Blackcurrant pomace powder	Pectin-based films	-	Increased water vapor permeability and antioxidant activity, and decreased tensile strength	[112]

Table 4. Studies on bioplastic materials developed from vegetable by-products for food packaging during 2017–2021.

Vegetable By-Products	Type of Bioplastic Materials	Target Microorganisms	Physical and Mechanical Properties	References
Whole potato peel	Active biodegradable films incorporated with bacterial cellulose and curcumin	-	Improved tensile strength, reduced water vapor, permeability, oxygen permeability, and moisture content	[113]
Tomato extract	PVOH films mixed with chitosan and itaconic acid	*Staphylococcus aureus, Pseudomonas aeruginosa, Salmonella enterica Enteritidis,* and *Salmonella enterica Typhimurium*	Improved physical properties	[114]
Lycopene from tomato extract	Poly-lactic acid films	-	Improved barrier against light and oxygen	[115]
Red cabbage extracts	Gelatin films	-	Increased water solubility, water vapor permeability	[116]
Red cabbage extracts	Active fish gelatin films	-	Improved water and mechanical resistance, and antioxidant activity	[117]
Red cabbage anthocyanins	PVOH and starch, propolis, anthocyanins, and rosemary extract composite films	*Escherichia coli, Staphylococcus aureus*	Improved mechanical strength	[118]
Solid sweet potato by-product	Poly(3-hydroxybutyrate-co-3-hydroxyvalerate) composites	-	Increased thermal stability	[119]
β-carotene from carrot	Films based on cassava starch	-	Increased thickness, and greater stability and solubility	[120]
Tomato-based pigments	PVOH-based biofilms	-	Reduced transparency and increased mechanical resistance	[121]
Okra mucilage	Carboxymethyl cellulose with ZnO nanoparticle nanocomposite films	*Staphylococcus aureus*	Reduced microbial growth, oxidation, and gas production.	[122]

The presented research studies from the last five years show the great potential of these types of materials. The following section will present the existing main types of utilizations for bioplastics in packaging materials, their main properties, and their applications at an industrial level.

5. Applications

In general, there are four main types of utilization of bioplastics in packaging materials, as follows:

1. Structural material—bioplastic is used as a mono-material, for polymer blends, or in composite materials [123–125];
2. Coating—bioplastic is used as a coating on the substrate material, forming a multi-layer material to increase barrier functions, enhance processability (printability and sealability), functionalize the surface, or serve another duty. Typically, a coating is accomplished by extrusion, film casting, or common lacquer application techniques [126];
3. Additive—biopolymers can be utilized as functional additives in plastics to control physico-mechanical properties, such as strength, stiffness, hardness, or barrier functions (e.g., nanocellulose diffusion barriers) [127,128];
4. Filler—bio-based materials serve as fillers that can reduce material costs and/or increase the ratio of renewable resources in bioplastic packaging materials [129].

Holistic approach for material selection

In general, there are a number of different, at least partially, plastic-based packaging systems. In this context, plastic can be represented either structurally or as a functional coating. The applicability of different plastics is primarily limited by mechanical material properties, which in turn can be derived from the molecular basis. These include, for example, rigid trays (T), bottles (B), pouches (P), coated cardboard (C), films, and wraps and bags (F) [97,128].

5.1. Processing

There are a number of different processes for manufacturing the various types of plastic-based packaging products mentioned above, each of which has specific requirements for different physical material properties with special emphasis on rheology [130]. Plastic melts are non-Newtonian shear-thinning (viscoelastic) fluids [131]. Due to the disentanglement and realignment of the molecules under high pressure, a drop-in viscosity and pseudo-plastic behavior are observed [132]. Typically, methods, such as melt flow index (MFI) measurement, are used to provide fast conclusions about chain length and melt viscosity (where lower values of the same polymer typically correspond to higher viscosity and higher chain length) [133].

For example, for injection molding, sufficient fluidity of the melt must be ensured to fully penetrate the mold [134,135], whereas for extrusion, due to the absence of such a mold, a higher viscosity is advantageous for stability. It is noteworthy that thermal and mechanical processing parameters, as well as throughput rate, may affect material degradation during processing [136]. Therefore, the desired type of packaging and the associated manufacturing method(s) play an essential role in material selection.

Limitations in applicability due to the molecular basis (caused by properties, such as brittleness) are addressed via variations in molecular weight or side-chain length, fillers, additives, plasticizers, blending with other types of polymers, and/or co-polymerization, resulting in different polymer grades and types of plastic tailor-made for various processing methods. It is important to point out that higher amounts of additional components may affect the recyclability of the material and that additives should be chosen carefully to minimize environmental harm after being littered [137].

In the following, the different processing approaches are described:

Extrusion coating and film production (casting and blown film)

In the extrusion process, previously compounded materials are fed into a screw barrel equipped with a screw conveyor, melted, compacted, and pressed via a die through a 2D shaping profile die to produce a continuous polymer strand whose cross-section corresponds to the applied die and which can optionally also contain cavities [138]. For packaging, cast film and tube extrusion are particularly relevant. In extrusion, the flow behavior is decisive for the quality of the product. The use of longer-chain and, therefore, higher-viscosity grades tends to reduce the risk of deformation in the obtained extrudate [139]. In addition, the molecular structure is decisive for the crystallization behavior and, thus, besides processing parameters, influences the sharpness of the melting range. Having control over crystallization behavior is an important aspect of polymer engineering [140]. Cast films typically have lower crystallinity due to rapid cooling and, thus, usually have better transparency and gloss [141]. The method is well-suited for thicker films that are subsequently further processed via thermoforming [142].

Injection molding (I)

Similar to extrusion, pre-compounded plastic is melted and compacted by a screw and conveyed to the injection nozzle. Instead of a profile mold, the material is pressed into an injection mold, allowing 3D structures to be made from plastic. For the process, with higher complexity of the injection mold, good flowability of the material is essential so that the mold is completely and uniformly filled. Furthermore, process parameters, such as mold temperature, significantly affect mechanical properties [143].

Thermoforming (T)

Here, 2D plastic films (semi-finished products) are continuously processed into a stable 3D shape by thermal softening in the elastic range above the glass transition temperature and with the aid of a cooling tool, whereby the process is usually supported by vacuum or compressed air. The films or sheets are clamped to ensure forming with wall thickness reduction [142]. After filling with a sealing film, thermoformed cups and trays are usually sealed by using pressure and spot heating above the melting temperature, whereby chemical compatibility and a similar melting range must be ensured for the material's combination as a basis for homogeneous bonding [144].

Blow molding (B)

In blow molding processes, preforms produced by injection molding are blown into a mold (e.g., PET bottle production) [145] or tubes are extruded and blown into films using ring dies coupled inline to an extruder (e.g., PE bag production) [146].

5.2. Properties

A huge variety of material properties need to be analyzed before a rational decision for a certain material can be made. This decision depends on specific barrier requirements of the packed food and other factors, such as ecological and economic criteria. Often, despite a favorable low price, no clear general pro or con can be formulated for different packaging types. Since mechanical properties affect the processability, materials that are applicable for injection molding may be unsuitable for extrusion and vice versa. Physical and mechanical properties are interconnected and a result of underlying chemical structures of the biopolymers, additives, and fillers, and their inter-and intramolecular interactions. Crystallinity correlates directly with properties, such as brittleness, tensile strength, and gas and aroma permeability. Furthermore, permeability is dependent on solubility that is a function of polarity [147].

5.2.1. Biodegradability

As previously presented in Section 3, the term biodegradable implies that microorganisms can completely degrade a material into elementary components or small molecules within a specific period and environment. Depending on environmental conditions (such as pH, temperature, and/or oxygen availability) resulting in differences in the microbial colonization of diverse habitats, different categories can be used to describe biodegradability. Additionally, standardized test methods are available and can be used to characterize it. However, some of these methods, such as solely analyzing weight loss over time, do not give a sufficient direct proof of biodegradation [148].

Four common categories are proposed that are typically used to describe the behavior of different materials in the context of biodegradability. Here, classification in a lower category automatically corresponds to "upward compatibility" for higher categories (except category 4). These categories are as follows:

Category 1—marine biodegradable (claimed to be biodegradable in the marine environment);
Category 2—home compostable (claimed to be biodegradable in soil without optimized composting conditions);
Category 3—industrially compostable (according to EN 13432);
Category 4—non-biodegradable (within the time frame specified by definition).

It is important to note that especially Categories 1 and 2 currently cannot be sufficiently backed up with standardized methods that allow reliable forecasts for estimating the degradation time in the natural environment. Determining the transferability of defined laboratory conditions is, in many cases, not possible or possible only to a limited extent due to the complexity and abundance of influencing parameters, as stated by Choe et al. [149] in their review which compared results from laboratory and environmental experiments.

Currently, little is known about the ecotoxicological impact of biodegradable micro- and nano-plastics. Increased degradation rates increase the amount of micro-bioplastics coming from biodegradable polymers that pose certain risks, such as shifts in microbial communities (that could destabilize delicate ecological balances). Microplastics from degradable polyesters, such as PLA and PHB (poly-3-hydroxybutyrate), were found to have negative effects on marine benthic communities [150]. A comprehensive recent review by Fan et al. shows that biodegradable microplastics can show more severe effects compared to conventional microplastics [151]. The release of micro- and nano-plastics into the environment during biodegradation will be discussed in more detail in Section 6.2.

5.2.2. Barrier Functions

Barrier functions against gases play a very important role in the selection of materials for food packaging. If a packaging material does not offer an adequate barrier, this can lead to untimely spoilage of the contents (for example, oxidation of sensitive fatty foods caused by an inadequate oxygen barrier [152] or premature wilting of lettuce due to an inadequate water vapor barrier [153]). As already mentioned in Section 4, the permeability is determined by the molecular basis of the material. In this context, permeability is dependent on sorption and diffusion, and there is an important entanglement between sorption and the polarity of a material. Moreover, crystallinity, for example, plays a role in the diffusion process within the phase. A wide palette of measuring methods is available as reviewed in detail by Baschetti et al. [154]. The permeability of a material is a key limitation in the substitution of typical petro-based plastics, such as polyolefins [155] and, where appropriate, it is shifted by combining the plastic with orthogonally effective materials (in the form of multilayer structures, compounds, or additives). In this case, an improvement in barrier properties comes at the potential price of reduced recyclability and/or degradability and is, therefore, a tightrope walk that should be made, taking into account additional considerations, such as life cycle assessment or local recycling infrastructure [156]. It should be noted that, in some cases, biopolymers may also have superior barrier properties, such as oxygen transmission rate (OTR), and that current packaging solutions may have higher barriers than necessary for certain products to secure their typical storage time and shelf life. To avoid potential over-packaging and to save resources in this area, re-evaluations based on storage trials are, therefore, useful in addition to material decisions based on the literature. A detailed permeability comparison between the most common bioplastics and conventional plastics was recently published by Wu et al. [157]. In addition to OTR and water vapor transmission rate (WVTR), other gas transmission rates, such as carbon dioxide transmission rate (CO_2TR), are also relevant for certain products, but these were not addressed in detail in the review. We propose a categorization on a scale basis (powers of ten) for our overview of existing materials, as follows:

OTR—A (<1), B (1–10), C (10–100), D (100–1000), E (>1000) [cm^3/m^2d]
WVTR—A (<1), B (1–10), C (10–100), D (100–1000), E (>1000) [$g/m^2/d$] at 25 °C

5.2.3. Feedstock

In the context of bioplastics, the question of the underlying resources is crucial, especially in terms of sustainability. Inherently, bioplastics are obtained from renewable raw materials and are the focus of research as an approach to the transition to a circular economic model. In this context, a comprehensive accompanying life cycle assessment [158] is essential to act as sustainably as possible in the choice of materials. The gap in knowledge on detailed life cycle assessment (LCA) data needed for properly assessing bioplastics has been discussed and has become the focus of research activities [159]. Tools, such as the "Product Environmental Footprints (PEF)" system developed by the EU Commission, serve to consider a large catalog of criteria, instead of one-dimensionally looking only at CO_2 footprints to avoid distorting the picture of the actual most sustainable solution [160]. The production of bioplastics requires resources, such as land and water, and can, therefore, compete with food or fodder production and lead to environmental pollution, for example through eutrophication [161]. Directly linked to this are food security and other SDGs that need to be considered. Therefore, it seems reasonable to present different possible feedstocks for the production of bioplastics [162,163], and the following categories were defined for the overview table:

Petrol-based (P);
Natural biomass (N);
Monomers from starch/food or feed competition (first-generation) (S);
Agricultural waste/nonfood competition land use (second-generation) (W);
CO_2 or other feedstocks decoupled from land use (third-generation) (C).

5.2.4. Price

One of the greatest current obstacles to the wider use of biopolymers as a substitute for conventional materials lies in their unattractiveness in terms of price, especially in the scenario where more expensive substitute materials do not meet the necessary barrier requirements to the same extent due to molecular differences. Especially in the case of food packaging, which belongs to the fast-moving consumer goods (FMCG) sector, profit margins are often low and, thus, the scope for increased packaging costs due to more expensive materials is correspondingly limited [164]. Nevertheless, there is strong customer demand for bio-based food packaging [165]. Four categories were defined and, as the category increases, the economic applicability shifts from potential substitute material in the FMCG sector to high-priced niche applications. The classification corresponds to the state of knowledge at the time of writing, i.e., a snapshot, and there may be transitions between different categories in either direction in the future. These categories are as follows:

Category A (0.5–2 €/kg);
Category B (2.1–5 €/kg);
Category C (6–10 €/kg);
Category D (>11 €/kg).

5.2.5. Production

Bioplastics account for a small but growing share of total plastics production (2019: around 1%; 2.11 million tons [166]). In addition to price aspects, the level of production capacities is also a main factor for the security of supply and, thus, affects the choice options for the materials in question, particularly for larger production volumes, since demand exceeds the current supply on the market [164]. For this reason, annual production capacities in this review are divided into the following four orders of magnitude (as with price, these are snapshots at the time the review was written):

Category A (>100 kt/a);
Category B (51–100 kt/a);
Category C (10–50 kt/a);
Category D (<10 kt/a).

5.2.6. Food-Contact Material

According to Regulation No. 1935/2004 [167], food contact materials must not transfer chemicals that are hazardous to health into food products. The approval of bioplastics for direct food contact is regulated in EU Commission Regulation No.10/2011 [168]. According to the classification, materials without direct contact can, for example, be used externally in a multilayer composite, provided that an intervening functional barrier ensures that a defined migration limit is not exceeded. Some novel materials require more detailed investigation and classification. In any case, supplementary migration measurements, mostly with simulants, on packaging prototypes are also necessary. These include, on the one hand, total migration, in which the total mass of migrated substances is quantified without detailed characterization, and, on the other hand, specific migration, in which specific contaminants, such as endocrine disruptors or carcinogens are tested for.

However, toxicological knowledge is still very limited. As an example, some recent studies suggest alterations in steroid hormone metabolism caused by acetyl tributyl citrate, a common replacement for phthalate plasticizers [169,170]. On a side note, non-intentionally added substances (NIAS) that can be a result of processing conditions or chemical reactions during food storage (e.g., under acidic conditions) should be of special concern when dealing with complex bio-based and novel materials [171,172]. Moreover, potential allergenic effects are worth investigating [173]. The following cases can be identified:

Not tested (~);
Declined (o);
Approved (+).

5.3. Examples

Bioplastics are rarely used as mono-materials but are typically applied as blends (in many cases compatibilizers are added to improve the miscibility) or multilayers to optimize the mechanical properties as well as barrier functions. For polar compounds, such as protic polyols (carbohydrates), modifiers, such as glycerol, are added to break hydrogen bonds and allow for thermoplastic behavior. Furthermore, additives are normally used to change the physical properties of materials. Therefore, Table 5 is based on application examples that contain the previously discussed polymers as the main structural component and do not always refer to pure material.

Foods **2022**, 11, 3087

Table 5. Overview of existing materials.

Class	Degradability				Barrier		Processability						Feedstock					Application				FC	Price	Prod	References
	M	H	I	N	O	W	I	C	E	T	B	P	N	W	S	C	T	B	P	C	F				
Proteins																									
Zein	X	X	X	-	B	D	X	X	X	X	X	-	X	X	X	-	?	?	?	X	X	+	B	?	[174–181]
Gluten	X	X	X	-	B-C	C-E	X	X	X	X	?	-	X	X	X	-	?	?	?	X	X	+	?	?	[182–187]
Soy	X	X	X	-	C	D	X	X	X	?	?	-	X	X	X	-	X	?	?	X	X	+	B	?	[188–191]
Whey	X	X	X	-	A-B	A-B	X	X	X	?	?	-	X	X	X	-	?	?	X	X	X	+	C	?	[192–194]
Casein	X	X	X	-	A	C	?	?	?	?	?	-	X	X	X	-	?	?	?	X	X	+	B-C	?	[195]
Collagen	X	X	X	-	-	C	?	?	?	?	?	-	X	X	X	-	?	?	?	X	X	+	C	?	[196]
Keratin	X	X	X	-	-	A-B	?	?	?	?	?	-	X	X	X	-	?	?	?	X	X	+	?	?	[197]
Carbohydrates																									
Cellulose-based	X	X	X		C	D	X	X	X	?	?	-	X	X		X	X	-	X	X	X	+	A-B	B	[198]
Starch-based	X	X	X	-	C	C-D	X	X	X	?	?	-	X	?	X	X	X	X	X	X	X	+	B	A	[199,200]
Chitosan	X	X	X	-	B-C	C-D	-	X	-	-	-	-	X	X	?	?	?	?	?	X	X	+	B-D	?	[201,202]
Alginate	X	X	X	-	B	D-E	-	X	-	-	-	-	X	X	?	?	?	?	?	X	X	+	B	-	[203–205]
Polyesters																									
PLA	-	-	X	-	D	D	X	X	X	X	X	-	X	X	X	-	X	X	X	X	X	+	A-B	A	[168,201,206,207]
PHA	X	X	X	-	C	C	X	X	X	X	X	-	X	X	X	X	X	X	?	X	X	+	D	C	[70,208–212]
PBS	-	X	X	-	D	B-C	X	X	X	X	?	X	X	-	X	-	X	X	?	X	X	+	B-C	B	[213]
PBAT	?	X	X	-	D	C	X	X	X	X	?	X	-	-	-	-	X	X	X	X	X	+	B	A	[214]
PEF	-	-	-	X	B	B-C	X	X	X	X	X	-	X	X	X	-	X	X	?	?	?	~	B	D	[164,215,216]
Bio-PET	-	-	-	X	D	C	X	X	X	X	X	X	X	X	X	-	X	X	X	X	X	+	A-B	A	[164,217,218]
PGA	?	X	X	-	B	B	X	X	?	?	?	X	X	?	?	?	?	?	?	?	?	-	B	?	[219]
Ethers																									
Lignin-based	?	X	X	-	E	D	?	?	?	?	?	-	X	X	-	-	?	?	?	X	X	~	A-B	?	[220]

Table 5. *Cont.*

Class	Degradability				Barrier		Processability					Feedstock					Application					FC	Price	Prod	References
	M	H	I	N	O	W	I	C	E	T	B	P	N	W	S	C	T	B	P	C	F				
Polyolefins																									
Bio-PE	-	-	-	X	E	B	X	X	X	X	X	X	X	X	X	-	X	X	X	X	X	X	A-B	A	[164,221]
Lipid-based																									
Waxes	?	X	X	-	E	B-C	X	?	?	?	?	X	X	X	-	-	?	?	?	X	X	X	B	?	[222–224]
Fatty Acid-based	?	X	X	-	?	?	?	?	?	?	?	-	X	X	-	-	?	?	?	X	X	+	?	?	[225]

General symbols are as follows: no data available (?), applicable (X), not applicable (-); degradability categories are as follows: marine (M), home compostable (H), industrial compostable (I), non-compostable (N); barrier categories are as follows: OTR (O), WVTR (W); barrier values are as follows: A (<1), B (1–10), C (10–100), D (100–1000), E (>1000) [$g/m^2/d$]; processability categories are as follows: injection molding (I), film casting (C), extrusion (E), thermoforming (T), blow extrusion/molding (B); feedstocks are as follows: petrol-based (P), natural biomass (N), monomers from starch/food or feed competition (S), agricultural byproducts/nonfood competition land-use (W), CO_2/decoupled from land-use (C); application categories are as follows: rigid trays (T), bottles (B), pouches (P), coated cardboard (C), films, wraps, and bags (F); food contact (FC) categories are as follows: approved (+), declined (-), not tested (~); price categories are as follows: A (0.5–2), B (2.1–5), C (6–10), D (>11) [€/kg]; production capacity (Prod) categories are as follows: A (>100), B (51–100), C (10–50), D (<10) [kt/a]; here, PBAT refers to poly(butylene adipate-coterephthalate).

5.4. Commercial Applications and Supply Chain

As outlined in the section above, a multitude of different materials has been developed through academic and industrial research. Most of the packaging include hot and cold cups, capsules, bowls, bags, overwrap and lamination films, pouches, and containers for different types of products, such as coffee and tea, beverages, salads, potato chips, bread, yogurt, fruits, vegetables, sweets, and pasta [86,97]. Specifically, starch-based materials are used as an alternative to polystyrene (PS) in disposable tableware and cutlery, coffee machine capsules, and bottles. Cellulose is used in bio-based trays wrapped with cellulose film, and cellulose-based packaging is used for bread, fruits, meat, and dried products. Additionally, PLA can be used as an alternative to low density polyethylene (LDPE), high density polyethylene (HDPE), PS, and PET in transparent, rigid containers, bags, jars, and films for yogurts, organic pretzels, potato chips, carbonated water, fresh juices, and dairy drinks. However, PEF has a better barrier function than PET and may be used in bottles, fibers, and films, while PBAT can be used in food disposable packaging and plastic films for fresh food. Moreover, as previously referenced, several producers also use other additives, such as plasticizers, to enhance the materials' final properties, e.g., mechanical stress and moisture [123,226,227]. However, the current bioplastic market makes up less than 1% of the entire plastic packaging market, although it is continuously growing and diversifying due to demand, R&D activities, increased environmental awareness with concerns about plastics (production and consumption), and implementation of strict environmental regulations [228]. More and more companies are looking for fully, rather than partially, bio-based alternatives, yet few performant options are available.

The main examples available are PLA (NatureWorks), PEF (Avantium), bio-PE (Braskem), bio-PET (currently only ethylene glycol, but soon expanding to terephthalic acid), PHAs, cellulose acetate, starch-based plastics (Novamont); other more niche examples are based on the latest generation lignocellulosic biomass from Stora Enso, Bloom Biorenewables, Lignopure, and Lignin Industries (e.g., lignin and hemicellulose), or are based on food wastes from traceless or UBQ.

The cost of bio-based plastics has been a major barrier to the development and growth of the market [228,229], but the prices are also decreasing since major food companies and well-known brands are launching or integrating bioplastic packaging products into their portfolios, contributing to the expansion of the production capacities, and the efficiency of the supply chains and all processing steps [230]. In addition, regulation and company-set goals of net-zero CO_2 emission in the near future also drive bio-based alternatives which were not plausible in the past. Nonetheless, the commercialization of novel (bio)polymers is an arduous task with many challenges to overcome. Notable ones were already discussed above, e.g., price, type, and processability. As for all materials, the properties of bioplastics present several advantages and disadvantages. Some bioplastics present a much higher water vapor permeability compared with normal plastics that, in some cases, can be useful for packed food to release excess vapor or steam [124]. Other disadvantages for food packaging applications include thermal instability, brittleness, poor mechanical properties, and difficulties with heat sealing [231]. On the other hand, these materials are sustainable alternatives with properties, such as biodegradability, and biocompatibility, and they are non-toxic and have a lower carbon footprint compared with oil-based plastics [231]. Furthermore, less obvious factors are the compatibility of the polymer's recyclability with existing polymers, the volumes at which such a polymer can be produced, and the seasonal and regional differences and availabilities of the starting material. Currently, only materials which can demonstrate their success in all of these aspects will be driven towards a commercial scale and, thus, become a real alternative to current petrol-based packaging.

The availability and seasonality of specific renewable resources needed for the above-mentioned polymers is a key bottleneck in the commercialization process. Successfully scaled bio-polymers have guaranteed this by typically relying on a fermentation stage of sugars from biomass, e.g., sugar cane, bagasse, and hemicellulose streams, as these are easily available all year round in different climates. Novel approaches on different types of renewable resources need to ensure similar resilience against seasonal and regional differences.

Several American and European companies are the top players in what concerns the commercialization of these types of packaging materials. The European Bioplastics Association, in cooperation with the nova-Institute, predicts that the global bioplastics production capacities will increase from around 2.11 million tons in 2020 to approximately 2.87 million tons in 2025 [230]. In addition to the above-mentioned regulations and company goals, supply chain, and resource availability crises, as are currently occurring, provide further pressures and incentives to facilitate increased bio-polymer production in the upcoming years.

6. Environmental Impact

In recent years, there is a dichotomy between "biodegradable products are all good" and "petrochemical-based products are all bad". The use of renewable sources (particularly from agricultural origin) and consumption of less energy are now requirements for the production of industrial products. Therefore, there is an increasing interest in bioplastics due to their renewable nature (raw materials from agriculture instead of crude oil) or biodegradable nature providing less landfilling. Plastics impact the environment and ecosystem during their production, during their service life, and after their disposal, producing contaminants and physical hazards. Bioplastics as potential replacements for petroleum-based polymers require less energy in their production steps and have significantly lower carbon emissions [232–235]. Therefore, replacing fossil-based polymers with renewable and lower carbon footprint bioplastics is seen to promote the transition to a green bioeconomy with less environmental impact.

For instance, PLA, as a biodegradable polymer, consumes two-thirds less energy in the production step when compared to conventional ones [236], provides no net increase in CO_2 gas during the biodegradation step [237,238], emits fewer greenhouse gasses when degrading in landfills [239], and reduces greenhouse gas emissions by 25% [240]. Thus, PLA can be considered one of the most suitable candidates for substituting conventional plastics. On the other hand, after composting a biodegradable polymer, the compost can be used as fertilizer or soil conditioner; however, the produced compost can also be a pollution source for soil, water, and groundwater [234]. At the end of their service life, used or wasted polymers are recycled with some losses due to degradation, are incinerated to produce energy with potential environmental pollution, are littered, resulting in environmental hazards, or are landfilled, resulting in carbon or methane emissions over time due to their uncontrolled degradation [70]. Even though, at this disposal cycle, biodegradable polymers are less harmful to the environment compared to petroleum-based polymers, biodegradable polymers are not generally suitable to be landfilled or digested anaerobically due to the potential methane production under anaerobic conditions [200]. The integration of bioplastics with disposal infrastructures includes various facilities, such as composting, anaerobic digestion, recycling, and waste to energy production, as well as their landfilling and debris to the environment. Bioplastics may be alternative materials to petroleum-based polymers; however, clear assessments of the environmental impacts of both petroleum-based polymers and their bio-based counterparts should be explained in greater detail.

In this paper, the environmental effects of bioplastics are examined at two different stages, i.e., "during the production" and "at the end of life", and the main reasons and key findings are highlighted.

6.1. "During the Production"

6.1.1. Land Use—Soil Erosion

Even though biomass is renewable, it requires responsible and optimal use for longer-term sustainability to avoid the overuse of water/fertilizers, soil erosion, reduced land availability, and changing biodiversity [234]. Because of the high competition for the use of biomass by several industries, such as energy (electricity, heat), food/feed (sugar-, starch-based), biofuel, and materials/carbon (wood and paper industry) [241], its use for bioplastic production may create a challenge to strike the balance among the industries. The impact of such use of plants for bioplastic production has gained attention because of direct and indirect land-use changes in agricultural areas or rainforests [234]. Further, the possible loss of soil, which is a non-renewable resource with its complex ecosystem, will result in considerable environmental and economic consequences. For example, the use of forests for agricultural purposes and intensive cultivation, and inappropriate land-use change for more bioplastic production, can result in more soil decomposition [242]. Including unavoidable agricultural or forestry wastes as biomass resources will minimize the competition with land-use for food production [161], which means that agricultural areas or plants remain available and accessible for food production and will be invaluable to the intended bioplastic production [243]. Several researchers have compared the energy use, greenhouse gas emissions, and direct/indirect land-use change for bio-PET [244], bio-LDPE, bio-PVC [245–247], and bio-HDPE [248] with their related petroleum-based counterparts. Eerhart et al. [244] studied the energy and greenhouse gas balance for polyethylene 2,5-furandicarboxylate (PEF) bioplastic and compared it to its petrochemical counterpart PET. The non-renewable energy use and greenhouse gas emissions for PEF production were reduced by 40–50% (440 and 520 PJ of non-renewable energy savings) and 45–55% (20 to 35 Mt of CO_2 equivalents), respectively. Similarly, Alvarenga et al. [246,247] concluded that bio-PVC showed better results than fossil-based PVC based on greenhouse gas emissions and energy savings. Liptow and Tillman [245] showed that bio-LDPE production requires more total energy compared to fossil-based LDPE, although the major share is renewable. For their potential impacts on acidification, eutrophication, and photochemical ozone creation, no significant difference between the two materials has been reported. However, with regard to global warming potential and the contribution of land-use change was reported as decisive. Accordingly, Piemonte et al. [248] studied the land-use carbon emission of corn-based bioplastics with their environmental impact while comparing the results with PE. It was found that the replacement of petroleum-based plastics with bioplastics from waste biomass might sustain the advantages of lowering greenhouse gas emissions. Likewise, Tsiropoulos et al. [249] found 140% lower greenhouse gas emissions for bio-PE than PE and approximately 65% energy savings for bio-PE production. The authors concluded that the combination of some of these measures and the use of biomass for the supply of process steam can further contribute to reducing greenhouse gas emissions.

6.1.2. Loss of Biodiversity

The reduction in global wild species populations, the decrease in crop yields and fish catches, and rising risk of extinction of species, especially farmland birds and insects, are some results of biodiversity loss. The growing interest in using bioplastics will increase land and water use due to bioplastic production [161], and the inappropriate use of pesticides/herbicides/fertilizers will increase deforestation. This trend will result in rising biodiversity loss [242]. Although there have been increasing studies comparing the energy consumption and global warming effects of bioplastics with petroleum-based plastics, more efforts are needed to assess the impacts of bioplastics on biodiversity [161].

6.2. "At the End of life"

6.2.1. Recycling of Bioplastics

Reusing the bioplastics, such as polyglycolide, PLA, PHA, bio-PE, and bio-PET, is recommended as a pre-step towards the recycling route, and mechanical recycling should be the following step for as long as possible, until they become low-grade [250]. For instance, bio-PET and bio-PE maintain their good mechanical properties for a decent number of recycles. Chemical recycling should be the route chosen once the polymers become low grade, where each bioplastic has an optimum route with the lowest activation energy [251]. For instance, PLA is recycled via alcoholysis, and bio-PET is recycled via glycolysis, as they produce value-added products [251,252], whereas bio-PE requires pyrolysis to be recycled due to its strong solvent resistance [253,254]. However, the environmental benefits of chemical recycling are deeply debated. Current processes for chemical recycling usually encounter the problems of high cost and high energy consumption and require additional steps for isolation and purification from excessive solvents and catalysts, creating environmental consequences [255]. On the other hand, the presence of biodegradable polymers in municipal waste streams and existing plastic recycling systems may cause problems. For instance, it was stated that the presence of natural fibers or starch might complicate recycling [256]. Even though mechanical recycling can be used a few times without losing the original properties of the biodegradable polymers, such as PLA, when recycled, the possible problems in supplying larger quantities of biodegradable polymer waste make it economically unattractive when compared to petroleum-based polymers [257,258]. The environmental impact of bioplastics can also include an economical angle; however, research has so far focused on the cost of bioplastic production instead of overall cost, including the impact of waste management. As a relatively accepted statement in the recycling systems of bio-based, yet non-biodegradable drop-in plastics, such as bio-PP, bio-PE, and bio-PET, such bioplastics are chemically identical to their fossil counterparts, and can be collected, sorted out, and introduced into the existing recycling streams same as their fossil counterparts. No additional processes or investment costs are expected to recycle these drop-in bio-based plastics [259].

6.2.2. Biodegradation of Bioplastics

The biodegradability and/or compostability of some polymers make a positive effect on the environment by generating carbon- and nutrient-rich compost. Methane gas can be produced via the biological waste treatment of biodegradable polymers at anaerobic conditions [260,261], contributing to global warming as a greenhouse gas [262–265] that is many times more potent than carbon dioxide [266]. In the aerobic biodegradation of bioplastics in soil systems, degradation products come into contact with soil, and affect the soil microbial environment, where the nutrient uptake by plants and soil physicochemical properties undergo a variety of changes [267]. On the other hand, in marine ecosystem, plastic debris may cause physical hazards for wildlife due to ingestion or becoming entangled in this debris or chemical hazards due to the formation of toxic compounds during oxidation [268].

Release of micro- and nano-plastics into the environment during biodegradation

Macro-, micro-, and/or nano-counterparts of polymers are released into the environment after the degradation or incomplete degradation of polymers. In recent years, the ecotoxicity and the possibility for those particles to enter the living organisms in the food chain are being treated with increasing concern [269]. The environmental persistence of biodegradable microplastics should be shorter than that of conventional plastics; however, they may have similar negative impacts on the environment [270] and their harm is more pronounced when their size decreases. The harmful effects of these particles are found on the biodiversity, growth, reproduction, and wellness of marine organisms. Green et al. [271] studied the effect of PLA microplastics on marine habitats/biodiversity and observed that such microplastics changed the bacteria population and their behavior in marine environments. The effects of biodegradable plastics and their micro counterparts after degradation in aquatic ecosystems has been very recently reviewed elsewhere [272]. On the contrary, Chu et al. [273] recently revealed that PLA-based bio-microplastics may not pose a serious risk for the agroecosystems in the short timeframe spanning from days to months. It was also reported that soil could hold more microplastics (>40,000 microplastic particles/kg of soil) compared to marine environment [274]. The potential environmental impact of microplastics coming from biodegradable polymers were assessed by Shruti et al. [275], and the authors concluded that microplastic formation was inevitable in biodegradable polymers and that their degradation to microplastics needs more research. Straub et al. [276] compared the uptake and effects of microplastic particles from petroleum-based counterparts and from a biodegradable polymer (PHB) in the freshwater amphipod and reported that there were no significant differences in their ingestion and excretion, but that they differed in biological effects. It is inevitable to note that microplastics from bioplastics can be formed faster than in the case of petroleum-based plastics in non-completed degradation systems [277]. Emadian et al. [268] showed that multiple biodegradation environments were not successful for complete biodegradation and, thus, most of the non-biodegraded material is fragmented into micro- or nano-plastics.

No standardized and accurate methodology is available to quantify the environmental impact of nano- or micro-plastics due to complications caused by a multitude of soil biotic and abiotic processes, the interaction of particles with various components of soil, strong matrix effects, and challenging extraction methods [278]. Even though there is a lack of analytical methods to determine biodegradable microplastics in water, soil, or compost [279], the presence of PHB bio-microplastics was observed by using microscopy [275]. On the other hand, Fojt et al. [280] studied a simple method for the quantification of PHB and PLA microplastics in soils and concluded that biodegradation of plastics might be incomplete and favor microplastic formation.

6.2.3. Incineration with Energy Recovery

Incineration with energy recovery from bioplastics is widely accepted and considered safe with no danger of releasing dioxins or heavy metals [200]. However, as biodegradability is the inherent property of bioplastics, energy recovery should be the least preferred end-of-life option after recycling and biodegradation. It is known that most renewable materials have low calorific values and consume significantly less energy in the production steps, which are positive for the environment [281,282]. However, the value of bioplastics for energy recovery by incineration has not been properly known due to the lack of calorific value determination of biodegradable polymers and the unknown impact of their moisture content on the process. Renewable resources are used for polymer production, which all have a defined circular end-of-life scenario. It is accepted that CO_2 produced from the incineration of fully bio-based plastics, aerobic composting, or incineration of CH_4 from anaerobic composting is a net-zero addition to the carbon cycle since, it is used in the photosynthesis to produce new biomass [164,283].

6.2.4. Disposal in Landfill

Even though it is accepted that the bioplastic disposal in landfill sites does not require preprocesses such as separation, cleaning, or pre-treatment [284], landfill disposal is considered as the least desirable approach due to the uncontrolled production of the highly potent greenhouse gas methane in landfilled areas. However, in the waste management systems, it has been proposed that such a 'landfill gas' can be captured as an energy source, and can then be used as a carbon source input (along with CO_2 produced during biodegradation) to biodegradation into CO_2 after its production during biodegradation [70,200]. The degradation of bioplastics in landfill areas consists of different stages [285] and different compounds are produced depending on the type of bioplastics. For example, sugars are produced during landfilling of TPS, and volatile fatty acids are produced during landfilling of PLA and PHB [286]. However, due to the continuous addition of bioplastics into landfills, the phases of degradation overlap and make the determination of the quantity and rate of biogas production in landfills quite complex [287]. During landfilling of bioplastics similar to petroleum-based plastics, the produced biogas will be the critical point that includes the potential uses of biogas for bioplastic production or as a substitute for natural gas [287]. Even though the use of biogas captured from landfills is still not cost-effective, the implementation of biogas capture and utilization is expected to increase by 50% by 2040 [288].

7. Consumer Research

The increased consumer demand for sustainable products is fundamental to reaching the proposed goals of minimizing the environmental impact of plastics.

Compared to the plethora of studies on the technical and scientific aspects of bio-based food packaging, contributions from social science consumer research are scarce. This might be due to the fact that, for consumers, the product itself and its price is in most cases more important than the packaging [289,290]. The packaging is rather seen as an information tool [291].

Among studies on how consumers respond to bio-based materials, food packaging-related research with 15 contributions comes first, while contributions on bio-based apparel, toys, furniture, and dinnerware, as well as other packaging (non-food) are not as frequent.

In this section, a systematic review on consumer research related to bio-based products based on the PRISMA protocol using Web of Science as our primary database was performed. Our literature search included forward and backward searches, and we added additional articles. Finally, this process yielded 36 studies in total, of which 15 covered food packaging.

Six studies (40%) looked at water bottles, three looked at Coca Cola or other cola products, and two looked at fruit, while other types of food were only represented by one study each (Table 6).

Table 6. Overview of packaged food products in the studies under review.

Packaged Products	Number of Studies	Studies
Water	6	[292–298]
Coca Cola/other colas	3	[296,299,300]
Fruit	2	[301,302]
Juice	1	[303]
Beer	1	[296]
Soup	1	[304]
Takeout food	1	[305]
Food in general (unspecified)	1	[306]

To start on a descriptive level, many authors did not explicitly state on which theory they based their study. Theories that were mentioned were the attitude network approach [293] and the cue utilization theory [304]. Except for two studies that used a mixed methods design [296,297], all other studies were quantitative studies and most of them relied on online surveys.

In line with a large part of consumer research in other areas, the studies under review in this paper often used a quantitative design aimed at explaining stated behavioral intentions, such as willingness-to-pay or intention to purchase by looking at factors that explain these intentions. The factors that were tested can be divided into two broad categories. First, factors pertaining to packaging and its attributes, such as material, recyclability, or labels were considered. Second, factors pertaining to consumers, such as attitudes, norms, and other psychographic or socio-demographic variables were considered.

The dependent variable that studies in our sample sought to explain was primarily willingness to pay (WTP) [292,293,295,298,301,305]. Furthermore, utility [301,302] and preferences [292,294] were closely related to WTP, as well as purchase intention [307]. Other dependent variables were perceived environmental friendliness or, more generally speaking, perceived sustainability [296,297]. One study also examined factors determining correct disposal of biodegradable packaging [297].

The WTP resulting in a surcharge for products packaged in bio-based materials is important information for companies seeking to use these materials in their packaging solutions. Likewise, it was a frequent object of research in our sample. Table 7 summarizes the price premium consumers were willing to pay for bio-based packaging compared to fossil-based packaging. Overall, the range of premiums is very wide, ranging from 8% to

30%. Most of the WTP studies were carried out for water bottles. Overall, it seems that 20% seems to be a premium that is at least a rough approximation for this product category.

Table 7. Overview of price premia in the studies under review.

Studies	Price premium	Method	Remarks
[305]	N/A	Choice-based conjoint	The study only tested bio-based alternatives, no fossil alternatives
[292]	0.07 Euro / bottle (PLA) 0.05 Euro / bottle (bio-PET)	Choice-based conjoint	Percentages were not shown and could not be calculated
[295]	25% PLA over PET (mean) 22–35% depending on treatment 13% PLA over PEF 6–17% depending on the treatment	Direct	Treatments: different messages on the environmental effects of different plastics
[301]	Control group: 23% Other groups: 19–51% (depending on the treatment)	Choice-based conjoint	Treatments: e.g., pictures, normative messages
[293] study 2	21%	Direct	
[293] study 3	18%	Direct	
[298] Study 2	30%	Direct	
[298] Study 3	20%	Direct	
[298] Study 4	8%	Direct	

One study also asked consumers how they thought about a local ban on expanded polystyrene (EPS) food containers, i.e., not a consumer choice but a regulatory measure [305].

Influencing Factors

All studies found that consumers harbor more positive attitudes towards bio-based plastic packaging than towards conventional plastics.

The most frequently tested attributes of bio-based food packaging were biodegradability, within six studies [292,293,297,301–303], and recyclability within four studies [297,301–303], both being seen positively by consumers. Biodegradability also scored positively in other studies not looking at WTP [306]. Furthermore, end-of-life related criteria were more important for consumers than production or transport [277]. Testa et al. [303] tested if third-party certification has an influence but found it to have no significant effect.

The influence of the material for producing bio-based packaging was tested as an influencing factor for WTP in several studies which will be discussed below. Barnes et al. [305], in their study of containers for takeout food, found different preferences in their latent classes, as some preferred sugar-cane, others paper, while corn was not popular among any of the latent classes. Moreover, the material was only the most important attribute for one group. De Marchi et al. [292] tested bio-PET and PLA, with PLA being clearly favored by consumers. Reinders et al. [300] showed that a 100% bio-based product scores much better with consumers than a product with a lower bio-based content.

Local production was tested in one study and, not surprisingly, found to have a positive influence on WTP. Other, less often tested attributes include microwaveability and water resistance [305].

Turning towards consumer attributes as influencing factors, two studies looked at socio-demographics [295,302]. Most other studies that considered consumer attributes examined the influence of various psychographic variables, such as attitudes about bio-based plastics [293,298], environmental attitudes [295], norms [300], trust [295], or knowledge [293,299].

Within their paper, Zwicker et al. [298] did not find attitudes towards bio-based plastics to predict WTP in studies 2 and 4 of their research. However, the attitude towards conventional plastic did, which hints at feelings of guilt. In study 3, both were found to influence WTP but with a very low explanatory power. Guilt was also found to be a driver of WTP [298].

Several studies tested the influence of interventions on choice behaviour, such as nudging and pro-environmental guidance [294], giving information on the environmental effects of different plastics [292,293,295], stimulating feelings of guilt [298], as well as stimulating norms or providing nature pictures or reflective questions [301]. All of these interventions positively influenced the participants' choice of bio-based packaging.

Finally, the differences between countries revealed in the few cross-country studies [300] make the importance of a differentiated internationalization strategy clear.

The studies under review identified the following barriers to an environmentally beneficial expansion of bio-based food packaging:

A lack of knowledge was frequently discussed to be a barrier. Even with labels clearly indicating a bio-based packaging's characteristics, consumers seem to have great difficulties in identifying these. In a study by Taufik et al. [297], participants were not able to tell apart bio-based recyclable water bottles and recyclable fossil-based bottles. The participants in the study by Zwicker et al. [298] believed that bio-based plastics are always biodegradable. This false belief can drive acceptance but can also backfire once consumers learn that they have been mistaken. Lynch et al. [299] and Testa et al. [303] pointed out the low level of familiarity with bio-based products in the Netherlands and Italy, while Dilkes-Hoffman et al. [306] and Boesen et al. [296], as well as Zwicker et al. [298] confirm the low level of Australian consumers' knowledge.

Consumers' perceptions of the origin of the biomass used to produce bio-based plastics is another potential barrier to further expansion. Zwicker et al. [298] showed that the majority of participants were neutral about whether bio-based plastics contribute to deforestation and food competition. However, nearly 20% (6 and 7 on a 7-point-scale) believed that these materials compete with those used in food production.

Environmental benefits include the correct disposal of the packaging. However, in a study by Taufik et al. [297], 63% of the participants disposed of the compostable bio-based bottle incorrectly. Participants with a higher bio-based product familiarity were more likely to dispose of the compostable bottle correctly. Apparently, the main reason was that participants could not think of plastic and compostable material together. Bio-based plastic was still plastic for them, with all the characteristics they attribute to this kind of material. Similarly, in the study by Dilkes-Hoffman et al. [306], 62% of the participants would dispose of biodegradable food packaging in a recycling bin rather than by composting it. Zwicker et al. [298] (studies 2 and 3 within the paper) showed that consumers find it more important to recycle fossil-based plastic bottles than bio-based bottles. They also showed that consumers in study 3 frequently believed bio-based plastics to be biodegradable, quite the opposite of the findings in the paper of Taufik et al. [297]. Further, in the study by Lynch et al. [299], focus group participants raised the issue of consumers possibly not knowing how to correctly dispose of a bio-based plastic bottle.

What can companies take away from extant consumer research? First, the studies under review have shown that biodegradability and recyclability are important product attributes for consumers. This can be directly applied in companies' choice of materials and product design, i.e., product strategy. Biodegradability is especially high on the consumer agenda, confirming findings from studies on bio-based packaging in general which have shown that consumers focus strongly on the end of packaging life, i.e., the disposal stage [291]. Furthermore, 100% bio-based products seem to be preferable compared with partially bio-based products. Second, analyses of influencing factors for WTP and differentiated treatments in experiments suggest promising approaches to communication strategy, namely that guilt (when using conventional plastics) seems to be a strong driver of WTP for bio-based products, and that companies can appeal to this emotion in their communication.

Along the same lines, norms were shown to be effective; therefore, evoking norms may be a promising element of communication strategy. Moreover, giving pro-environmental guidance in the buying process and pointing out the environmental effects of different types of plastics also have clear effects. However, companies and governments clearly need to educate consumers on how to dispose of bio-based plastics correctly, especially with regard to their biodegradability. Third-party certification did not prove effective; however, since this was tested only in one study, companies should probably consult more studies or include this question into their market research. These hints on communication strategy can not only be applied by companies but also by governments and NGOs in their efforts to persuade consumers to reduce plastics consumption.

Additionally, the pricing strategy can be informed by extant research. The results in Table 7 suggest that a price premium of around 20% could be a good starting point for deliberations on pricing strategies. However, for a final decision, other factors, such as competition and cost, have to be considered.

Looking at the above analysis of consumer research on bio-based food packaging, there are several avenues for further research that seem promising. From a methodological perspective, there is clearly a dearth of qualitative research. Understanding in more detail why consumers prefer certain materials over others and the influences of various attributes, i.e., consumers' subjective logic, would certainly help to inform both policymakers and marketeers. The study on attitude networks by Zwicker et al. [293] demonstrated how useful this can be. Second, if WTP is to be examined using a quantitative design, it is surprising that direct WTP elicitation methods are still used despite their well-known shortcomings [307]. Choice-based conjoint, which is well-established, and neuroscience-based methods offer interesting alternatives.

However, the consumer–citizen gap must also be considered. While, as citizens, consumers support sustainable packaging, in real shopping situations, the WTP is often much lower, as the citizens then act as consumers, and they have to pay a surplus for more sustainable packaging. This phenomenon has already been studied in depth in the field of animal welfare (cf. e.g., [308,309]).

Concerning potential communication strategies, it would be helpful for companies and governments alike if researchers tested more communication measures, varying both messages and ways of communication, such as text, labeling, or pictures.

8. Conclusions

The interest of researchers has turned in the last two decades to the research of bioplastics, as they are quite promising materials with good properties, such as biodegradability and biocompatibility [310]. The use of biological resources is going to contribute significantly to the production of innovative materials. The advantages of these materials regarding the environmentally friendly solutions are expected to be significant and, to some extent, address the future bioeconomy [83], although mechanical and barrier properties, thermal stability, and water resistance are major problems for many materials, preventing their use in many cases [96]. The application of bioplastics in food packaging compared to conventional materials remains small for reasons related to specific regulations, requirements, price, safety, and their post-use management [86]. This review shows that further research is needed to improve the production of bioplastics and their potential applications, according to different properties, mechanisms of biodegradation, environmental impact, their market and how consumers perceive bioplastics. Governmental economic incentives for these materials and specific rules to limit the use of non-bioplastic materials are mandatory in the future to contribute to the development and commercialization of bioplastics for food packaging and to reduce our dependency on limited petroleum resources. Together with motivated consumers, industry, and also governments, environmental awareness and a willingness to focus on sustainability will definitely contribute to an ecological and circular economy.

Author Contributions: Conceptualization, R.M.S.C., V.K., S.K. (Simon Krauter), S.A., R.W., C.H., I.U.-U. and T.V.; writing—original draft preparation, R.M.S.C., V.K., S.K. (Simon Krauter), S.A., R.W., C.H., P.B.V.S., I.U.-U., E.S., J.L. and T.V.; writing—review and editing, R.M.S.C., V.K., S.K. (Simon Krauter), S.A., R.W., C.H., P.B.V.S., I.U.-U., E.S., S.K. (Samir Kopacic), J.L., R.R. and T.V.; supervision, R.M.S.C., V.K., S.K. (Simon Krauter), S.A., R.W., C.H., P.B.V.S., I.U.-U., E.S., J.L., R.R. and T.V; project administration, V.K.; funding acquisition, V.K. All authors have read and agreed to the published version of the manuscript.

Funding: This article/publication is based upon work from COST Action Circul-a-bility (CA19124), supported by COST (European Cooperation in Science and Technology), www.cost.eu (accessed on 28 March 2022).

Conflicts of Interest: The author Philip B. V. Scholten is an employee of the company "Bloom Biorenewables". The company "Bloom Biorenewables" has no active role in this manuscript. The authors declare no conflict of interest.

References

1. HLPE. Nutrition and food systems. A report by the High-Level Panel of Experts on Food Security and Nutrition of the Committee on World Food Security, Rome. 2017. Available online: https://www.fao.org/3/i7846e/i7846e.pdf (accessed on 12 April 2022).
2. Van Alfen, N.K. *Encyclopedia of Agriculture and Food Systems*; Academic Press: Cambridge, MA, USA, 2014; pp. 232–249.
3. Robertson, G.L. *Food Packaging—Principles and Practice*, 3rd ed.; CRC Press: Boca Raton, FL, USA, 2012; p. 736.
4. Singh, P.; Wani, A.A.; Langowski, H.-C. *Food Packaging Materials: Testing & Quality Assurance*; CRC Press: New York, NY, USA, 2017; p. 356.
5. Soroka, W. *Fundamentals of Packaging Technology*, 5th ed.; Institute of Packaging Professional: Herndon, VA, USA, 2014.
6. European Commission. Communication from the Commission to the European Parliament, the Council, the European Economic and Social Committee and the Committee of the Regions: A new Circular Economy Action Plan. For a Cleaner and More Competitive Europe COM (2020) 98 Final, Brussels. 2020. Available online: https://eur-lex.europa.eu/legal-content/EN/TXT/?qid=1583933814386&uri=COM:2020:98:FIN (accessed on 2 February 2022).
7. Circularity Gap Report 2022: Five Years of Analysis by Circle Economy, European Circular Economy Stakeholder 1878 Platform. Available online: https://circulareconomy.europa.eu/platform/en/knowledge/circularity-gap-report-2022-five-years-analysis-circle-economy (accessed on 27 March 2022).
8. European Commission. Communication from the Commission to the European Parliament, the Council, the European Economic and Social Committee and the Committee of the Regions: A European Strategy for Plastics in a Circular Economy COM/2018/028 Final, Brussels. 2018. Available online: https://eur-lex.europa.eu/resource.html?uri=cellar:2df5d1d2-fac7-11e7-b8f5-01aa75ed7 1a1.0001.02/DOC_1&format=PDF (accessed on 2 February 2022).
9. United Nations. Resolution Adopted by the General Assembly on 25 September 2015: Transforming Our World: The 2030 Agenda for Sustainable Development. 2015. Available online: https://www.un.org/ga/search/view_doc.asp?symbol=A/RES/70/1 &Lang=E (accessed on 2 February 2022).
10. Bakri, A.A.; Nordenfelt, M.; Ajdic, T.; Nail, E.L. *Recycling Food Packaging and Food Waste in Plastic Revolution*; The European Economic and Social Committee: Bruxelles, Belgium, 2021. [CrossRef]
11. Bonilla, J.; Talón, E.; Atarés, L.; Vargas, M.; Chiralt, A. Effect of the Incorporation of Antioxidants on Physicochemical and Antioxidant Properties of Wheat Starch–Chitosan Films. *J. Food Eng.* **2013**, *118*, 271–278. [CrossRef]
12. Eurostat. Packaging Waste Statistics. Consulted in October. 2021. Available online: https://ec.europa.eu/eurostat/statistics-explained/index.php?title=Packaging_waste_statistics (accessed on 10 May 2022).
13. Duque-Acevedo, M.; Belmonte-Ureña, L.J.; Cortés-García, F.J.; Camacho-Ferre, F. Agricultural Waste: Review of the Evolution, Approaches and Perspectives on Alternative Uses. *Glob. Ecol. Conserv.* **2020**, *22*, e00902. [CrossRef]
14. United Nations (UN) General Assembly. Transforming Our World: The 2030 Agenda for Sustainable Development; Resolution Adopted by the General Assembly A/RES/70/1. 2015. Available online: https://www.un.org/en/development/desa/population/migration/generalassembly/docs/globalcompact/A_RES_70_1_E.pdf (accessed on 10 May 2022).
15. Jariyasakoolroj, P.; Leelaphiwat, P.; Harnkarnsujarit, N. Advances in Research and Development of Bioplastic for Food Packaging. *J. Sci. Food Agric.* **2020**, *100*, 5032–5045. [CrossRef] [PubMed]
16. de Boer, J.; Aiking, H. Prospects for pro-environmental protein consumption in Europe: Cultural, culinary, economic and psychological factors. *Appetite* **2018**, *121*, 29–40. [CrossRef] [PubMed]
17. European Bioplastics. What Are Bioplastics? Available online: https://www.european-bioplastics.org/bioplastics (accessed on 22 August 2021).
18. Iacovidou, E.; Gerassimidou, S. Sustainable Packaging and the Circular Economy: An EU Perspective. In *Reference Module in Food Science*; Elsevier: Amsterdam, The Netherlands, 2018.
19. Hahladakis, J.N.; Iacovidou, E.; Gerassimidou, S. Plastic Waste in a Circular Economy. In *Plastic Waste and Recycling*; Elsevier: Oxford, UK, 2020; pp. 481–512.

20. Plastics—The Facts 2021. Available online: https://plasticseurope.org/knowledge-hub/plastics-the-facts-2021/ (accessed on 10 May 2022).
21. Guengerich, F.P.; Yoshimoto, F.K. Formation and Cleavage of C-C Bonds by Enzymatic Oxidation-Reduction Reactions. *Chem. Rev.* **2018**, *118*, 6573–6655. [CrossRef]
22. Chen, F.; Wang, T.; Jiao, N. Recent Advances in Transition-Metal-Catalyzed Functionalization of Unstrained Carbon-Carbon Bonds. *Chem. Rev.* **2014**, *114*, 8613–8661. [CrossRef] [PubMed]
23. Kobayashi, S.; Uyama, H.; Kimura, S. Enzymatic Polymerization. *Chem. Rev.* **2001**, *101*, 3793–3818. [CrossRef]
24. Beringer, M.; Rodnina, M.V. The Ribosomal Peptidyl Transferase. *Mol. Cell* **2007**, *26*, 311–321. [CrossRef]
25. Luley-Goedl, C.; Nidetzky, B. Glycosides as Compatible Solutes: Biosynthesis and Applications. *Nat. Prod. Rep.* **2011**, *28*, 875–896. [CrossRef]
26. Rehm, B.H.A. Polyester Synthases: Natural Catalysts for Plastics. *Biochem. J.* **2003**, *376*, 15–33. [CrossRef]
27. Behabtu, N.; Kralj, S. Enzymatic Polymerization Routes to Synthetic–Natural Materials: A Review. *ACS Sustain. Chem. Eng.* **2020**, *8*, 9947–9954. [CrossRef]
28. Douka, A.; Vouyiouka, S.; Papaspyridi, L.-M.; Papaspyrides, C.D. A Review on Enzymatic Polymerization to Produce Poly-condensation Polymers: The Case of Aliphatic Polyesters, Polyamides and Polyesteramides. *Prog. Polym. Sci.* **2018**, *79*, 1–25. [CrossRef]
29. Samuels, A.L.; Rensing, K.H.; Douglas, C.J.; Mansfield, S.D.; Dharmawardhana, D.P.; Ellis, B.E. Cellular Machinery of Wood Production: Differentiation of Secondary Xylem in Pinus contorta Var. *latifolia*. *Planta* **2002**, *216*, 72–82. [CrossRef]
30. Ganster, J.; Fink, H.-P. Cellulose and Cellulose Acetate. In *Bio-Based Plastics*; John Wiley & Sons Ltd: Chichester, UK, 2013; pp. 35–62.
31. Meister, J.J. Modification of Lignin. *J. Macromol. Sci. Part C Polym. Rev.* **2002**, *42*, 235–289. [CrossRef]
32. Alma, M.H.; Kelley, S.S. Conversion of barks of several tree species into bakelite-like thermosetting materials by their phenolysis. *J. Polym. Eng.* **2000**, *20*, 365–380. [CrossRef]
33. Bai, Y.; Gao, Z. Water-Resistant Plywood Adhesives Prepared from Phenolated Larch Bark in the Presence of Various Combination Catalysts. *J. Appl. Polym. Sci.* **2012**, *126*, E374–E382. [CrossRef]
34. Zadeh, E.M.; O'Keefe, S.F.; Kim, Y.-T. Utilization of Lignin in Biopolymeric Packaging Films. *ACS Omega* **2018**, *3*, 7388–7398. [CrossRef]
35. Bioplastics, European. Frequently Asked Questions on Bioplastics. Available online: https://docs.european-bioplastics.org/publications/EUBP_FAQ_on_bioplastics.pdf (accessed on 10 May 2022).
36. USDA. *Voluntary Labeling Program for Bio-Based Products*; United States Department of Agriculture (USDA): Washington, DC, USA, 2011; pp. 3789–3813.
37. Gerassimidou, S.; Martin, O.V.; Chapman, S.P.; Hahladakis, J.N.; Iacovidou, E. Development of an Integrated Sustainability Matrix to Depict Challenges and Trade-Offs of Introducing Bio-Based Plastics in the Food Packaging Value Chain. *J. Clean. Prod.* **2021**, *286*, 125378. [CrossRef]
38. European Commission. Communication from the Commission to the European Parliament, the Council, the European Economic and Social Committee and the Committee of the Regions: The Role of Waste-To-Energy in the Circular Economy COM/2017/034 Final, Brussels. 2017. Available online: https://eur-lex.europa.eu/legal-content/en/TXT/?uri=CELEX%3A52017DC0034 (accessed on 11 May 2022).
39. European Commission. Communication from the Commission to the European Parliament, the Council, the European Economic and Social Committee and the Committee of the Regions: Closing the Loop—An EU Action Plan for the Circular Economy COM/2015/0614 Final, Brussels. 2015. Available online: https://eur-lex.europa.eu/legal-content/EN/ALL/?uri=CELEX%3A5 2015DC0614 (accessed on 11 May 2022).
40. European Parliament and Council of the European Union. *Directive (EU) 2019/904 of the European Parliament and of the Council of 5 June 2019 on the Reduction of the Impact of Certain Plastic Products on the Environment*; Official Journal of the European Union: Brussels, Belgium, 2019; Volume 155.
41. Ellen MacArthur Foundation. Global Commitment. A Circular Economy for Plastic in which It Never Becomes Waste. 2020. Available online: https://www.newplasticseconomy.org/projects/global-commitment (accessed on 16 September 2020).
42. European Plastics Pact. The European Plastics Pact. Bringing Together Frontrunner Companies and Governments to Accelerate the Transition toward Circular Plastics Economy. 2020. Available online: https://europeanplasticspact.org/ (accessed on 16 September 2020).
43. United Nations. *The Sustainable Development Goals Report*; United Nations: New York, NY, USA, 2020.
44. Nikiema, J.; Asiedu, Z. A Review of the Cost and Effectiveness of Solutions to Address Plastic Pollution. *Environ. Sci. Pollut. Res. Int.* **2022**, *29*, 24547–24573. [CrossRef]
45. Shin, S.-K.; Um, N.; Kim, Y.-J.; Cho, N.-H.; Jeon, T.-W. New Policy Framework with Plastic Waste Control Plan for Effective Plastic Waste Management. *Sustainability* **2020**, *12*, 6049. [CrossRef]
46. Nikiema, J.; Mateo-Sagasta, J.; Asiedu, Z.; Saad, D.; Lamizana, B. *Water Pollution by Macroplastics and Microplastics: A Review of Technical Solutions from Source to Sea*; United Nations Environment Programme (UNEP): Nairobi, Kenia, 2020. Available online: https://www.unep.org/resources/report/water-pollution-plastics-and-microplastics-review-technical-solutions-source-sea (accessed on 27 July 2022).

47. EU. A European Strategy for Plastics in a Circular Economy. 2018. Available online: https://www.europarc.org/wp-content/uploads/2018/01/Eu-plastics-strategy-brochure.pdf (accessed on 30 July 2022).

48. PlasticAtlas. *Plastic Atlas 2019: Facts and Figures about the World of Synthetic Polymers*; Heinrich Boll Foundation: Berlin, Germany, 2019.

49. Filiciotto, L.; Rothenberg, G. Biodegradable Plastics: Standards, Policies, and Impacts. *ChemSusChem* **2021**, *14*, 56–72. [CrossRef] [PubMed]

50. Brooks, A.L.; Wang, S.; Jambeck, J.R. The Chinese Import Ban and Its Impact on Global Plastic Waste Trade. *Sci. Adv.* **2018**, *4*, eaat0131. [CrossRef] [PubMed]

51. European Environment Agency (EEA). *Plastics, the Circular Economy and Europe's Environment—A Priority for Action*; European Environment Agency: Copenhagen, Denmark, 2021.

52. Hira, A.; Pacini, H.; Attafuah-Wadee, K.; Vivas-Eugui, D.; Saltzberg, M.; Yeoh, T.N. Plastic Waste Mitigation Strategies: A Review of Lessons from Developing Countries. *J. Dev. Soc.* **2022**, *38*, 336–359. [CrossRef]

53. UNEP (United Nations Environment Programme). Single-Use Plastics: A Roadmap for Sustainability. 2018. Available online: https://www.unep.org/resources/report/single-use-plastics-roadmap-sustainability (accessed on 27 July 2022).

54. Lam, C.-S.; Ramanathan, S.; Carbery, M.; Gray, K.; Vanka, K.S.; Maurin, C.; Bush, R.; Palanisami, T. A Comprehensive Analysis of Plastics and Microplastic Legislation Worldwide. *Water Air Soil Pollut.* **2018**, *229*, 345. [CrossRef]

55. Eriksen, M.; Thiel, M.; Prindiville, M.; Kiessling, T. Microplastic: What Are the Solutions? In *The Handbook of Environmental Chemistry*; Springer: Cham, Switzerland, 2018; pp. 273–298. ISBN 9783319616148.

56. United Nations. Report of the Conference of the Parties to the Basel Convention on the Control of Transboundary Movements of Hazardous Wastes and Their Disposal on the Work of Its Fourteenth Meeting. In Proceedings of the 14th Meeting of the Conference of the Parties to the Basel Convention (COP14), Geneva, Switzerland, 29 May–10 April 2019. UNEP/CHW.14/28.

57. European Commission. Communication from the Commission to the European Parliament, the European Council, the Council, the European Economic and Social Committee and the Committee of the Regions: The European Green Deal COM(2019) 640 Final, Brussels. 2019. Available online: https://eur-lex.europa.eu/resource.html?uri=cellar:b828d165-1c22-11ea-8c1f-01aa75ed7 1a1.0002.02/DOC_1&format=PDF (accessed on 22 May 2022).

58. European Commission. Proposal for a Regulation of the European Parliament and of the Council on the Establishment of a Framework to Facilitate Sustainable Investment COM(2018) 253 Final, Brussels. 2018. Available online: https://ec.europa.eu/transparency/regdoc/rep/1/2018/EN/COM-2018-353-F1-EN-MAIN-PART-1.PDF (accessed on 22 May 2022).

59. Legislative Train Schedule. Available online: https://www.europarl.europa.eu/legislative-train (accessed on 22 May 2022).

60. Monier, V.; Hestin, M.; Cavé, J.; Laureysens, I.; Watkins, E.; Reisinger, H.; Porsch, L. Development of Guidance on Extended Producer Responsibility (EPR). 2014. Available online: https://ec.europa.eu/environment/archives/waste/eu_guidance/pdf/Guidance%20on%20EPR%20-%20Final%20Report.pdf (accessed on 22 May 2022).

61. European Chemicals Agency (ECHA). Annex XV Restriction Report: Proposal of Restriction for Intentionally Added Microplastics. 2019. Available online: https://echa.europa.eu/documents/10162/05bd96e3-b969-0a7c-c6d0-441182893720 (accessed on 22 May 2022).

62. European Chemicals Agency (ECHA). Authorisation-ECHA. 2019. Available online: https://echa.europa.eu/substances-of-very-high-concernidentification-explained (accessed on 22 May 2022).

63. Kubicki, M. *Guidance on the Preparation of an Application for Authorisation*; European Chemicals Agency: Helsinki, Finland, 2011.

64. Singh, M. Most Ambitious US Law to Tackle Single-Use Plastics Faces Make-Or-Break Moment. 2019. Available online: http://www.theguardian.com/us-news/2019/sep/13/california-plastics-legislation-singleuse (accessed on 22 May 2022).

65. OECD. Policies for Bioplastics in the Context of a Bioeconomy. 2013. Available online: https://www.oecd-ilibrary.org/science-and-technology/policies-for-bioplastics-in-the-context-of-a-bioeconomy_5k3xpf9rrw6den (accessed on 22 May 2022).

66. Ugorji, C.; van der Ven, C. Trade and the Circular Economy: A Deep Dive into Plastics Action in Ghana. 2021. Available online: https://globalplasticaction.org/wp-content/uploads/Ghana_NPAP_Trade_and_Circular_Economy.pdf (accessed on 30 July 2022).

67. Briassoulis, D.; Dejean, C.; Picuno, P. Critical Review of Norms and Standards for Biodegradable Agricultural Plastics Part II: Composting. *J. Polym. Environ.* **2010**, *18*, 364–383. [CrossRef]

68. European Commission. Bio-Based, Biodegradable and Compostable Plastics. Environment. 2022. Available online: https://ec.europa.eu/environment/topics/plastics/bio-based-biodegradable-and-compostable-plastics_en (accessed on 6 June 2022).

69. Lucas, N.; Bienaime, C.; Belloy, C.; Queneudec, M.; Silvestre, F.; Nava-Saucedo, J.E. Polymer Biodegradation: Mechanisms and Estimation Techniques—A Review. *Chemosphere* **2008**, *73*, 429–442. [CrossRef]

70. Meereboer, K.W.; Misra, M.; Mohanty, A.K. Review of Recent Advances in the Biodegradability of Polyhydroxyalkanoate (PHA) Bioplastics and Their Composites. *Green Chem.* **2020**, *22*, 5519–5558. [CrossRef]

71. Xu, L.; Crawford, K.; Gorman, C.B. Effects of Temperature and pH on the Degradation of Poly(Lactic Acid) Brushes. *Macromolecules* **2011**, *44*, 4777–4782. [CrossRef]

72. Volova, T.G.; Gladyshev, M.I.; Trusova, M.Y.; Zhila, N.O. Degradation of Polyhydroxyalkanoates in Eutrophic Reservoir. *Polym. Degrad. Stab.* **2007**, *92*, 580–586. [CrossRef]

73. Liu, M.; Zhuo, J.K.; Xiong, S.J.; Yao, Q. Catalytic Degradation of High-Density Polyethylene over a Clay Catalyst Compared with Other Catalysts. *Energy Fuels* **2014**, *28*, 6038–6045. [CrossRef]

74. Michalak, M.; Kwiecień, M.; Kawalec, M.; Kurcok, P. Oxidative Degradation of Poly (3-Hydroxybutyrate). *A New Method of Synthesis for the Malic Acid Copolymers. RSC Adv.* **2016**, *6*, 12809–12818.
75. Verhoogt, H.; Ramsay, B.A.; Favis, B.D. Polymer Blends Containing Poly(3-Hydroxyalkanoate)s. *Polymer* **1994**, *35*, 5155–5169. [CrossRef]
76. Kijchavengkul, T.; Auras, R.; Rubino, M.; Alvarado, E.; Camacho Montero, J.R.; Rosales, J.M. Atmospheric and Soil Degradation of Aliphatic–Aromatic Polyester Films. *Polym. Degrad. Stab.* **2010**, *95*, 99–107. [CrossRef]
77. Proikakis, C.S.; Mamouzelos, N.J.; Tarantili, P.A.; Andreopoulos, A.G. Swelling and Hydrolytic Degradation of Poly(d,l-Lactic Acid) in Aqueous Solutions. *Polym. Degrad. Stab.* **2006**, *91*, 614–619. [CrossRef]
78. Helbling, C.; Abanilla, M.; Lee, L.; Karbhari, V.M. Issues of Variability and Durability under Synergistic Exposure Conditions Related to Advanced Polymer Composites in the Civil Infrastructure. *Compos. Part A Appl. Sci. Manuf.* **2006**, *37*, 1102–1110. [CrossRef]
79. İpekoğlu, B.; Böke, H.; Çizer, Ö. Assessment of Material Use in Relation to Climate in Historical Buildings. *Build. Environ.* **2007**, *42*, 970–978. [CrossRef]
80. Kijchavengkul, T.; Auras, R.; Rubino, M.; Ngouajio, M.; Fernandez, R.T. Assessment of Aliphatic–Aromatic Copolyester Biodegradable Mulch Films. Part I: Field Study. *Chemosphere* **2008**, *71*, 942–953. [CrossRef]
81. Karan, H.; Funk, C.; Grabert, M.; Oey, M.; Hankamer, B. Green Bioplastics as Part of a Circular Bioeconomy. *Trends Plant Sci.* **2019**, *24*, 237–249. [CrossRef]
82. Aggarwal, J.; Sharma, S.; Kamyab, H.; Kumar, A. The Realm of Biopolymers and Their Usage:An Overview The Realm of Biopolymers and Their Usage: An Overview. *J. Environ. Treat. Tech.* **2020**, *8*, 1005–1016.
83. Brodin, M.; Vallejos, M.; Opedal, M.T.; Area, M.C.; Chinga-Carrasco, G. Lignocellulosics as Sustainable Resources for Production of Bioplastics—A Review. *J. Clean. Prod.* **2017**, *162*, 646–664. [CrossRef]
84. Kumar, S.; Thakur, K.S. Bioplastics—Classification, Production and Their Potential Food Applications. *J. Hill Agric.* **2017**, *8*, 118. [CrossRef]
85. Acquavia, M.A.; Pascale, R.; Martelli, G.; Bondoni, M.; Bianco, G. Natural Polymeric Materials: A Solution to Plastic Pollution from the Agro-Food Sector. *Polymers* **2021**, *13*, 158. [CrossRef] [PubMed]
86. Nilsen-Nygaard, J.; Fernández, E.N.; Radusin, T.; Rotabakk, B.T.; Sarfraz, J.; Sharmin, N.; Sivertsvik, M.; Sone, I.; Pettersen, M.K. Current Status of Bio-based and Biodegradable Food Packaging Materials: Impact on Food Quality and Effect of Innovative Processing Technologies. *Compr. Rev. Food Sci. Food Saf.* **2021**, *20*, 1333–1380. [CrossRef] [PubMed]
87. Jafarzadeh, S.; Jafari, S.M.; Salehabadi, A.; Nafchi, A.M.; Uthaya Kumar, U.S.; Khalil, H.P.S.A. Biodegradable Green Packaging with Antimicrobial Functions Based on the Bioactive Compounds from Tropical Plants and Their By-Products. *Trends Food Sci. Technol.* **2020**, *100*, 262–277. [CrossRef]
88. Singha, S.; Mahmutovic, M.; Zamalloa, C.; Stragier, L.; Verstraete, W.; Svagan, A.J.; Das, O.; Hedenqvist, M.S. Novel Bioplastic from Single Cell Protein as a Potential Packaging Material. *ACS Sustain. Chem. Eng.* **2021**, *9*, 6337–6346. [CrossRef]
89. Coltelli, M.-B.; Wild, F.; Bugnicourt, E.; Cinelli, P.; Lindner, M.; Schmid, M.; Weckel, V.; Müller, K.; Rodriguez, P.; Staebler, A.; et al. State of the Art in the Development and Properties of Protein-Based Films and Coatings and Their Applicability to Cellulose Based Products: An Extensive Review. *Coatings* **2015**, *6*, 1. [CrossRef]
90. Zubair, M.; Ullah, A. Recent Advances in Protein Derived Bionanocomposites for Food Packaging Applications. *Crit. Rev. Food Sci. Nutr.* **2020**, *60*, 406–434. [CrossRef]
91. Zárate-Ramírez, L.S.; Romero, A.; Bengoechea, C.; Partal, P.; Guerrero, A. Thermo-Mechanical and Hydrophilic Properties of Polysaccharide/Gluten-Based Bioplastics. *Carbohydr. Polym.* **2014**, *112*, 24–31. [CrossRef]
92. Reshmy, R.; Philip, E.; Madhavan, A.; Sindhu, R.; Pugazhendhi, A.; Binod, P.; Sirohi, R.; Awasthi, M.K.; Tarafdar, A.; Pandey, A. Advanced Biomaterials for Sustainable Applications in the Food Industry: Updates and Challenges. *Environ. Pollut.* **2021**, *283*, 117071. [CrossRef]
93. Wang, S.; Lu, A.; Zhang, L. Recent Advances in Regenerated Cellulose Materials. *Prog. Polym. Sci.* **2016**, *53*, 169–206. [CrossRef]
94. Qasim, U.; Osman, A.I.; Al-Muhtaseb, A.H.; Farrell, C.; Al-Abri, M.; Ali, M.; Vo, D.-V.N.; Jamil, F.; Rooney, D.W. Renewable Cellulosic Nanocomposites for Food Packaging to Avoid Fossil Fuel Plastic Pollution: A Review. *Environ. Chem. Lett.* **2021**, *19*, 613–641. [CrossRef]
95. Bayer, I.S.; Guzman-Puyol, S.; Heredia-Guerrero, J.A.; Ceseracciu, L.; Pignatelli, F.; Ruffilli, R.; Cingolani, R.; Athanassiou, A. Direct Transformation of Edible Vegetable Waste into Bioplastics. *Macromolecules* **2014**, *47*, 5135–5143. [CrossRef]
96. Vilarinho, F.; Sanches Silva, A.; Vaz, M.F.; Farinha, J.P. Nanocellulose in Green Food Packaging. *Crit. Rev. Food Sci. Nutr.* **2018**, *58*, 1526–1537. [CrossRef]
97. Jabeen, N.; Majid, I.; Nayik, G.A. Bioplastics and Food Packaging: A Review. *Cogent Food Agric.* **2015**, *1*, 1117749. [CrossRef]
98. Nemes, S.A.; Szabo, K.; Vodnar, D.C. Applicability of Agro-Industrial by-Products in Intelligent Food Packaging. *Coatings* **2020**, *10*, 550. [CrossRef]
99. Freeland, B.; McCarthy, E.; Balakrishnan, R.; Fahy, S.; Boland, A.; Rochfort, K.D.; Dabros, M.; Marti, R.; Kelleher, S.M.; Gaughran, J. A Review of Polylactic Acid as a Replacement Material for Single-Use Laboratory Components. *Materials* **2022**, *15*, 2989. [CrossRef] [PubMed]
100. Zhang, M.; Biesold, G.M.; Choi, W.; Yu, J.; Deng, Y.; Silvestre, C.; Lin, Z. Recent advances in polymers and polymer composites for food packaging. *Materials Today.* **2022**, *53*, 134–161. [CrossRef]

101. Priyadarshi, R.; Sauraj; Kumar, B.; Deeba, F.; Kulshreshtha, A.; Negi, Y.S. Chitosan Films Incorporated with Apricot (*Prunus armeniaca*) Kernel Essential Oil as Active Food Packaging Material. *Food Hydrocoll.* **2018**, *85*, 158–166. [CrossRef]
102. Kim, J.-H.; Hong, W.-S.; Oh, S.-W. Effect of Layer-by-Layer Antimicrobial Edible Coating of Alginate and Chitosan with Grapefruit Seed Extract for Shelf-Life Extension of Shrimp (Litopenaeus vannamei) Stored at 4 °C. *Int. J. Biol. Macromol.* **2018**, *120*, 1468–1473. [CrossRef]
103. Dilucia, F.; Lacivita, V.; Conte, A.; Del Nobile, M.A. Sustainable Use of Fruit and Vegetable By-Products to Enhance Food Packaging Performance. *Foods* **2020**, *9*, 857. [CrossRef]
104. Torres-León, C.; Vicente, A.A.; Flores-López, M.L.; Rojas, R.; Serna-Cock, L.; Alvarez-Pérez, O.B.; Aguilar, C.N. Edible Films and Coatings Based on Mango (Var. Ataulfo) by-Products to Improve Gas Transfer Rate of Peach. *LWT* **2018**, *97*, 624–631. [CrossRef]
105. Adilah, Z.A.M.; Jamilah, B.; Nur Hanani, Z.A. Functional and Antioxidant Properties of Protein-Based Films Incorporated with Mango Kernel Extract for Active Packaging. *Food Hydrocoll.* **2018**, *74*, 207–218. [CrossRef]
106. Riaz, A.; Lei, S.; Akhtar, H.M.S.; Wan, P.; Chen, D.; Jabbar, S.; Abid, M.; Hashim, M.M.; Zeng, X. Preparation and Characterization of Chitosan-Based Antimicrobial Active Food Packaging Film Incorporated with Apple Peel Polyphenols. *Int. J. Biol. Macromol.* **2018**, *114*, 547–555. [CrossRef] [PubMed]
107. Choi, I.; Chang, Y.; Shin, S.-H.; Joo, E.; Song, H.J.; Eom, H.; Han, J. Development of Biopolymer Composite Films Using a Microfluidization Technique for Carboxymethylcellulose and Apple Skin Particles. *Int. J. Mol. Sci.* **2017**, *18*, 1278. [CrossRef] [PubMed]
108. Zhang, W.; Li, X.; Jiang, W. Development of Antioxidant Chitosan Film with Banana Peels Extract and Its Application as Coating in Maintaining the Storage Quality of Apple. *Int. J. Biol. Macromol.* **2020**, *154*, 1205–1214. [CrossRef] [PubMed]
109. Kumar, N.; Pratibha; Neeraj; Ojha, A.; Upadhyay, A.; Singh, R.; Kumar, S. Effect of Active Chitosan-Pullulan Composite Edible Coating Enrich with Pomegranate Peel Extract on the Storage Quality of Green Bell Pepper. *LWT* **2021**, *138*, 110435. [CrossRef]
110. Hanani, Z.A.N.; Yee, F.C.; Nor-Khaizura, M.A.R. Effect of Pomegranate (*Punica granatum* L.) Peel Powder on the Antioxidant and Antimicrobial Properties of Fish Gelatin Films as Active Packaging. *Food Hydrocoll.* **2019**, *89*, 253–259. [CrossRef]
111. Mushtaq, M.; Gani, A.; Gani, A.; Punoo, H.A.; Masoodi, F.A. Use of Pomegranate Peel Extract Incorporated Zein Film with Improved Properties for Prolonged Shelf Life of Fresh Himalayan Cheese (Kalari/Kradi). *Innov. Food Sci. Emerg. Technol.* **2018**, *48*, 25–32. [CrossRef]
112. Kurek, M.; Benbettaieb, N.; Ščetar, M.; Chaudy, E.; Repajić, M.; Klepac, D.; Valić, S.; Debeaufort, F.; Galić, K. Characterization of Food Packaging Films with Blackcurrant Fruit Waste as a Source of Antioxidant and Color Sensing Intelligent Material. *Molecules* **2021**, *26*, 2569. [CrossRef]
113. Xie, Y.; Niu, X.; Yang, J.; Fan, R.; Shi, J.; Ullah, N.; Feng, X.; Chen, L. Active Biodegradable Films Based on the Whole Potato Peel Incorporated with Bacterial Cellulose and Curcumin. *Int. J. Biol. Macromol.* **2020**, *150*, 480–491. [CrossRef]
114. Szabo, K.; Teleky, B.-E.; Mitrea, L.; Călinoiu, L.-F.; Mărtău, G.-A.; Simon, E.; Varvara, R.-A.; Vodnar, D.C. Active Packaging—Poly(Vinyl Alcohol) Films Enriched with Tomato by-Products Extract. *Coatings* **2020**, *10*, 141. [CrossRef]
115. Stoll, L.; Rech, R.; Flôres, S.H.; Nachtigall, S.M.B.; de Oliveira Rios, A. Poly(Acid Lactic) Films with Carotenoids Extracts: Release Study and Effect on Sunflower Oil Preservation. *Food Chem.* **2019**, *281*, 213–221. [CrossRef] [PubMed]
116. Musso, Y.S.; Salgado, P.R.; Mauri, A.N. Smart Gelatin Films Prepared Using Red Cabbage (*Brassica oleracea* L.) Extracts as Solvent. *Food Hydrocoll.* **2019**, *89*, 674–681. [CrossRef]
117. Uranga, J.; Etxabide, A.; Guerrero, P.; de la Caba, K. Development of Active Fish Gelatin Films with Anthocyanins by Compression Molding. *Food Hydrocoll.* **2018**, *84*, 313–320. [CrossRef]
118. Mustafa, P.; Niazi, M.B.K.; Jahan, Z.; Samin, G.; Hussain, A.; Ahmed, T.; Naqvi, S.R. PVA/Starch/Propolis/Anthocyanins Rosemary Extract Composite Films as Active and Intelligent Food Packaging Materials. *J. Food Saf.* **2020**, *40*, e12725. [CrossRef]
119. Vannini, M.; Marchese, P.; Sisti, L.; Saccani, A.; Mu, T.; Sun, H.; Celli, A. Integrated Efforts for the Valorization of Sweet Potato By-Products within a Circular Economy Concept: Biocomposites for Packaging Applications Close the Loop. *Polymers* **2021**, *13*, 1048. [CrossRef]
120. Assis, R.Q.; Pagno, C.H.; Costa, T.M.H.; Flôres, S.H.; Rios, A.d.O. Synthesis of Biodegradable Films Based on Cassava Starch Containing Free and Nanoencapsulated β-Carotene. *Packag. Technol. Sci.* **2018**, *31*, 157–166. [CrossRef]
121. Mitrea, L.; Călinoiu, L.-F.; Mărtău, G.-A.; Szabo, K.; Teleky, B.-E.; Mureșan, V.; Rusu, A.-V.; Socol, C.-T.; Vodnar, D.-C. Poly(Vinyl Alcohol)-Based Biofilms Plasticized with Polyols and Colored with Pigments Extracted from Tomato by-Products. *Polymers* **2020**, *12*, 532. [CrossRef]
122. Mohammadi, H.; Kamkar, A.; Misaghi, A.; Zunabovic-Pichler, M.; Fatehi, S. Nanocomposite Films with CMC, Okra Mucilage, and ZnO Nanoparticles: Extending the Shelf-Life of Chicken Breast Meat. *Food Packag. Shelf Life* **2019**, *21*, 100330. [CrossRef]
123. Peelman, N.; Ragaert, P.; De Meulenaer, B.; Adons, D.; Peeters, R.; Cardon, L.; Van Impe, F.; Devlieghere, F. Application of Bioplastics for Food Packaging. *Trends Food Sci. Technol.* **2013**, *32*, 128–141. [CrossRef]
124. Ibrahim, N.I.; Shahar, F.S.; Sultan, M.T.H.; Shah, A.U.M.; Safri, S.N.A.; Mat Yazik, M.H. Overview of Bioplastic Introduction and Its Applications in Product Packaging. *Coatings* **2021**, *11*, 1423. [CrossRef]
125. Nanda, S.; Patra, B.R.; Patel, R.; Bakos, J.; Dalai, A.K. Innovations in Applications and Prospects of Bioplastics and Biopolymers: A Review. *Environ. Chem. Lett.* **2022**, *20*, 379–395. [CrossRef] [PubMed]
126. Asgher, M.; Qamar, S.A.; Bilal, M.; Iqbal, H.M.N. Bio-Based Active Food Packaging Materials: Sustainable Alternative to Conventional Petrochemical-Based Packaging Materials. *Food Res. Int.* **2020**, *137*, 109625. [CrossRef] [PubMed]

127. Dang, B.-T.; Bui, X.-T.; Tran, D.P.H.; Hao Ngo, H.; Nghiem, L.D.; Hoang, T.-K.-D.; Nguyen, P.-T.; Nguyen, H.H.; Vo, T.-K.-Q.; Lin, C.; et al. Current Application of Algae Derivatives for Bioplastic Production: A Review. *Bioresour. Technol.* **2022**, *347*, 126698. [CrossRef]
128. Brodin, F.W.; Gregersen, Ø.W.; Syverud, K. Cellulose Nanofibrils: Challenges and Possibilities as a Paper Additive or Coating Material: A Review. *Nord. Pulp Paper Res. J.* **2014**, *29*, 156–166. [CrossRef]
129. Yang, J.; Ching, Y.C.; Chuah, C.H. Applications of Lignocellulosic Fibers and Lignin in Bioplastics: A Review. *Polymers* **2019**, *11*, 751. [CrossRef]
130. Polychronopoulos, N.D.; Vlachopoulos, J. Polymer Processing and Rheology. In *Functional Polymers. Polymers and Polymeric Composites: A Reference Series*; Springer: Cham, Switzerland, 2019. [CrossRef]
131. Liang, J.-Z. Characteristics of Melt Shear Viscosity during Extrusion of Polymers. *Polym. Test.* **2002**, *21*, 307–311. [CrossRef]
132. Irgens, F. *Rheology and Non-Newtonian Fluids*; Springer: Cham, Switzerland, 2014.
133. Shenoy, A.V.; Saini, D.R. Melt Flow Index: More than Just a Quality Control Rheological Parameter. Part, I. *Adv. Polym. Technol.* **1986**, *6*, 1–58. [CrossRef]
134. Hieber, C.A. Melt-Viscosity Characterization and Its Application to Injection Molding. In *Injection and Compression Molding Fundamentals*; Routledge: London, UK, 2017; pp. 1–136.
135. Lord, H.A. Flow of Polymers with Pressure-Dependent Viscosity in Injection Molding Dies. *Polym. Eng. Sci.* **1979**, *19*, 469–473. [CrossRef]
136. Capone, C.; Di Landro, L.; Inzoli, F.; Penco, M.; Sartore, L. Thermal and Mechanical Degradation during Polymer Extrusion Processing. *Polym. Eng. Sci.* **2007**, *47*, 1813–1819. [CrossRef]
137. Hermabessiere, L.; Dehaut, A.; Paul-Pont, I.; Lacroix, C.; Jezequel, R.; Soudant, P.; Duflos, G. Occurrence and Effects of Plastic Additives on Marine Environments and Organisms: A Review. *Chemosphere* **2017**, *182*, 781–793. [CrossRef] [PubMed]
138. Rauwendaal, C. *Polymer Extrusion*; Carl Hanser Verlag: Munich, Germany, 2014.
139. Abeykoon, C.; Pérez, P.; Kelly, A.L. The Effect of Materials' Rheology on Process Energy Consumption and Melt Thermal Quality in Polymer Extrusion. *Polym. Eng. Sci.* **2020**, *60*, 1244–1265. [CrossRef]
140. Mileva, D.; Tranchida, D.; Gahleitner, M. Designing Polymer Crystallinity: An Industrial Perspective. *Polym. Cryst.* **2018**, *1*, e10009. [CrossRef]
141. Billham, M.; Clarke, A.H.; Garrett, G.; Mcnally, G.M.; Murphy, W.R. The Effect of Extrusion Processing Conditions on the Properties of Blown and Cast Polyolefin Packaging Films. *Dev. Chem. Eng. Miner. Process.* **2008**, *11*, 137–146. [CrossRef]
142. Gruenwald, G. *Thermoforming: A Plastics Processing Guide*; Routledge: London, UK, 2018.
143. Farotti, E.; Natalini, M. Injection Molding. *Influence of Process Parameters on Mechanical Properties of Polypropylene Polymer. A First Study. Procedia Struct. Integr.* **2018**, *8*, 256–264. [CrossRef]
144. Hishinuma, K. *Heat Sealing Technology and Engineering: Principles and Packaging Applications*; Miyagawa, H., Translator; DEStech Publications: Lancaster, CA, USA, 2009.
145. Lee, N.C. *Understanding Blow Molding*, 2nd ed.; Carl Hanser Verlag: Munich, Germany, 2007.
146. Cantor, K. *Blown Film Extrusion*; Hanser Publications: Liberty Twp, OH, USA, 2018.
147. Siracusa, V. Food Packaging Permeability Behaviour: A Report. *Int. J. Polym. Sci.* **2012**, *2012*, 302029. [CrossRef]
148. Shah, A.A.; Hasan, F.; Hameed, A.; Ahmed, S. Biological Degradation of Plastics: A Comprehensive Review. *Biotechnol. Adv.* **2008**, *26*, 246–265. [CrossRef]
149. Choe, S.; Kim, Y.; Won, Y.; Myung, J. Bridging Three Gaps in Biodegradable Plastics: Misconceptions and Truths about Biodegradation. *Front. Chem.* **2021**, *9*, 671750. [CrossRef]
150. Moshood, T.D.; Nawanir, G.; Mahmud, F.; Mohamad, F.; Ahmad, M.H.; AbdulGhani, A. Biodegradable Plastic Applications towards Sustainability: A Recent Innovations in the Green Product. *Clean. Eng. Technol.* **2022**, *6*, 100404. [CrossRef]
151. Fan, P.; Yu, H.; Xi, B.; Tan, W. A Review on the Occurrence and Influence of Biodegradable Microplastics in Soil Ecosystems: Are Biodegradable Plastics Substitute or Threat? *Environ. Int.* **2022**, *163*, 107244. [CrossRef]
152. Dey, A.; Neogi, S. Oxygen Scavengers for Food Packaging Applications: A Review. *Trends Food Sci. Technol.* **2019**, *90*, 26–34. [CrossRef]
153. Pascall, M.A.; Siddiq, F. Fresh Vegetables and Vegetable Products Packaging. In *Handbook of Vegetables and Vegetable Processing*; John Wiley & Sons, Ltd: Chichester, UK, 2018; pp. 265–286.
154. Baschetti, M.G.; Minelli, M. Test Methods for the Characterization of Gas and Vapor Permeability in Polymers for Food Packaging Application: A Review. *Polym. Test.* **2020**, *89*, 106606. [CrossRef]
155. Pandit, P.; Nadathur, G.T.; Maiti, S.; Regubalan, B. Functionality and Properties of Bio-Based Materials. In *Bio-Based Materials for Food Packaging*; Springer: Singapore, 2018; pp. 81–103.
156. Pauer, E.; Tacker, M.; Gabriel, V.; Krauter, V. Sustainability of Flexible Multilayer Packaging: Environmental Impacts and Recyclability of Packaging for Bacon in Block. *Cleaner Environ. Syst.* **2020**, *1*, 100001. [CrossRef]
157. Wu, F.; Misra, M.; Mohanty, A.K. Challenges and New Opportunities on Barrier Performance of Biodegradable Polymers for Sustainable Packaging. *Prog. Polym. Sci.* **2021**, *117*, 101395. [CrossRef]
158. *ISO 14044:2006*; Environmental Management—Life Cycle Assessment—Requirements and Guidelines. International Organization for Standardization: Geneva, Switzerland, 2006.

159. Grabowski, A.; Selke, S.E.M.; Auras, R.; Patel, M.K.; Narayan, R. Life Cycle Inventory Data Quality Issues for Bioplastics Feedstocks. *Int. J. Life Cycle Assess.* **2015**, *20*, 584–596. [CrossRef]
160. *2013/179/EU*; Commission Recommendation of 9 April 2013 on the Use of Common Methods to Measure and Communicate the Life Cycle Environmental Performance of Products and Organisations. European Commission: Brussels, Belgium, 9 April 2013.
161. Brizga, J.; Hubacek, K.; Feng, K. The Unintended Side Effects of Bioplastics: Carbon, Land, and Water Footprints. *One Earth* **2020**, *3*, 45–53. [CrossRef]
162. Wellenreuther, C.; Wolf, A. *Innovative Feedstocks in Biodegradable Bio-Based Plastics: A Literature Review*; HWWI Research Papers 194; Hamburg Institute of International Economics (HWWI): Hamburg, Germany, 2020.
163. De Souza Vandenberghe, L.P.; de Oliveira, P.Z.; Bittencourt, G.A.; de Mello, A.F.M.; Vásquez, Z.S.; Karp, S.G.; Soccol, C.R. The 2G and 3G Bioplastics: An Overview. *Biotechnol. Res. Innov.* **2021**, *5*, e2021004. [CrossRef]
164. Rosenboom, J.-G.; Langer, R.; Traverso, G. Bioplastics for a Circular Economy. *Nat. Rev. Mater.* **2022**, *7*, 117–137. [CrossRef]
165. Filho, W.L.; Barbir, J.; Abubakar, I.R.; Paço, A.; Stasiskiene, Z.; Hornbogen, M.; Christin Fendt, M.T.; Voronova, V.; Klõga, M. Consumer Attitudes and Concerns with Bioplastics Use: An International Study. *PLoS ONE* **2022**, *17*, e0266918. [CrossRef]
166. Bioplastics: Facts and figures. Available online: https://docs.european-bioplastics.org/publications/EUBP_Facts_and_figures.pdf (accessed on 22 June 2022).
167. Regulation (EC) No 1935/2004 of the European Parliament and of the Council of 27 October 2004 on materials and articles intended to come into contact with food and repealing Directives 80/590/EEC and 89/109/EEC. Available online: https://eur-lex.europa.eu/eli/reg/2004/1935/oj (accessed on 22 June 2022).
168. Commission Regulation (EU) No 10/2011 of 14 January 2011 on plastic materials and articles intended to come into contact with food. Text with EEA relevance. Available online: https://eur-lex.europa.eu/legal-content/EN/ALL/?uri=celex%3A32011R0010 (accessed on 22 June 2022).
169. Takeshita, A.; Igarashi-Migitaka, J.; Nishiyama, K.; Takahashi, H.; Takeuchi, Y.; Koibuchi, N. Acetyl Tributyl Citrate, the Most Widely Used Phthalate Substitute Plasticizer, Induces Cytochrome P450 3a through Steroid and Xenobiotic Receptor. *Toxicol. Sci.* **2011**, *123*, 460–470. [CrossRef]
170. Sheikh, I.A.; Beg, M.A. Structural Characterization of Potential Endocrine Disrupting Activity of Alternate Plasticizers Di-(2-Ethylhexyl) Adipate (DEHA), Acetyl Tributyl Citrate (ATBC) and 2,2,4-Trimethyl 1,3-Pentanediol Diisobutyrate (TPIB) with Human Sex Hormone-Binding Globulin. *Reprod. Toxicol.* **2019**, *83*, 46–53. [CrossRef] [PubMed]
171. Rainer, B.; Pinter, E.; Czerny, T.; Riegel, E.; Kirchnawy, C.; Marin-Kuan, M.; Schilter, B.; Tacker, M. Suitability of the Ames Test to Characterise Genotoxicity of Food Contact Material Migrates. *Food Addit. Contam. Part A Chem. Anal. Control Expo. Risk Assess.* **2018**, *35*, 2230–2243. [CrossRef] [PubMed]
172. Kato, L.S.; Conte-Junior, C.A. Safety of Plastic Food Packaging: The Challenges about Non-Intentionally Added Substances (NIAS) Discovery, Identification and Risk Assessment. *Polymers* **2021**, *13*, 2077. [CrossRef] [PubMed]
173. Cavazza, A.; Mattarozzi, M.; Franzoni, A.; Careri, M. A Spotlight on Analytical Prospects in Food Allergens: From Emerging Allergens and Novel Foods to Bioplastics and Plant-Based Sustainable Food Contact Materials. *Food Chem.* **2022**, *388*, 132951. [CrossRef]
174. Tihminlioglu, F.; Atik, İ.D.; Özen, B. Water Vapor and Oxygen-Barrier Performance of Corn–Zein Coated Polypropylene Films. *J. Food Eng.* **2010**, *96*, 342–347. [CrossRef]
175. Koh, H.-Y.; Chinnan, M.S. Characteristics of Corn Zein and Methyl Cellulose Bilayer Edible Films according to Preparation Protocol. *Food Sci. Biotechnol.* **2002**, *11*, 310–315.
176. Bayer, I.S. Zein in Food Packaging. In *Sustainable Food Packaging Technology*; Wiley: Hoboken, NJ, USA, 2021; pp. 199–224. [CrossRef]
177. Álvarez-Castillo, E.; Felix, M.; Bengoechea, C.; Guerrero, A. Proteins from Agri-Food Industrial Biowastes or Co-Products and Their Applications as Green Materials. *Foods* **2021**, *10*, 981. [CrossRef]
178. Verbeek, C.J.R.; van den Berg, L.E. Extrusion Processing and Properties of Protein-Based Thermoplastics. *Macromol. Mater. Eng.* **2010**, *295*, 10–21. [CrossRef]
179. Di Gioia, L.; Guilbert, S. Corn Protein-Based Thermoplastic Resins: Effect of Some Polar and Amphiphilic Plasticizers. *J. Agric. Food Chem.* **1999**, *47*, 1254–1261. [CrossRef]
180. Wang, Y.; Padua, G.W. Tensile Properties of Extruded Zein Sheets and Extrusion Blown Films. *Macromol. Mater. Eng.* **2003**, *288*, 886–893. [CrossRef]
181. Chen, H.; Wang, J.; Cheng, Y.; Wang, C.; Liu, H.; Bian, H.; Pan, Y.; Sun, J.; Han, W. Application of Protein-Based Films and Coatings for Food Packaging: A Review. *Polymers* **2019**, *11*, 2039. [CrossRef] [PubMed]
182. Türe, H.; Gällstedt, M.; Johansson, E.; Hedenqvist, M.S. Wheat-Gluten/Montmorillonite Clay Multilayer-Coated Paperboards with High Barrier Properties. *Ind. Crops Prod.* **2013**, *51*, 1–6. [CrossRef]
183. Klüver, E.; Meyer, M. Thermoplastic Processing, Rheology, and Extrudate Properties of Wheat, Soy, and Pea Proteins. *Polym. Eng. Sci.* **2015**, *55*, 1912–1919. [CrossRef]
184. Cho, S.-W.; Gällstedt, M.; Johansson, E.; Hedenqvist, M.S. Injection-Molded Nanocomposites and Materials Based on Wheat Gluten. *Int. J. Biol. Macromol.* **2011**, *48*, 146–152. [CrossRef] [PubMed]
185. Tunc, S.; Angellier, H.; Cahyana, Y.; Chalier, P.; Gontard, N.; Gastaldi, E. Functional Properties of Wheat Gluten/Montmorillonite Nanocomposite Films Processed by Casting. *J. Memb. Sci.* **2007**, *289*, 159–168. [CrossRef]

186. Redl, A.; Morel, M.H.; Bonicel, J.; Vergnes, B.; Guilbert, S. Extrusion of Wheat Gluten Plasticized with Glycerol: Influence of Process Conditions on Flow Behavior, Rheological Properties, and Molecular Size Distribution. *Cereal Chem.* **1999**, *76*, 361–370. [CrossRef]
187. Pallos, F.M.; Robertson, G.H.; Pavlath, A.E.; Orts, W.J. Thermoformed Wheat Gluten Biopolymers. *J. Agric. Food Chem.* **2006**, *54*, 349–352. [CrossRef]
188. Swain, S.N.; Biswal, S.M.; Nanda, P.K.; Nayak, P.L. Biodegradable Soy-Based Plastics: Opportunities and Challenges. *J. Polym. Environ.* **2004**, *12*, 35–42. [CrossRef]
189. Gökkaya Erdem, B.; Dıblan, S.; Kaya, S. A Comprehensive Study on Sorption, Water Barrier, and Physicochemical Properties of Some Protein- and Carbohydrate-Based Edible Films. *Food Bioproc. Tech.* **2021**, *14*, 2161–2179. [CrossRef]
190. Garrido, T.; Peñalba, M.; de la Caba, K.; Guerrero, P. Injection-Manufactured Biocomposites from Extruded Soy Protein with Algae Waste as a Filler. *Compos. B Eng.* **2016**, *86*, 197–202. [CrossRef]
191. Mohareb, E.; Mittal, G.S. Formulation and Process Conditions for Biodegradable/Edible Soy-Based Packaging Trays. *Packag. Technol. Sci.* **2007**, *20*, 1–15. [CrossRef]
192. Schmid, M.; Dallmann, K.; Bugnicourt, E.; Cordoni, D.; Wild, F.; Lazzeri, A.; Noller, K. Properties of Whey-Protein-Coated Films and Laminates as Novel Recyclable Food Packaging Materials with Excellent Barrier Properties. *Int. J. Polym. Sci.* **2012**, *2012*, 562381. [CrossRef]
193. Hernandez-Izquierdo, V.M.; Krochta, J.M. Thermoplastic Processing of Proteins for Film Formation—A Review. *J. Food Sci.* **2008**, *73*, R30–R39. [CrossRef] [PubMed]
194. Phupoksakul, T.; Leuangsukrerk, M.; Somwangthanaroj, A.; Tananuwong, K.; Janjarasskul, T. Storage Stability of Packaged Baby Formula in Poly(Lactide)-Whey Protein Isolate Laminated Pouch. *J. Sci. Food Agric.* **2017**, *97*, 3365–3373. [CrossRef] [PubMed]
195. Pankaj, S.K.; Bueno-Ferrer, C.; Misra, N.N.; O'Neill, L.; Tiwari, B.K.; Bourke, P.; Cullen, P.J. Physicochemical Characterization of Plasma-Treated Sodium Caseinate Film. *Food Res. Int.* **2014**, *66*, 438–444. [CrossRef]
196. Nurdiani, R.; Ma'rifah, R.D.A.; Busyro, I.K.; Jaziri, A.A.; Prihanto, A.A.; Firdaus, M.; Talib, R.A.; Huda, N. Physical and Functional Properties of Fish Gelatin-Based Film Incorporated with Mangrove Extracts. *PeerJ* **2022**, *10*, e13062. [CrossRef] [PubMed]
197. Mi, X.; Xu, H.; Yang, Y. Submicron Amino Acid Particles Reinforced 100% Keratin Biomedical Films with Enhanced Wet Properties via Interfacial Strengthening. *Colloids Surf. B Biointerfaces* **2019**, *177*, 33–40. [CrossRef]
198. Nair, S.S.; Zhu, J.Y.; Deng, Y.; Ragauskas, A.J. High Performance Green Barriers Based on Nanocellulose. *Sustain. Chem. Process.* **2014**, *2*, 1–7. [CrossRef]
199. Fazeli, M.; Keley, M.; Biazar, E. Preparation and Characterization of Starch-Based Composite Films Reinforced by Cellulose Nanofibers. *Int. J. Biol. Macromol.* **2018**, *116*, 272–280. [CrossRef]
200. Song, J.H.; Murphy, R.J.; Narayan, R.; Davies, G.B.H. Biodegradable and Compostable Alternatives to Conventional Plastics. *Philos. Trans. R. Soc. Lond. B Biol. Sci.* **2009**, *364*, 2127–2139. [CrossRef]
201. Park, S.-H.; Lee, H.S.; Choi, J.H.; Jeong, C.M.; Sung, M.H.; Park, H.J. Improvements in Barrier Properties of Poly(Lactic Acid) Films Coated with Chitosan or Chitosan/Clay Nanocomposite. *J. Appl. Polym. Sci.* **2012**, *125*, E675–E680. [CrossRef]
202. Wiles, J.L.; Vergano, P.J.; Barron, F.H.; Bunn, J.M.; Testin, R.F. Water Vapor Transmission Rates and Sorption Behavior of Chitosan Films. *J. Food Sci.* **2000**, *65*, 1175–1179. [CrossRef]
203. Gu, C.-H.; Wang, J.-J.; Yu, Y.; Sun, H.; Shuai, N.; Wei, B. Biodegradable Multilayer Barrier Films Based on Alginate/Polyethyleneimine and Biaxially Oriented Poly(Lactic Acid). *Carbohydr. Polym.* **2013**, *92*, 1579–1585. [CrossRef]
204. Kopacic, S.; Walzl, A.; Zankel, A.; Leitner, E.; Bauer, W. Alginate and Chitosan as a Functional Barrier for Paper-Based Packaging Materials. *Coatings* **2018**, *8*, 235. [CrossRef]
205. Łopusiewicz, Ł.; Macieja, S.; Śliwiński, M.; Bartkowiak, A.; Roy, S.; Sobolewski, P. Alginate Biofunctional Films Modified with Melanin from Watermelon Seeds and Zinc Oxide/Silver Nanoparticles. *Materials* **2022**, *15*, 2381. [CrossRef]
206. Wang, S.; Liu, B.; Qin, Y.; Guo, H. Effects of Processing Conditions and Plasticizing-Reinforcing Modification on the Crystallization and Physical Properties of PLA Films. *Membranes* **2021**, *11*, 640. [CrossRef]
207. Bioplastics market. Development update 2021. Available online: https://docs.european-bioplastics.org/publications/market_data/Report_Bioplastics_Market_Data_2021_short_version.pdf (accessed on 1 June 2022).
208. Thellen, C.; Coyne, M.; Froio, D.; Auerbach, M.; Wirsen, C.; Ratto, J.A. A Processing, Characterization and Marine Biodegradation Study of Melt-Extruded Polyhydroxyalkanoate (PHA) Films. *J Polym Environ* **2008**, *16*, 1–11. [CrossRef]
209. Hänggi, U.J. Requirements on Bacterial Polyesters as Future Substitute for Conventional Plastics for Consumer Goods. *FEMS Microbiol. Rev.* **1995**, *16*, 213–220. [CrossRef]
210. Al Rowaihi, I.S.; Kick, B.; Grötzinger, S.W.; Burger, C.; Karan, R.; Weuster-Botz, D.; Eppinger, J.; Arold, S.T. A Two-Stage Biological Gas to Liquid Transfer Process to Convert Carbon Dioxide into Bioplastic. *Bioresour. Technol. Rep.* **2018**, *1*, 61–68. [CrossRef]
211. Tanaka, K.; Mori, S.; Hirata, M.; Matsusaki, H. Autotrophic Growth of Paracoccus denitrificans in Aerobic Condition and the Accumulation of Biodegradable Plastics from CO2. *Environ. Ecol. Res.* **2016**, *4*, 231–236. [CrossRef]
212. Ghysels, S.; Mozumder, M.S.I.; De Wever, H.; Volcke, E.I.P.; Garcia-Gonzalez, L. Targeted Poly(3-Hydroxybutyrate-Co-3-Hydroxyvalerate) Bioplastic Production from Carbon Dioxide. *Bioresour. Technol.* **2018**, *249*, 858–868. [CrossRef] [PubMed]
213. Vytejčková, S.; Vápenka, L.; Hradecký, J.; Dobiáš, J.; Hajšlová, J.; Loriot, C.; Vannini, L.; Poustka, J. Testing of Polybutylene Succinate Based Films for Poultry Meat Packaging. *Polym. Test.* **2017**, *60*, 357–364. [CrossRef]

214. Threepopnatkul, P.; Wongsuton, K.; Jaiaue, C.; Rakkietwinai, N.; Kulsetthanchalee, C. Thermal and Barrier Properties of Poly(Butylene Adipate-Coterephthalate) Incorporated with Zeolite Doped Potassium Ion for Packaging Film. *IOP Conf. Ser. Mater. Sci. Eng.* **2020**, *773*, 012042. [CrossRef]
215. Guinault, A.; Messin, T.; Anderer, G.; Krawielitzki, S.; Sollogoub, C.; Roland, S.; Grandmontagne, A.; Dole, P.; Julien, J.-M.; Loriot, C.; et al. Relationship between crystallization, mechanical and gas barrier properties of poly(ethylene furanoate) (PEF) in multinanolayered PLA-PEF and PET-PEF films. In Proceedings of the ESAFORM 2021, 24th International Conference on Material Forming, Liège, Belgium, 14–16 April 2021. [CrossRef]
216. Reichert, C.L.; Bugnicourt, E.; Coltelli, M.-B.; Cinelli, P.; Lazzeri, A.; Canesi, I.; Braca, F.; Martínez, B.M.; Alonso, R.; Agostinis, L.; et al. Bio-Based Packaging: Materials, Modifications, Industrial Applications and Sustainability. *Polymers* **2020**, *12*, 1558. [CrossRef]
217. Lange, J.; Wyser, Y. Recent Innovations in Barrier Technologies for Plastic Packaging? *A Review. Packag. Technol. Sci.* **2003**, *16*, 149–158. [CrossRef]
218. Tullo, A. New Route Planned to Bio-based Ethylene Glycol. *C&EN Glob. Enterp.* **2017**, *95*, 10. [CrossRef]
219. Yamane, K.; Sato, H.; Ichikawa, Y.; Sunagawa, K.; Shigaki, Y. Development of an Industrial Production Technology for High-Molecular-Weight Polyglycolic Acid. *Polym. J.* **2014**, *46*, 769–775. [CrossRef]
220. Hult, E.-L.; Ropponen, J.; Poppius-Levlin, K.; Ohra-Aho, T.; Tamminen, T. Enhancing the Barrier Properties of Paper Board by a Novel Lignin Coating. *Ind. Crops Prod.* **2013**, *50*, 694–700. [CrossRef]
221. Bhatia, K.; Asrey, R.; Jha, S.K.; Singh, S.; Kannaujia, P.K. Influence of packaging material on quality characteristics of minimally processed Mridula pomegranate (*Punica granatum*) arils during cold storage. *Indian J. Agric. Sci.* **2013**, *83*, 872–876.
222. Lindström, T.; Österberg, F. Evolution of Bio-based and Nanotechnology Packaging—A Review. *Nord. Pulp Paper Res. J.* **2020**, *35*, 491–515. [CrossRef]
223. Bucio, A.; Moreno-Tovar, R.; Bucio, L.; Espinosa-Dávila, J.; Anguebes-Franceschi, F. Characterization of Beeswax, Candelilla Wax and Paraffin Wax for Coating Cheeses. *Coatings* **2021**, *11*, 261. [CrossRef]
224. Hassan, B.; Chatha, S.A.S.; Hussain, A.I.; Zia, K.M.; Akhtar, N. Recent Advances on Polysaccharides, Lipids and Protein-Based Edible Films and Coatings: A Review. *Int. J. Biol. Macromol.* **2018**, *109*, 1095–1107. [CrossRef] [PubMed]
225. Hult, E.-L.; Koivu, K.; Asikkala, J.; Ropponen, J.; Wrigstedt, P.; Sipilä, J.; Poppius-Levlin, K. Esterified Lignin Coating as Water Vapor and Oxygen Barrier for Fiber-Based Packaging. *Holzforschung* **2013**, *67*, 899–905. [CrossRef]
226. Geueke, B. FPF Dossier: Bioplastics. 2015. Zenodo. Available online: https://doi.org/10.5281/ZENODO.33517 (accessed on 3 July 2022).
227. Shlush, E.; Davidovich-Pinhas, M. Bioplastics for Food Packaging. *Trends Food Sci. Technol.* **2022**, *125*, 66–80. [CrossRef]
228. Zhao, X.; Cornish, K.; Vodovotz, Y. Narrowing the Gap for Bioplastic Use in Food Packaging: An Update. *Environ. Sci. Technol.* **2020**, *54*, 4712–4732. [CrossRef]
229. PlasticsBusinessMagazine. Available online: http://www.plasticsbusinessmag.com/enews/stories/050713/biodegradable.shtml#.VjgZZZZca6Ao\T1\textgreater{} (accessed on 27 June 2022).
230. European Bioplastics. FAQ-European Bioplastics. 2021. Available online: http://www.european-bioplastics.org/market (accessed on 27 June 2022).
231. Rhim, J.-W.; Hong, S.-I.; Ha, C.-S. Tensile, Water Vapor Barrier and Antimicrobial Properties of PLA/Nanoclay Composite Films. *LWT* **2009**, *42*, 612–617. [CrossRef]
232. Gironi, F.; Piemonte, V. Bioplastics and Petroleum-Based Plastics: Strengths and Weaknesses. *Energy Sources Recovery Util. Environ. Eff.* **2011**, *33*, 1949–1959. [CrossRef]
233. Harding, K.G.; Dennis, J.S.; von Blottnitz, H.; Harrison, S.T.L. Environmental Analysis of Plastic Production Processes: Comparing Petroleum-Based Polypropylene and Polyethylene with Biologically-Based Poly-Beta-Hydroxybutyric Acid Using Life Cycle Analysis. *J. Biotechnol.* **2007**, *130*, 57–66. [CrossRef]
234. Vanderreydt, I.; Rommens, T.; Tenhunen, A.; Mortensen, L.F.; Tange, I. Greenhouse Gas Emissions and Natural Capital Implications of Plastics (Including Bio-based Plastics). *Eur. Top. Cent. Waste Mater. a Green Econ.* 2021.
235. Kircher, M. Bioeconomy: Markets, Implications, and Investment Opportunities. *Economies* **2019**, *7*, 73. [CrossRef]
236. Woodford, C. Bioplastics and Biodegradable Plastics-How Do They Work? *Explain that St.* 2020. Available online: https://www.explainthatstuff.com/bioplastics.html (accessed on 28 June 2022).
237. Elsawy, M.A.; Kim, K.-H.; Park, J.-W.; Deep, A. Hydrolytic Degradation of Polylactic Acid (PLA) and Its Composites. *Renew. Sustain. Energy Rev.* **2017**, *79*, 1346–1352. [CrossRef]
238. Shamsuddin, I.M.; Jafar, J.A.; Shawai, A.S.A.; Yusuf, S.; Lateefah, M.; Aminu, I. Bioplastics as Better Alternative to Petroplastics and Their Role in National Sustainability: A Review. *Adv. Biosci. Bioeng.* **2017**, *5*, 63. [CrossRef]
239. Iwata, T. Biodegradable and Bio-Based Polymers: Future Prospects of Eco-Friendly Plastics. *Angew. Chem. Int. Ed Engl.* **2015**, *54*, 3210–3215. [CrossRef] [PubMed]
240. Porta, R.; Sabbah, M. Plastic Pollution and the Challenge of Bioplastics. *J. Appl. Biotechnol. Bioeng.* **2017**, *2*, 111. [CrossRef]
241. Bazzanella, A.M.; Ausfelder, F. *Low Carbon Energy and Feedstock for the European Chemical Industry*; DECHEMA, Gesellschaft für Chemische Technik und Biotechnologie eV: Frankfurt Am Main, Germany, 2017.
242. Ketene, M.B. Bioeconomy: Shaping the Transition to a Sustainable, Bio-based Economy, Edited by Iris Lewandowski (Springer, 2018). *Corvinus J. Sociol. Soc. Policy* **2019**, *10*, 165–172. [CrossRef]

243. Popp, J.; Lakner, Z.; Harangi-Rákos, M.; Fári, M. The Effect of Bioenergy Expansion: Food, Energy, and Environment. *Renew. Sustain. Energy Rev.* **2014**, *32*, 559–578. [CrossRef]
244. Eerhart, A.J.J.E.; Faaij, A.P.C.; Patel, M.K. Replacing fossil based PET with biobased PEF; process analysis, energy and GHG balance. *Energy Environ. Sci.* **2012**, *5*, 6407. [CrossRef]
245. Liptow, C.; Tillman, A.-M. A Comparative Life Cycle Assessment Study of Polyethylene Based on Sugarcane and Crude Oil. *J. Ind. Ecol.* **2012**, *16*, 420–435. [CrossRef]
246. Alvarenga, R.A.F.; Dewulf, J.; De Meester, S.; Wathelet, A.; Villers, J.; Thommeret, R.; Hruska, Z. Life Cycle Assessment of Bioethanol-Based PVC: Part 2: Consequential Approach. *Biofuel. Bioprod. Biorefin.* **2013**, *7*, 396–405. [CrossRef]
247. Alvarenga, R.A.F.; Dewulf, J.; De Meester, S.; Wathelet, A.; Villers, J.; Thommeret, R.; Hruska, Z. Life Cycle Assessment of Bioethanol-Based PVC: Part 1: Attributional Approach. *Biofuel. Bioprod. Biorefin.* **2013**, *7*, 386–395. [CrossRef]
248. Piemonte, V.; Gironi, F. Land-Use Change Emissions: How Green Are the Bioplastics? *Environ. Prog. Sustain. Energy* **2011**, *30*, 685–691. [CrossRef]
249. Tsiropoulos, I.; Faaij, A.P.C.; Lundquist, L.; Schenker, U.; Briois, J.F.; Patel, M.K. Life Cycle Impact Assessment of Bio-Based Plastics from Sugarcane Ethanol. *J. Clean. Prod.* **2015**, *90*, 114–127. [CrossRef]
250. Lamberti, F.M.; Román-Ramírez, L.A.; Wood, J. Recycling of Bioplastics: Routes and Benefits. *J. Polym. Environ.* **2020**, *28*, 2551–2571. [CrossRef]
251. Raheem, A.B.; Noor, Z.Z.; Hassan, A.; Abd Hamid, M.K.; Samsudin, S.A.; Sabeen, A.H. Current Developments in Chemical Recycling of Post-Consumer Polyethylene Terephthalate Wastes for New Materials Production: A Review. *J. Clean. Prod.* **2019**, *225*, 1052–1064. [CrossRef]
252. Alzuhairi, M.A.H.; Khalil, B.I.; Hadi, R.S. Nano ZnO Catalyst for Chemical Recycling of Polyethylene Terephthalate (PET). *Eng. Technol. J.* **2017**, *35*, 831–837.
253. Wang, Z.; Shen, D.; Wu, C.; Gu, S. Thermal Behavior and Kinetics of Co-Pyrolysis of Cellulose and Polyethylene with the Addition of Transition Metals. *Energy Convers. Manag.* **2018**, *172*, 32–38. [CrossRef]
254. Lee, K.-H. Composition of Aromatic Products in the Catalytic Degradation of the Mixture of Waste Polystyrene and High-Density Polyethylene Using Spent FCC Catalyst. *Polym. Degrad. Stab.* **2008**, *93*, 1284–1289. [CrossRef]
255. Hong, M.; Chen, E.Y.X. Chemically recyclable polymers: A circular economy approach to sustainability. *Green Chem.* **2017**, *19*, 3692–3706. [CrossRef]
256. Scott, G. Photo-Biodegradable Plastics. In *Degradable Polymers*; Springer: Dordrecht, The Netherlands, 1995; pp. 169–185.
257. Claesen, C. Hycail-More than the Other PLA Producer. In Proceedings of the RSC Symposium Sustainable Plastic: Biodegradability vs. Recycling, Manchester, UK, 8–9 March 2005; pp. 8–9.
258. Hopewell, J.; Dvorak, R.; Kosior, E. Plastics Recycling: Challenges and Opportunities. *Philos. Trans. R. Soc. Lond. B Biol. Sci.* **2009**, *364*, 2115–2126. [CrossRef]
259. Odegard, I.; Nusselder, S.; Lindgreen, E.R.; Bergsma, G.; Graaff, L. Biobased Plastics in a Circular Economy. 2017. Available online: https://cedelft.eu/publications/biobased-plastics-in-a-circular-economy/ (accessed on 30 June 2022).
260. Ramsay, B.A.; Langlade, V.; Carreau, P.J.; Ramsay, J.A. Biodegradability and Mechanical Properties of Poly-(Beta-Hydroxybutyrate-Co-Beta-Hydroxyvalerate)-Starch Blends. *Appl. Environ. Microbiol.* **1993**, *59*, 1242–1246. [CrossRef]
261. Mohee, R.; Unmar, G.D.; Mudhoo, A.; Khadoo, P. Biodegradability of Biodegradable/Degradable Plastic Materials under Aerobic and Anaerobic Conditions. *Waste Manag.* **2008**, *28*, 1624–1629. [CrossRef] [PubMed]
262. Knoblauch, C.; Beer, C.; Liebner, S.; Grigoriev, M.N.; Pfeiffer, E.-M. Methane Production as Key to the Greenhouse Gas Budget of Thawing Permafrost. *Nat. Clim. Chang.* **2018**, *8*, 309–312. [CrossRef]
263. Stevens, E.S. *Green Plastics*; Princeton University Press: Princeton, NJ, USA, 2020; ISBN 0691214174.
264. Mohanty, A.K.; Misra, M.; Hinrichsen, G. Biofibres, Biodegradable Polymers and Biocomposites: An Overview. *Macromol. Mater. Eng.* **2000**, *276–277*, 1–24. [CrossRef]
265. Lashof, D.A.; Ahuja, D.R. Relative Contributions of Greenhouse Gas Emissions to Global Warming. *Nature* **1990**, *344*, 529–531. [CrossRef]
266. Atiwesh, G.; Mikhael, A.; Parrish, C.C.; Banoub, J.; Le, T.A.T. Environmental impact of bioplastic use: A review. *Heliyon* **2021**, *7*, e07918. [CrossRef] [PubMed]
267. Liu, L.; Xu, M.; Ye, Y.; Zhang, B. On the degradation of (micro) plastics: Degradation methods, influencing factors, environmental impacts. *Sci. Total Environ.* **2022**, *806*, 151312. [CrossRef] [PubMed]
268. Emadian, S.M.; Onay, T.T.; Demirel, B. Biodegradation of Bioplastics in Natural Environments. *Waste Manag.* **2017**, *59*, 526–536. [CrossRef]
269. Astner, A.F.; Hayes, D.G.; O'Neill, H.; Evans, B.R.; Pingali, S.V.; Urban, V.S.; Young, T.M. Mechanical Formation of Micro- and Nano-Plastic Materials for Environmental Studies in Agricultural Ecosystems. *Sci. Total Environ.* **2019**, *685*, 1097–1106. [CrossRef]
270. Haider, T.P.; Völker, C.; Kramm, J.; Landfester, K.; Wurm, F.R. Plastics of the Future? *The Impact of Biodegradable Polymers on the Environment and on Society.* Angew. Chem. Int. Ed Engl. **2019**, *58*, 50–62. [CrossRef]
271. Green, D.S.; Boots, B.; O'Connor, N.E.; Thompson, R. Microplastics affect the ecological functioning of an important biogenic habitat. *Environ. Sci. Technol.* **2017**, *51*, 68–77. [CrossRef]
272. Ribba, L.; Lopretti, M.; de Oca-Vásquez, G.M.; Batista, D.; Goyanes, S.; Vega-Baudrit, J.R. Biodegradable plastics in aquatic ecosystems: Latest findings, research gaps, and recommendations. *Environ. Res. Lett.* **2022**, *17*, 033003. [CrossRef]

273. Chu, J.; Zhou, J.; Wang, Y.; Jones, D.L.; Yang, Y.; Brown, R.W.; Zang, H.; Zeng, Z. Field Application of Biodegradable Microplastics Has No Significant Effect on Plant and Soil Health in the Short Term. *SSRN Electron. J.* **2022**. [CrossRef]
274. de Souza Machado, A.A.; Lau, C.W.; Kloas, W.; Bergmann, J.; Bachelier, J.B.; Faltin, E.; Becker, R.; Görlich, A.S.; Rillig, M.C. Microplastics Can Change Soil Properties and Affect Plant Performance. *Environ. Sci. Technol.* **2019**, *53*, 6044–6052. [CrossRef] [PubMed]
275. Shruti, V.C.; Kutralam-Muniasamy, G. Bioplastics: Missing Link in the Era of Microplastics. *Sci. Total Environ.* **2019**, *697*, 134139. [CrossRef] [PubMed]
276. Straub, S.; Hirsch, P.E.; Burkhardt-Holm, P. Biodegradable and Petroleum-Based Microplastics Do Not Differ in Their Ingestion and Excretion but in Their Biological Effects in a Freshwater Invertebrate Gammarus fossarum. *Int. J. Environ. Res. Public Health* **2017**, *14*, 774. [CrossRef]
277. Sintim, H.Y.; Bary, A.I.; Hayes, D.G.; English, M.E.; Schaeffer, S.M.; Miles, C.A.; Zelenyuk, A.; Suski, K.; Flury, M. Release of Micro- and Nanoparticles from Biodegradable Plastic during in Situ Composting. *Sci. Total Environ.* **2019**, *675*, 686–693. [CrossRef]
278. Fojt, J.; David, J.; Přikryl, R.; Řezáčová, V.; Kučerík, J. A Critical Review of the Overlooked Challenge of Determining Micro-Bioplastics in Soil. *Sci. Total Environ.* **2020**, *745*, 140975. [CrossRef]
279. Crawford, C.B.; Quinn, B. *Microplastic Pollutants*; Elsevier: Amsterdam, The Netherlands, 2016; ISBN 0128094060.
280. Fojt, J.; Románeková, I.; Procházková, P.; David, J.; Brtnický, M.; Kučerík, J. A Simple Method for Quantification of Polyhydroxybutyrate and Polylactic Acid Micro-Bioplastics in Soils by Evolved Gas Analysis. *Molecules* **2022**, *27*, 1898. [CrossRef]
281. Davis, G.; Song, J.H. Biodegradable Packaging Based on Raw Materials from Crops and Their Impact on Waste Management. *Ind. Crops Prod.* **2006**, *23*, 147–161. [CrossRef]
282. Patel, M.; Bastioli, C.; Marini, L.; Würdinger, E. Life-cycle Assessment of Bio-based Polymers and Natural Fiber Composites. *Biopolym. Online* **2002**, *10*. [CrossRef]
283. Stagner, J. Methane Generation from Anaerobic Digestion of Biodegradable Plastics—A Review. *Int. J. Environ. Stud.* **2016**, *73*, 462–468. [CrossRef]
284. Abraham, A.; Park, H.; Choi, O.; Sang, B.-I. Anaerobic Co-Digestion of Bioplastics as a Sustainable Mode of Waste Management with Improved Energy Production—A Review. *Bioresour. Technol.* **2021**, *322*, 124537. [CrossRef] [PubMed]
285. Rodrigo-Ilarri, J.; Rodrigo-Clavero, M.-E. Mathematical Modeling of the Biogas Production in MSW Landfills. *Impact of the Implementation of Organic Matter and Food Waste Selective Collection Systems. Atmosphere* **2020**, *11*, 1306. [CrossRef]
286. Donovan, S.M.; Bateson, T.; Gronow, J.R.; Voulvoulis, N. Modelling the Behaviour of Mechanical Biological Treatment Outputs in Landfills Using the GasSim Model. *Sci. Total Environ.* **2010**, *408*, 1979–1984. [CrossRef] [PubMed]
287. Van Roijen, E.C.; Miller, S.A. A Review of Bioplastics at End-of-Life: Linking Experimental Biodegradation Studies and Life Cycle Impact Assessments. *Resour. Conserv. Recycl.* **2022**, *181*, 106236. [CrossRef]
288. Hobbs, S.R.; Harris, T.M.; Barr, W.J.; Landis, A.E. Life Cycle Assessment of Bioplastics and Food Waste Disposal Methods. *Sustainability* **2021**, *13*, 6894. [CrossRef]
289. Heide, M.; Olsen, S.O. Influence of Packaging Attributes on Consumer Evaluation of Fresh Cod. *Food Qual. Prefer.* **2017**, *60*, 9–18. [CrossRef]
290. Schuch, A.F.; da Silva, A.C.; Kalschne, D.L.; da Silva-Buzanello, R.A.; Corso, M.P.; Canan, C. Chicken Nuggets Packaging Attributes Impact on Consumer Purchase Intention. *Food Sci. Technol.* **2019**, *39*, 152–158. [CrossRef]
291. Herbes, C.; Beuthner, C.; Ramme, I. Consumer Attitudes towards Bio-based Packaging—A Cross-Cultural Comparative Study. *J. Clean. Prod.* **2018**, *194*, 203–218. [CrossRef]
292. De Marchi, E.; Pigliafreddo, S.; Banterle, A.; Parolini, M.; Cavaliere, A. Plastic Packaging Goes Sustainable: An Analysis of Consumer Preferences for Plastic Water Bottles. *Environ. Sci. Policy* **2020**, *114*, 305–311. [CrossRef]
293. Zwicker, M.V.; Nohlen, H.U.; Dalege, J.; Gruter, G.-J.M.; van Harreveld, F. Applying an Attitude Network Approach to Consumer Behaviour towards Plastic. *J. Environ. Psychol.* **2020**, *69*, 101433. [CrossRef]
294. Grebitus, C.; Roscoe, R.D.; Van Loo, E.J.; Kula, I. Sustainable Bottled Water: How Nudging and Internet Search Affect Consumers' Choices. *J. Clean. Prod.* **2020**, *267*, 121930. [CrossRef]
295. Orset, C.; Barret, N.; Lemaire, A. How Consumers of Plastic Water Bottles Are Responding to Environmental Policies? *Waste Manag.* **2017**, *61*, 13–27. [CrossRef] [PubMed]
296. Boesen, S.; Bey, N.; Niero, M. Environmental Sustainability of Liquid Food Packaging: Is There a Gap between Danish Consumers' Perception and Learnings from Life Cycle Assessment? *J. Clean. Prod.* **2019**, *210*, 1193–1206. [CrossRef]
297. Taufik, D.; Reinders, M.J.; Molenveld, K.; Onwezen, M.C. The Paradox between the Environmental Appeal of Bio-Based Plastic Packaging for Consumers and Their Disposal Behaviour. *Sci. Total Environ.* **2020**, *705*, 135820. [CrossRef]
298. Zwicker, M.V.; Brick, C.; Gruter, G.-J.M.; van Harreveld, F. (Not) Doing the Right Things for the Wrong Reasons: An Investigation of Consumer Attitudes, Perceptions, and Willingness to Pay for Bio-Based Plastics. *Sustainability* **2021**, *13*, 6819. [CrossRef]
299. Lynch, D.H.J.; Klaassen, P.; Broerse, J.E.W. Unraveling Dutch Citizens' Perceptions on the Bio-Based Economy: The Case of Bioplastics, Bio-Jetfuels and Small-Scale Bio-Refineries. *Ind. Crops Prod.* **2017**, *106*, 130–137. [CrossRef]
300. Reinders, M.J.; Onwezen, M.C.; Meeusen, M.J.G. Can Bio-Based Attributes Upgrade a Brand? *How Partial and Full Use of Bio-Based Materials Affects the Purchase Intention of Brands. J. Clean. Prod.* **2017**, *162*, 1169–1179. [CrossRef]
301. Wensing, J.; Caputo, V.; Carraresi, L.; Bröring, S. The Effects of Green Nudges on Consumer Valuation of Bio-Based Plastic Packaging. *Ecol. Econ.* **2020**, *178*, 106783. [CrossRef]

302. Koutsimanis, G.; Getter, K.; Behe, B.; Harte, J.; Almenar, E. Influences of Packaging Attributes on Consumer Purchase Decisions for Fresh Produce. *Appetite* **2012**, *59*, 270–280. [CrossRef]
303. Testa, F.; Di Iorio, V.; Cerri, J.; Pretner, G. Five Shades of Plastic in Food: Which Potentially Circular Packaging Solutions Are Italian Consumers More Sensitive To. *Resour. Conserv. Recycl.* **2021**, *173*, 105726. [CrossRef]
304. Steenis, N.D.; van Herpen, E.; van der Lans, I.A.; Ligthart, T.N.; van Trijp, H.C.M. Consumer Response to Packaging Design: The Role of Packaging Materials and Graphics in Sustainability Perceptions and Product Evaluations. *J. Clean. Prod.* **2017**, *162*, 286–298. [CrossRef]
305. Barnes, M.; Chan-Halbrendt, C.; Zhang, Q.; Abejon, N. Consumer Preference and Willingness to Pay for Non-Plastic Food Containers in Honolulu, USA. *J. Environ. Prot.* **2011**, *2*, 1264–1273. [CrossRef]
306. Dilkes-Hoffman, L.; Ashworth, P.; Laycock, B.; Pratt, S.; Lant, P. Public Attitudes towards Bioplastics—Knowledge, Perception and End-of-Life Management. *Resour. Conserv. Recycl.* **2019**, *151*, 104479. [CrossRef]
307. Herbes, C.; Friege, C.; Baldo, D.; Mueller, K.-M. Willingness to Pay Lip Service? *Applying a Neuroscience-Based Method to WTP for Green Electricity. Energy Policy* **2015**, *87*, 562–572. [CrossRef]
308. Harvey, D.; Hubbard, C. Reconsidering the Political Economy of Farm Animal Welfare: An Anatomy of Market Failure. *Food Policy* **2013**, *38*, 105–114. [CrossRef]
309. Weinrich, R.; Kühl, S.; Spiller, A.; Zühlsdorf, A. Consumer attitudes in Germany towards different dairy housing systems and their implications for the marketing of pasture-raised milk. *Int. Food Agribus. Manag. Rev.* **2014**, *4*, 205–222.
310. Tsang, Y.F.; Kumar, V.; Samadar, P.; Yang, Y.; Lee, J.; Ok, Y.S.; Song, H.; Kim, K.-H.; Kwon, E.E.; Jeon, Y.J. Production of Bioplastic through Food Waste Valorization. *Environ. Int.* **2019**, *127*, 625–644. [CrossRef]

Review

Physical, Chemical and Biochemical Modification Approaches of Potato (Peel) Constituents for Bio-Based Food Packaging Concepts: A Review

Katharina Miller [1,2], Corina L. Reichert [2], Markus Schmid [2] and Myriam Loeffler [1,*]

1 Research Group: Meat Technology & Science of Protein-Rich Foods (MTSP), Department of Microbial and Molecular Systems, Leuven Food Science and Nutrition Research Centre, KU Leuven Ghent Technology Campus, B-9000 Ghent, Belgium
2 Sustainable Packaging Institute SPI, Faculty of Life Sciences, Albstadt-Sigmaringen University, 72488 Sigmaringen, Germany
* Correspondence: myriam.loeffler@kuleuven.be; Tel.: +32-9-3102553

Abstract: Potatoes are grown in large quantities and are mainly used as food or animal feed. Potato processing generates a large amount of side streams, which are currently low value by-products of the potato processing industry. The utilization of the potato peel side stream and other potato residues is also becoming increasingly important from a sustainability point of view. Individual constituents of potato peel or complete potato tubers can for instance be used for application in other products such as bio-based food packaging. Prior using constituents for specific applications, their properties and characteristics need to be known and understood. This article extensively reviews the scientific literature about physical, chemical, and biochemical modification of potato constituents. Besides short explanations about the modification techniques, extensive summaries of the results from scientific articles are outlined focusing on the main constituents of potatoes, namely potato starch and potato protein. The effects of the different modification techniques are qualitatively interpreted in tables to obtain a condensed overview about the influence of different modification techniques on the potato constituents. Overall, this article provides an up-to-date and comprehensive overview of the possibilities and implications of modifying potato components for potential further valorization in, e.g., bio-based food packaging.

Keywords: potato starch; potato protein; potato peel-based films; biopolymer modification

Citation: Miller, K.; Reichert, C.L.; Schmid, M.; Loeffler, M. Physical, Chemical and Biochemical Modification Approaches of Potato (Peel) Constituents for Bio-Based Food Packaging Concepts: A Review. *Foods* **2022**, *11*, 2927. https://doi.org/10.3390/foods11182927

Academic Editors: Theodoros Varzakas and Rui M.S. Cruz

Received: 30 May 2022
Accepted: 5 September 2022
Published: 19 September 2022

Publisher's Note: MDPI stays neutral with regard to jurisdictional claims in published maps and institutional affiliations.

1. Introduction

Potatoes are one of the four major food crops worldwide (first rice, second wheat, third corn, fourth potato) [1]. In the European Union (EU-28), the total potato harvest was 56.4 million tons in 2019. Especially in Europe, potato cultivation and the potato industry are very important in the domestic food culture and diet, and the increasing consumption of processed potatoes in the form of potato chips, French fries, mashed potatoes, etc., represents a worldwide trend [2]. In comparison, the annual potato harvest in China in 2019 was 91.8 million tons. China is thus the largest potato-producing country in the world [3].

Within the potato industry, approximately half of the harvested potatoes become side streams or residual material. Thus, the entire supply chain (including farmers, wholesalers, the processing industry, retailers and consumers) produces tubers, peels and pulp that are discarded [4], as illustrated in Figure 1 for non-organic processing potatoes.

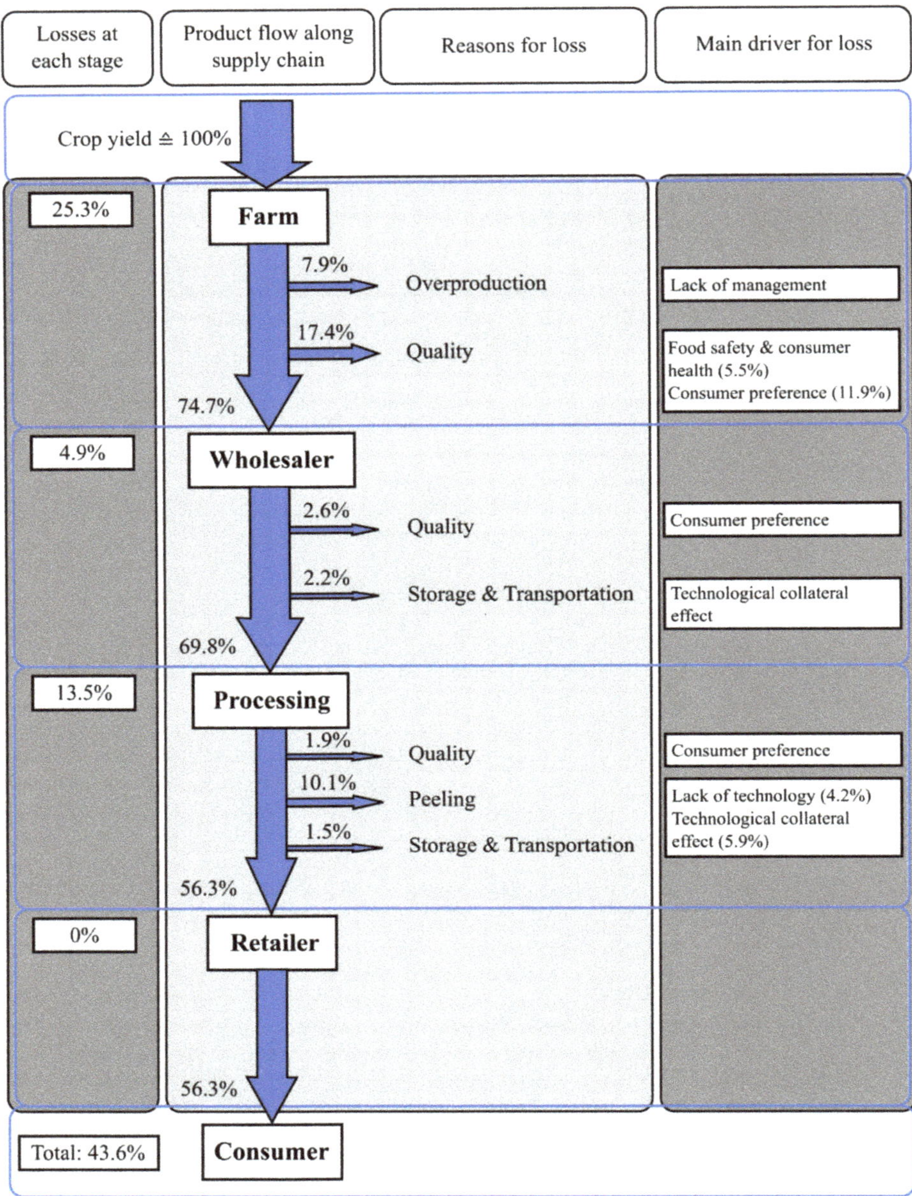

Figure 1. Mean food loss rates of non-organic processing potatoes in Switzerland at each stage of the total agricultural potato production, total supply chain losses and cause of losses. All percentages are related to an initial crop yield of 100%. ↓ Thickness of arrows represent percentage of total product flow. Cause of loss "quality" refers to potatoes, which did not suffice according to the standard quality specifications of the "Swiss trade customs for potatoes". Cause of driver "Food safety & consumer health" refers to green and rotten potatoes, and "Consumer preference" refers to "Consumer preferences for certain aesthetic standards or typologies of food" (own illustration based on data published by Willersinn et al. [4]).

To summarize the findings, the product losses along the supply chain of non-organic processing potatoes in Switzerland accounts for about 44%, which means that from an initial input of approximately 1.8 kg potatoes one receives an output of 1 kg processed potato product such as French fries [5]. Along the supply chain of processing potatoes, main losses can be attributed to quality standards (21.9%), followed by the potato peel side stream (10.1%) [4]. Similar data on the amount of side streams and losses along the potato supply chain can be found in Germany [6] and Europe in general [7]. Quality standards lead to a high amount of potato-based side streams or residuals, which in part are determined by the potato processing industry. For instance, PepsiCo has specific processing quality parameters in their potato crisp processing manufacturing in India to control, amongst others, tuber size, dry matter content, sugar content, starch content, damage and discoloration [8]. Besides the amount of potatoes that are lost based on quality aspects, there is also a high amount of potato peel. In Figure 1, the loss attributed to peeling is 10.1% based on the initial crop yield. If attributed to the input quantity at the processing stage (69.8% of initial crop yield) losses attributed to peeling would amount to 15.5%. In the literature, potato peel side stream losses of up to 40% have been reported which varied depending on the peeling method used [9]. The large amount of potato side streams led to the establishment of different strategies to valorize them [10].

Microbial spoilage of potato side stream products and widespread production locations result in low utilization of these byproducts. In Switzerland, most of the potato side streams (90%) are used as animal feed due to the high protein content [4] corroborating with data from other countries [1]. In accordance with the principles of a circular bio-economy, the use of potato side streams is of great interest for the production of high quality products with zero-waste generation [11,12]. As these large amounts of side streams currently possess low or even negative economic value potential, valorization approaches of potato side streams are becoming important [13].

One valorization approach for potato side streams is the production of bioplastics [14]. Due to environmental challenges caused by large amounts of petrochemical-based and not biodegradable packaging, the utilization of bio-based alternatives is demanded. For instance, potato side streams such as potato peel or potato pulp can be used for potato starch or potato protein extraction which can be further processed to bio-based plastics [15–21]. However, extraction processes are energy and time consuming and result in additional side stream productions. A material efficient approach represents the use of the whole side stream without extracting single constituents. Several studies demonstrated that mechanical and barrier properties of potato flour- or potato peel-based films are not inferior to the properties of films composed of potato starch or potato protein suggesting that extensive industrial purification may not be necessary for film production [15,17,18,22–24].

To successfully implement a circular bio-economy concept, the quality and characteristics of alternative materials (in this case biopolymers made from potato side streams) must be at least as good as their commercial equivalent (in this case petrochemical plastics) [11]. At the moment, potato side stream-based biopolymers are characterized by an overall low technology readiness level and in part insufficient processing- and packaging related properties including processability and mechanical properties [23,25]. One approach to increase the technology readiness level and optimize processing- and packaging related properties of potato side stream-based films and coatings can be physical, chemical or biochemical modification which are outlined and discussed in this review article.

Based on previously conducted studies, some modification methods have been used to influence physicochemical properties of potato peel-based films including high-pressure homogenization, ultrasonication, gamma irradiation and acid hydrolysis. This review aims to summarize and discuss the influence of different physical, chemical and biochemical modification methods on the structure and physicochemical properties of potato constituents (mainly potato starch and potato protein) and their films. In this review, different modification methods of potato constituents are evaluated upon their suitability to change a certain property especially in terms of the modification effect on the film-forming and packaging related properties such as rheological, thermal, and barrier characteristics. Besides a comprehensive review, we highlight recent research gaps and future research potentials of potato constituents for packaging applications.

2. Potato Tubers

Potato tubers are the stems of the potato plant that develop under the soil from the propagation of previously planted mother tubers by sprouting. Sprouting occurs under suitable conditions from the dormant/lateral buds of the potato tubers (also known as "potato eyes"). Although shape, color and size vary between different cultivars, the morphology of potato tubers is generally characterized by an oval to round shape, white flesh and light brown skin. The potato flesh consists of the pith, perimedullary zone, vascular ring and cortex (Figure 2), which are usually consumed by humans. The periderm (skin) contains the apical bud, stem end, lenticels and lateral buds, as shown in Figure 2. In addition to the type of potato cultivars, composition of potato tubers also depends on several other factors including soil type and temperature, location, cultural practices, maturity and postharvest storage conditions. On average, potato tubers mostly consist of water (approximately 80%) and 20% dry matter. Related to the dry matter content, carbohydrates, especially starch (see Section 2.1), are the main fraction, followed by proteins (see Section 2.2), which can be both extracted from potato tubers or their side streams [26–28].

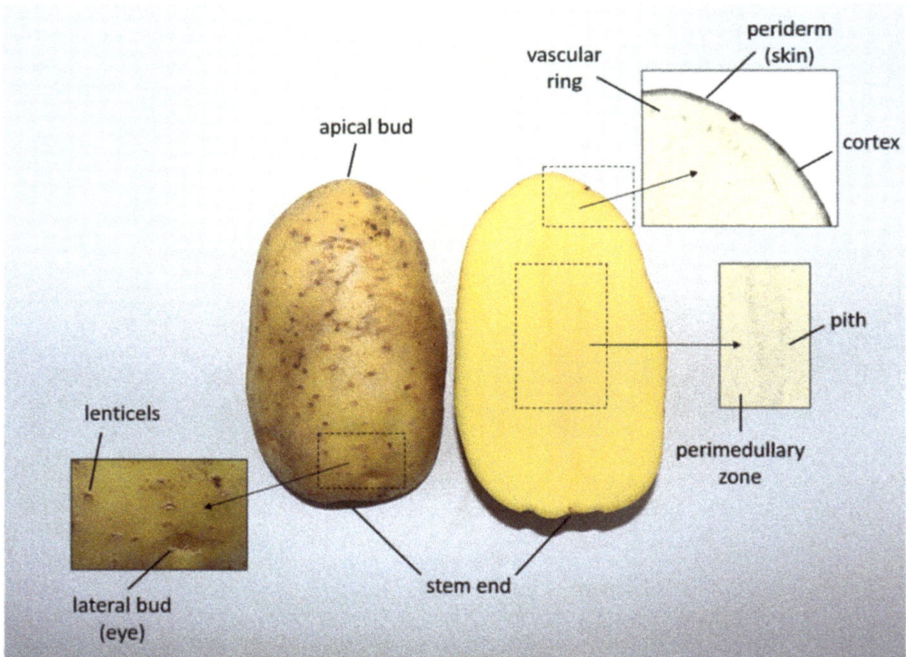

Figure 2. Structure of a potato tuber of the bernina variety (own illustration, terminology based on [28]).

Based on its high dry-matter starch content (Section 2.1), potato tubers are favorably used by the industry for starch extraction. In laboratory scale experiments, starch extraction is often conducted via sedimentation or centrifugation method [15,29,30]. Therefore, potato tubers are usually peeled and washed to remove defective parts and contaminations, and then grounded to destroy cell structures and release starch granules. The obtained slurry is filtered and washed to further remove contaminants. Based on the higher density and low cold-water solubility most of the starch can be removed as a sediment using decantation of the supernatant prior drying to obtain starch powder. Alternatively, the separation between starch and supernatant can be performed by centrifugation. Potato starch can also be extracted from potato side streams including potato peels [15,29,30].

Accordingly, potato protein (Section 2.2) can be extracted either from potato tubers or potato side streams. For instance, after potato starch and tuber cell (pulp) removal, the supernatant (potato juice) of the centrifuged sample, can be further used for protein recovery [31]. To increase the concentration of potato protein, purification by ultrafiltration, reverse osmosis or fractional precipitation can be conducted. Precipitation methods generally include salting out, heat denaturation and isoelectric precipitation [32–34]. In laboratory scale experiments, potato protein is recovered at pH 3.0 or 5.5 under thermal treatment [35]. However, dependent on the extraction and purification methods, the recovered yield of potato protein and its composition (e.g., amount of patatin, protease inhibitors and high molecular weight proteins) vary [36].

2.1. Potato Starch

With 70 to 80% w/w of dry matter content, starch is the major component of potato tubers. The structure of starch on molecular to macromolecular level is schematically illustrated in its main levels in Figure 3. The structure of potato starch typically consists of 20 to 30% (w/w) amylose, which is a linear to slightly branched (degree of polymerization > 60) polymer chain consisting of α-d-(1,4)-linked glucose units (Figure 3a), and 70 to 80% (w/w) of amylopectin, which is a highly branched polymer based on relatively short chains of α-d-(1,4)-linked glucose units with 4 to 6% (w/w) additional α-d-(1,6)-linkages (Figure 3b). Low molecular weight amylose (0.13–0.5 MDa) is assumed to form a helical structure (Figure 3c). Amylopectin is characterized by its high molecular weight (10–1000 MDa) and its branched structure, with branched chains being classified by their substitution into A chains (not substituted with other chains), and B chains (substituted with other chains) (Figure 3d). These branched chains derive from the main chain, which is called C chain (Figure 3d). The reducing end of the amylopectin polymer can be found at the end of the C chain (Figure 3d). Due to the tight packing of A and B chains, they can form double helices with six glucose units building one turn (Figure 3e). The arrangement of interlinked clusters results in lamellar structures, where alternating amorphous (branching points/interconnecting chains) and (semi-) crystalline (A and B chain clusters) parts are present (Figure 3f). On a macromolecular level, several of the alternating lamellar structures are organized in so called growth rings forming the starch granules (Figure 3g). The amorphous center of the granule (hilum) is formed mainly by amylose chains (Figure 3h) [37–39].

Potato starch granules have a spherical to oval shape with a diameter of 10 to 100 μm and a smooth surface. Furthermore, potato starch is usually characterized by its B type arrangement of double helices, which can be observed in X-ray diffraction analysis based on peaks at 5.6°, 17°, 22° and 24° (2θ) and a relative crystallinity of approx. 30%. In contrast to A type starch (peaks at 15°, 17°, 18° and 23° (2θ)), B type polymorphs basically have a more open structure and contain more water molecules [40]. In addition to A and B type polymorphs, there are C type polymorphs, which represent a mixture of the A and B type. V-type structures are composed of complexes between amylose single, left-handed helices and complexing compounds, which fill the helical cavities of the amylose in the presence of water [41,42]. One example are amylose–lipid complexes, which can be found in native starch. In those, lipid chains fill in the central cavity of amylose helices by interacting with the hydrophobic moiety of the amylose chain. These complexes can affect different properties of the starch and their films [43]. V-type crystallinity can also occur during starch retrogradation [44]. Another characteristic of starches is their phosphorous content, which was found to correlate in reverse with the amylose content and influences pasting and retrogradation properties. In comparison to other starches, potato starch contains high amounts of phosphorous referring to a high amylopectin content and long amylopectin chains. Mostly, the phosphorous is covalently bound to the amylopectin fraction (ester of phosphoric acid and hydroxyl group) of potato starch [45,46]. Regarding functional properties, the onset gelatinization temperature of potato starch is around 58–63 °C, dependent on its cultivar and crystalline order [47,48]. Amongst various starch sources, potato starches are known for its high swelling power and pasting viscosity, enabling the formation of thick viscoelastic gels [49].

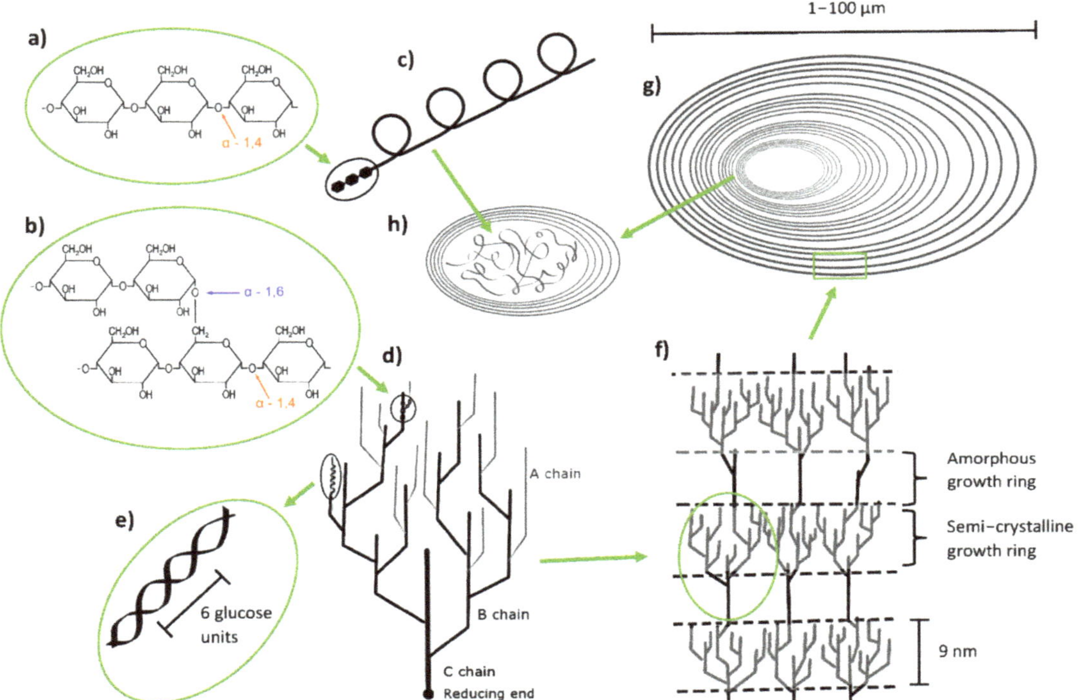

Figure 3. Schematic illustration of some key elements of the macro- and microstructure of potato starch (**a**) molecular structure of amylose, (**b**) molecular structure of amylopectin, (**c**) helical arrangement of amylose, (**d**) arrangement of amylopectin, (**e**) double helix structure of neighboring A and B chains, (**f**) cluster model of amylopectin with alternating amorphous and semi-crystalline parts, (**g**) potato starch granule, (**h**) starch hilum containing mainly amylose. (Own illustration, based on references [37–39]).

2.2. Potato Protein

Although potato proteins account for only 1 to 3% of the fresh weight of potato tubers, they are the second largest fraction of the dry matter of potato tubers. Potato proteins consist of water insoluble and water-soluble proteins, which account for 25 and 75% of potato tuber proteins, respectively. Based on their function, soluble potato proteins are classified into three groups, namely patatins (30–40%), protease inhibitors (40–50%) and other high molecular weight proteins (10–15%) [39,50].

Patatins (Figure 4a) are storage proteins, consisting of homologous glycoprotein isoforms. They are characterized by molecular weights of 40–45 kDa and vary in their isoelectric points between 4.5–5.2. The secondary structure of patatin consists of approximately 33% α-helices, 45% β-sheets and 15% random turns [51]. Stability of patatin decreases with temperatures above 45 °C and the denaturation temperature was measured to be between 59–60 °C. The best gelatinization of patatin was reported to be at pH values between 4.8–5.5 [52]. At a pH of 4, the solubility of patatin is minimal, and at a pH below 4, irreversible unfolding of the tertiary and secondary structure of patatin can occur. Patatins are known for their emulsifying, foaming, gel forming or antioxidant properties, making them attractive for food and non-food applications [53–55].

Protease inhibitors (Figure 4b) are a heterogeneous group with varying reactive groups. Protease inhibitors are classified into four major sub-groups, namely serine-, cysteine-, metalloid- and aspartic protease inhibitors. Protease inhibitors have molecular weights of approximately 5–25 kDa and differ in their isoelectric points between 5.1–9.0 [56]. The gelatinization optimum for protease inhibitors was reported to be at pH 3.2–4.3 [52]. With approximately 24%, serine protease inhibitors are the biggest fraction within the group of protease inhibitors in potato proteins. The denaturation temperature of serine protease inhibitors is about 69 °C, and their secondary structure is composed of about 2% α-helices, 38% β-sheets, 23% β-turns and 37% random coils [57]. Compared to patatin, protease inhibitors are more hydrophilic and were reported to have superior antifungal and antimicrobial properties [36,39,50,56].

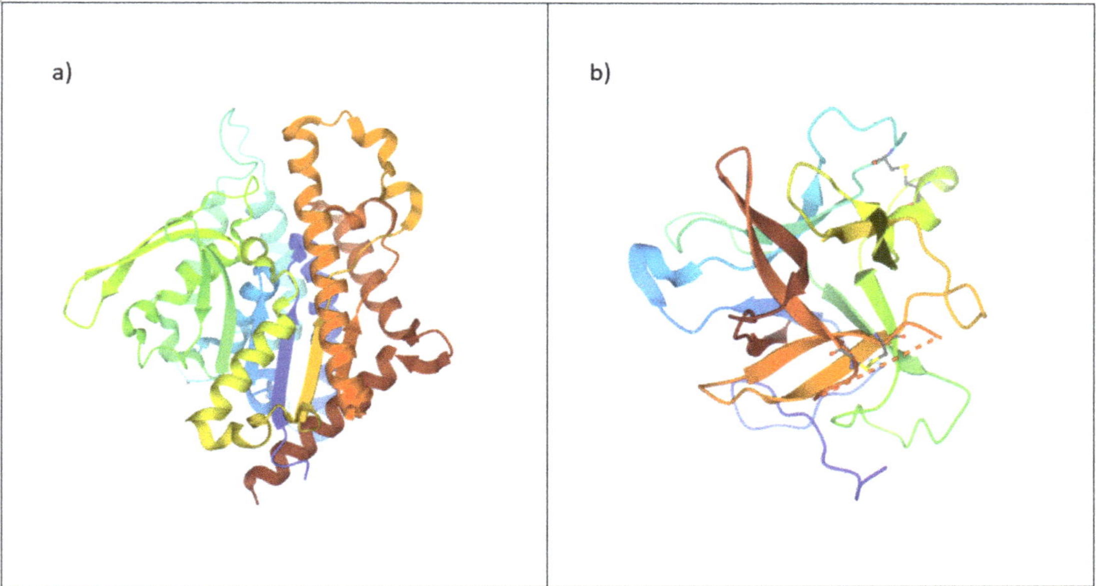

Figure 4. Images from the RCSB PDB (rcsb.org) of the tertiary structures of (**a**) native patatin, PDB ID 4PK9 [58] and (**b**) serine protease inhibitor, PBD ID 3TC2 [59], reprinted with permission from Ref. [59]. 2012, International Union of Crystallography.

Other high molecular weight proteins (>40 kDa) are the third group of proteins present in potato tubers. Based on their relative low percentage as mentioned above, the group of other high molecular weight potato proteins is not as well described as patatin and protease inhibitors. However, it has been reported that this group contains among others polyphenol oxidases, lipoxygenases and starch synthesis associated enzymes [39,50,55].

In addition to the mentioned potato protein fractions, potatoes also contain free amino acids, which account for up to 40% of the total nitrogen content. Amino acid composition of potato protein does not only play a crucial role in dietary/nutrition [39] but also determines modification potentials, as the composition impacts available functional groups of the amino acid side chains [60]. Table 1 shows the amino acid composition of potato proteins with their associated functional groups, ordered from the smallest to the biggest fraction.

Table 1. Amino acid composition of potato protein with the reactive groups of their side chains, ordered from the smallest to biggest fraction (adopted and extended from [39,61]).

Amino Acid	Abbreviation	Range (%)	Reactive Groups
Cysteine	Cys	0.2–1.3	Sulfhydryl
Tryptophan	Try	0.3–1.9	Indole
Methionine	Met	1.2–2.25	Thioester
Histidine	His	2.0–2.5	Imidazole
Isoleucine	Iso	3.7–5.8	
Glycine	Gly	4.3–6.1	
Tyrosine	Tyr	4.5–5.7	Hydroxyl
Threonine	Thr	4.6–6.5	Hydroxyl
Alanine	Ala	4.6–5.6	
Proline	Pro	4.7–5.6	
Arginine	Arg	4.7–5.7	Guanidino
Phenylalanine	Phe	4.8–6.5	
Valine	Val	4.9–7.4	
Serine	Ser	4.9–5.9	Hydroxyl
Lysine	Lys	6.7–10.1	Amino
Leucine	Leu	9.6–10.7	
Glutamic acid	Glu	9.6–11.8	Carboxyl
Aspartic acid	Asp	11.7–13.9	Carboxyl

3. Modifications

Depending on the nature of changes induced, modification of molecules or polymers can be divided into physical (Section 3.1), chemical (Section 3.2) or biochemical (Section 3.3) modifications, which are subsequently described for potato constituents. Besides individual modification techniques, combinations of modifications can be applied. These modifications are often referred to as dual modifications, and are also described below (Section 3.4).

3.1. Physical Modification

One approach to alter the physicochemical properties of powders including starches and proteins is via physical treatment (e.g., hydrothermal treatment, irradiation, ultra-sonication and high-pressure treatment, see Sections 3.1.1 and 3.1.2). Besides chemical modification, starches and proteins, which have been modified via physical methods do not have to be claimed as "modified". Physical modification is also viewed as cost-efficient and environmentally friendly, because no hazardous substances (chemicals) are used. Among the various physical modifications, a distinction is made between thermal (Section 3.1.1) and non-thermal treatment methods (Section 3.1.2) [62].

The physical modification of starch has already been reviewed by several authors [62–66]. However, due to fundamental differences in structure, morphology and physicochemical properties of physically modified starches from different botanical sources (e.g., corn, maize, tapioca, rice, pea, potato), it is important to specifically review the literature on physical modifications of potato starch [67–70]. So far, the physical modification of potato starch has been reviewed elsewhere describing the effect from emerging technologies, including high pressure, ultrasound and microwaves, cold plasma and electric pulse [71].

3.1.1. Thermal Physical Modification

In this review article, thermal physical modification of a substance is defined as a physical treatment, which generates or transfers heat or cold onto a substance in the presence or absence of moisture to create structural and/or functional changes. Therefore, thermal methods to modify potato starch include hydrothermal methods such as heat moisture treatment and annealing, dielectric treatments such as microwave heating and radio frequency heating, as well as other methods including freeze-thaw treatment, dry heat treatment, flash drying, autoclaving and modification via superheated steam.

Heat Moisture Treatment

Heat moisture treatment (HMT) refers to the treatment of a product with heat in combination with a certain amount of moisture. Thus, the parameters for temperature, time, moisture and cycles/repetitions are crucial for this treatment method.

In HMT the amount of water in the treated samples is limited to 10–35%, to prevent starch gelatinization [72]. Therefore, granular morphology of starch is preserved during HMT [73]. However, several authors noted that HMT leads to some granular damage [67,73]. The temperature applied during HMT is characterized being above the gelatinization temperature, but below the glass transition temperature. For HMT of potato starch temperatures of 80–121 °C, durations of 30 min to 16 h, and iterations of up to six have been applied in the last years [29,74,75].

Overall, the thermal properties, such as gelatinization temperatures (onset (To), peak (Tp) and conclusion (Tc)) of potato starch significantly increased and melting enthalpy (ΔH) decreased by HMT. This increase in thermal properties was accelerated and/or intensified with increasing temperature [67], increasing holding time [74] and decreasing moisture content [75]. Shi et al. [67] linked the changes in gelatinization temperatures with increasing HMT temperatures to intramolecular interactions towards more amylose–amylose and amylose–lipid interactions at higher temperatures, which hindered the mobility of the amorphous region thus increasing the gelatinization temperature. Decreased mobility leads to decreased water uptake and swelling, which is necessary to facilitate melting of crystals and double helices. Therefore, stability of the crystalline regions of potato starch increased and higher temperatures for gelatinization were required after HMT. Accordingly, amylose content increased, degree of crystallinity decreased and type of crystallinity varied upon HMT of potato starch, using different temperatures and moisture contents [67,73,75,76]. Based on the reviewed studies it seems that the change in type of crystallinity of potato starch from B to C or A, which occurs at 120 °C, depends on the moisture content of the material: C-type occurred at 30% [73], but A-type at 10–27% [29,67,75] moisture content. At lower temperature (100 °C) a change from B to A-type crystallinity occurred at 24% moisture content [76].

Hydration and pasting properties of potato starch were also affected by HMT. Cold water solubility of potato starch increased whereas swelling power, water absorption capacity and hot water solubility decreased [20,29,73,74,76]. Upon HMT, pasting temperature increased with increasing temperature [29]. Through HMT, the final viscosity decreased with increasing temperature [29] and increasing holding time [74], but final viscosity increased at short holding times [74] and temperatures < 120 °C [29].

Concerning potato starch as packaging material, HMT did not significantly affect packaging related properties of potato starch-based bioplastic [15].

Annealing

Annealing (ANN) is another method to physically modify starch, without destroying its granular morphology. In contrast to HMT, moisture content of the samples is not limited (usually being > 60%, w/w) and temperature used during ANN is below the sample To (onset gelatinization temperature). Common processing conditions during ANN of potato starch are 30 to 55 °C for 12 h to 96 h. Here, iterations of ANN treatment cycles can be applied [20,77,78].

Overall, ANN treatment of potato starch increased gelatinization temperatures (To, Tp, Tc), had no effect on type of crystallinity and granule surface, but increased pasting properties (pasting temperature and final viscosity) [20,77,78] (see Table 2), which were favored by higher temperatures [77] and longer treatment times/iterations [78] used for ANN treatment. Furthermore, increased hot water solubility and hot water swelling power of ANN treated potato starch granules were reported [78]. Through processing of starch in water having temperatures between 50 and 70 °C, solubility and swelling power of ANN potato starch was lower than for native potato starch [78]. Wang et al. [77] and Xu et al. [78] reported an increase in relative crystallinity, increase in ΔH, as well as changes in absorbance of potato starch, which might be attributed to the higher ANN temperature, treatment times or suspension concentration used (55 °C vs. 30–50 °C, 12–96 h vs. 24 h, 1:3 *w/v* vs. 1:5 *w/v*).

Microwave Treatment

Microwave (MW) treatment is a thermal heating method using microwaves (300 MHz to 300 GHz), that can be absorbed by polar materials, such as potato starch. Upon MW heating, molecular vibrations and frictions occur due to their alignment to the vibrating electromagnetic field, resulting in the generation of heat within the product. Therefore, MW heating requires less treatment time and can be more efficient and less energy consuming than conventional heating [79,80]. The ability of a product to be heated via MW treatment is dependent on moisture content, frequency, as well as compaction density and temperature of the sample [81–83]. The influence of MW treatment on starch structure and starch properties has been recently reviewed [84,85]. In addition, Jiang et al. [86] reviewed the effects of MW treatment on different food components, including proteins. In contrast to other thermal modification methods such as HMT or ANN, MW treatment is not limited to certain moisture or temperature and can be therefore performed not only for potato starch powders [68,87,88], but also in potato tubers [89] or for starch–water suspensions [30,90].

Miernik and Jakubowski [89] found that MW treatment of potato tubers show a strong correlation between MW dose and starch granule size, which resembled a parabola with its minimal turning point at 53.19 J/g. This means that granule size increased, when MW doses of 0–47.32 J/g or 66.14–508.47 J/g were applied.

Compared to native potato starch granules, MW treated potato starch granules showed damaged structure or severe cracking, which correlated to an increased water absorption capacity [88]. For potato starch with low moisture content (16.5%) decreasing gelatinization temperatures and decreasing final viscosities have been reported after MW treatment [87]. Whereas increasing To, decreasing relative crystallinity and increasing pasting properties (pasting temperature and final viscosity) have been observed for MW treated potato starch with higher moisture contents (21% and 30%) [68,88]. Furthermore, changes in relative crystallinity, gelatinization temperatures (To, Tp, Tc) and pasting properties seem to be accelerated with increasing MW intensity and treatment time [68,87,88]. Xu et al. [68] reported a change in crystallinity from B to A-type at high power (6.63 W/g) MW treatment.

In potato starch solutions, decreased amylose content, increased relative crystallinity, increased gelatinization properties, increased pasting properties, decreased water solubility and increased swelling power were observed after MW treatment [30]. However, Xia et al. [90] did not observe any changes in gelatinization and pasting temperatures in potato starch solutions, but changes in hydration and pasting properties were also observed in potato starch solutions containing added protein [91].

Based on the reviewed studies, once can conclude that the effect of MW treatment on potato constituent structure and properties can vary, depending on the processing conditions (moisture content, power, treatment time, cool-off period, etc.) applied [30,68,87–91].

Radio Frequency Treatment

Similar to MW heating, radio frequency (RF) are electromagnetic waves in the frequency range of 1 to 300 MHz. In comparison to MW, RF has been reported to penetrate products even more deeply and distribute heat more uniformly.

In addition to MW treatment, which is reviewed above, Xia et al. [90] also treated potato starch–water suspensions with RF, varying the water content. Upon RF treatment small fractures on starch granule surfaces were observed, which were less severe than for MW treated starch granules. Similar to MW treated samples, RF treated samples showed lower amylose contents, but relative crystallinity decreased after RF treatment and with increasing water content. In contrast to MW treatment, type of crystallinity of potato starch samples changed from B-type to C-type after RF treatment. Compared to native potato starch, gelatinization temperatures, pasting temperature and ΔH slightly decreased and final viscosity slightly increased after RF treatment [90]. An increase in pasting properties was also reported by Zhu et al. [92], who investigated the influence of RF drying on potato flour properties. However, no changes in sample color, relative crystallinity, thermal properties or type of crystallinity were observed in this study.

Others

In addition to hydrothermal (HMT, ANN) and dielectric (MW, RF) treatment methods, we found other thermal modification methods including freezing and thawing (FT), flash drying, autoclaving or steam heating, which can be used to alter potato starch, flour and protein properties. In general, there are methods that can be used in the presence or absence of water to generate/transfer heat.

FT can be applied to induce structural and functional changes in potato constituents. For instance, Zhang et al. [93] investigated the effect of FT treatment on potato starch showing disruption and aggregation of granule fragments after FT treatment. Furthermore, pasting temperature slightly increased and final viscosity of potato starch slightly decreased. A decrease in gelatinization properties (To, Tp and ΔH) after FT treatment was reported by Wang et al. [94], indicating a weakening of the crystalline structure. Another study demonstrated the influence of the thawing method on structural and functional properties of pre-gelatinized potato flour samples, showing an overall increase in gelatinization temperatures and pasting properties upon FT treatment [80].

Similarly, an increase in gelatinization properties (To, Tp, Tc), final viscosity, and a reduction in granule damage and gelatinization were achieved in potato flour upon flash drying (130–135 °C, 1–2 min) as compared to commercial potato flakes and granules [95].

The influence of superheated steam (100–160 °C) on potato starch properties was investigated by Hu et al. [96], showing minor effects on granule surface, compared to native starch granules. The type of crystalline structure was not affected by the superheated steam treatment, but relative crystallinity decreased progressively with increasing temperature. Furthermore, thermal properties (To, Tp, Tc and ΔH), solubility and swelling power decreased, and pasting properties increased upon superheated steam treatment [96].

Drozłowska et al. [97] gelatinized potato protein–water solutions (10% w/w) at 100 °C for 15 or 30 min prior autoclaving the samples at 126 °C for 60 or 120 min to induce thermal hydrolysis. No influence on solubility was observed for potato protein samples after thermal hydrolysis treatment. Viscosity of modified samples significantly increased with decreasing pre-treatment time and increasing autoclaving time compared to native and only pre-treated samples. Finally, emulsion properties of potato protein were slightly increased due to modification.

3.1.2. Non-Thermal Physical Modification

In this subsection, non-thermal physical modification methods are reviewed. Non-thermal physical modification refers to physical treatments that affect structural and/or functional properties of materials without the intentional use of heat or cold. In non-thermal treatment, heat is sometimes generated as a "side stream" (e.g., during high pressure treatment or ultrasonication due to friction, etc.). For instance, high pressure, ultrasound, ionizing irradiation, milling and electric field treatment are non-thermal techniques that are outlined in the following for potato constituents.

High Pressure Treatment

High pressure (HP) treatment (30–1000 MPa) is a non-thermal modification method, which can cause starch gelatinization and protein denaturation/gelation [98,99]. Thereby, HP can be applied in different ways, which are commonly referred to as high hydrostatic pressure/high isostatic pressure, high pressure homogenization and dynamic high pressure microfluidization. During high hydrostatic pressure/high isostatic pressure modification of starch and protein, pressure is transmitted uniformly through a pressure-transmitting medium (e.g., water) to the sample (sample-water suspensions), which is usually packed in vacuum bags and placed inside the high pressure vessel [69,100]. The high hydrostatic pressure treatment of starch in its granular form has been previously reported [101]. During high pressure homogenization a liquid sample is pressed through a small orifice causing shear forces. In dynamic high pressure microfluidization the pressure is transferred in an interaction chamber [102,103]. Similar to other modification methods, the effect of HP treatment can vary for starch from different plant sources as reviewed elsewhere [104,105].

Different effects and partially controversial findings of HP treatment on potato starch properties (thermal properties, pasting properties, crystallinity, granular/surface damaging) and potato protein properties (solubility, foam stability) were reported in the reviewed literature [69,70,100,101,106–108], which might be attributed to the different types of pressure (high hydrostatic pressure, high isostatic pressure, microfluidization) used. For instance, granular structure of potato starch as well as granule surface were only slightly affected by high isostatic/hydrostatic pressure application [69,70], but significantly damaged by HP microfluidization [106]. Furthermore, within similar HP application and starch–water solution ranges, different effects of HP treatment on potato starch crystallinity, gelatinization properties and pasting properties have been reported in different studies [69,70,101]. Wang et al. [69] did not observe any significant changes in gelatinization temperatures and ΔH, whereas Rahman et al. [70] found increasing gelatinization temperatures with increasing pressure applied to potato starch–water suspensions (1:5 w/w) with HP (100–500 MPa vs. 200–600 MPa). In contrast, Słomińska et al. [101] observed decreasing gelatinization temperatures for potato starch granules treated at 50–1000 MPa HP, which was accelerated with increasing pressure applied. Rahman et al. [70] additionally investigated hydration properties. A decrease in solubility and swelling power of potato starch was observed upon HP treatment.

For potato proteins, differences in purity and pH did likely contribute to observed qualitative and quantitative differences in some functional properties throughout the reviewed studies. For instance, it seems that physical changes by HP treatment cannot be induced at pH 7 but likely at pH < 7 [100,108]. This was shown by Baier and Knorr [100] who observed a significant decrease in solubility and an increase in surface hydrophobicity for patatin fractions that were HP treated at pH 6, but no effect were observed at pH 7. In addition, Katzav et al. [108] reported that potato protein gelation was only possible when samples were HP treated at pH 3 but not at pH 7, independent of the amount of pressure (300–500 MPa) applied.

Upon HP treatment at pH 3.6, the secondary structure of potato protein was significantly affected as an increase in α-helix content and decrease in β-sheet structures has been determined by Hu et al. [107]. Furthermore, particle size and emulsion stability of potato protein stabilized emulsions decreased and foaming properties slightly increased. In contrast to the patatin fraction studied by Baier and Knorr [100], Hu et al. [107] reported increased solubility of potato protein upon HP treatment.

Kang et al. [25] investigated the effect of HP treatment on the film-forming properties of potato peel which are used for film production. Compared to native film-forming potato peel solutions, HP treated solutions showed higher elastic modulus and viscosity at higher temperatures (>60 °C). Increasing viscosity values with increasing pressure and number of passes were observed for HP treated film-forming solutions. This indicated the improvement of film formation (smoothness) by HP treatment, which has been previously reported by [109].

Ultrasonication

During ultrasound (US) treatment, acoustic waves with a frequency > 16 kHz are transmitted through solid, liquid or gaseous systems. US can be classified into different types based on frequency and intensity [110,111]. As for the modification of potato constituents, US treatment is usually carried out in a liquid medium (usually in a water bath), with starch- or protein-based samples treated in their granular form or in the form of suspensions before or after the gelatinization/gelation process. US can be conducted at room or elevated temperatures. During US treatment cavitation occurs, which is the formation and implosion of gas bubbles that can break polymer chains by mechanical shear forces, which occur when these bubbles collapse. As a side effect, local temperature increases contributing to the modification effect [112]. For different carbohydrate and protein-based films or coatings US treatment has shown to be able to improve gelling properties and tensile properties, increase solubility and surface hydrophobicity, and reduce water vapor permeability [113,114].

For potato starch granules, damaging occurred after US treatment (especially in water [115,116]), which increased with increasing temperature [117], US power [118], US frequency [119], as well as by the application of dual-frequency US-treatment [117]. Furthermore, a decrease in average molecular weight [120] and slight decrease in relative crystallinity [117,118] as well as an increase in amylose content [117] was observed after US treatment. Among these studies, a change in type of crystallinity from B to V type was only reported by Nie et al. [120]. However, other functional properties including thermal, and pasting properties were not significantly influenced by US treatment according to Hu et al. [117], though hydration properties were slightly increased, when compared to native potato starch [115,117].

Mao et al. [121] demonstrated that solubility of potato protein can be increased by US treatment, but only in combination with a pH shift to pH 12. In addition, α-helix content decreased by approx. 10%, while β-sheet content increased by approx. 10%, suggesting a partial unfolding of potato proteins upon US treatment at pH 12. However, Hussain et al. [122] reported an increase in potato protein and patatin solubility by +25% through US treatment at pH 7. Other functional properties including radical scavenging activity, foaming ability, emulsifying ability and emulsifying stability also increased with increasing treatment time for both, potato protein and patatin fraction by two times or more, whereas particle size and foaming stability decreased [122].

In case of potato peel-based films, US treatment can enhance packaging related properties, as tensile properties were increased and hydration properties decreased by US treatment of potato peel-based films [123]. Similarly, an increase in tensile strength and decrease in water vapor permeability was reported by Wang et al. [124] for potato starch-based films after US treatment.

Ionizing Irradiation

Upon interaction with ionizing irradiation, cleavage of water molecules (in the product) into free radicals and high energy electrons occur, which can facilitate cross-linking and hydrolysis reaction of the polymer chains. The affinity of a polymer towards crosslinking and chain scission is described by the $G(x)$ and $G(s)$ value (quantification of the chemical yield obtained from ionizing irradiation treatment), respectively, which can highly differ among polymers. Thereby, the ratio between $G(s):G(x)$ indicates the overall prevailing reaction (cross-linking < 1, hydrolysis > 1) which, however, depends on the irradiation dose and temperature. Accordingly, the molecular weight is either increased (cross-linking is prevailing), decreased (chain scission is prevailing) or not significantly changed (cross-linking and chain scission occur equally). For starches (including potato starch) mostly decreasing average molecular weights upon irradiation have been reported in several studies which were reviewed by Bashir and Aggarwal [125], indicating the prevalence of hydrolysis [125]. The induced structural changes by ionizing irradiation can thus alter product properties [126–128].

Investigating the effect of electron beam or gamma irradiation on potato starch morphology, none or only slight changes in starch granule surface and color have been observed throughout the reviewed literature [129–131]. However, the occurrence of free radicals on starch surface was observed by Rao et al. [129]. A decrease in amylose content and increase in water solubility index with increasing irradiation dose was determined by Atrous et al. [130], who presumed that these findings were attributed to the occurrence of depolymerization hindering the formation of iodine complexes and increasing water affinity. Interestingly, swelling power of potato starch was up to 1.5 times higher when irradiated at $\leq$20 kGy, similar to native starch when irradiated at 35 kGy (1 times) and significantly lower (0.4 times), when irradiated at 50 kGy [130]. According to the authors [130], this binomial relationship might be attributed to starch granule morphology/degree of disruption, enabling water penetration at first, but then decreasing water holding capability after gelatinization at higher disruption levels.

Overall, gelatinization temperatures, ΔH, viscosity and relative crystallinity of potato starch decreased upon irradiation treatment and with increasing irradiation dose [129]. The decrease in potato starch viscosity by ionizing irradiation treatment and increasing irradiation dose was supported by the findings of Atrous et al. [130] and Teixeira et al. [131].

In potato starch-based films, ionizing irradiation treatment resulted in decreased hydration properties (solubility, swelling, water absorption) [132] and increased hydrophobicity [133]. Investigating the effect of gamma irradiation on tensile properties of potato starch-based films, no change was observed for irradiation treatments at 0–15 kGy [131], but an increase was observed at 30 kGy [133], indicating an irradiation dose dependency. However, Teixeira et al. [132] reported a decrease in tensile properties of electron beam (0–60 kGy) treated potato starch/hibiscus extract-based films, indicating weakening of the films due to depolymerization into shorter chains, upon irradiation [132]. This behavior was in contrast to the findings reported by Cieśla and Sartowska [133] and Teixeira et al. [131], which might be attributed to the different irradiation source (electron beam vs. ^{60}CO gamma) and/or a higher amount of incorporated hibiscus extract into potato starch-based films by Teixeira et al. [132].

Others

In addition to HP, US treatment and ionizing irradiation, there are other non-thermal modification methods to change the structural and functional properties of potato constituents, such as milling and electric field treatment.

Milling is commonly applied in the food industry to reduce the particle size and produce flour/powder. Ball-milling of potato starch does not only reduce the particle size of the samples but can also highly damage granule morphology, induce partial gelatinization and destroy B-type crystallinity, which is accelerated with increasing milling time [134]. Furthermore, for jet-milled potato starch a decrease in molecular weight with increasing milling speed was observed [135]. Relative crystallinity, gelatinization temperatures and pasting properties were decreased and thermal stability increased upon jet-milling of potato starch. These effects were gradually accelerated with increasing jet-milling speed [135]. An increase in solubility, swelling and viscosity was observed for high-energy ball-milled potato starch at 0.5–2.2 J/g. Applying higher energy ($\geq$2.8 J/g), a complete loss of crystallinity and viscosity occurred [136]. Based on the results, Juarez-Arellano et al. [136] categorized the effect of ball-milling into three stages: Modification-stage, mechanolysis-stage and over-destruction-stage, depending on the amount of energy applied and the corresponding behavior of the characteristic properties [136].

Electric field treatment is the generic term for a number of different applications, including high voltage electric field (HVEF), or induced electric field (IEF). HVEF application is regarded as a method to physically modify starch properties in the absence of thermal influence, as no significant changes in product temperature occur during this treatment. In HVEF application, the sample is exposed to an electric field in liquid or gaseous medium. High voltage electric field can be classified into high electrostatic field (uniform electric field with no currents or varying voltages) and into high voltage electrical discharge (current flow causing ionizing/plasma) [137].

Cao and Gao [138] investigated the effect of HVEF on potato starch properties, by varying treatment time (10–50 min) and dose (10–40 kV). SEM images revealed some granular deformation, which increased with increasing processing time. Compared to native potato starch, apparent amylose content of HVEF treated starches significantly increased with increasing treatment time. Furthermore, relative crystallinity, water solubility and swelling power gradually decreased upon HVEF treatment with increasing voltage and treatment time [138].

In IEF treatment, electromagnetic induction is induced according to Faraday's law, where energy is conducted from a primary coil to the sample (solution), acting as a secondary coil [139]. In contrast to HVEF, no significant influence of IEF treatment (30 or 96 h) on relative crystallinity of potato starch was observed by Li et al. [140]. However, gelatinization and pasting temperatures increased independent of IEF treatment time and the final viscosity increased by approx. 33% (IEF for 30 h) and 40% (IEF for 96 h). Swelling power of potato starch at 65–95 °C slightly decreased upon all treatment constellations, showing a tendency of decreasing values upon increasing treatment time and temperature [140].

Conclusion Physical Modification

In this part of the article, different thermal and non-thermal physical modification methods were reviewed in terms of their influence on the structure and properties of potato constituents. From the evaluated studies, it can be concluded that different physical treatment methods require different preconditions for the sample. For instance, for HMT modification, a granular form of the sample with low moisture content is needed, whereas other methods such as MW treatment or ionizing irradiation can modify samples in form of granules, suspensions/hydrogels or films. Moreover, the required treatment time among different physical modification methods did highly vary, reaching from 5 min for MW treatment up to 24 h or more used for ANN and IEF treatments. Furthermore, different physical modifications generally resulted in different sample properties; the trends are summarized in Table 2. However, it should be considered that qualitative and quantitative changes in sample structure and properties can differ among physical modification methods, due to differences in modification and/or sample parameters (e.g., intensity, temperature, time, moisture content and pH). Especially for potato protein modification, pH is one of the most contributing factors, as the change in some functional properties is only susceptible under acidic or basic conditions.

Table 2. General overview of changes induced by different physical modification methods on potato starch, potato protein, and potato-based films structure and properties (specific processing parameters or changes in processing parameters are not considered here, but can be looked up above). Abbrevations and references: **HMT** (heat moisture treatment) [15,29,67,73–75], **ANN** (annealing) [20,77,78], **MW** (microwave) [30,68,87,88,90,91], **RF** (radio frequency) [90,92], **FT** (freezing and-thawing) [93,94], **Superheated steam** [97], **HP** (high pressure) [25,100,107–109], **US** (ultrasound) [115–124], **Ionizing irradiation** [129–133], **Milling** [134–136], **EF** (electric field) [138,140], **G′** (storage modulus), **WVP** (Water vapor permeability). Meaning of symbols: ↑ (increase), ↓ (decrease), → (no sig. influence), ↗ (slight increase), ↘ (slight decrease), - (no data available). Meaning of superscript numbers: **1** (for patatin at pH 6), **2** (at pH 7), **3** (at pH 12), **4** (for potato starch-lauric acid-complexes), **5** (for potato starch-hibiscus extract-based films).

Starch

| | Structure | | Crystallinity | | Hydration properties | | | Thermal properties | | Pasting properties | |
| | | | | | | | | | | | |
Modification	Damaging	Amylose content	Degree	Type	Solubility	Swelling power	Water absorption capacity	Gelatinization temperatures	Melting enthalpy	Pasting temperature	Final viscosity
HMT	↗	↑	↓	B → C/A	↑	↓	↓	↑	↓	↑	↓/↑
ANN	→	-	→/↗	→	↓/↑	↓/↑	-	↑	→/↗	↑	↑
MW	↑	↓	↓/↑	B → A	↓	↑	↑	↘/→/↗	↗	→/↑	↘/↑
RF	↗	↓	↓	B → C	-	-	-	↘	↘	↘	↗
FT	↑	-	-	-	-	-	-	↘	↘	↘	↘
Superheated steam	→	-	↓	→	↓	↓	-	↘	↘	↑	↑
HP	↗/↑	↗	↓/→	B → B + V	↓	↘	-	↘/→/↗	→	→	↘/→/↗
US	↗	↑	↘	→/B → V	↗	↗	↗	→	→	→	→
Ionizing irradiation	→/↗	↓	↘	→	↗	↘/↗	-	↓	↓	-	-
Milling	↑	-	↓	loss	↑	↑	↑	↓	-	↓	↓
EF	↗	↑	↓/→	-	↓	↘/↓	-	↑	→	→/↑	↑

Protein

| | Structure | | Film forming properties | Hydration properties | | Foaming properties | | Emulsifying properties | |
| | | | | | | | | | |
Modification	α-helix content	β-sheet content	Particle size	Viscosity	Solubility	Surface hydrophobicity	Ability/Capability	Stability	Activity	Stability
Autoclaving	-	-		↑	→	-	-	-	↗	↗
HP	↓	↑	↓	-	↓[1]/→[2]/↑	↗	↗	↗	-	↓
US	↓	↑	↓	-	→[2]/↑[2/3]	↘	↑	↓	↑	↑

Table 2. *Cont.*

Modification	Crystallinity	Film forming properties			Hydration properties					Thermal properties	Tensile properties		
	Degree	Viscosity	G′	Smoothness	Solubility	Moisture content	Water absorption	WVP	Surface hydrophobicity	Glass transition/Thermal stability	Elongation modulus	Tensile strength	Elongation
HMT	-	-	-	-	↘	-	-	↘	-	-	-	↗	↗
HP	-	↑	↑	↑	-	-	-	-	-	-	-	-	-
US	↑[4]	-	-	-	↓	-	↓	↘[4]	-	↗[4]	-	↑	↓[4]/↑
Ionizing irradiation	-/↑	↓	-	-	↘	-	↓	-	↑	-	-	↘/→[5]/↗	→/↑

3.2. Chemical Modification

In this section, the effects of chemical modification on potato constituents are outlined and the most important findings presented in Table 3. In general, chemical modifications refer to the substitution, cross-linking or degradation of a polymer via chemical reaction.

For instance, in substitution reaction, new functional groups/existing functional groups in a polymer are introduced/blocked to alter polymer functionality. In cross-linking modification of polymers new intermolecular linkages are created [141], and in degradation modification of polymers, polymer chains are either cleaved via hydrolysis [142] or functional groups such as hydroxyl groups are oxidized to carbonyl and carboxyl groups [143].

Basically, starch contains a large number of hydroxyl groups, and proteins contain a variety of different functional groups (hydroxyl-, carboxyl-, amine groups, etc.). These functional groups can be used as reactive sides for chemical modification reactions such as acylation, esterification, etherification, cross-linking, grafting, acid hydrolysis and oxidation. A common technique to evaluate chemical changes represents FT-IR analysis [142,144–147]. Some general information on the chemical modification of starches and proteins can be also found in other review articles [60,148–151]. In this review article, studies on chemical modifications of potato constituents over the last years are reviewed in detail.

3.2.1. Chemical Substitution

Hydroxyl groups of potato starch and different functional groups of potato proteins can be substituted via chemical reaction. In this review article, substitution reactions are divided into subcategories, dependent on the type of agent used and starch/protein derivate obtained (i.e., acetylation, phosphorylation, fatty acid esterification, octenyl succinate, citric acid esterification and etherification). The amount of derivatization via substitution reaction often influences resulting properties of the product and is commonly described by the degree of substitution, which reflects the amount of theoretically possible substitutions per polymer unit. For instance, the degree of substitution is maximum three for potato starch as each anhydride glucose unit possess three hydroxyl groups, which can be substituted by other groups [152]. Amongst others, the degree of substitution reactions can be adjusted by the acylating agent concentration, treatment time and treatment temperature. For instance, a correlation between the degree of substitution and plasticizer migration was found in starch-based films [153]. Acylation of potato protein with different acylating agents showed that the acylation effect depended on the used acylating agent and reaction temperature, resulting in decreased thermal stability and increased swelling towards water [154]. When acylation agents with multiple functional groups are used, chemical modifications (e.g., phosphorylation, citric acid esterification) can be either used as (single) substitution reaction or as cross-linking reaction, if functional groups of different chains are substituted by the same acylation agent [155–157].

Acetylation

Acetylation refers to the chemical reaction of potato starch or potato protein with acetic anhydride [158–160].

Acetylation of potato starch resulted in an increase in moisture and amylose content as well as increased relative crystallinity, solubility and paste clarity. However, type of crystallinity and gelatinization temperatures of potato starch were not significantly influenced by acetylation [158].

Miedzianka et al. [160] investigated the influence of acetylation on potato protein concentrate and isolate, varying acetylation concentration. Upon acetylation, solubility and water holding capacity of the potato protein concentrate and isolate increased by up to 100%. Foaming properties (capacity and stability) were lowest for potato concentrate and isolate acetylated with the medium concentration of acetic anhydride (1 mL/g). However, regarding the oil binding capacity, emulsion properties and essential amino acid composition, the effect of acetylation on potato protein was highly influenced by protein purity. For instance, upon acetylation the oil binding capacity increased for potato protein concentrate but decreased for potato protein isolate. For potato protein concentrate, the decrease in emulsion activity and increase in emulsion stability was much higher than for potato protein isolate. Upon acetylation, the amount of essential amino acids decreased in potato protein concentrate and isolate. For potato protein concentrates, highest decrease in essential amino acids was determined at the lowest acetylating agent concentration (0.4 mL/g) and for potato protein isolates, highest decrease in essential amino acids was measured at the highest acetic anhydride concentration (2 mL/g) [160].

Phosphorylation

Phosphorylation of starch can be performed using different agents including phosphoryl chloride, sodium trimetaphosphate, sodium tripolyphosphate and sodium or potassium orthophosphate to obtain mono- and/or di-starch phosphates. In di-starch phosphates, orthophosphate groups can act as intramolecular or intermolecular bridges between the C2, C3 and C6 atoms of glucose units from the same or different (cross-linking) chain (see Section 3.2.2) [155].

Phosphorylation of potato starch resulted in slightly decreased moisture content, no change in amylose content and a slight change in color, which was however not visible to the human eye. Compared to native starch, gelatinization temperatures were slightly decreased when phosphorylation was performed at 15 °C, but slightly increased when phosphorylation was performed at 45 °C. The authors interpreted these effects as loosening and strengthening of the potato starch structure at 15 °C and 45 °C, respectively. Accordingly, an increase in viscosity was observed in potato starch pastes, which were phosphorylated at 45 °C [155].

For the phosphorylation of potato proteins, phosphorus oxychloride, phosphoric acid and sodium trimetaphosphate can be used [161]. Miedzianka and Pęksa [161] phosphorylated potato protein at different pH values (5.2, 6.2, 8.0 and 10.5) using sodium trimetaphosphate as the phosphorylating reagent. Properties of potato protein were greatly affected by phosphorylation under alkaline conditions (pH 8.0) showing the most significant increase in water absorption capacity (approx. 20%), oil absorption capacity (2.5 times), emulsifying activity (approx. four times) and foaming capacity (more than two times), when compared to native potato protein and potato protein modified at other pH values. However, protein solubility was not significantly affected by phosphorylation [161].

Fatty Acid Esterification

Another approach to modify potato starch by chemical substitution is the esterification of hydroxyl groups by fatty acid chains. Vanmarcke et al. [144] investigated the effect of fatty acid chain length (C8, C12 and C16) on potato starch fatty acid ester cast films. Results showed that chain length did not influence esterification reactivity and FT-IR spectra confirmed the almost complete substitution of hydroxyl groups by disappearance of the 3300 cm^{-1} band. However, with increasing chain length, weight of the samples increased and the X-ray diffraction peak in the low angle region shifted towards lower angles, indicating an increase in nanometer scale ordered structure based on the esterification of longer fatty acid chains. Thermographs showed that the thermal stability of potato starch increased due to fatty acid esterification, as hydrophobicity increased, which was indicated by the fact that the mass did not decrease due to water evaporation and the starch degradation shifted to higher temperatures. However, no significant differences between different chain lengths were observed. Fatty acid chain length highly influenced tensile properties of the modified starch films. With increasing chain length from C8, C12 to C16, elongation at break decreased by 150%, 133% and 13%, respectively. Tensile strength and elongation modulus showed no linear trend but highly depended on the fatty acid chain length. For C8, C12 and C16 fatty acid esterified potato starch films, the tensile strength was measured to be 3.8, 2.5 and 6.5 MPa and for the elongation modulus 68, 40 and 122 MPa was determined, respectively [144].

Octenyl Succinylation

The hydrophobic group of alkenyl succinic anhydrides can be esterified with starch, to weaken internal bonding and disorder the internal structure to obtain more fluid and clear pastes. Due to its food approval at low substitution levels, the influence of octenyl succinic anhydride on starch properties is widely studied [162–164].

Wang et al. [162] demonstrated that potato starch granule size has an effect on the efficiency of octenyl succinylation. For octenyl succinic anhydride modified starch, higher degree of substitution (DS) was reported with decreasing granule size varying between 0.0147–0.0219 (degree of substitution), indicating that smaller particles with higher surface areas are more susceptible to modification. However, this was only observed when octenyl succinic anhydride modification was carried out prior to particle fractioning [162].

Investigating the effect of octenyl succinic anhydride modification on potato starch structure, Won et al. [164] observed an increase in surface roughness, porosity, granule size and the occurrence of cavities and slight deformation after octenyl succinic anhydride modification. However, no differences between different degrees of substitutions were observed and crystallinity of potato starch was not affected by octenyl succinic anhydride modification [164].

Compared to native potato starch, octenyl succinic anhydride modified starch showed lower amylose leaching, gel strength and viscosity, which gradually decreased with increasing degree of substitution. Won et al. [163] explained these results by hindered interactions between amylose chains caused by octenyl succinic anhydride modification, which resulted in decreased amylose leaching, retrogradation and therefore weaker gel formation. The weakening of inter- and intramolecular bonding was also reflected by the decrease in final viscosity. However, gelatinization temperatures were not influenced by octenyl succinic anhydride modification [163].

In another study, slightly increased amylose content, swelling power, gelatinization temperatures and slightly decreased water binding capacity and pasting temperature were observed upon octenyl succinylation of potato starch [165].

Citric Acid Esterification

Based on its three carboxyl groups and acidity, it has been reported that citric acid can induce esterification, cross-linking and hydrolysis, when reacting under elevated temperatures with potato starch [156,157].

SEM images of citric acid esterified potato starch revealed the occurrence of some surface damage due to citric acid treatment, which was, however, not intensified using a higher citric acid concentration (up to 10%). Furthermore, amylose content of potato starch slightly in-/or decreased and moisture content and relative crystallinity decreased upon citric acid modification [156,166]. A change in type of crystallinity was not reported [156]. In addition to structural changes, citric acid esterification of potato starch also influenced different functional/physicochemical properties. Upon citric acid esterification, the solubility of potato starch decreased, whereas swelling power, water binding capacity and gelatinization temperatures increased. The reported effects of citric acid modification on potato starch pasting properties were slightly different, although pasting temperature did not appear to be significantly affected and final viscosity appeared to increase with citric acid modification compared to native potato starch [156,166]. In addition, Kapelko-Żeberska et al. [167] showed that increasing processing temperatures increased the degree of substitution as well. Accordingly, decreased solubility, swelling power and gelatinization temperatures were observed upon citric acid modification that further deceased with increasing processing temperature.

The effect of citric acid modification (5–20% w/w potato starch) to reduce hydrophilic properties of potato starch/chitosan composite films was studied by Wu et al. [146]. Cross-linking of composite films resulted in decreased smoothness but also decreased hydration properties (moisture content, water solubility, and swelling degree) as well as decreased water vapor permeability and moisture sorption. Furthermore, mechanical properties (tensile strength and elongation at break) and thermal properties increased after cross-linking reaction. All changes in film properties gradually increased with increasing citric acid concentration up to 15%. Excessive use of citric acid (20%) led to solidification and crystal formation on the film surface.

Etherification

The etherification reaction of potato starch with various reagents, including those containing carboxymethyl, hydroxypropyl and/or hydroxyethyl groups, requires alkaline catalysts, of which sodium hydroxide (NaOH) is usually used. A common etherification reaction for starches is the hydroxypropylation, which has been extensively reviewed elsewhere [168].

In the reviewed literature, carboxylation [169], hydroxypropylation [145] and hydroxyethylation [170] of potato starch were found to alter product structure and properties. In addition to reagent concentration, other reaction and solvent related variables such as reaction temperature, reaction time, catalyst concentration, type of solvent and solvent concentration were found to affect the degree of substitution and the product properties [169]. Therefore, optimal etherification conditions need to be determined individually, as demonstrated by Prajapati et al. [169] for carboxypropylated potato starch. However, in general etherification of potato starch resulted in granular surface damaging [170], increased solubility [145], higher final viscosity [170], decreased crystallinity [145], decreased pasting temperatures [145] and decreased thermal stability [170]. Similar to other substitution reactions such as citric acid esterification, properties of potato starch were found to decline when optimal reaction conditions are exceeded [170].

3.2.2. Chemical Cross-Linking

Cross-linking of potato starch describes the formation of intermolecular covalent bonds (between different chains) via esterification or etherification reaction with a cross-linking agent. Thereby, the cross-linking agents possess bi- or multi- functional groups, which enable the reaction on different sides with multiple chains.

In the case of starch, it has been reported that cross-linking only occurs between two amylopectin molecules or between amylopectin and amylose, but not between two amylose molecules [171]. The reduction in amylose content by cross-linking reaction, as cross-linked molecules can be considered as amylopectin molecules, can (among others) be used to determine the degree of cross-linking by the starch–iodine method [171].

Cross-linking of potato starch can be performed using cross-linking agents such as acetylmalic acid chloroanhydride [172], sodium trimetaphosphate (STMP)/sodium tripolyphosphate (STPP) [173] and deep eutectic solvents [174]. According to Shulga et al. [172] cross-linking of potato starch resulted in destruction of its granular form and reduced relative crystallinity as well as reduced thermal stability. Furthermore, Heo et al. [173] reported that the increased formation of covalent bridges between starch molecules due to cross-linking hindered granular swelling and thermally induced gelatinization. In addition, pasting temperature and (final) viscosity of potato starch pastes increased upon cross-linking.

For potato starch-based films, the formation of new covalent bonds through cross-linking generally strengthened the starch gel network, resulting in an increase in rheological, thermal and tensile (elongation modulus, tensile strength) properties. Solubility of potato starch-based films was not affected and water sorption degree increased upon deep eutectic solvent addition [174].

3.2.3. Degradation
Acid Hydrolysis

Acid hydrolysis describes the process of chemical bond cleavage via a nucleophilic substitution reaction with water under thermal conditions and in an acidic environment.

Common acids that are used in acid hydrolysis include hydrochloric acid (HCl), acetic acid, citric acid and sulfuric acid (H_2SO_4).

Generally, hydrolysis of potato starch in water results in a decrease in molecular weight and some granular damage. Furthermore, an increase in thermal properties (gelatinization temperatures and stability), gel strength and viscosity of hydrolyzed potato starch was observed [142,166,175].

Reviewing several studies, it seems that the qualitative and quantitative impact of acid hydrolysis on potato starch did not only depend on the amount but also on the type of acid as well as the type of solvent used [142,175]. For instance, Celikci et al. [142] showed that 0.25–1.25 M of HCl or H_2SO_4 influenced the quantitative increase in bonding strength of potato starch-based adhesives. However, when alcohol–water mixtures were used as a solvent instead of water, bonding strength of the hydrolyzed starch-based adhesives decreased, when compared to the native reference [142]. Furthermore, the use of water as solvent led to surface cracking, an increase in thermal properties and to a slight decrease in relative crystallinity, whereas the use of alcohol or alcohol–water mixtures resulted in the occurrence of granular pitting, a decrease in thermal properties, no change in relative crystallinity and an increase in solubility [175].

Varying the amount of acid used for potato starch hydrolysis, the degree of hydrolysis is usually affected, which can in turn also influence qualitative and quantitative changes in potato starch properties. On the one hand, a lower degree of hydrolysis, (using milder conditions) seem to accelerate changes in potato starch including a further decrease in amylose content and hydration properties, and increase in thermal stability and viscosity [142,166]. On the other hand, the use of high acid concentrations (≥ 1 M) can result in opposite effects such as a decrease in viscosity [142].

A favorable effect of mild acid hydrolysis was also observed for film formation of potato peel-based films by Merino et al. [176] as the films became smoother, more homogeneous, more transparent and more flexible after acid hydrolysis treatment (1 M acetic acid), which was performed prior to cast film plasticization.

Oxidation

Oxidation of potato starch can be carried out using oxidizing agents under controlled pH to oxidize hydroxyl groups into carbonyl and carboxyl groups. Common oxidizing agents used, include sodium hypochlorite, hydrogen peroxide and ozone [143]. Degree of oxidation is usually expressed as the sum of carbonyl and carboxyl content. In contrast to the use of chemical oxidizing agents, ozone can be decomposed after the oxidation reaction into oxygen by an ozone destructor. Thus, ozone oxidation is often described as a safe and environmental friendly method [143], which can also be applied to fresh potato tubers to reduce the spread of potato diseases during storage [177].

Basically, it was found that with increasing concentration of oxidizing agent and increasing treatment time, the degree of oxidation of oxidized potato starch increased [147,178,179] and the amylose content decreased, indicating depolymerization [179]. Higher degrees of oxidation, which means the conversion of hydroxyl groups into carbonyl and carboxyl groups, resulted in increasing granule surface cracking and shape deformation and in a decrease in Maltese cross intensity, determined by polarized light microscopy [178,179]. B-type crystallinity of potato starch was not affected by oxidation [179], but change in relative crystallinity varied upon the amount of oxidant agent concentration used [178]. Authors suggested that oxidation occurred first in the amorphous (increasing crystallinity at low oxidant concentration), and second in the crystalline structures (decreasing crystallinity at higher oxidant concentrations) [178]. However, no change in relative crystallinity was reported by Castanha et al. [179] for ozonized potato starch, but particle size was slightly decreased [180]. Overall, an increase in solubility, decrease in swelling power, increase in pasting temperature and decrease in final viscosity upon potato starch oxidation was observed in several studies [178,179]. The induced changes were amplified with increasing oxidant agent concentration and treatment time [178,179].

Oxidation treatment of potato starch prior cast film formation revealed positive effects on the potato starch films such as improved smoothness and transparency [181]. B-type crystallinity of potato starch-based films was not affected by oxidation [182] but relative crystallinity of films plasticized with glycerol or sorbitol decreased according to Fonseca et al. [182] using sodium hypochlorite, whereas it decreased for glycerol plasticized films upon ozone treatment [181]. Similarly, no change in moisture content or water vapor permeability and slight increase in solubility were observed in ozonized films [181], but slightly decreased values were reported by Fonseca et al. [182]. Furthermore, according to La Fuente et al. [181] elongation at break of potato starch-based films decreased and elongation modulus increased upon oxidation. Similar results were found by Fonseca et al. [183] who reported a decrease in elongation at break upon oxidation but an oxidant agent concentration dependency of changes in elongation modulus and tensile strength of oxidized potato starch-based films. In a follow-up study, the high influence of plasticizer type and amount used in potato starch-based films on the qualitative and quantitative changes in tensile properties induced by oxidation was demonstrated [182]. Overall, the different effects on different film-related properties such as hydration properties, water vapor permeability and tensile properties were determined for potato starch-based films upon oxidation which might be attributed to the different oxidizing agents and/or plasticizer (concentrations) used [181,182].

3.2.4. Conclusion Chemical Modification

Based on the reviewed literature, several substitution, cross-linking and degradation reactions have been studied to chemically modify potato constituents. In most cases potato starch was analyzed. Overall, chemical modifications can change physicochemical and functional properties as summarized in Table 3. Through chemical modification, constituents obtain different functional groups depending on the degree of substitution and change their molecular weight by either cross-linking, hydrolysis or oxidation.

There are qualitative and quantitative differences in sample structure and properties depending on the chemical modification method used. Parameters that influence the method are e.g., treatment time, temperature, type and amount of chemical(s), while e.g., solvent, purity, moisture content and pH-value influence the raw material, i.e., the potato constituents. As an example, the purity of the potato protein (either concentrate or isolate) has shown to not only quantitatively, but also qualitatively influence hydration and emulsifying properties upon chemical modification.

Table 3. General overview of changes induced by different chemical modification methods on potato starch, potato protein and potato-based films structure and properties (specific processing parameters or changes in processing parameters are not considered here, but can be looked up above). Abbreviations and references: **Acetylation** [158,160], **Phosphorylation** [155,161], **FA** (fatty acid) **esterification** [144], **OSA** (octenyl succinic anhydride) **modification** [162–165], **Citric acid esterification** [146,156,166,167], **Etherification** [145,169,170], **Cross-linking** [172–174], **Acid hydrolysis** [142,166,175,176], **Oxidation** [147,178,179,181–183], **Acylation** [154], **G'** (storage modulus), **WVP** (Water vapor permeability). Meaning of symbols: ↑ (increase), ↓ (decrease), → (no sig. influence), ↗ (slight increase), ↘ (slight decrease), - (no data available).

Starch

Properties

Modification	Structure		Crystallinity		Film forming properties	Hydration properties				Thermal properties			Pasting properties	
	Damaging	Amylose content	Degree	Type	Viscosity	Solubility	Swelling power	Water absorption capacity	Moisture content	Thermal stability	Gelatinization temperatures	Melting enthalpy	Pasting temp.	Final viscosity
Acetylation	-	↑	↑	→	-	↑	-	-	↑	-	→	↓	-	-
Phosphorylation	-	→	-	-	↑	-	-	-	↘	-	↘/→/↗	→/↗	-	-
FA esterification	-	-	-	-	-	-	-	-	-	↑	-	-	-	-
OSA modification	↗	↗	-	→	↓	-	↗	↘	-	-	→/↗	↘/→	↘	↓
Citric acid esterification	↗	↘/↗	↓	→	-	↓	↗	↗	↓	-	↗	↓	↘/→	↘/↑
Etherification	↗	-	↓	-	-	↑	-	-	-	↓	-	-	↓	↑
Cross-linking	↗/↑	-	↓	-	↑	-	↓	-	-	↓	-	-	↗	↑
Acid hydrolysis	↑	↓	↘	→	↓/↑	↑	↓	-	↓	↑	↘/↑	↓	→	↑
Oxidation	→/↗	↓	↓/→/↑	→	-	↑	↓	↓	-	-	-	-	↗	↓

Protein

Properties

Modification	Structure		Hydration properties			Thermal properties	Foaming properties		Emulsifying properties	
	α-helix content	β-sheet content	Solubility	Water absorption capacity	Oil binding/absorption capacity	Thermal stability	Ability/Capability	Stability	Activity	Stability
Acylation	-	-	-	↑	-	↓	-	-	-	-
Acetylation	-	-	↗/↑	↑	↘/↗	-	↘	↘/↗	↓/↘	↑/→
Phosphorylation	-	-	→	↗	↑	-	↑	-	↑	-

Potato-based films

Properties

Modification	Crystallinity	Film forming properties			Hydration properties			Thermal properties			Tensile properties		
	Degree	Viscosity	G'	Smoothness	Solubility	Moisture content	Water absorption	WVP	Surface hydrophobicity	Glass transition/Thermal stability	Elongation modulus	Tensile strength	Elongation
Citric acid esterification	-	-	-	↘	↓	↓	↓	↓	↑	↑	-	↑	↑
Cross-linking	-	↑	↑	-	→	-	↑	-	-	↑	↑	↑	↓/→
Acid hydrolysis	-	-	-	↑	-	-	-	-	-	-	-	-	-
Oxidation	↘/↑	-	-	-	↓/→/↗	-	-	↘/→	↑	-	↓/↑	↘/→/↗	↓

3.3. Biochemical Modification

In general, biochemical modifications of starches and proteins including enzymatic substitution (Section 3.3.1), cross-linking (Section 3.3.2) or hydrolysis (Section 3.3.4) are usually regarded as a clean or green alternative to chemical modification (Section 3.2). Throughout the different biochemical modification methods, substrate specific enzymes can be used such as in enzymatic de-/branching modification (Section 3.3.3), where the branched structure of potato starch can be altered to effect starch crystallinity and thus its properties.

3.3.1. Enzymatic Substitution

Substitution of hydroxyl groups of potato starch building blocks can be either performed by chemical substitution as outlined in Section 3.2.1 or by enzymatic modification. One example of enzymatic substitution is the esterification/transesterification of fatty acids by lipases. Based on fatty acid solubility, organic solvents or ionic liquids need to be used. Dependent on the lipase used, degree of esterification and esterification site can vary [184].

According to Zarski et al. [185] degree of substitution gradually decreased with increasing reaction temperature (60–80 °C) and time (4–8 h), when enzymatic potato starch esterification with oleic acid by lipase in an ionic liquid was performed. The authors reported that a longer reaction time resulted in an increase in water formation, which changed pH, increased hydrolysis and possibly decreased lipase activity. The results showed that esterification of potato starch with oleic acid led to a loss in crystallinity and SEM images revealed the destruction of granule shape and the smooth surface due to the enzymatic treatment. The authors suggested that these results are due to hydrogen bond disruption and substitution by ionic liquids and esterification. A decrease in thermal stability due to enzymatic esterification was reported in terms of the initial and onset degradation temperatures, which decreased from 251.8 and 300.1 °C to 198.4 and 150.7 °C, respectively [185].

3.3.2. Enzymatic Cross-Linking

Enzymatic cross-linking of potato protein can be performed in a variety of different ways including acyl-transfer reaction, oxidation/radical formation and 1,4-addition reaction. For potato protein, enzymatic cross-linking using transglutaminase, laccase, tyrosinase and peroxidase has been reported [186–189].

Of these four enzymes, transglutaminase and peroxidase showed higher quantitative impacts on potato protein properties than tyrosinase and laccase [189]. For instance, thermal stability of potato protein was increased to a higher extent using transglutaminase and peroxidase than using laccase or tyrosinase as the cross-linking enzyme. Similarly, rheological properties increased using transglutaminase and peroxidase but decreased using laccase or tyrosinase as the cross-linking enzyme [189–191]. The extent of enzymatic cross-linking using laccase could be enhanced using a mediator such as ferulic acid [190]. Furthermore, Glusac et al. [192] demonstrated that tyrosinase-cross-linking could improve emulsion properties of potato protein.

Upon enzymatic cross-linking, an increase in structural order of potato protein was observed in different studies due to a decrease in random coil content [189] and increase in β-sheet content as well as relative crystallinity [189,191]. In a follow-up study, Gui et al. [193] demonstrated that the combination of potato starch and enzymatically cross-linked potato protein resulted in increased intermolecular interactions in the modified mixture, but in decreased intermolecular interactions in the native mixture. In turn, the increase in molecular interactions enhanced gel formation, pasting and rheological properties [192,193], which was also observed in cross-linked potato flour [191].

3.3.3. Enzymatic De-/Branching

During enzymatic branching of potato starch, α-1,4-glycosidic linkages are cleaved and new α-1,6-glycosidic bonds are formed via a transglycosylation reaction using enzymes such as branching enzymes and/or transglucosidase. Therefore, polymer chain length distributions shift from longer chains towards shorter chains, indicating the occurrence of enzymatic hydrolysis. This means that starch polymers become more branched, resulting in a decrease in amylose and increase in amylopectin content [194,195]. Overall, branching of potato starch induced granular damaging, affected type of crystallinity (B $\rightarrow$ B + C or C), decreased relative crystallinity, gelatinization temperatures and shear viscosity and increased solubility [195].

The opposite happens during enzymatic debranching of potato starch, where debranching enzymes such as pullulanase [194] or isoamylase [196] can be used. Here α-1,6-glycosidic bonds are cleaved and new α-1,4-glycosidic bonds are formed. As a result, starch polymers become less branched, resulting in a decrease in amylopectin and increase in amylose content [196]. Accordingly, average molecular weight of the samples gradually decreased with increasing isoamylase concentration and granular damaging occurred, which intensified with increasing enzyme concentration. However, B-type crystallinity, thermal properties and solubility were not significantly affected upon isoamylase treatment, but gel strength and viscosity of debranched potato starch pastes decreased [196].

Investigating the influence of debranching on potato starch–lauric acid-based films, increasing debranching time up to 1.5 h gradually decreased surface roughness, water vapor permeability and elongation of the films, whereas tensile strength gradually increased. However, a longer debranching time (2 h) resulted in a reverse effect on potato starch–lauric acid-based films properties [194].

3.3.4. Enzymatic Hydrolysis

Enzymatic hydrolysis of potato starch and potato protein refers to the enzymatic cleavage of glycosidic and peptide bonds, respectively, to decrease chain length and molecular weight and therefore, alter product properties.

Different enzymes including α-amylase [197–199], glucoamylase [198], glycosyltransferase [198], or mixtures of different enzymes [199] can be used to hydrolyze potato starch. In general, the enzymatic hydrolysis of potato starch results in a decrease in average molecular weight, average chain length and therefore amylose content [197–199]. In starch granules, this was expressed by surface roughness and cracking, which increased with increasing enzyme concentration used, which in turn increased the degree of hydrolysis [197].

Furthermore, enzymatic potato starch hydrolysis resulted in increasing water solubility and absorption capability and in a decrease in gel strength and (final) viscosity [197,199]. No changes in moisture content [199], or pasting temperature [197] were observed upon enzymatic hydrolysis. However, there were differences reported by Asiri et al. [197] and Vafina et al. [199] regarding the changes in thermal properties of potato starch induced by enzymatic hydrolysis using different or similar enzymes. While Asiri et al. [197] did not observe any significant influence of hydrolysis using α-amylase on gelatinization temperatures of potato starch, Vafina et al. [199] reported a decrease in gelatinization temperature and thermal stability upon hydrolysis using a commercial enzyme-mixture or α-amylase.

For potato protein, the influence of hydrolysis time and different enzymes including protease and alcalase was investigated in several studies [200–202]. Upon enzymatic hydrolysis, secondary structure of potato protein was affected, as the amount of α-helix increased and β-sheet content decreased with increasing enzyme concentration [200]. In all studies, degree of hydrolysis gradually increased with increasing reaction time [200–202]. DSC analysis performed by Galves et al. [201] indicated that protease treatment of potato protein causes protein denaturation, as the endothermic peak correlated with protein denaturation disappears in native potato protein samples. According to Akbari et al. [200] the number of carboxyl and amino groups increased upon enzymatic hydrolysis, as molecular weight decreased, which in turn increased protein–water interactions. This was expressed in an increase in protein solubility. According to the authors [200], the observed increase in foam capacity and decrease in foam stability with increasing degree of hydrolysis could be related to the increase in solubility enabling shorter protein chains to migrate to air bubble interfaces for faster stabilization, but decrease in protein–protein interactions, which are necessary for long-term stabilization against environmental conditions [200]. Similar changes in protein solubility and foaming properties were also found by Miedzianka et al. [202]. In addition increased emulsifying properties of potato protein were observed upon enzymatic hydrolysis [200].

3.3.5. Conclusion Biochemical Modification

Changes in potato starch structure and properties by enzymatic substitution, hydrolysis, branching and debranching are depicted in Table 4. Furthermore, Table 4 outlines the changes in structure and properties of potato protein, induced by enzymatic hydrolysis and cross-linking, and changes in potato starch-based films induced by debranching. Overall, reviewed biochemical modifications are highly pH and temperature specific, as the used enzymes require different optimal conditions. Compared to chemical modifications, enzymatic modifications often require more time to achieve a similar outcome. However, enzymatic treatment resulted in a high impact on starch and protein morphology and structure, causing changes in crystallinity, as well as pasting, thermal, hydration, foaming and emulsion properties, which depend on the type of enzyme, enzyme concentration and treatment time used.

Table 4. General overview of changes induced by different biochemical modification methods on potato starch, potato protein and potato-based films structure and properties (specific processing parameters or changes in processing parameters are not considered here, but can be looked up above). Abbrevations and references: Substitution [185], Branching [195], Debranching [194,196], Hydrolysis [197–202], Cross-linking [189–193], **G'** (storage modulus), **WVP** (Water vapor permeability). Meaning of symbols: ↑ (increase), ↓ (decrease), → (no sig. influence), ↗ (slight increase), ↘ (slight decrease), - (no data available). Meaning of superscript numbers: **1** (for potato starch–lauric acid-complexes).

Starch

Properties

Modification	Structure		Crystallinity		Film forming properties	Hydration properties					Thermal properties			Pasting properties	
	Damaging	Amylose content	Degree	Type	Viscosity	Solubility	Swelling power	Water absorption capacity	Moisture content	Thermal stability	Gelatinization temperatures	Melting enthalpy	Pasting temp.	Final viscosity
Substitution	↑	-	loss	loss	-	-	-	-	-	↓	-	-	-	-
Branching	↑	↓	↓	B → C	↑	↑	-	-	-	-	↓	↓	-	-
Debranching	↑	-	-	→	→	→	-	-	-	→	→	-	-	-
Hydrolysis	↗/↑	↓	↑	-	↓	↑	-	-	→	↓	↓/→	→	→	↓

Protein

Properties

Modification	Enzyme	Structure		Crystallinity	Rheological properties	Hydration properties	Thermal properties	Foaming properties		Emulsifying properties	
		α-helix content	β-sheet content	Degree		Solubility	Stability	Ability/Capability	Stability	Activity	Stability
Cross-linking	Transglucosidase	→	↑	↓	↑	-	↑	-	-	-	-
Cross-linking	Peroxidase	↘	↑	↓	↑	-	↑	-	-	-	-
Cross-linking	Tyrosinase	↘	↑	↓	↑	-	→/↗	-	-	↑	↑
Cross-linking	Laccase	↘	↑	↓	↑	-	→	-	-	-	-
Hydrolysis	Alcalase	↑	↓	-	-	↑	-	↑	↓	↑	↑
Hydrolysis	Protease	-	-	-	-	-	-	-	-	↗	-

Potato-based films

Properties

Modification	Crystallinity	Film forming properties			Hydration properties					Thermal properties	Tensile properties		
	Degree	Viscosity	G'	Smoothness	Solubility	Moisture content	Water absorption	WVP	Surface hydrophobicity	Tg/Thermal stability	Elongation modulus	Tensile strength	Elongation
Debranching	-	-	-	↑[1]	-	-	-	↓[1]	-	-	-	↑[1]	↓[1]

3.4. Dual Modification

Dual modifications are combinations of different physical, chemical and/or biochemical modifications, which are applied simultaneously or successively.

3.4.1. Physical–Physical

While some studies extensively investigated the relationship between modification parameters and the corresponding effect on structural and functional properties of potato constituents applying a certain type of modification, other studies investigated the effect of combined modification in comparison to native samples and/or the effect of an individual modification. Common dual modifications include the combination of thermal treatments such as HMT, ANN, MW, FT with non-thermal treatments such as HP, US and milling [203–206], but different heating methods have been combined as well [207,208].

As demonstrated by Wang et al. [203] and Cao and Gao [209], the order in which the two different physical treatment methods are performed can also influence the resulting qualitative and quantitative properties of the dual-modified samples, compared to native samples. For instance, ANN treatment prior to HP treatment resulted in a decrease in potato starch relative crystallinity, whereas an increase was reported when physical treatments were applied vice versa [203]. In potato starch, solubility and swelling power decreased, and gel hardness increased, when treated with US prior to electric field treatment, compared to untreated native potato starch. Treating potato starch first with an electric field and then with US resulted in reverse effects meaning increase in solubility and swelling power. Interestingly, the simultaneous treatment of electric field and US resulted in similar changes as the electric field prior to US treatment, except for a decrease in swelling power and increase in gel adhesiveness [209]. Some additional general information on dual-modification of starch can be found elsewhere [210].

3.4.2. Physical-Chemical

To enhance the effect of chemical modification on product properties, several studies were performed in the last years, combining different physical treatment methods with chemical modification methods. For instance, the assistance of acetylation or octenyl succinylation of potato starch by high voltage electric field, ultrasonication, pulsed electric field or microwave treatment was reported to increase the degree of substitution compared to a single chemical modification [211–214]. Moreover, some structural, pasting, rheological, thermal and other functional properties were affected by US treatment assisted acid hydrolysis [215], dry heat and $CaCl_2$ dual-treated [216] and annealing treated acetylated [217] potato starch samples.

3.4.3. Physical–Biochemical

Among biochemical modifications, enzymatic hydrolysis is often reported to be amplified by a physical pre-treatment. For instance, enzymatic hydrolysis of potato protein was increased by US treatment [218], and the enzymatic hydrolysis of potato starch by HP [219] and HMT [220]. This can be mostly attributed to an increase in susceptibility due to increasing structural damaging and/or surface cracking. With increasing degree of hydrolysis, different sample properties can be further in-/decreased, as demonstrated by Mu et al. [219] with increasing enzyme concentration and pressure.

3.4.4. Chemical–Chemical

Similar to dual physical modification, which has been described above (Section 3.4.1), dual chemical modification can be performed as well. For potato starch different dual modifications, including etherification + esterification, etherification + acid hydrolysis, and $CaCl_2$ treatment + succinylation, have been reported [221–223]. One commonly applied combination is acetylation + cross-linking. In potato starch granules, this dual modification resulted in a decrease in granule size and relative crystallinity and in an increase in thermal and pasting properties. Regarding potato starch-based films, acetylation prior to cross-linking caused lower water vapor permeability, solubility, moisture sorption and relative crystallinity, while flexibility of the films was increased [224].

3.4.5. Chemical–Biochemical

No combined treatment of chemical and biochemical modification of potato constituents was found by the authors during the literature search.

3.4.6. Biochemical–Biochemical

Dual biochemical modification can accelerate the impact on potato constituent properties, compared to the single treatments. As reviewed above, the combined use of branching enzymes and transglucosidase had a synergistic effect that more strongly influenced the properties of potato starch [195].

4. Overall Conclusions

In this review article, various physical, chemical and biochemical modifications of potato constituents were identified and the resulting structural and property changes were presented in the text as well as qualitatively illustrated in Tables 2–4. Overall, most of the scientific literature on physical and chemical modification was found for potato starch, and only a few selective studies on potato protein, potato flour or potato peel. Regarding biochemical modification, most studies focused on potato protein modification, followed by potato starch. This indicates a contained research interest on physical, chemical and biochemical modification of other potato constituents besides starch. This relates especially to potato protein, potato starch–protein interactions and bio-based plastics derived from different potato constituents, as some of the studies suggested synergistic effects on potato starch–protein interactions and resulting functional properties. As discussed earlier, the need for high value utilization of potato constituents, in form of isolated starch, proteins or unpurified side streams, is based on the large amounts of side streams occurring along the fresh and processed potato supply chain. However, it became evident that modification of potato constituents is highly complex, due to the great influence of processing parameters (including temperature, time, solvent, concentration and pH) and material properties (including chemical composition and moisture content) resulting in quantitative and qualitative differences throughout modifications. For enhancing the use of potato constituents also for other applications than for food products, modifications are required. For instance, using modified potato constituents can lead to films applicable for packaging applications, as they can have a positive effect on film forming, hydration, barrier and mechanical properties.

Author Contributions: K.M. conceived the focus of the review, conducted the literature search and drafted the manuscript. C.L.R., M.S. and M.L. reviewed and edited the manuscript upon critical revision of the text. All authors have read and agreed to the published version of the manuscript.

Funding: Parts of this work were financially supported by the Heinrich-Stockmeyer Stiftung as part of a PhD scholarship (scholarship holder: Katharina Miller) and the project PLA4Map (BMEL 2219NR448).

Data Availability Statement: Data is contained within the article.

Acknowledgments: The authors would like to thank Hanna Lia Reinhardt for her help in creating Figures 1 and 3.

Conflicts of Interest: The authors declare no conflict of interest. The funders had no role in the design, writing or publishing of the manuscript.

References

1. Bundesanstalt für Landwirtschaft und Ernährung. *Bericht zur Markt- und Versorgungslage Kartoffeln*; Bundesanstalt für Landwirtschaft und Ernährung: Bonn, Germany, 2020.
2. Gebrechristos, H.Y.; Chen, W. Utilization of potato peel as eco-friendly products: A review. *Food Sci. Nutr.* **2018**, *6*, 1352–1356. [CrossRef] [PubMed]
3. Landesanstalt für Entwicklung der Landwirtschaft und der ländlichen Räume, Schwäbisch Gmünd. Kartoffeln. Available online: https://lel.landwirtschaft-bw.de/pb/site/pbs-bw-mlr/get/documents_E973294336/MLR.LEL/PB5Documents/lel/Abteilung_4/Agrarm%C3%A4rkte%202020/04%20Kartoffeln_%28BW%29_2020.pdf (accessed on 30 November 2021).
4. Willersinn, C.; Mack, G.; Mouron, P.; Keiser, A.; Siegrist, M. Quantity and quality of food losses along the Swiss potato supply chain: Stepwise investigation and the influence of quality standards on losses. *Waste Manag.* **2015**, *46*, 120–132. [CrossRef] [PubMed]
5. Mouron, P.; Willersinn, C.; Möbius, S.; Lansche, J. Environmental profile of the Swiss supply chain for French fries: Effects of food loss reduction, loss treatments and process modifications. *Sustainability* **2016**, *8*, 1214. [CrossRef]
6. Kranert, M.; Hanfer, G.; Barabosz, J.; Schuller, H.; Leverez, D.; Kölbig, A.; Schneider, F.; Lebersorger, S.; Scherhaufer, S. Ermittlung der Weggeworfenen Lebensmittelmengen und Vorschläge zur Verminderung der Wegwerfrate bei Lebensmitteln in Deutschland. Available online: https://www.bmel.de/SharedDocs/Downloads/DE/_Ernaehrung/Lebensmittelverschwendung/Studie_Lebensmittelabfaelle_Langfassung.pdf?__blob=publicationFile&v=3 (accessed on 23 February 2022).
7. Gustavsson, J.; Cederberg, C.; Sonesson, U.; Otterdijk, R.V.; Meybeck, A. Global Food Losses and Food Waste: Extent, Causes and Prevention. Available online: https://www.semanticscholar.org/paper/Global-food-losses-and-food-waste%3A-extent%2C-causes-Gustavsson-Cederberg/19c0065b1ad3f83f5ce7b0b16742d137d0f2125e (accessed on 20 April 2022).
8. Punjabi, M. The Potato Supply Chain to PepsiCo's Frito Lay: India. Available online: https://www.fao.org/fileadmin/user_upload/ivc/PDF/Asia/15_Punjabi_potato_contract_farming_for_Pepsi_India_formatted.pdf (accessed on 8 December 2021).
9. Sepelev, I.; Galoburda, R. Industrial potato peel waste application in food production: A review. *Res. Rural. Dev.* **2015**, *1*, 130–136.
10. Maroušek, J.; Rowland, Z.; Valášková, K.; Král, P. Techno-economic assessment of potato waste management in developing economies. *Clean Technol. Environ. Policy* **2020**, *22*, 937–944. [CrossRef]
11. Brandão, A.S.; Gonçalves, A.; Santos, J.M. Circular bioeconomy strategies: From scientific research to commercially viable products. *J. Clean. Prod.* **2021**, *295*, 126407. [CrossRef]
12. Sharma, P.; Gaur, V.K.; Sirohi, R.; Varjani, S.; Hyoun Kim, S.; Wong, J.W.C. Sustainable processing of food waste for production of bio-based products for circular bioeconomy. *Bioresour. Technol.* **2021**, *325*, 124684. [CrossRef]
13. Pathak, P.D.; Mandavgane, S.A.; Puranik, N.M.; Jambhulkar, S.J.; Kulkarni, B.D. Valorization of potato peel: A biorefinery approach. *Crit. Rev. Biotechnol.* **2018**, *38*, 218–230. [CrossRef]
14. Virtanen, S.; Chowreddy, R.R.; Irmak, S.; Honkapää, K.; Isom, L. Food industry co-streams: Potential raw materials for biodegradable mulch film applications. *J. Polym. Environ.* **2017**, *25*, 1110–1130. [CrossRef]
15. Karki, D.B.; KC, Y.; Khanal, H.; Bhattarai, P.; Koirala, B.; Khatri, S.B. Analysis of biodegradable films of starch from potato waste. *Asian Food Sci. J.* **2020**, *14*, 28–40. [CrossRef]
16. Artkan, E.B.; Bilgen, H.D. Production of bioplastic from potato peel waste and investigation of its biodegradability. *Int. Adv. Res. Eng. J.* **2019**, *3*, 93–97. [CrossRef]
17. Rommi, K.; Rahikainen, J.; Vartiainen, J.; Holopainen, U.; Lahtinen, P.; Honkapää, K.; Lantto, R. Potato peeling costreams as raw materials for biopolymer film preparation. *J. Appl. Polym. Sci.* **2016**, *133*, 42862. [CrossRef]
18. Schäfer, D.; Reinelt, M.; Stäbler, A.; Schmid, M. Mechanical and barrier properties of potato protein isolate-based films. *Coatings* **2018**, *8*, 58. [CrossRef]
19. Waglay, A.; Achouri, A.; Karboune, S.; Zareifard, M.R.; L'Hocine, L. Pilot plant extraction of potato proteins and their structural and functional properties. *LWT* **2019**, *113*, 108275. [CrossRef]
20. Chen, X.; Luo, J.; Liang, Z.; Zhu, J.; Li, L.; Wang, Q. Structural and physicochemical/digestion characteristics of potato starch-amino acid complexes prepared under hydrothermal conditions. *Int. J. Biol. Macromol.* **2020**, *145*, 1091–1098. [CrossRef]
21. Chen, W.; Oldfield, T.L.; Cinelli, P.; Righetti, M.C.; Holden, N.M. Hybrid life cycle assessment of potato pulp valorisation in biocomposite production. *J. Clean. Prod.* **2020**, *269*, 122366. [CrossRef]
22. Talja, R.A.; Helén, H.; Roos, Y.H.; Jouppila, K. Effect of various polyols and polyol contents on physical and mechanical properties of potato starch-based films. *Carbohyd. Polym.* **2007**, *67*, 288–295. [CrossRef]
23. Miller, K.; Silcher, C.; Lindner, M.; Schmid, M. Effects of glycerol and sorbitol on optical, mechanical, and gas barrier properties of potato peel-based films. *Packag. Technol. Sci.* **2021**, *34*, 11–23. [CrossRef]
24. Fu, Z.; Guo, S.; Sun, Y.; Wu, H.; Huang, Z.; Wu, M. Effect of glycerol content on the properties of potato flour films. *Starch—Stärke* **2021**, *73*, 2000203. [CrossRef]
25. Kang, H.J.; Won, M.Y.; Lee, S.J.; Min, S.C. Plasticization and moisture sensitivity of potato peel-based biopolymer films. *Food Sci. Biotechnol.* **2015**, *24*, 1703–1710. [CrossRef]
26. Singh, J.; Kaur, L. Chemistry, processing, and nutritional attributes of potatoes–An introduction. In *Advances in Potato Chemistry and Technology*; Singh, J., Kaur, L., Eds.; Elsevier: Amsterdam, The Netherlands, 2016; pp. 23–26. ISBN 9780128000021.
27. Leonel, M.; do Carmo, E.L.; Fernandes, A.M.; Soratto, R.P.; Ebúrneo, J.A.M.; Garcia, É.L.; Dos Santos, T.P.R. Chemical composition of potato tubers: The effect of cultivars and growth conditions. *J. Food Sci. Technol.* **2017**, *54*, 2372–2378. [CrossRef]

28. Lieberei, R.; Reisdorff, C. *Nutzpflanzen*, 8th ed.; Thieme: Stuttgart, Germany, 2012; ISBN 9783131516381.
29. Subroto, E.; Indiarto, R.; Marta, H.; Shalihah, S. Effect of heat-moisture treatment on functional and pasting properties of potato (Solanum tuberosum L. var. Granola) starch. *Food Res.* **2018**, *3*, 469–476. [CrossRef]
30. Awokoya, K.N.; Odeleye, I.E.; Muhammed, Y.A.; Ndukwe, N.A.; Ibikunle, A.A. Impact of microwave irradiation energy levels on molecular rotation, structural, physicochemical, proximate and functional properties of potato (Ipomoea batatas) starch. *Ghana J. Sci.* **2021**, *61*, 57–72. [CrossRef]
31. Van Koningsveld, G.A.; Gruppen, H.; Jongh, H.H.; de Wijngaards, G.; van Boekel, M.A.; Walstra, P.; Voragen, A.G. The solubility of potato proteins from industrial potato fruit juice as influenced by pH and various additives. *J. Sci. Food Agric.* **2002**, *82*, 134–142. [CrossRef]
32. Løkra, S.; Strætkvern, K.O. Industrial proteins from potato juice. A review. *Food* **2009**, *3*, 88–95.
33. Lee, C.H. A simple outline of methods for protein isolation and purification. *Endocrinol. Metab.* **2017**, *32*, 18–22. [CrossRef]
34. Li, H.; Zeng, X.; Shi, W.; Zhang, H.; Huang, S.; Zhou, R.; Qin, X. Recovery and purification of potato proteins from potato starch wastewater by hollow fiber separation membrane integrated process. *Innov. Food Sci. Emerg. Technol.* **2020**, *63*, 102380. [CrossRef]
35. Knorr, D.; Kohler, G.O.; Betschart, A.A. Potato protein concentrates: The influence of various methods of recovery upon yield, compositional and functional characteristics. *J. Food Process. Preserv.* **1977**, *1*, 235–247. [CrossRef]
36. Waglay, A.; Karboune, S.; Alli, I. Potato protein isolates: Recovery and characterization of their properties. *Food Chem.* **2014**, *142*, 373–382. [CrossRef] [PubMed]
37. Cornejo-Ramírez, Y.I.; Martínez-Cruz, O.; Del Toro-Sánchez, C.L.; Wong-Corral, F.J.; Borboa-Flores, J.; Cinco-Moroyoqui, F.J. The structural characteristics of starches and their functional properties. *CyTA J. Food* **2018**, *16*, 1003–1017. [CrossRef]
38. Bertoft, E.; Blennow, A. Structure of Potato Starch. In *Advances in Potato Chemistry and Technology*; Singh, J., Kaur, L., Eds.; Elsevier: Amsterdam, The Netherlands, 2016; pp. 57–73. ISBN 9780128000021.
39. Raigond, P.; Singh, B.; Dutt, S.; Chakrabarti, S.K. *Potato*; Springer: Singapore, 2020; ISBN 978-981-15-7661-4.
40. Wang, S. *Starch Structure, Functionality and Application in Foods*; Springer: Singapore, 2020; ISBN 978-981-15-0621-5.
41. Van Soest, J.J.G.; Vliegenthart, J.F.G. Crystallinity in starch plastics: Consequences for material properties. *Trends Biotechnol.* **1997**, *15*, 208–213. [CrossRef]
42. Van Soest, J.J.; van Hullemann, S.H.; de Wit, D.; Vliegenthart, J.F. Crystallinity in starch bioplastics. *Ind. Crops Prod.* **1996**, *5*, 11–22. [CrossRef]
43. Kang, X.; Liu, P.; Gao, W.; Wu, Z.; Yu, B.; Wang, R.; Cui, B.; Qiu, L.; Sun, C. Preparation of starch-lipid complex by ultrasonication and its film forming capacity. *Food Hydrocoll.* **2020**, *99*, 105340. [CrossRef]
44. Frost, K.; Kaminski, D.; Kirwan, G.; Lascaris, E.; Shanks, R. Crystallinity and structure of starch using wide angle X-ray scattering. *Carbohyd. Polym.* **2009**, *78*, 543–548. [CrossRef]
45. Karim, A.A.; Toon, L.C.; Lee, V.P.L.; Ong, W.Y.; Fazilah, A.; Noda, T. Effects of phosphorus contents on the gelatinization and retrogradation of potato starch. *J. Food Sci.* **2007**, *72*, C132–C138. [CrossRef]
46. Thomasik, P. Specific physical and chemical properties of potato starch. *Food* **2009**, *3*, 45–56.
47. Jagadeesan, S.; Govindaraju, I.; Mazumder, N. An insight into the ultrastructural and physiochemical characterization of potato starch: A review. *Am. J. Potato Res.* **2020**, *97*, 464–476. [CrossRef]
48. Alvani, K.; Qi, X.; Tester, R.F.; Snape, C.E. Physico-chemical properties of potato starches. *Food Chem.* **2011**, *125*, 958–965. [CrossRef]
49. Thakur, R.; Pristijono, P.; Scarlett, C.J.; Bowyer, M.; Singh, S.P.; Vuong, Q.V. Starch-based films: Major factors affecting their properties. *Int. J. Biol. Macromol.* **2019**, *132*, 1079–1089. [CrossRef]
50. Waglay, A.; Karboune, S. Potato proteins: Functional food ingredients. In *Advances in Potato Chemistry and Technology*; Singh, J., Kaur, L., Eds.; Elsevier: Amsterdam, The Netherlands, 2016; pp. 75–104. ISBN 9780128000021.
51. Pots, A.M.; De Jongh, H.H.J.; Gruppen, H.; Hamer, R.J.; Voragen, A.G.J. Heat-induced conformational changes of patatin, the major potato tuber protein. *Eur. J. Biochem.* **1998**, *252*, 66–72. [CrossRef]
52. Giuseppin, M.L.F.; van der Sluis, C.; Laus, M.C. Native Potato Protein Isolates. U.S. Patent 8,465,911, 10 November 2006.
53. Fu, Y.; Liu, W.-N.; Soladoye, O.P. Towards potato protein utilisation: Insights into separation, functionality and bioactivity of patatin. *Int. J. Food Sci. Technol.* **2020**, *55*, 2314–2322. [CrossRef]
54. Schmidt, J.M.; Damgaard, H.; Greve-Poulsen, M.; Larsen, L.B.; Hammershøj, M. Foam and emulsion properties of potato protein isolate and purified fractions. *Food Hydrocoll.* **2018**, *74*, 367–378. [CrossRef]
55. Schmidt, J.M.; Damgaard, H.; Greve-Poulsen, M.; Sunds, A.V.; Larsen, L.B.; Hammershøj, M. Gel properties of potato protein and the isolated fractions of patatins and protease inhibitors—Impact of drying method, protein concentration, pH and ionic strength. *Food Hydrocoll.* **2019**, *96*, 246–258. [CrossRef]
56. Bártová, V.; Bárta, J.; Jarošová, M. Antifungal and antimicrobial proteins and peptides of potato (Solanum tuberosum L.) tubers and their applications. *Appl. Microbiol.* **2019**, *103*, 5533–5547. [CrossRef]
57. Pouvreau, L.; Kroef, T.; Gruppen, H.; van Koningsveld, G.; van den Broek, L.A.M.; Voragen, A.G.J. Structure and stability of the potato cysteine protease inhibitor group (cv. Elkana). *J. Agric. Food Chem.* **2005**, *53*, 5739–5746. [CrossRef]
58. Wijeyesakere, S.J.; Stuckey, J.A.; Richardson, R.J. The Crystal Structure of Native Patatin. *PLoS ONE* **2014**, *9*, e108245.
59. Meulenbroek, E.M.; Thomassen, E.; Pannu, N.S. Crystal Structure of Potato Serine Protease Inhibitor. *Crystallogr. Sect. D Biol. Crystallogr.* **2012**, *68*, 794–799. [CrossRef] [PubMed]

60. Zink, J.; Wyrobnik, T.; Prinz, T.; Schmid, M. Physical, chemical and biochemical modifications of protein-based films and coatings: An extensive review. *Int. J. Mol. Sci.* **2016**, *17*, 1376. [CrossRef] [PubMed]

61. Pęksa, A.; Miedzianka, J. Amino acid composition of enzymatically hydrolysed potato protein preparations. *Czech J. Food Sci.* **2014**, *32*, 265–272. [CrossRef]

62. BeMiller, J.N.; Huber, K.C. Physical modification of food starch functionalities. *Annu. Rev. Food Sci. Technol.* **2015**, *6*, 19–69. [CrossRef] [PubMed]

63. Grgić, I.; Ačkar, Đ.; Barišić, V.; Vlainić, M.; Knežević, N.; Medverec Knežević, Z. Nonthermal methods for starch modification—A review. *J. Food Process. Preserv.* **2019**, *43*, 1. [CrossRef]

64. Han, Z.; Shi, R.; Sun, D.-W. Effects of novel physical processing techniques on the multi-structures of starch. *Trends Food Sci. Technol.* **2020**, *97*, 126–135. [CrossRef]

65. Zia-Ud-Din; Xiong, H.; Fei, P. Physical and chemical modification of starches: A review. *Crit. Rev. Food Sci. Nutr.* **2017**, *57*, 2691–2705. [CrossRef]

66. Maniglia, B.C.; Castanha, N.; Rojas, M.L.; Augusto, P.E.D. Emerging technologies to enhance starch performance. *Curr. Opin. Food Sci.* **2021**, *37*, 26–36. [CrossRef]

67. Shi, M.; Gao, Q.; Liu, Y. Corn, potato, and wrinkled pea starches with heat-moisture treatment: Structure and digestibility. *Cereal Chem.* **2018**, *95*, 603–614. [CrossRef]

68. Xu, X.; Chen, Y.; Luo, Z.; Lu, X. Different variations in structures of A- and B-type starches subjected to microwave treatment and their relationships with digestibility. *LWT* **2019**, *99*, 179–187. [CrossRef]

69. Wang, J.; Zhu, H.; Li, S.; Wang, S.; Wang, S.; Copeland, L. Insights into structure and function of high pressure-modified starches with different crystalline polymorphs. *Int. J. Biol. Macromol.* **2017**, *102*, 414–424. [CrossRef]

70. Rahman, M.H.; Mu, T.-H.; Zhang, M.; Ma, M.-M.; Sun, H.-N. Comparative study of the effects of high hydrostatic pressure on physicochemical, thermal, and structural properties of maize, potato, and sweet potato starches. *J. Food Process. Preserv.* **2020**, *44*, 2854. [CrossRef]

71. Guillén, J.S. Modification of potato starch by emerging technologies. *Trop. Agric. Res.* **2021**, *32*, 480. [CrossRef]

72. Fonseca, L.M.; Halal, S.L.M.E.; Dias, A.R.G.; Zavareze, E.D.R. Physical modification of starch by heat-moisture treatment and annealing and their applications: A review. *Carbohyd. Polym.* **2021**, *274*, 118665. [CrossRef]

73. Zhang, B.; Saleh, A.S.M.; Su, C.; Gong, B.; Zhao, K.; Zhang, G.; Li, W.; Yan, W. The molecular structure, morphology, and physicochemical property and digestibility of potato starch after repeated and continuous heat-moisture treatment. *J. Food Sci.* **2020**, *85*, 4215–4224. [CrossRef] [PubMed]

74. Lin, C.-L.; Lin, J.-H.; Lin, J.-J.; Chang, Y.-H. Progressive alterations in crystalline structure of starches during heat-moisture treatment with varying iterations and holding times. *Int. J. Biol. Macromol.* **2019**, *135*, 472–480. [CrossRef] [PubMed]

75. Oliveira, C.S.; de Bet, C.D.; Bisinella, R.Z.B.; Waiga, L.H.; Colman, T.A.D.; Schnitzler, E. Heat-moisture treatment (HMT) on blends from potato starch (PS) and sweet potato starch (SPS). *J. Therm. Anal. Calorim.* **2018**, *133*, 1491–1498. [CrossRef]

76. Bartz, J.; da Rosa Zavareze, E.; Dias, A.R.G. Study of heat-moisture treatment of potato starch granules by chemical surface gelatinization. *J. Sci. Food Agric.* **2017**, *97*, 3114–3123. [CrossRef] [PubMed]

77. Wang, S.; Wang, J.; Wang, S.; Wang, S. Annealing improves paste viscosity and stability of starch. *Food Hydrocoll.* **2017**, *62*, 203–211. [CrossRef]

78. Xu, M.; Saleh, A.S.M.; Gong, B.; Li, B.; Jing, L.; Gou, M.; Jiang, H.; Li, W. The effect of repeated versus continuous annealing on structural, physicochemical, and digestive properties of potato starch. *Food Res. Int.* **2018**, *111*, 324–333. [CrossRef]

79. Zhu, Z.; Guo, W. Frequency, moisture content, and temperature dependent dielectric properties of potato starch related to drying with radio-frequency/microwave energy. *Sci. Rep.* **2017**, *7*, 9311. [CrossRef]

80. Shen, G.; Zhang, L.; Hu, T.; Li, Z.; Chen, A.; Zhang, Z.; Wu, H.; Li, S.; Hou, X. Preparation of potato flour by freeze-thaw pretreatment: Effect of different thawing methods on hot-air drying process and physicochemical properties. *LWT* **2020**, *133*, 110157. [CrossRef]

81. Ozturk, S.; Kong, F.; Trabelsi, S.; Singh, R.K. Dielectric properties of dried vegetable powders and their temperature profile during radio frequency heating. *J. Food Eng.* **2016**, *169*, 91–100. [CrossRef]

82. Fan, D.; Wang, L.; Shen, H.; Huang, L.; Zhao, J.; Zhang, H. Ultrastructure of potato starch granules as affected by microwave treatment. *Int. J. Food Prop.* **2017**, *20*, S3189–S3194. [CrossRef]

83. Zhu, H.; Li, D.; Li, S.; Wang, S. A novel method to improve heating uniformity in mid-high moisture potato starch with radio frequency assisted treatment. *J. Food Eng.* **2017**, *206*, 23–36. [CrossRef]

84. Tao, Y.; Yan, B.; Fan, D.; Zhang, N.; Ma, S.; Wang, L.; Wu, Y.; Wang, M.; Zhao, J.; Zhang, H. Structural changes of starch subjected to microwave heating: A review from the perspective of dielectric properties. *Trends Food Sci. Technol.* **2020**, *99*, 593–607. [CrossRef]

85. Oyeyinka, S.A.; Akintayo, O.A.; Adebo, O.A.; Kayitesi, E.; Njobeh, P.B. A review on the physicochemical properties of starches modified by microwave alone and in combination with other methods. *Int. J. Biol. Macromol.* **2021**, *176*, 87–95. [CrossRef]

86. Jiang, H.; Liu, Z.; Wang, S. Microwave processing: Effects and impacts on food components. *Crit. Rev. Food Sci. Nutr.* **2018**, *58*, 2476–2489. [CrossRef]

87. Przetaczek-Rożnowska, I.; Fortuna, T.; Wodniak, M.; Łabanowska, M.; Pająk, P.; Królikowska, K. Properties of potato starch treated with microwave radiation and enriched with mineral additives. *Int. J. Biol. Macromol.* **2019**, *124*, 229–234. [CrossRef]

88. Kumar, Y.; Singh, L.; Sharanagat, V.S.; Patel, A.; Kumar, K. Effect of microwave treatment (low power and varying time) on potato starch: Microstructure, thermo-functional, pasting and rheological properties. *Int. J. Biol. Macromol.* **2020**, *155*, 27–35. [CrossRef]
89. Miernik, A.; Jakubowski, T. Response of potato starch granules to microwave radiation. *J. Phys. Conf. Ser.* **2021**, *1782*, 12018. [CrossRef]
90. Xia, T.; Gou, M.; Zhang, G.; Li, W.; Jiang, H. Physical and structural properties of potato starch modified by dielectric treatment with different moisture content. *Int. J. Biol. Macromol.* **2018**, *118*, 1455–1462. [CrossRef]
91. Villanueva, M.; Lamo, B.; de Harasym, J.; Ronda, F. Microwave radiation and protein addition modulate hydration, pasting and gel rheological characteristics of rice and potato starches. *Carbohyd. Polym.* **2018**, *201*, 374–381. [CrossRef]
92. Zhu, H.; Yang, L.; Fang, X.; Wang, Y.; Li, D.; Wang, L. Effects of intermittent radio frequency drying on structure and gelatinization properties of native potato flour. *Food Res. Int.* **2021**, *139*, 109807. [CrossRef]
93. Zhang, C.; Lim, S.-T.; Chung, H.-J. Physical modification of potato starch using mild heating and freezing with minor addition of gums. *Food Hydrocoll.* **2019**, *94*, 294–303. [CrossRef]
94. Wang, L.; Xie, B.; Xiong, G.; Wu, W.; Wang, J.; Qiao, Y.; Liao, L. The effect of freeze–thaw cycles on microstructure and physicochemical properties of four starch gels. *Food Hydrocoll.* **2013**, *31*, 61–67. [CrossRef]
95. Li, J.; Shen, C.; Ge, B.; Wang, L.; Wang, R.; Luo, X.; Chen, Z. Preparation and application of potato flour with low gelatinization degree using flash drying. *Dry. Technol.* **2018**, *36*, 374–382. [CrossRef]
96. Hu, X.; Guo, B.; Liu, C.; Yan, X.; Chen, J.; Luo, S.; Liu, Y.; Wang, H.; Yang, R.; Zhong, Y.; et al. Modification of potato starch by using superheated steam. *Carbohyd. Polym.* **2018**, *198*, 375–384. [CrossRef]
97. Drozłowska, E.; Weronis, M.; Bartkowiak, A. The influence of thermal hydrolysis process on emulsifying properties of potato protein isolate. *J. Food Sci. Technol.* **2020**, *57*, 1131–1137. [CrossRef]
98. Liu, Y.; Chao, C.; Yu, J.; Wang, S.; Wang, S.; Copeland, L. New insights into starch gelatinization by high pressure: Comparison with heat-gelatinization. *Food Chem.* **2020**, *318*, 126493. [CrossRef]
99. Kunugi, S.; Tanaka, N. Cold denaturation of proteins under high pressure. *Biochim. Biophys. Acta* **2002**, *1595*, 329–344. [CrossRef]
100. Baier, A.K.; Knorr, D. Influence of high isostatic pressure on structural and functional characteristics of potato protein. *Food Res. Int.* **2015**, *77*, 753–761. [CrossRef]
101. Słomińska, L.; Zielonka, R.; Jarosławski, L.; Krupska, A.; Szlaferek, A.; Kowalski, W.; Tomaszewska-Gras, J.; Nowicki, M. High pressure impact on changes in potato starch granules. *Pol. J. Chem. Technol.* **2015**, *17*, 65–73. [CrossRef]
102. Dos Santos Aguilar, J.G.; Cristianini, M.; Sato, H.H. Modification of enzymes by use of high-pressure homogenization. *Food Res. Int.* **2018**, *109*, 120–125. [CrossRef]
103. Guo, X.; Chen, M.; Li, Y.; Dai, T.; Shuai, X.; Chen, J.; Liu, C. Modification of food macromolecules using dynamic high pressure microfluidization: A review. *Trends Food Sci. Technol.* **2020**, *100*, 223–234. [CrossRef]
104. Castro, L.M.; Alexandre, E.M.; Saraiva, J.A.; Pintado, M. Impact of high pressure on starch properties: A review. *Food Hydrocoll.* **2020**, *106*, 105877. [CrossRef]
105. Yang, Z.; Chaib, S.; Gu, Q.; Hemar, Y. Impact of pressure on physicochemical properties of starch dispersions. *Food Hydrocoll.* **2017**, *68*, 164–177. [CrossRef]
106. He, X.; Luo, S.; Chen, M.; Xia, W.; Chen, J.; Liu, C. Effect of industry-scale microfluidization on structural and physicochemical properties of potato starch. *Innov. Food Sci. Emerg. Technol.* **2020**, *60*, 102278. [CrossRef]
107. Hu, C.; Xiong, Z.; Xiong, H.; Chen, L.; Zhang, Z. Effects of dynamic high-pressure microfluidization treatment on the functional and structural properties of potato protein isolate and its complex with chitosan. *Food Res. Int.* **2021**, *140*, 109868. [CrossRef]
108. Katzav, H.; Chirug, L.; Okun, Z.; Davidovich-Pinhas, M.; Shpigelman, A. Comparison of thermal and high-pressure gelation of potato protein isolates. *Foods* **2020**, *9*, 1041. [CrossRef]
109. Kang, H.J.; Min, S.C. Potato peel-based biopolymer film development using high-pressure homogenization, irradiation, and ultrasound. *LWT* **2010**, *43*, 903–909. [CrossRef]
110. Patist, A.; Bates, D. Ultrasonic innovations in the food industry: From the laboratory to commercial production. *Innov. Food Sci. Emerg. Technol.* **2008**, *9*, 147–154. [CrossRef]
111. Cárcel, J.A.; García-Pérez, J.V.; Benedito, J.; Mulet, A. Food process innovation through new technologies: Use of ultrasound. *J. Food Eng.* **2012**, *110*, 200–207. [CrossRef]
112. Maniglia, B.C.; Castanha, N.; Le-Bail, P.; Le-Bail, A.; Augusto, P.E.D. Starch modification through environmentally friendly alternatives: A review. *Crit. Rev. Food Sci. Nutr.* **2021**, *61*, 2482–2505. [CrossRef]
113. Wang, X.; Majzoobi, M.; Farahnaky, A. Ultrasound-assisted modification of functional properties and biological activity of biopolymers: A review. *Ultrason. Sonochem.* **2020**, *65*, 105057. [CrossRef]
114. Brodnjak, U.V. Influence of ultrasonic treatment on properties of bio-based coated paper. *Prog. Org. Coat.* **2017**, *103*, 93–100. [CrossRef]
115. Sujka, M.; Jamroz, J. Ultrasound-treated starch: SEM and TEM imaging, and functional behaviour. *Food Hydrocoll.* **2013**, *31*, 413–419. [CrossRef]
116. Sujka, M. Ultrasonic modification of starch—Impact on granules porosity. *Ultrason. Sonochem.* **2017**, *37*, 424–429. [CrossRef]
117. Hu, A.; Li, Y.; Zheng, J. Dual-frequency ultrasonic effect on the structure and properties of starch with different size. *LWT* **2019**, *106*, 254–262. [CrossRef]

118. Zhu, J.; Li, L.; Chen, L.; Li, X. Study on supramolecular structural changes of ultrasonic treated potato starch granules. *Food Hydrocoll.* **2012**, *29*, 116–122. [CrossRef]
119. Bai, W.; Hébraud, P.; Ashokkumar, M.; Hemar, Y. Investigation on the pitting of potato starch granules during high frequency ultrasound treatment. *Ultrason. Sonochem.* **2017**, *35*, 547–555. [CrossRef]
120. Nie, H.; Li, C.; Liu, P.-H.; Lei, C.-Y.; Li, J.-B. Retrogradation, gel texture properties, intrinsic viscosity and degradation mechanism of potato starch paste under ultrasonic irradiation. *Food Hydrocoll.* **2019**, *95*, 590–600. [CrossRef]
121. Mao, C.; Wu, J.; Zhang, X.; Ma, F.; Cheng, Y. Improving the solubility and digestibility of potato protein with an online ultrasound-assisted pH shifting treatment at medium temperature. *Foods* **2020**, *9*, 1908. [CrossRef]
122. Hussain, M.; Qayum, A.; Zhang, X.; Hao, X.; Liu, L.; Wang, Y.; Hussain, K.; Li, X. Improvement in bioactive, functional, structural and digestibility of potato protein and its fraction patatin via ultra-sonication. *LWT* **2021**, *148*, 111747. [CrossRef]
123. Borah, P.P.; Das, P.; Badwaik, L.S. Ultrasound treated potato peel and sweet lime pomace based biopolymer film development. *Ultrason. Sonochem.* **2017**, *36*, 11–19. [CrossRef]
124. Wang, R.; Liu, P.; Cui, B.; Kang, X.; Yu, B. Effects of different treatment methods on properties of potato starch-lauric acid complex and potato starch-based films. *Int. J. Biol. Macromol.* **2019**, *124*, 34–40. [CrossRef]
125. Bashir, K.; Aggarwal, M. Physicochemical, structural and functional properties of native and irradiated starch: A review. *J. Food Sci. Technol.* **2019**, *56*, 513–523. [CrossRef]
126. Zivanovic, S. Electron beam processing to improve the functionality of biodegradable food packaging. In *Electron Beam Pasteurization and Complementary Food Processing Technologies*; Pillai, S.D., Shayanfar, S., Eds.; Elsevier: Amsterdam, The Netherlands, 2015; pp. 279–294. ISBN 9781782421009.
127. Makuuchi, K.; Cheng, S. *Radiation Processing of Polymer Materials and Its Industrial Applications*; John Wiley & Sons, Inc.: Hoboken, NJ, USA, 2012; ISBN 978-0-470-58769-0.
128. Zhu, F. Impact of γ-irradiation on structure, physicochemical properties, and applications of starch. *Food Hydrocoll.* **2016**, *52*, 201–212. [CrossRef]
129. Rao, N.R.; Rao, B.S.; Reddy, S.V.S.R.; Rao, T.V. The effect of electron beam irradiation on physicochemical properties of potato starch. In Proceedings of the International Conference on Renewable Energy Research and Education (RERE-2018), Andhra Pradesh, India, 8–10 February 2018; p. 40030.
130. Atrous, H.; Benbettaieb, N.; Chouaibi, M.; Attia, H.; Ghorbel, D. Changes in wheat and potato starches induced by gamma irradiation: A comparative macro and microscopic study. *Int. J. Food Prop.* **2017**, *20*, 1532–1546. [CrossRef]
131. Teixeira, B.S.; Garcia, R.H.; Takinami, P.Y.; del Mastro, N.L. Comparison of gamma radiation effects on natural corn and potato starches and modified cassava starch. *Radiat. Phys. Chem.* **2018**, *142*, 44–49. [CrossRef]
132. Teixeira, B.S.; Chierentin, G.S.; Del Mastro, N.L. Impact of electron beam irradiation in potato starch films containing hibiscus aquous extract. *Braz. J. Radiat. Sci.* **2021**, *9*. [CrossRef]
133. Cieśla, K.; Sartowska, B. Modification of the microstructure of the films formed by gamma irradiated starch examined by SEM. *Radiat. Phys. Chem.* **2016**, *118*, 87–95. [CrossRef]
134. Fu, Z.; Wu, M.; Zhang, H.; Wang, J. Retrogradation of partially gelatinised potato starch prepared by ball milling. *Int. J. Food Sci. Technol.* **2018**, *53*, 1065–1071. [CrossRef]
135. Wang, L.; Liu, S.; Hou, Y.; Lang, S.; Wang, C.; Zhang, D. Changes in particle size, structure, and physicochemical properties of potato starch after jet-milling treatments. *J. Food Process. Preserv.* **2020**, *44*, 1. [CrossRef]
136. Juarez-Arellano, E.A.; Urzua-Valenzuela, M.; Peña-Rico, M.A.; Aparicio-Saguilan, A.; Valera-Zaragoza, M.; Huerta-Heredia, A.A.; Navarro-Mtz, A.K. Planetary ball-mill as a versatile tool to controlled potato starch modification to broaden its industrial applications. *Food Res. Int.* **2021**, *140*, 109870. [CrossRef] [PubMed]
137. Dalvi-Isfahan, M.; Hamdami, N.; Le-Bail, A.; Xanthakis, E. The principles of high voltage electric field and its application in food processing: A review. *Food Res. Int.* **2016**, *89*, 48–62. [CrossRef] [PubMed]
138. Cao, M.; Gao, Q. Effects of high-voltage electric field treatment on physicochemical properties of potato starch. *J. Food Meas. Charact.* **2019**, *13*, 3069–3076. [CrossRef]
139. Zhu, F. Modifications of starch by electric field based techniques. *Trends Food Sci. Technol.* **2018**, *75*, 158–169. [CrossRef]
140. Li, D.; Yang, N.; Jin, Y.; Zhou, Y.; Xie, Z.; Jin, Z.; Xu, X. Changes in crystal structure and physicochemical properties of potato starch treated by induced electric field. *Carbohyd. Polym.* **2016**, *153*, 535–541. [CrossRef]
141. Garavand, F.; Rouhi, M.; Razavi, S.H.; Cacciotti, I.; Mohammadi, R. Improving the integrity of natural biopolymer films used in food packaging by crosslinking approach: A review. *Int. J. Biol. Macromol.* **2017**, *104*, 687–707. [CrossRef]
142. Celikci, N.; Dolaz, M.; Colakoglu, A.S. An environmentally friendly carton adhesive from acidic hydrolysis of waste potato starch. *Int. J. Polym. Anal. Charact.* **2021**, *26*, 97–110. [CrossRef]
143. Vanier, N.L.; El Halal, S.L.M.; Dias, A.R.G.; da Rosa Zavareze, E. Molecular structure, functionality and applications of oxidized starches: A review. *Food Chem.* **2017**, *221*, 1546–1559. [CrossRef]
144. Vanmarcke, A.; Leroy, L.; Stoclet, G.; Duchatel-Crépy, L.; Lefebvre, J.-M.; Joly, N.; Gaucher, V. Influence of fatty chain length and starch composition on structure and properties of fully substituted fatty acid starch esters. *Carbohyd. Polym.* **2017**, *164*, 249–257. [CrossRef]
145. Almonaityte, K.; Bendoraitiene, J.; Babelyte, M.; Rosliuk, D.; Rutkaite, R. Structure and properties of cationic starches synthesized by using 3-chloro-2-hydroxypropyltrimethylammonium chloride. *Int. J. Biol. Macromol.* **2020**, *164*, 2010–2017. [CrossRef]

146. Wu, H.; Lei, Y.; Lu, J.; Zhu, R.; Xiao, D.; Jiao, C.; Xia, R.; Zhang, Z.; Shen, G.; Liu, Y.; et al. Effect of citric acid induced crosslinking on the structure and properties of potato starch/chitosan composite films. *Food Hydrocoll.* **2019**, *97*, 105208. [CrossRef]
147. Barbosa, J.V.; Martins, J.; Carvalho, L.; Bastos, M.M.; Magalhães, F.D. Effect of peroxide oxidation on the expansion of potato starch foam. *Ind. Crops Prod.* **2019**, *137*, 428–435. [CrossRef]
148. Shah, U.; Naqash, F.; Gani, A.; Masoodi, F.A. Art and science behind modified starch edible films and coatings: A review. *Compr. Rev. Food Sci. Food Saf.* **2016**, *15*, 568–580. [CrossRef]
149. Haq, F.; Yu, H.; Wang, L.; Teng, L.; Haroon, M.; Khan, R.U.; Mehmood, S.; Bilal-Ul-Amin; Ullah, R.S.; Khan, A.; et al. Advances in chemical modifications of starches and their applications. *Carbohyd. Res.* **2019**, *476*, 12–35. [CrossRef]
150. Haroon, M.; Wang, L.; Yu, H.; Abbasi, N.M.; Zain-ul-Abdin, Z.-A.; Saleem, M.; Khan, R.U.; Ullah, R.S.; Chen, Q.; Wu, J. Chemical modification of starch and its application as an adsorbent material. *RSC Adv.* **2016**, *6*, 78264–78285. [CrossRef]
151. Wang, X.; Huang, L.; Zhang, C.; Deng, Y.; Xie, P.; Liu, L.; Cheng, J. Research advances in chemical modifications of starch for hydrophobicity and its applications: A review. *Carbohyd. Polym.* **2020**, *240*, 116292. [CrossRef]
152. Xu, J.; Andrews, T.D.; Shi, Y.-C. Recent advances in the preparation and characterization of intermediately to highly esterified and etherified starches: A review. *Starch-Stärke* **2020**, *72*, 1900238. [CrossRef]
153. Li, X.; He, Y.; Huang, C.; Zhu, J.; Lin, A.H.-M.; Chen, L.; Li, L. Inhibition of plasticizer migration from packaging to foods during microwave heating by controlling the esterified starch film structure. *Food Control.* **2016**, *66*, 130–136. [CrossRef]
154. Capezza, A.J.; Glad, D.; Özeren, H.D.; Newson, W.R.; Olsson, R.T.; Johansson, E.; Hedenqvist, M.S. Novel sustainable superabsorbents: A one-pot method for functionalization of side-stream potato proteins. *ACS Sustain. Chem. Eng.* **2019**, *7*, 17845–17854. [CrossRef]
155. Rożnowski, J.; Juszczak, L.; Szwaja, B.; Przetaczek-Rożnowska, I. Effect of esterification conditions on the physicochemical properties of phosphorylated potato starch. *Polymers* **2021**, *13*, 2548. [CrossRef]
156. Remya, R.; Jyothi, A.N.; Sreekumar, J. Effect of chemical modification with citric acid on the physicochemical properties and resistant starch formation in different starches. *Carbohyd. Polym.* **2018**, *202*, 29–38. [CrossRef]
157. Van Hung, P.; Vien, N.L.; Lan Phi, N.T. Resistant starch improvement of rice starches under a combination of acid and heat-moisture treatments. *Food Chem.* **2016**, *191*, 67–73. [CrossRef]
158. Mbougueng, P.D.; Tenin, D.; Scher, J.; Tchiégang, C. Influence of acetylation on physicochemical, functional and thermal properties of potato and cassava starches. *J. Food Eng.* **2012**, *108*, 320–326. [CrossRef]
159. Zięba, T.; Kapelko, M.; Szumny, A. Effect of preparation method on the properties of potato starch acetates with an equal degree of substitution. *Carbohyd. Polym.* **2013**, *94*, 193–198. [CrossRef]
160. Miedzianka, J.; Pęksa, A.; Aniołowska, M. Properties of acetylated potato protein preparations. *Food Chem.* **2012**, *133*, 1283–1291. [CrossRef]
161. Miedzianka, J.; Pęksa, A. Effect of pH on phosphorylation of potato protein isolate. *Food Chem.* **2013**, *138*, 2321–2326. [CrossRef]
162. Wang, C.; Tang, C.-H.; Fu, X.; Huang, Q.; Zhang, B. Granular size of potato starch affects structural properties, octenylsuccinic anhydride modification and flowability. *Food Chem.* **2016**, *212*, 453–459. [CrossRef]
163. Won, C.; Jin, Y.I.; Chang, D.-C.; Kim, M.; Lee, Y.; Gansean, P.; Lee, Y.-K.; Chang, Y.H. Rheological, pasting, thermal and retrogradation properties of octenyl succinic anhydride modified potato starch. *Food Sci. Technol.* **2017**, *37*, 321–327. [CrossRef]
164. Won, C.; Jin, Y.; Kim, M.; Lee, Y.; Chang, Y.H. Structural and rheological properties of potato starch affected by degree of substitution by octenyl succinic anhydride. *Int. J. Food Prop.* **2017**, *20*, 3076–3089. [CrossRef]
165. Remya, R.; Jyothi, A.N.; Sreekumar, J. Comparative study of RS4 type resistant starches derived from cassava and potato starches via octenyl succinylation. *Starch-Stärke* **2017**, *69*, 1600264. [CrossRef]
166. Martins, P.C.; Gutkoski, L.C.; Martins, V.G. Impact of acid hydrolysis and esterification process in rice and potato starch properties. *Int. J. Biol. Macromol.* **2018**, *120*, 959–965. [CrossRef] [PubMed]
167. Kapelko-Żeberska, M.; Zięba, T.; Pietrzak, W.; Gryszkin, A. Effect of citric acid esterification conditions on the properties of the obtained resistant starch. *Int. J. Food Sci. Technol.* **2016**, *51*, 1647–1654. [CrossRef]
168. Fu, Z.; Zhang, L.; Ren, M.-H.; BeMiller, J.N. Developments in hydroxypropylation of starch: A review. *Starch-Stärke* **2019**, *71*, 1800167. [CrossRef]
169. Prajapati, N.K.; Patel, N.K.; Sinha, V.K. Optimization of reaction condition for carboxylicpropyl starch from potato starch. *Int. J. Plast. Technol.* **2018**, *22*, 104–114. [CrossRef]
170. Öztürk, Y.S.; Dolaz, M. Synthesis and characterization of hydroxyethyl starch from chips wastes under microwave irradiation. *J. Polym. Environ.* **2021**, *29*, 948–957. [CrossRef]
171. Kou, T.; Gao, Q. New insight in crosslinking degree determination for crosslinked starch. *Carbohyd. Res.* **2018**, *458–459*, 13–18. [CrossRef]
172. Shulga, O.; Simurova, N.; Shulga, S.; Smirnova, J. Modification of potato starch by acetylmalic acid chloroanhydride and physicochemical research of the new product. *Int. J. Polym. Sci.* **2018**, *2018*, 1–7. [CrossRef]
173. Heo, H.; Lee, Y.-K.; Chang, Y.H. Rheaological, pasting, and structural properties of potato starch by cross-linking. *Int. J. Food Prop.* **2017**, *2*, 1–13. [CrossRef]
174. Zdanowicz, M.; Johansson, C. Mechanical and barrier properties of starch-based films plasticized with two- or three component deep eutectic solvents. *Carbohyd. Polym.* **2016**, *151*, 103–112. [CrossRef]

175. Sun, B.; Tian, Y.; Wei, B.; Chen, L.; Bi, Y.; Jin, Z. Effect of reaction solvents on the multi-scale structure of potato starch during acid treatment. *Int. J. Biol. Macromol.* **2017**, *97*, 67–75. [CrossRef]
176. Merino, D.; Paul, U.C.; Athanassiou, A. Bio-based plastic films prepared from potato peels using mild acid hydrolysis followed by plasticization with a polyglycerol. *Food Packag. Shelf Life* **2021**, *29*, 100707. [CrossRef]
177. Hutla, P.; Kolaříková, M.; Hájek, D.; Doležal, P.; Hausvater, E.; Petráčková, B. Ozone treatment of stored potato tubers. *Agron. Res.* **2020**, *18*, 100–112. [CrossRef]
178. Zhou, F.; Liu, Q.; Zhang, H.; Chen, Q.; Kong, B. Potato starch oxidation induced by sodium hypochlorite and its effect on functional properties and digestibility. *Int. J. Biol. Macromol.* **2016**, *84*, 410–417. [CrossRef]
179. Castanha, N.; Matta Junior, M.D.D.; Augusto, P.E.D. Potato starch modification using the ozone technology. *Food Hydrocoll.* **2017**, *66*, 343–356. [CrossRef]
180. Castanha, N.; Santos, D.N.E.; Cunha, R.L.; Augusto, P.E.D. Properties and possible applications of ozone-modified potato starch. *Food Res. Int.* **2019**, *116*, 1192–1201. [CrossRef]
181. La Fuente, C.I.A.; Castanha, N.; Maniglia, B.C.; Tadini, C.C.; Augusto, P.E.D. Biodegradable films produced from ozone-modified potato starch. *J. Package. Technol. Res.* **2020**, *4*, 3–11. [CrossRef]
182. Fonseca, L.M.; Henkes, A.K.; Bruni, G.P.; Viana, L.A.N.; Moura, C.M.; de Flores, W.H.; Galio, A.F. Fabrication and characterization of native and oxidized potato starch biodegradable films. *Food Biophys.* **2018**, *13*, 163–174. [CrossRef]
183. Fonseca, L.M.; Roque, V.R.; Roquete, A.P.J.; Mossmann, A.; Viana, L.A.N.; Henkes, A.K.; Luiz, C.D.C.; Galio, A.F.; Da Zavareze, E.R. Mechanical analysis of biodegradable films from native and chemically modified potato starches. *Mater. Sci. Forum* **2016**, *869*, 830–834. [CrossRef]
184. Boruczkowska, H.; Boruczkowski, T.; Gubica, T.; Anioł, M.; Tomaszewska-Ciosk, E. Analysis of the chemical structure of insoluble products of enzymatic esterification of starch and transesterification of acetylated starch with oleic acid by solid-state CP/MAS 13 C NMR. *Starch-Stärke* **2016**, *68*, 1180–1186. [CrossRef]
185. Zarski, A.; Ptak, S.; Siemion, P.; Kapusniak, J. Esterification of potato starch by a biocatalysed reaction in an ionic liquid. *Carbohyd. Polym.* **2016**, *137*, 657–663. [CrossRef]
186. Li, M.; Karboune, S.; Light, K.; Kermasha, S. Oxidative cross-linking of potato proteins by fungal laccases: Reaction kinetics and effects on the structural and functional properties. *Innov. Food Sci. Emerg. Technol.* **2021**, *71*, 102723. [CrossRef]
187. Isaschar-Ovdat, S.; Fishman, A. Crosslinking of food proteins mediated by oxidative enzymes—A review. *Trends Food Sci. Technol.* **2018**, *72*, 134–143. [CrossRef]
188. Glusac, J.; Isaschar-Ovdat, S.; Fishman, A.; Kukavica, B. Partial characterization of bean and maize root peroxidases and their ability to crosslink potato protein. *Arch. Biol. Sci.* **2019**, *71*, 293–303. [CrossRef]
189. Gui, Y.; Li, J.; Zhu, Y.; Guo, L. Roles of four enzyme crosslinks on structural, thermal and gel properties of potato proteins. *LWT* **2020**, *123*, 109116. [CrossRef]
190. Li, M.; Blecker, C.; Karboune, S. Molecular and air-water interfacial properties of potato protein upon modification via laccase-catalyzed cross-linking and conjugation with sugar beet pectin. *Food Hydrocoll.* **2021**, *112*, 106236. [CrossRef]
191. Zhu, Y.; Tao, H.; Janaswamy, S.; Zou, F.; Cui, B.; Guo, L. The functionality of laccase- or peroxidase-treated potato flour: Role of interactions between protein and protein/starch. *Food Chem.* **2021**, *341*, 128082. [CrossRef]
192. Glusac, J.; Isaschar-Ovdat, S.; Kukavica, B.; Fishman, A. Oil-in-water emulsions stabilized by tyrosinase-crosslinked potato protein. *Food Res. Int.* **2017**, *100*, 407–415. [CrossRef]
193. Gui, Y.; Zou, F.; Zhu, Y.; Li, J.; Wang, N.; Guo, L.; Cui, B. The structural, thermal, pasting and gel properties of the mixtures of enzyme-treated potato protein and potato starch. *LWT* **2022**, *154*, 112882. [CrossRef]
194. Wang, R.; Liu, P.; Cui, B.; Kang, X.; Yu, B.; Qiu, L.; Sun, C. Effects of pullulanase debranching on the properties of potato starch-lauric acid complex and potato starch-based film. *Int. J. Biol. Macromol.* **2020**, *156*, 1330–1336. [CrossRef]
195. Guo, L.; Deng, Y.; Lu, L.; Zou, F.; Cui, B. Synergistic effects of branching enzyme and transglucosidase on the modification of potato starch granules. *Int. J. Biol. Macromol.* **2019**, *130*, 499–507. [CrossRef]
196. Ulbrich, M.; Asiri, S.A.; Bussert, R.; Flöter, E. Enzymatic modification of granular potato starch using isoamylase—Investigation of morphological, physicochemical, molecular, and techno-functional properties. *Starch-Stärke* **2021**, *73*, 2000080. [CrossRef]
197. Asiri, S.A.; Ulbrich, M.; Flöter, E. Partial hydrolysis of granular potato starch using α-amylase—Effect on physicochemical, molecular, and functional properties. *Starch-Stärke* **2019**, *71*, 1800253. [CrossRef]
198. Guo, L.; Li, J.; Li, H.; Zhu, Y.; Cui, B. The structure property and adsorption capacity of new enzyme-treated potato and sweet potato starches. *Int. J. Biol. Macromol.* **2020**, *144*, 863–873. [CrossRef] [PubMed]
199. Vafina, A.; Proskurina, V.; Vorobiev, V.; Evtugin, V.G.; Egkova, G.; Nikitina, E. Physicochemical and morphological characterization of potato starch modified by bacterial amylases for food industry applications. *J. Chem.* **2018**, *2018*, 1627540. [CrossRef]
200. Akbari, N.; Mohammadzadeh Milani, J.; Biparva, P. Functional and conformational properties of proteolytic enzyme-modified potato protein isolate. *J. Sci. Food Agric.* **2020**, *100*, 1320–1327. [CrossRef]
201. Galves, C.; Galli, G.; Miranda, C.G.; Kurozawa, L.E. Improving the emulsifying property of potato protein by hydrolysis: An application as encapsulating agent with maltodextrin. *Innov. Food Sci. Emerg. Technol.* **2021**, *70*, 102696. [CrossRef]
202. Miedzianka, J.; Pęksa, A.; Pokora, M.; Rytel, E.; Tajner-Czopek, A.; Kita, A. Improving the properties of fodder potato protein concentrate by enzymatic hydrolysis. *Food Chem.* **2014**, *159*, 512–518. [CrossRef]

203. Wang, S.; Guo, P.; Xiang, F.; Wang, J.; Yu, J.; Wang, S. Effect of dual modification by annealing and ultrahigh pressure on properties of starches with different polymorphs. *Carbohyd. Polym.* **2017**, *174*, 549–557. [CrossRef]
204. Colussi, R.; Kringel, D.; Kaur, L.; da Rosa Zavareze, E.; Dias, A.R.G.; Singh, J. Dual modification of potato starch: Effects of heat-moisture and high pressure treatments on starch structure and functionalities. *Food Chem.* **2020**, *318*, 126475. [CrossRef]
205. Liu, C.; Song, M.; Liu, L.; Hong, J.; Guan, E.; Bian, K.; Zheng, X. Effect of heat-moisture treatment on the structure and physicochemical properties of ball mill damaged starches from different botanical sources. *Int. J. Biol. Macromol.* **2020**, *156*, 403–410. [CrossRef]
206. Wang, S.; Hu, X.; Wang, Z.; Bao, Q.; Zhou, B.; Li, T.; Li, S. Preparation and characterization of highly lipophilic modified potato starch by ultrasound and freeze-thaw treatments. *Ultrason. Sonochem.* **2020**, *64*, 105054. [CrossRef]
207. Chi, C.; Li, X.; Lu, P.; Miao, S.; Zhang, Y.; Chen, L. Dry heating and annealing treatment synergistically modulate starch structure and digestibility. *Int. J. Biol. Macromol.* **2019**, *137*, 554–561. [CrossRef]
208. Li, Y.-D.; Xu, T.-C.; Xiao, J.-X.; Zong, A.-Z.; Qiu, B.; Jia, M.; Liu, L.-N.; Liu, W. Efficacy of potato resistant starch prepared by microwave-toughening treatment. *Carbohyd. Polym.* **2018**, *192*, 299–307. [CrossRef]
209. Cao, M.; Gao, Q. Effect of dual modification with ultrasonic and electric field on potato starch. *Int. J. Biol. Macromol.* **2020**, *150*, 637–643. [CrossRef]
210. Ashogbon, A.O. Dual modification of various starches: Synthesis, properties and applications. *Food Chem.* **2021**, *342*, 128325. [CrossRef]
211. Cao, M.; Gao, Q. Internal structure of high degree substitution acetylated potato starch by chemical surface gelatinization. *Int. J. Biol. Macromol.* **2020**, *145*, 133–140. [CrossRef]
212. Chen, B.-R.; Wen, Q.-H.; Zeng, X.-A.; Abdul, R.; Roobab, U.; Xu, F.-Y. Pulsed electric field assisted modification of octenyl succinylated potato starch and its influence on pasting properties. *Carbohyd. Polym.* **2021**, *254*, 117294. [CrossRef]
213. Hong, J.; Zeng, X.-A.; Han, Z.; Brennan, C.S. Effect of pulsed electric fields treatment on the nanostructure of esterified potato starch and their potential glycemic digestibility. *Innov. Food Sci. Emerg. Technol.* **2018**, *45*, 438–446. [CrossRef]
214. Zhao, K.; Li, B.; Xu, M.; Jing, L.; Gou, M.; Yu, Z.; Zheng, J.; Li, W. Microwave pretreated esterification improved the substitution degree, structural and physicochemical properties of potato starch esters. *LWT* **2018**, *90*, 116–123. [CrossRef]
215. Ulbrich, M.; Bai, Y.; Flöter, E. The supporting effect of ultrasound on the acid hydrolysis of granular potato starch. *Carbohyd. Polym.* **2020**, *230*, 115633. [CrossRef]
216. Zhang, H.; He, F.; Wang, T.; Chen, G. Thermal, pasting, and rheological properties of potato starch dual-treated with CaCl2 and dry heat. *LWT* **2021**, *146*, 111467. [CrossRef]
217. Zięba, T.; Wilczak, A.; Kobryń, J.; Musiał, W.; Kapelko-Żeberska, M.; Gryszkin, A.; Meisel, M. The annealing of acetylated potato starch with various substitution degrees. *Molecules* **2021**, *26*, 2096. [CrossRef]
218. Cheng, Y.; Liu, Y.; Wu, J.; Ofori Donkor, P.; Li, T.; Ma, H. Improving the enzymolysis efficiency of potato protein by simultaneous dual-frequency energy-gathered ultrasound pretreatment: Thermodynamics and kinetics. *Ultrason. Sonochem.* **2017**, *37*, 351–359. [CrossRef]
219. Mu, T.-H.; Zhang, M.; Raad, L.; Sun, H.-N.; Wang, C. Effect of α-amylase degradation on physicochemical properties of pre-high hydrostatic pressure-treated potato starch. *PLoS ONE* **2015**, *10*, e0143620. [CrossRef]
220. Qi, X.; Tester, R.F. Heat and moisture modification of native starch granules on susceptibility to amylase hydrolysis. *Starch-Stärke* **2016**, *68*, 816–820. [CrossRef]
221. Wang, J.; Ren, F.; Huang, H.; Wang, Y.; Copeland, L.; Wang, S.; Wang, S. Effect of CaCl2 pre-treatment on the succinylation of potato starch. *Food Chem.* **2019**, *288*, 291–296. [CrossRef]
222. Li, L.; Hong, Y.; Gu, Z.; Cheng, L.; Li, Z.; Li, C. Synergistic effect of sodium dodecyl sulfate and salts on the gelation properties of acid-hydrolyzed-hydroxypropylated potato starch. *LWT* **2018**, *93*, 556–562. [CrossRef]
223. He, X.; Gong, X.; Li, W.; Cao, W.; Yan, J.; Guo, R.; Niu, B.; Jia, L. Preparation and characterization of amphiphilic composites made with double-modified (etherified and esterified) potato starches. *Starch-Stärke* **2019**, *71*, 1900089. [CrossRef]
224. González-Soto, R.A.; Núñez-Santiago, M.C.; Bello-Pérez, L.A. Preparation and partial characterization of films made with dual-modified (acetylation and crosslinking) potato starch. *J. Sci. Food Agric.* **2019**, *99*, 3134–3141. [CrossRef]

 *foods*

Article

(Not) Communicating the Environmental Friendliness of Food Packaging to Consumers—An Attribute- and Cue-Based Concept and Its Application

Krisztina Rita Dörnyei [1,†], Anna-Sophia Bauer [2,†], Victoria Krauter [2,*] and Carsten Herbes [3]

1   Institute of Marketing, Corvinus University of Budapest, 8 Fovam ter, 1093 Budapest, Hungary; krisztina.dornyei@uni-corvinus.hu
2   Packaging and Resource Management, Department Applied Life Sciences, FH Campus Wien, University of Applied Sciences, Helmut-Qualtinger-Gasse 2/2/3, 1030 Vienna, Austria; anna-sophia.bauer@fh-campuswien.ac.at
3   Institute for International Research on Sustainable Management and Renewable Energy, Nuertingen Geislingen University, Neckarsteige 6-10, 72622 Nuertingen, Germany; carsten.herbes@hfwu.de
*   Correspondence: victoria.krauter@fh-campuswien.ac.at; Tel.: +43-1-606-68-77-3592
†   These authors contributed equally to this work.

Abstract: While consumer understanding of and preferences for environmentally friendly packaging options have been well investigated, little is known about the environmentally friendly packaging attributes communicated to consumers by suppliers via packaging cues. We thus propose a literature-based attribute-cue matrix as a tool for analyzing packaging solutions. Using a 2021 snapshot of the wafer market in nine European countries, we demonstrate the tool's utility by analyzing the cues found that signal environmentally friendly packaging attributes. While the literature suggests that environmentally friendly packaging is increasingly used by manufacturers, our analysis of 164 wafer packages shows that communication is very limited except for information related to recyclability and disposal. This is frequently communicated via labels (e.g., recycling codes, Green Dot) and structural cues that implicitly signal reduced material use (e.g., less headspace and few packaging levels). Our attribute–cue matrix enables researchers, companies, and policymakers to analyze and improve packaging solutions across countries and product categories. Our finding that environmentally friendly packaging attributes are not being communicated to consumers underscores a pressing need for better communication strategies. Both direct on-pack and implicit communication should help consumers choose more environmentally friendly packaging. Governments are encouraged to apply our tool to identify communication gaps and adopt labeling regulations where needed.

Keywords: packaging; environmentally friendly; eco-friendly; sustainable; consumer; strategy; attribute; cue; marketing; wafer

Citation: Dörnyei, K.R.; Bauer, A.-S.; Krauter, V.; Herbes, C. (Not) Communicating the Environmental Friendliness of Food Packaging to Consumers—An Attribute- and Cue-Based Concept and Its Application. *Foods* **2022**, *11*, 1371. https://doi.org/10.3390/foods11091371

Academic Editor: Bernardo Pace and Sergio Torres-Giner

Received: 4 April 2022
Accepted: 5 May 2022
Published: 9 May 2022

Publisher's Note: MDPI stays neutral with regard to jurisdictional claims in published maps and institutional affiliations.

1. Introduction

From an environmental perspective, food packaging is both boon and bane. As a boon, it preserves food and supports its efficient transport; thus limiting the waste of food and resources [1–4]. However, the bane of packaging can seem overwhelming: nearly 200 kg of packaging waste is generated each year in the European Union per inhabitant [5]. A large part of that waste goes to incinerators or landfill [6], but much of the packaging ends up in the environment [7]. As the packaging market is expected to grow [8] and many of today's packaging solutions are less environmentally friendly than they could be, both waste management and packaging systems call for redesign [9,10].

Packaging has become an environmental villain, a necessary evil, or even an unnecessary cost position that ought to be minimized [11,12]. The European Commission's action plan for the circular economy aims at developing a sustainable, low carbon, resource efficient, and competitive European economy. Developing environmentally friendly packaging is one of the key items on its agenda [13], and thus a keen area of interest to scholars and practitioners [14,15].

Developing environmentally friendly packaging, however, is a difficult task. It is a balancing act between competing demands. Packaging must satisfy environmental requirements, food protection and logistics requirements, production and marketing requirements, and strategic and operational requirements. Such solutions cannot be developed by one company in isolation but only in the context of multidisciplinary product-packaging development teams [14]. These need to involve the entire supply chain: suppliers of raw material, products and packaging, brand owners, retailers, collectors, and recyclers [12].

It further complicates packaging development that consumers are not always eager to embrace environmentally friendly designs. Consumers perceive compromises between environmental friendliness and functional performance that they are unwilling to make [16,17]. Environmentally friendly packages are usually negatively associated with convenience, which leads to lower perceived functionality and a reduced willingness to purchase [16]. Consumers also harbor false beliefs about the benefits of alternative packaging materials (e.g., recycled, bio-based, or bio-degradable plastic) [17,18]. The willingness to purchase environmentally friendly packaging is further limited by time pressures and the cognitive overload caused by much information and a disinclination to process it [10]. In their defense, though, communication on sustainability is often misleading, which creates confusion and discomfort among consumers who are unable to differentiate between environmentally friendly packaging and packaging that just claims to be [19].

Still, for an environmentally friendly packaging solution to succeed in a consumer market, it has to meet with consumer acceptance. To better understand what that entails, this study adopts a consumer-based perspective on environmentally friendly packaging. That means we examine attributes and cues explicitly or implicitly perceived as sustainable by consumers.

Consumer perception does not necessarily agree with life cycle assessments (LCA), nor does it recognize the economic and social pillars of sustainability. However, it is critical for acceptance of a packaging solution. The fact is, consumers often harbor a simplified understanding of a packaging's environmental impact and rely on behavioral routines and simple heuristics such as colors, material, or recycling options [10,20]. Additionally, sometimes they are outright wrong in their assumptions or evaluations of packaging [14,20–22]. That is why the consumer view of what makes a package environmentally friendly and how that friendliness can be recognized will not necessarily align with what is known by science of the environmental impact of a given solution [23].

This leads to a dilemma for packaging designers. If they base their design decisions solely on environmental assessments such as LCA, the design might fail in the market. However, if they base their designs on consumer perceptions, they might end up with environmentally inferior solutions. Hence, designers need to fulfill two objectives. First, packaging must fulfill engineering (or scientific) requirements. Designs must fit existing infrastructure (including machinery, available material, and food product needs), comply with changing regulatory environments, and have a comparatively low environmental impact as assessed by tools such as LCAs [22]. Second, packaging should communicate its benefits to consumers in a way that will be understood and recognized [10]. Thus, companies need to understand what consumers think makes a packaging solution environmentally friendly, i.e., the attributes and what they think and how they can recognize these attributes, i.e., the perceptual cues that signal environmental friendliness [23].

While research on environmentally friendly packaging has gained momentum with all stakeholders along the food and packaging supply chain [24], it has often focused on assessing environmental friendliness from a scientific view or from the demand-side perspective, i.e., consumer attitudes and perceptions or acceptance of environmentally friendly packaging solutions (e.g., ocean plastic). To the best of our knowledge, research has overlooked communication on the supply side, i.e., packaging choices available to consumers in the market. A well-structured mapping of packaging solutions in the market, one reflecting consumer perceptions of environmental friendliness, is needed to provide more effective messaging to consumers.

The present research aims at providing such a tool—an attribute–cue matrix combining two hitherto separate theoretical perspectives—to analyze packaging solutions across countries and product categories. To demonstrate the tool's utility, we apply it to a snapshot of the wafer market as found in nine countries: Austria, Denmark, Finland, Greece, Hungary, Poland, Portugal, Slovakia, and Turkey. To the best of our knowledge, this field study is also the first comprehensive market analysis of environmentally friendly packaging communication in a specific product category across multiple countries.

How is our research useful for stakeholders in the packaging sector? First, while we apply the attribute–cue matrix to a sub-market of the food sector, it can be used to analyze any packaging solution in business-to-consumer (B2C) industries. That gives companies a practical tool to analyze their own packaging solutions and benchmark them against specific competitors or the industry. Any gaps derived from these analyses can be used as a point of departure for improving communication to the consumer at the point of sale.

Second, the attribute–cue matrix provides governments and regulators with a tool to survey how companies implement packaging solutions that claim to be environmentally friendly. Any inconsistencies found can then be addressed, if needed, by changes in regulations. The matrix also points out attributes and cues that a company does not yet use in its packaging solutions. These missed opportunities can serve as a starting point for new strategies, just as the matrix can help policymakers draft new regulations or project the impact of potential future regulations. The tool can also help environmental pressure groups and non-governmental organizations (NGOs) to document the state of packaging and communication approaches in specific industries and build their strategies from there.

Finally, for researchers in marketing and strategy, our combined conceptualization of attributes and cues provides a launch pad for cross-national and cross-industry studies of packaging strategies from the consumer perspective. This is especially important since a strong theoretical understanding of consumer perception of environmentally friendly packaging is still lacking [9,10,25] and existing knowledge is rather fragmented [15]. Therefore, we combine two theoretical perspectives: attributes that capture what consumers think makes a packaging solution environmentally friendly and cues, the core concept of cue utilization theory, that show how consumers think they can recognize these attributes. Moreover, we contribute to cue utilization theory by adding new cues that can be used in future research to analyze communication via packaging.

Our matrix, however, not only supports comparative research into strategies across countries, industries, or market segments on the supply side, but it also enables evaluating consumer attitudes and behavior against company strategies. This can potentially reveal gaps between the focus of a company's messaging and what is important to consumers. Moreover, because we expand the concept of cues to include one sensory and three structural signals new to the literature, our attribute–cue matrix extends the strategic range of messaging to consumers.

The remainder of the paper unfolds as follows: in Section 2, we explain the theoretical foundations underlying our attribute–cue matrix. We then describe how we acquired empirical data from the multiple wafer markets and how we applied the matrix. In Section 3, we present our results. In Section 4, we discuss potential reasons for the packaging strategies found, after which we present implications for companies and policymakers. We conclude with avenues for further research.

2. Materials and Methods

2.1. Attribute–Cue Matrix

The attribute–cue matrix aims at identifying the messages cued by packaging that companies can use to communicate the environmentally friendly attributes of their packaging.

The framework can be used both for analyzing consumer perceptions and behavior and for analyzing company packaging strategies. The concept of stimuli contents vs. formats that Ketelsen et al. have used for analyzing past studies (not products) ties in with our approach [15].

In its essence, the matrix combines packaging attributes that consumers perceive as environmentally friendly, e.g., biodegradability, with cues that explicitly or implicitly communicate the given attribute, e.g., a label indicating biodegradability. We derived both attributes and cues from previous consumer research on packaging perceived as environmentally friendly, using attributes proven to matter to consumers in their decision making. We also include attributes related to the efficient use of packaging, e.g., space-saving packaging, for the same reason.

From a consumer perspective, attributes are those characteristics that make a package environmentally friendly. These attributes can relate to various phases of packaging life: in raw material production, for example, consumers regard the use of recycled material or renewable material as environmentally friendly. In the post-use phase, consumers pay attention to biodegradability and recyclability. Table 1 summarizes these attributes, grouped by packaging life stages. The third column indicates previous studies that have shown the relevance of the respective attribute to consumer decision making. In the compilation of the attributes, we drew on Herbes et al. [21].

Table 1. Pro-environmental attributes of packaging solutions.

Stage in Packaging Life	Pro-Environmental Attribute	Source
	Reused packaging	[26,27]
Material production	Recycled materials	[18,19,28–30]
	Renewable materials (bio-based)	[18,20,22,23,25,31–33]
	Less packaging	[34–38]
Packaging production	Local/regional production	*
	Environmentally friendly production	[25]
Transport and use	Lightweight	*
	Space-saving	[39]
	Reusable	[23,25,26,40,41]
Post-use	Recyclable	[19,23,25,29,30,35,42,43]
	Bio-degradable	[23,25,29,30,40,44]
General (no specific stage)	Environmentally friendly in general	[40]

* newly proposed attributes.

From a producer perspective, attributes describe packaging design choices, for example, the choice to use bio-based plastics for producing a pouch or to design the polymers for the pouch so they are bio-degradable. To communicate these attributes so they can enter into consumer purchasing decisions, designers need appropriate cues.

Cues are about communication. They are how companies communicate pro-environmental attributes of their packaging. This might be done by describing what part of the packaging is from a certain material, say ocean plastic. Cues describe how consumers recognize, or think they can recognize, pro-environmental attributes. Cues are necessary, because consumers often cannot experience directly the pro-environmental attributes of packaging. How, for instance, would a consumer know that the polymers for a pouch were bio-degradable? This is where cue utilization theory [45] comes in, when product characteristics cannot be objectively evaluated by observation. To reduce complexity, consumers make conclusions about products from the configuration of cues available [46]. Attributes that cannot be directly observed are called credence attributes [47]; for these, consumers have to trust the information provided by manufacturers on the package [48,49]. For example, the biodegradability of packaging is an attribute neither visible nor otherwise sense-perceptible. A consumer has to trust a manufacturer's claim.

One attribute may be recognized through several cues. For example, consumers might think they can recognize renewable or recycled material by its color, but they may also look for a label or text on the packaging confirming the material's origins.

Cues, however, can be treacherous if consumers have wrong ideas about packaging. Companies may deliberately mislead consumers by capitalizing on these wrong ideas, for example, using brown tones and coarse surfaces for packaging that is not from recycled or renewable material [15]. Some consumers, on the other hand, interpret pro-environmental cues as greenwashing, especially when claims diverge from expectations for environmentally friendly packaging design [25,28]. The multiple meanings of environmentally friendly packaging and the unclear packaging messages (e.g., labels) can create ambiguity, especially when environmental information is incorporated into a single metric or cue [50].

We chose to group environmental friendliness cues as experienced along the consumer journey: from first seeing the package at the point of sale, to then looking at the package closely, touching it and later, after the purchase, opening and using it (consumption). In the compilation of cues we drew on Herbes et al. [10].

We then added one new sensory cue and two new structural cues that consumers experience when using a product. They include, first, the sensory cue of how loosely or tightly a product is packaged, signaling how much packaging volume could have been saved. Next the product-to-packaging weight ratio, a structural cue, which though never measured directly by consumers does leave an impression. If the ratio is too low, consumers will read the cue as "overpackaged." The second new structural cue we added is the number of packaging levels, which along with packaging waste pieces, is experienced directly by consumers when opening a product. The calculus of perception is as follows: the more levels, the more waste pieces, the less environmentally friendly.

We would like to point out that, in contrast to most other cues, these cues do not require a conscious marketing decision on the part of the manufacturer. Manufacturers may design lightweight packages (e.g., few packaging levels, few packaging pieces) for other reasons than consumer communication, such as savings in material or in logistic costs.

Table 2 presents the cues used in our analysis. These can all be found on or in the packaging itself, a constraint we imposed on our analysis since only these cues can be directly influenced by the manufacturer. Other cues consumers have been shown to use are the so-called social cues, information provided by retailers, friends, and family [10].

Table 2. Cues on pro-environmental attributes of packaging solutions.

Consumer Journey	Cue Type	Cue	Source
Point of Sale	Visual (from distance)	Color	[10,39,40]
		Label/logo	[10,39,51,52]
		Image/picture	[22,53,54]
	Sensory (touching/picking up)	Haptics/texture/material	[10,12,22,25,55,56]
		Loose/tight packaging	*
	Informational (reading)	Text	[10,27,28,39,53,54,57]
Consumption	Structural (use-phase)	Product-to-packaging ratio	*
		Number of packaging levels	*
		Number of packaging waste pieces	[10]

* newly proposed cues.

Figure 1 presents the attribute–cue matrix, combining the attributes and cues described in Tables 1 and 2. The matrix contains a total of 108 possible attribute–cue combinations, of which 49 are identified as practically applicable (colored white in Figure 1). For example, the fact that packaging is from renewable materials can be explicitly communicated through a label and text, and implicitly through images, surface texture, and color. Certain cues, such as labels, images, and text could be called all-purpose-cues, because they can be used to provide attribute-specific communication for all attributes. Other cues are more limited in their communication power; haptics for example, can be taken as a cue for renewable materials but not much else.

		CUES								
		CONSUMER JOURNEY								
		View Visual Cues			Touch Sensory Cues		Read Informational Cues	Consume Structural Cues		
ATTRIBUTES / LIFE CYCLE STAGES PACKAGING		Color	Label	Image / picture	Haptics / texture	Tightly packed	Text	P2P ratio	Packaging levels	Waste pieces
No stage	-					X		X	X	X
Material production	Reused	X				X		X	X	X
Material production	Recycled					X		X	X	X
Material production	Renewable					X		X	X	X
Packaging production	Less packaging	X			X					
Packaging production	Local production	X			X	X		X	X	X
Packaging production	Environmentally friendly production	X			X	X		X	X	X
Transport and use	Light-weight	X			X	X			X	X
Transport and use	Space-saving	X			X			X	X	X
Post-use	Recyclable	X			X	X		X	X	X
Post-use	Reuseable	X			X	X		X	X	X
Post-use	Bio-degradeable	X			X	X		X	X	X

Figure 1. Attribute–cue matrix. Abbreviation: P2P ratio (product-to-packaging ratio).

2.2. Sampling

To examine the environmentally friendly packaging options available to consumers in the market and to provide a snapshot of which messages about which attributes companies send to consumers through their packaging, a field study was conducted (with similarities to the field study of Deng and Srinivasan [58]). Wafer products were purchased from retail outlets to serve as data for the analysis.

Wafers are in the product group of cereals and confectionary; they were chosen for the study as a prime example of the impact that packaging can have on consumer decisions at the point of sale (POS). Among wafers, many different packaging options for similar products are available. The product category includes multiple sizes and packaging formats (types, material, shapes), as well as flexible packaging solutions such as fold wraps, flow packs, stand-up pouches and laminated paper bags, rigid plastic trays and boxes, metal-based boxes, and cardboard boxes.

Moreover, sustainable production and packaging of confectionery goods is a main area of interest for packaging redesign [59]. Sweets in general depend heavily on packaging [60] to take advantage of seasonal trade through colorful special editions. The main quality-related criteria for packaging confectionary products are protection against light, oxygen, and water vapor transmission [61]. To provide these high barriers, packaging designers often use material combinations that might yield non-recyclable packaging solutions [62]. However, the industry aspires to make progress in sustainable packaging. Indeed, an increasing number of news articles have appeared recently about the environmentally friendly aspirations of the confectionery industry [63].

The data collection portion of our field study ran from January to May 2021 in nine different countries—Austria, Denmark, Finland, Greece, Hungary, Poland, Portugal, Slovakia, and Turkey to cover as many products as possible. In each country, available packaging solutions in the wafer category were collected. Collections were made by a local researcher following these instructions: (1) define one shopping area (street, district, etc.); (2) within one week, visit all shops selling confectionary products in that area; (3) purchase all available wafer products (uncoated, chocolate, or nut-based filled wafers with at least two layers of wafers and one layer of filling); (4) repeat the shopping trip after 4–6 weeks to search for new products; and (5) send all (unopened) products accompanied by the shopping trip information to the research team members in Vienna for analysis. If researchers found the exact same packaging solutions in different "product series" of one brand with different sizes or fillings/flavors that matched the criteria, they were asked to purchase the cheaper option. This procedure resulted in a sample of 189 wafer products overall, of which 25 were excluded for being duplicates or not meeting the defined criteria for, i.e., flavor selection.

2.3. Analysis and Coding

Analysis of this data meant the careful examination of cues and the attributes companies communicate. The packaging examination was designed to best imitate the consumer journey and be as realistic as possible, so the analysis included not only the visual examination [64,65] but also the description (e.g., material, packaging type) and physical examination [66] of packages, including manually opening the packages. First, the content analysis [67] of packaging information was conducted; all environment-related textual and visual attributes were compiled in an Excel database. Second, the physical examination of packages was conducted, which included the opening, emptying, and exploring of disposal information of each package in a way that most closely resembles average product usage.

Coding used a combination of deductive and inductive approaches [67,68], since it started with environmental attributes and cues identified by previous research (deductive approach). Then during the analytical phase, new codes were added (inductive coding) to the category system—one sensory and two structural cues. One researcher coded the packages while two researchers assisted and revised coding to ensure objectivity and reduce rater bias. Codes were also re-examined by a fourth researcher, before the final coding scheme was developed (see Table A1 (Appendix A) for examples of coding rules).

After coding, the wafer data was processed through the attribute–cue matrix to obtain the frequency of use of each practically applicable attribute–cue combination. Based on these frequencies, we identified three main groups of cue usage. We then prepared the data for visual analysis using a heat map where cues used by the majority of products ($\geq$50%) were marked red, cues used by a sizeable percentage ($\geq$20%) were marked orange, and cues rarely used (<20%) were marked yellow. Other combinations, which were applicable but not used at all, remained white.

3. Results

In total, 164 different wafer products were included in the analysis (see Figure A1 (Appendix B) for pictures of all collected packages, n = 189). The top three contributing countries for packages were Austria (33%), Turkey (20%), and Poland (13%). Other countries in the sample had shares of 10% or lower (n = 164, after discarding 25 as non-qualifying).

3.1. Descriptives

Flexible solutions were used by 88% of the products analyzed, whereas 11% of the products combined flexible (i.e., flow packs, fold wraps) and rigid elements, mostly plastic, rarely cardboard trays. Only one solution contained wafers as a bulk product in a solely rigid packaging solution, similar to a bucket with a lid and handle. Packaging made solely from plastic (excluding labels and clips) dominated the sample, making up 87% of the solutions. Information about the packaging being made from polypropylene (PP) and/or the recycling code/number five was frequently found. Only 13% of the packaging solutions included paper or cardboard elements, irrespective of labels including multilayer material (fold wraps, stand-up-pouches with paper layers) as well as boxes, trays, and inlays.

Referring to the surface haptics, 17% of the packaging surfaces were found to be coarse and/or matte as opposed to sleek and shiny. Investigating another sensory cue, the perception of excess air (headspace), found 79% of the solutions to be packed tightly, meaning the product could not move around in the package. Some solutions, such as trays in flow packs, were found to be intermediate (3%), i.e., between packed tightly and loosely. About one-fifth were packed loosely (18%).

The packages in the sample showed a variety of labels. Most of them related to the products, fewer to the packaging. One could find regional labels referring to local production, local certification schemes, as well as international certification standards commonly applied in the food production industry. Labels referring to certain ingredients, giving information about the cultivation or production of mostly cocoa, were frequently present. Labels relating to the packaging solution, e.g., the composition of the used materials and, less often, information about certified production standards in fiber-based solutions (paper, cardboard), for example, were found less often. Only one packaging solution in the 22 samples including paper carried a label related to agroforestry certification. Independent of the communicated material, the use of arrows arranged in a triangle or circle, with and without recycling code/number and the Green Dot, indicating collection or recycling context (76%), were found as well. Although symbols with recycling context/logos could help with correct post-use treatment by consumers, 39 of the collected packages did not contain the recycling code/number or a triangle/circle with arrows or the Green Dot on the outer packaging.

Surprisingly, text-based information referring to packaging was also quite rare (19 samples). Even though it is an all-purpose cue, text related to the packaging solution appeared on very few packages, stating, i.e., that the packaging solution is recyclable or that it is important to separate waste. On some packaging solutions one could find specific collection systems mentioned, i.e., for specific regions. Partly, the text-based information was available in combination or within a symbol, for example, stating in words which container to use for collection. These cases are reflected in the text-based share, not in the percentage of labels. More often, one could find, next to legally required labeling, information about the production, the ingredients and flavors, promotions or, for example, the brand values. As for the packaging solutions, the production or supplying company was communicated, but with logos rather than text. This was also the case for materials communicated as certified for food contact (FCM, fork, and glass). Moreover, none of the packages claimed to be bio-plastic/bio-based or of an environmentally friendly origin.

In terms of design, a total of 49 (30%) packages applied green as one of three main (most dominant) colors in the font of the brand name or the background color. If no brand name was found on the front of the pack, the product name was taken instead. Counting packages that were coded as being solely green, merely 7 (4%) of the wafers were found to have such a packaging design. Addressing images and pictures, one could find a multitude of different designs in backgrounds, brands, and product names on the wafer packages. Many of these images and pictures were, however, not found to be nature related (i.e., buildings, people, furniture, kitchen appliances, etc.) or, secondly, found to directly present the specific products (i.e., wafers), represent related processed ingredients (i.e., cocoa powder, chocolate, milk, cream, flour, etc.) and ingredient-related plants (i.e., hazelnuts, leaves of hazelnut trees, cocoa beans, cocoa plants, leaves of cocoa plants, vanilla blossom, ears of wheat, etc.). One could also find images and pictures of animals, but mostly cartoon style. All other additional images and pictures that were found to be nature-related (excluding the ones representing ingredients, animals, drop, and petal shapes), were rather limited and included trees, leaves and flowers, grass, mountains, landscapes, sun, moon, stars, clouds, etc. Counting only these, 18 (11%) packaging solutions carried one or more of such images or pictures.

The structural cue "product-to-packaging ratio" (written product weight versus emptied packaging) showed a broad range. The least efficient sample had a ratio of 1.75:1 whereas the most efficient solution had a rounded product-to-packaging ratio of 109:1. The most efficient solution was one package of 500 g wafers in a 4.6 g transparent flow pack. The sample's average product-to-packaging-ratio rounded was 38:1, what was taken as a benchmark to identify the more efficient ones within the sample. In total, 79 (48%)

packaging solutions had a higher ratio than this, meaning even higher efficiency, while the remaining 85 packages were less efficient.

Two other structural cues were investigated—the number of packaging levels (elements) that have to be opened to access the wafers, and the number of waste pieces of packaging that accumulate after consumption. Of the purchased products, 21% were multipacks with single packaged units (15% with 2–15; 4% with 6–10; and 2% with 11–25 pieces). However, only 15% of the purchased packages counted as having at least two levels to open. The difference between these two shares results from multipacks with single units that were held together by stickers, and therefore not considered as one level to open. The remaining 85% of packaging solutions required opening only one packaging element to access the wafers. Some solutions also included tear tapes/strips as well as text and/or graphic arrows to indicate where best to open the package.

The number of single packs and packaging levels goes hand-in-hand with the number of waste pieces generated by consuming the products. In 73% of the cases, only one piece of packaging waste accrued. Clearly, this number is smaller than that of levels to open, because partly open elements (such as trays) were counted as waste pieces, but not necessarily ones to open. Furthermore, opening multipacks was calculated as accessing one unit, which also accounts for the difference between waste pieces and levels to open. Only 2% of the packaging solutions produced more than 15 pieces of packaging waste; these cases were very small packages of less than 15 g of product.

3.2. Heatmap Based on the Attribute–Cue Matrix

Analyzing the wafer packaging data through the attribute–cue matrix yields the heatmap shown in Figure 2. Attribute–cue combinations that are not applicable appear as dashed cells, while practically applicable combinations not used appear in white. Of the 49 practically applicable attribute–cue combinations, only 12 (24%) were used by at least one product. Only four cues were hot (red $\geq$ 50%), with two cues lukewarm (orange $\geq$ 20%).

			CUES								
			CONSUMER JOURNEY								
			View Visual Cues			Touch Sensory Cues		Read Informational Cues	Consume Structural Cues		
ATTRIBUTES (LIFE CYCLE STAGES PACKAGING)			Color	Label	Image / picture	Haptics / texture	Tightly packed	Text	P2P ratio	Packaging levels	Waste pieces
		No stage / -	30 %		11 %	17 %		12 %			
	Material production	Reused									
		Recycled									
		Renewable									
	Packaging production	Less packaging					79 %		48 %	85 %	73 %
		Local production									
		Environmentally friendly production									
	Transport and use	Light-weight							48 %		
		Space-saving					79 %				
	Post-use	Recyclable		76 %							
		Reuseable									
		Bio-degradeable									

Figure 2. Heatmap of environmentally friendly cues that were utilized. Abbreviation: P2P ratio (product-to-packaging ratio).

Traveling left to right in Figure 2, along the consumer journey, "color" was partially ($\geq$20%) used, so it shows up orange. Labels were used more often, but primarily to indicate post-use: 76% of sampled packages carried labels relating to sorting or recyclability, reflecting recycling codes/numbers and the Green Dot. Images and pictures that communicate naturalness without any link to a specific stage in the packaging life were sufficiently present to move this cue from cold to cool, but still yellow in Figure 2.

Moving to the physical experience of the packaging, coarse and matte packaging textures, evoking a sense of naturalness, appeared as a cue with the same frequency category as images and pictures, leading to a similar yellow coding. The second sensory cue, "tightly packed", was a hot signal for two different attributes ("less packaging" and "space-saving"). Text as an informational cue was used sparingly, leading also to its yellow coding.

Moving to the consumption phase, shown in the leftmost columns in Figure 2, more structural cues appear than in the other phases. This leads to more and hotter fields. A low number of packaging levels were used by around 85% of the packages. An optimized product-to-packaging ratio was found in more than 48% of the samples (the more efficient ones above average), producing the two orange fields and reflecting less packaging use in production and transportation.

The last structural cue along the consumer journey as well as the packaging's life cycle stage, is given by accumulated waste pieces after consumption. This cue led to a hot field, as 73% of the packaging solutions only generated one piece of packaging waste.

Other cues were either not used or limited to a fraction of the products in the sample.

4. Discussion and Conclusions

This study developed a tool, the attribute–cue matrix, for analyzing the effectiveness of packaging solutions in communicating their environmental credentials to decision-making consumers. Only when consumers can recognize environmentally friendly packaging options will they be able to choose them. Without that demand-side perspective, even the best packaging solutions can go for naught.

We demonstrated the matrix through a field study of the wafer market in nine European countries—Austria, Denmark, Finland, Greece, Hungary, Poland, Portugal, Slovakia, and Turkey. While the matrix provides a powerful and versatile heuristic for academics, marketing managers, and policymakers, the results of the field study, based on 164 wafer packages, highlight more current topics relevant to communication and environmental specialists. The results show that even in the ever-popular wafer market, the supply side rarely communicates the potentially perceivable environmental attributes of its packaging solutions, compared to what would be possible.

These results are surprising, since environmentally friendly packaging is at the forefront of both academic and applied research. That it is not (yet) observable at the point-of-sale is thought-provoking, since here consumer perception of environmental friendliness and not the objective facts enter into a purchasing decision [69]. Our results are particularly sad given the gap between consumer perception of environmental friendliness and objective assessments of the life-cycle costs of a package [20,70]. This gap could narrow were effective guidance by unmistakable on-pack communication available to support pro-environmental product choices. That it is not in a popular mass market is puzzling.

In the next section we consider potential reasons for the puzzle. We then consider implications for companies and policymakers before outlining avenues for future research.

4.1. Potential Reasons for (Not) Communicating the Environmental Friendliness of Packaging Solutions

The first potential reason behind non-communicating lies in the properties of the product and the practical requirements of its packaging. Wafers are susceptible to water uptake (e.g., loss of crispness), are sensitive to oxidation (e.g., rancidity, unwanted color, and/or taste changes), can take up flavors and suffer structural damage [71–73] while having low water activity and therefore low susceptibility to the growth of pathogenic microorganisms. To extend the shelf life and the overall acceptability of wafers, producers opt for packaging solutions with high barriers against moisture, oxygen, light, and flavor loss. In many cases, it is difficult to meet all the packaging design requirements using a single material, so producers frequently opt for multilayer flexible food packaging solutions. These are built up of different materials that combine to meet functional requirements for i.e., resealability, barrier protection, strength, and lightweight, along with economic requirements for cost efficiency [74]. The latter also dictates minimal use of materials and often a reduced carbon footprint, both of which are environmental benefits. However, these materials show poor recyclability, a disadvantage heavily discussed as a trade-off among scientists and the public [62]. Therefore, even if a packaging solution is environmentally optimal for the product category, that fact is not easy to communicate.

The second potential reason for the dearth of effective on-pack communication can be found in the role the product plays for consumers and the context in which wafers are consumed. The consumption of confectionary products is often driven by hedonic motives [75], and it still relies on classic impulse triggers. Being reminded of one's responsibility for the waste generated by the package, being beleaguered by details on the environmental impact of the packaging could have a sobering effect on a consumer, perhaps prompting second thoughts that would undermine the sale.

The contradiction between hedonic motives and moral choices is well-known in the literature [76] and most probably not from a perspective appealing to manufacturers of confectionaries. Manufacturers may not want to suppress hedonic impulses with environmental friendly packaging claims or to place moral principles over pleasure [76], because sustainability-linked attributes can affect hedonic properties negatively [77]. However, it is also possible for consumers to derive pleasure from doing something positive for the planet (see the concepts of 'alternative hedonism' [78,79] or 'warm glow' [71]), but this concept is probably difficult (though not impossible) to apply to environmentally friendly packaging of confectionaries. Still, despite extensive academic discourse on the dichotomy between hedonism and morality in consumption practices, we do not know what role these concerns played in the decisions made by the companies. How companies go about meeting both business and ethical obligations becomes a question for further research.

A third way to look at (the lack of) manufacturers' on-pack communication strategies is through the model of ecological responsiveness [80], which names three motives for companies to behave pro-environmentally: to improve competitiveness, to create legitimation, and to fulfill a sense of responsibility to the earth. All three goals can be advanced by environmentally friendly packaging, a straightforward example being the competitive edge gained by saving resources and waste and streamlining logistics [81,82]. However, the development of such packaging entails high production costs, slow time-to-market, technical difficulties, and complex cross-team alignments [14]. Many times, companies lack the business expertise or long-term planning horizon needed to pursue eco-friendly packaging [21]. This is especially true in a product category not under criticism. As it is, businesses are often compelled by law to adopt environmentally friendly packaging initiatives (the legitimation motive) [14,25], but maybe not yet pressing over all product categories.

The fourth potential reason behind scarce on-pack communication is the novelty of the topic. Communicating the environmental friendliness of packaging is just beginning, especially when compared to product related on-pack information (e.g., organic labels or health claims), which have been hotly debated for decades and have evolved from the nonregulatory action policy of a few selected companies to a heavily regulated area [83].

4.2. Implications for Companies and Policymakers

How can scholars, managers, and policymakers use our research and what can be gained from it? This section advances implications aimed at addressing the key issues in relation to environmental packaging management, to stimulate greater attention to this important topic and to expand the scope of discussion.

The tool we have demonstrated provides guidance to companies considering environmentally friendly packaging communication. The attribute–cue matrix summarizes and visualizes the attribute–cue combinations that manufacturers may use. The matrix can help evaluate the status quo, compare competitive offerings, analyze potential communication directions, and improve existing packaging solutions.

Furthermore, the matrix can be used to improve packaging design: both communication changes and structural design changes can emerge from applying it. While considering packaging redesign, companies need to consider questions such as: How are consumers making sense of the current on-pack communication? Do they want to make a well-founded choice decision prioritizing certain eco-friendly attributes over others? How do consumers make sure that they recognize these attributes from the cues on the packaging across products from different manufacturers? And how can consumer perceptions be aligned with objective environmental impact?

Besides that, our results also indicate that both direct on-pack and implicit communication should be used more often to inform consumers and allow them to choose environmentally friendlier packaging solutions. Companies can use the matrix to identify better ways to provide this information [70] and explicitly signal the package attributes that qualify as environmentally friendly—especially compared to competitors. Using multiple signals of environmental friendliness is supported by cue congruence theory.

This study also provides guidance to policymakers. Our results show that with absent regulation, packaging communications can run the gamut, presenting the consumer with a cacophony of different messages from different producers, each highlighting different attributes with different cues. This more often creates misunderstanding and confusion for the consumer than providing real help in making pro-environmental purchase decisions. As in other markets for eco-friendly products, such as the markets for green electricity or for eco-friendly food, there is a potential positive role for a standardized, easy-to-understand information system, possibly administered by the state. However, the agonizing discourse and stubborn resistance from manufacturers over the nutriscore front-of-pack labeling [84] of food in Germany, France, and other countries [85–88] shows how difficult it is for policymakers to establish such a system. However, with sustainability-related credence attributes gaining more and more importance and consumers being less and less able to judge products with their five senses, accurate and informative labeling becomes a key task for third party actors such as industry associations or the state.

Both policymakers and manufacturers should consider the lack of communication about the end of life of packages. Not only is there almost no on-pack information to help consumers dispose of the package, but even if there were, the collection system in Europe varies from country-to-country and in some countries by region. Perceivable cues on products sold in multiple European countries would have to include regional labeling, which simply is not feasible. Therefore, it appears that action is still needed to reach the recyclability goals of the European Plastic Strategy by 2030 [89] and to ensure that improvements align with the overall goal of sustainability.

4.3. Avenues for Further Research and Limitations

This study is not without limitations and our work hints at multiple avenues for future research. First, we demonstrate a versatile and powerful tool, but do so considering only packaging from one product category in nine countries. Undoubtedly, a larger and more heterogeneous sample would provide a richer understanding of current on-pack communications and might even expand the tool, as applied packaging solutions could differ from the ones found in the category of confectionary products. The validation or further development of the matrix with different sample sets would be beneficial to check for differences across product groups. Therefore, we recommend the attribute–cue matrix be used in the analysis of packaging strategies across product categories and markets, where large differences can be expected due to different consumption factors or packaging solutions.

Second, it would be helpful to understand why companies design packaging solutions the way they do and why they do (not) communicate the way we might think they should.

Do restrictions stemming from technical properties of packaging material and machinery as well as requirements of packaged products largely govern packaging solutions? How do companies position the environmental friendliness of packaging solutions in their marketing strategies? Which stakeholders inside and outside the company are involved in packaging design decisions? How do companies see their potential customers and how do they think customers factor environmental issues of packaging into their buying decisions? These questions call for a qualitative study of decision-making processes involved in packaging design in companies.

Third, let us turn from the supply side to the demand side. Largely absent from the literature are comparative studies of consumer preferences for environmentally friendly packaging across product categories. Do consumers have different preferences regarding pro-environmental attributes of packaging and are they receptive in different ways to cues communicating these attributes depending on the product category and the consumption context? The discourse on the relationship between hedonism and sustainable consumption suggests that environmental impact may be less of a concern for consumers when the product and its consumption are embedded in hedonism.

Another question is which cues are especially credible and effective in communicating pro-environmental attributes. We hypothesize that some attributes would best be communicated by text, others best by nontextual cues. Lastly, it would be helpful to understand how consumers examine a package to determine its environmental friendliness. Observations and eye tracking could be suitable methods to explore this.

Answering these questions would help companies better understand how they can build pro-environmental considerations into their packaging strategies and how they might better help consumers make sound pro-environmental choices. Pursuing these questions would also help policymakers understand where consumer preferences, even if understood well by companies, cannot drive improvements in the overall sustainability of packaging solutions and where, therefore, a positive role for regulation may exist.

Author Contributions: Conceptualization, K.R.D., A.-S.B., C.H., V.K.; methodology, K.R.D., A.-S.B., C.H., V.K.; formal analysis, A.-S.B. and C.H.; writing—original draft preparation, K.R.D., A.-S.B., C.H., V.K.; writing—review and editing, K.R.D., A.-S.B., C.H. and V.K.; visualization, A.-S.B. and C.H. All authors have read and agreed to the published version of the manuscript.

Funding: This article/publication is based upon work from COST Action Circul-a-bility, supported by COST (European Cooperation in Science and Technology). www.cost.eu (accessed on 28 March 2022). The APC was funded by FH Campus Wien.

Acknowledgments: The authors would like to thank the COST Community for inspiration and especially Ibrahim Gülseren, Grzegorz Ganczewksi, Maristiina Nurmi, Rui M. S. Cruz, Sofia Agriopoulou, Ilke Uysal Unalan, and Veronika Gežík for dedicating time selecting, buying, and sending the products to Vienna.

Appendix A

Table A1. Examples of coding rules for the analysis.

Cue	Coding Rule: Coded as Signaling Environmental Friendliness if …	Example
Color	Packaging was any shade of green	
Label	Recycling code and/or symbol and/or green dot was present	
Image/picture	Nature-related images were present. We excluded any nature-related pictures or graphics that had a direct link to the product and its ingredients, e.g., a cocoa tree	
Haptics/texture	Material was coarse or matte	
Tightly packed	Product was tightly packed (minimum headspace)	
Text	Information on environmental attributes of the packaging was present, e.g., general ecological benefits, appeals for waste treatment or reduction in greenhouse gas emissions	

Table A1. *Cont.*

Cue	Coding Rule: Coded as Signaling Environmental Friendliness if . . .	Example
Weight of the product relative to packaging weight	High product-to-packaging ratio	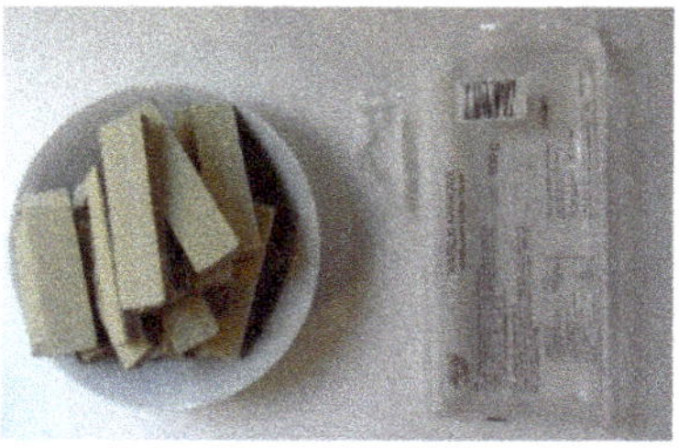
Number of packaging levels	No more than one level to open	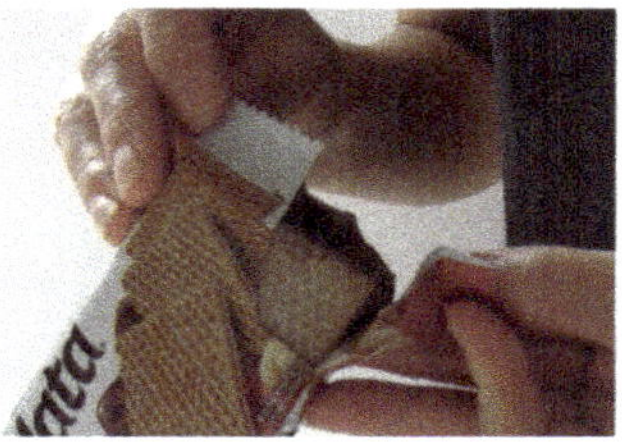
Number of packaging waste pieces	Only one waste piece	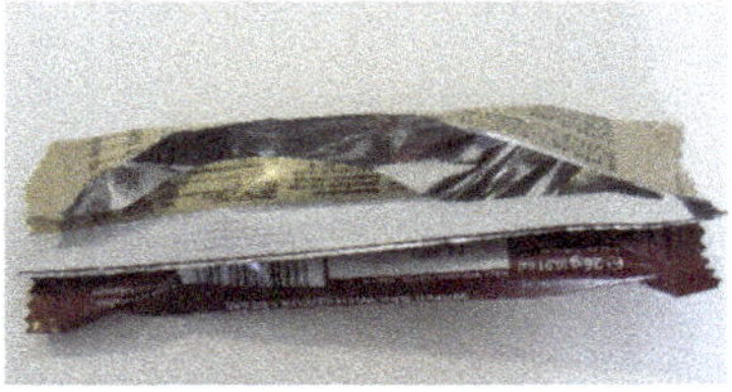

Appendix B

Figure A1. Pictures of all collected packages.

References

1. Yokokawa, N.; Kikuchi-Uehara, E.; Amasawa, E.; Sugiyama, H.; Hirao, M. Environmental analysis of packaging-derived changes in food production and consumer behavior. *J. Ind. Ecol.* **2019**, *23*, 1253–1263. [CrossRef]
2. Surucu-Balci, E.; Tuna, O. Investigating logistics-related food loss drivers: A study on fresh fruit and vegetable supply chain. *J. Clean. Prod.* **2021**, *318*, 128561. [CrossRef]
3. Food and Agriculture Organization of the United Nations. *The State of Food and Agriculture: Moving forward on Food Loss and Waste Reduction*; Food and Agriculture Organization of the United Nations: Rome, Italy, 2019. Available online: http://www.fao.org/3/ca6030en/ca6030en.pdf (accessed on 2 February 2022).
4. Food and Agriculture Organization of the United Nations. *Global Food Losses and Food Waste: Extent, Causes and Prevention. Study conducted for the International Congress SAVE FOOD! At Interpack2011 Düsseldorf, Germany*; Food and Agriculture Organization of the United Nations: Rome, Italy, 2011. Available online: http://www.fao.org/3/mb060e/mb060e.pdf (accessed on 2 February 2022).
5. EUROSTAT. Packaging Waste Statistics. Available online: https://ec.europa.eu/eurostat/statistics-explained/index.php?title=Packaging_waste_statistics (accessed on 21 September 2021).
6. European Commission. Waste Statistics—Waste Treatment. Available online: https://ec.europa.eu/eurostat/statistics-explained/index.php?title=Waste_statistics#Waste_treatment (accessed on 2 April 2022).
7. Directive (EU) 2019/904 of the European Parliament and of the Council of 5 June 2019 on the Reduction of the Impact of Certain Plastic Products on the Environment (Text with EEA Relevance). 2019. Available online: https://eur-lex.europa.eu/legal-content/DE/ALL/?uri=celex%3A32019L0904 (accessed on 2 April 2022).
8. Markets and Markets. Industrial Packaging Market Global Forecast to 2025 | MarketsandMarkets. Available online: https://www.marketsandmarkets.com/Market-Reports/industrial-packaging-market-10341323.html (accessed on 23 September 2021).
9. Testa, F.; Iovino, R.; Iraldo, F. The circular economy and consumer behaviour: The mediating role of information seeking in buying circular packaging. *Bus. Strat. Environ.* **2020**, *29*, 3435–3448. [CrossRef]
10. Herbes, C.; Beuthner, C.; Ramme, I. How green is your packaging—A comparative international study of cues consumers use to recognize environmentally friendly packaging. *Int. J. Consum. Stud.* **2020**, *44*, 258–271. [CrossRef]
11. Williams, H.; Lindström, A.; Trischler, J.; Wikström, F.; Rowe, Z. Avoiding food becoming waste in households—The role of packaging in consumers' practices across different food categories. *J. Clean. Prod.* **2020**, *265*, 121775. [CrossRef]
12. Lindh, H.; Olsson, A.; Williams, H. Consumer Perceptions of Food Packaging: Contributing to or Counteracting Environmentally Sustainable Development? *Packag. Technol. Sci.* **2016**, *29*, 3–23. [CrossRef]
13. Escursell, S.; Llorach-Massana, P.; Roncero, M.B. Sustainability in e-commerce packaging: A review. *J. Clean. Prod.* **2020**, *280*, 124314. [CrossRef]
14. Wandosell, G.; Parra-Meroño, M.C.; Alcayde, A.; Baños, R. Green Packaging from Consumer and Business Perspectives. *Sustainability* **2021**, *13*, 1356. [CrossRef]
15. Ketelsen, M.; Janssen, M.; Hamm, U. Consumers' response to environmentally-friendly food packaging—A systematic review. *J. Clean. Prod.* **2020**, *254*, 120123. [CrossRef]
16. Steenis, N.D.; van der Lans, I.A.; van Herpen, E.; van Trijp, H.C. Effects of sustainable design strategies on consumer preferences for redesigned packaging. *J. Clean. Prod.* **2018**, *205*, 854–865. [CrossRef]
17. Magnier, L.; Mugge, R.; Schoormans, J. Turning ocean garbage into products—Consumers' evaluations of products made of recycled ocean plastic. *J. Clean. Prod.* **2019**, *215*, 84–98. [CrossRef]
18. De Marchi, E.; Pigliafreddo, S.; Banterle, A.; Parolini, M.; Cavaliere, A. Plastic packaging goes sustainable: An analysis of consumer preferences for plastic water bottles. *Environ. Sci. Policy* **2020**, *114*, 305–311. [CrossRef]
19. Jerzyk, E. Design and Communication of Ecological Content on Sustainable Packaging in Young Consumers' Opinions. *J. Food Prod. Mark.* **2016**, *22*, 707–716. [CrossRef]
20. Boesen, S.; Bey, N.; Niero, M. Environmental sustainability of liquid food packaging: Is there a gap between Danish consumers' perception and learnings from life cycle assessment? *J. Clean. Prod.* **2019**, *210*, 1193–1206. [CrossRef]
21. Boz, Z.; Korhonen, V.; Koelsch Sand, C. Consumer Considerations for the Implementation of Sustainable Packaging: A Review. *Sustainability* **2020**, *12*, 2192. [CrossRef]
22. Steenis, N.D.; van Herpen, E.; van der Lans, I.A.; Ligthart, T.N.; van Trijp, H.C. Consumer response to packaging design: The role of packaging materials and graphics in sustainability perceptions and product evaluations. *J. Clean. Prod.* **2017**, *162*, 286–298. [CrossRef]
23. Herbes, C.; Beuthner, C.; Ramme, I. Consumer attitudes towards biobased packaging—A cross-cultural comparative study. *J. Clean. Prod.* **2018**, *194*, 203–218. [CrossRef]
24. Vila-Lopez, N.; Küster-Boluda, I. A bibliometric analysis on packaging research: Towards sustainable and healthy packages. *Br. Food J.* **2021**, *123*, 684–701. [CrossRef]
25. Nguyen, A.T.; Parker, L.; Brennan, L.; Lockrey, S. A consumer definition of eco-friendly packaging. *J. Clean. Prod.* **2020**, *252*, 119792. [CrossRef]
26. Neill, C.L.; Williams, R.B. Consumer preference for alternative milk packaging: The case of an inferred environmental attribute. *J. Agric. Appl. Econ.* **2016**, *48*, 241–256. [CrossRef]

27. Bhardwaj, A. A Study on Consumer Preference Towards Sustainability and Post-Use Consumption of Product Package in Chandigarh. *IUP J. Bus. Strategy* **2019**, *16*, 127–146.
28. Magnier, L.; Schoormans, J. Consumer reactions to sustainable packaging: The interplay of visual appearance, verbal claim and environmental concern. *J. Environ. Psychol.* **2015**, *44*, 53–62. [CrossRef]
29. Testa, F.; Di Iorio, V.; Cerri, J.; Pretner, G. Five shades of plastic in food: Which potentially circular packaging solutions are Italian consumers more sensitive to. *Resour. Conserv. Recycl.* **2021**, *173*, 105726. [CrossRef]
30. Orset, C.; Barret, N.; Lemaire, A. How consumers of plastic water bottles are responding to environmental policies? *Waste Manag.* **2017**, *61*, 13–27. [CrossRef]
31. Koenig-Lewis, N.; Palmer, A.; Dermody, J.; Urbye, A. Consumers' evaluations of ecological packaging—Rational and emotional approaches. *J. Environ. Psychol.* **2014**, *37*, 94–105. [CrossRef]
32. Koutsimanis, G.; Getter, K.; Behe, B.; Harte, J.; Almenar, E. Influences of packaging attributes on consumer purchase decisions for fresh produce. *Appetite* **2012**, *59*, 270–280. [CrossRef]
33. Sijtsema, S.J.; Onwezen, M.C.; Reinders, M.J.; Dagevos, H.; Partanen, A.; Meeusen, M. Consumer perception of bio-based products—An exploratory study in 5 European countries. *NJAS-Wagening. J. Life Sci.* **2016**, *77*, 61–69. [CrossRef]
34. Monnot, E.; Parguel, B.; Reniou, F. Consumer responses to elimination of overpackaging on private label products. *Int. J. Retail Distrib. Manag.* **2015**, *43*, 329–349. [CrossRef]
35. Lu, S.; Yang, L.; Liu, W.; Jia, L. User preference for electronic commerce overpackaging solutions: Implications for cleaner production. *J. Clean. Prod.* **2020**, *258*, 120936. [CrossRef]
36. Elgaaïed-Gambier, L. Who Buys Overpackaged Grocery Products and Why? Understanding Consumers' Reactions to Overpackaging in the Food Sector. *J. Bus. Ethics* **2016**, *135*, 683–698. [CrossRef]
37. Monnot, E.; Reniou, F.; Parguel, B.; Elgaaied-Gambier, L. "Thinking Outside the Packaging Box": Should Brands Consider Store Shelf Context When Eliminating Overpackaging? *J. Bus. Ethics* **2019**, *154*, 355–370. [CrossRef]
38. Elgaaied-Gambier, L.; Monnot, E.; Reniou, F. Using descriptive norm appeals effectively to promote green behavior. *J. Bus. Res.* **2018**, *82*, 179–191. [CrossRef]
39. Magnier, L.; Crié, D. Communicating packaging eco-friendliness. *Int. J. Retail Distrib. Manag.* **2015**, *43*, 350–366. [CrossRef]
40. Scott, L.; Vigar-Ellis, D. Consumer understanding, perceptions and behaviours with regard to environmentally friendly packaging in a developing nation. *Int. J. Consum. Stud.* **2014**, *38*, 642–649. [CrossRef]
41. Jeżewska-Zychowicz, M.; Jeznach, M. Consumers' behaviours related to packaging and their attitudes towards environment. *J. Agribus. Rural. Dev.* **2015**, *37*, 447–457.
42. Klaiman, K.; Ortega, D.L.; Garnache, C. Consumer preferences and demand for packaging material and recyclability. *Resour. Conserv. Recycl.* **2016**, *115*, 1–8. [CrossRef]
43. Rokka, J.; Uusitalo, L. Preference for green packaging in consumer product choices—Do consumers care? *Int. J. Consum. Stud.* **2008**, *32*, 516–525. [CrossRef]
44. Arboretti, R.; Bordignon, P. Consumer preferences in food packaging: CUB models and conjoint analysis. *Br. Food J.* **2016**, *118*, 527–540. [CrossRef]
45. Olson, J.C.; Jacoby, J. Cue utilization in the quality perception process. In *SV—Proceedings of the Third Annual Conference of the Association for Consumer Research*; Venkatesan, M., Ed.; Association for Consumer Research: Chicago, IL, USA, 1972.
46. Lee, E.J.; Bae, J.; Kim, K.H. The effect of environmental cues on the purchase intention of sustainable products. *J. Bus. Res.* **2020**, *120*, 425–433. [CrossRef]
47. Darby, M.R.; Karni, E. Free competition and the optimal amount of fraud. *J. Law Econ.* **1973**, *16*, 67–88. [CrossRef]
48. Zeng, T.; Durif, F.; Robinot, E. Can eco-design packaging reduce consumer food waste? an experimental study. *Technol. Forecast. Soc. Chang.* **2021**, *162*, 120342. [CrossRef]
49. Maesano, G.; Di Vita, G.; Chinnici, G.; Pappalardo, G.; D'Amico, M. The Role of Credence Attributes in Consumer Choices of Sustainable Fish Products: A Review. *Sustainability* **2020**, *12*, 10008. [CrossRef]
50. Rees, W.; Tremma, O.; Manning, L. Sustainability cues on packaging: The influence of recognition on purchasing behavior. *J. Clean. Prod.* **2019**, *235*, 841–853. [CrossRef]
51. Seo, S.; Ahn, H.-K.; Jeong, J.; Moon, J. Consumers' Attitude toward Sustainable Food Products: Ingredients vs. Packaging. *Sustainability* **2016**, *8*, 1073. [CrossRef]
52. Songa, G.; Slabbinck, H.; Vermeir, I.; Russo, V. How do implicit/explicit attitudes and emotional reactions to sustainable logo relate? A neurophysiological study. *Food Qual. Prefer.* **2019**, *71*, 485–496. [CrossRef]
53. Wensing, J.; Caputo, V.; Carraresi, L.; Bröring, S. The effects of green nudges on consumer valuation of bio-based plastic packaging. *Ecol. Econ.* **2020**, *178*, 106783. [CrossRef]
54. Spack, J.A.; Board, V.E.; Crighton, L.M.; Kostka, P.M.; Ivory, J.D. It's Easy Being Green: The Effects of Argument and Imagery on Consumer Responses to Green Product Packaging. *Environ. Commun.* **2012**, *6*, 441–458. [CrossRef]
55. Eberhart, A.K.; Naderer, G. Quantitative and qualitative insights into consumers' sustainable purchasing behaviour: A segmentation approach based on motives and heuristic cues. *J. Mark. Manag.* **2017**, *33*, 1149–1169. [CrossRef]
56. Karana, E. Characterization of 'natural' and 'high-quality' materials to improve perception of bio-plastics. *J. Clean. Prod.* **2012**, *37*, 316–325. [CrossRef]

57. Ertz, M.; François, J.; Durif, F. How consumers react to environmental information: An experimental study. *J. Int. Consum. Mark.* **2017**, *29*, 162–178. [CrossRef]
58. Deng, X.; Srinivasan, R. When Do Transparent Packages Increase (or Decrease) Food Consumption? *J. Mark.* **2013**, *77*, 104–117. [CrossRef]
59. Confectionery Production. ProSweets 2022 Promises Major Sustainable Packaging Focus—Confectionery Production. Available online: https://www.confectioneryproduction.com/news/37144/prosweets-2022-promises-major-sustainable-packaging-focus/ (accessed on 21 January 2022).
60. Dörnyei, K.R. Limited edition packaging: Objectives, implementations and related marketing mix decisions of a scarcity product tactic. *J. Consum. Mark.* **2020**, *37*, 617–627. [CrossRef]
61. Bauer, A.-S.; Leppik, K.; Galić, K.; Anestopoulos, I.; Panayiotidis, M.I.; Agriopoulou, S.; Milousi, M.; Uysal-Unalan, I.; Varzakas, T.; Krauter, V. Cereal and Confectionary Packaging: Background, Application and Shelf-Life Extension. *Foods* **2022**, *11*, 697. [CrossRef]
62. Bauer, A.-S.; Tacker, M.; Uysal-Unalan, I.; Cruz, R.M.S.; Varzakas, T.; Krauter, V. Recyclability and Redesign Challenges in Multilayer Flexible Food Packaging-A Review. *Foods* **2021**, *10*, 2702. [CrossRef]
63. Packaging Insights. Confectionery Packaging Gears up for Circular Economy with Recyclable Plastic, Fiber-Based and Compostable Innovation. Available online: https://www.packaginginsights.com/news/confectionery-packaging-gears-up-for-circular-economy-with-recyclable-plastic-fiber-based-and-compostable-innovation.html (accessed on 21 January 2022).
64. Dörnyei, K.R.; Lunardo, R. When limited edition packages backfire: The role of emotional value, typicality and need for uniqueness. *J. Bus. Res.* **2021**, *137*, 233–243. [CrossRef]
65. Chrysochou, P.; Festila, A. A content analysis of organic product package designs. *J. Consum. Mark.* **2019**, *36*, 441–448. [CrossRef]
66. Javier de la Fuente; Stephanie Gustafson; Colleen Twomey; Laura Bix. An Affordance-Based Methodology for Package Design. *Packag. Technol. Sci.* **2015**, *28*, 157–171. [CrossRef]
67. Mayring, P. Qualitative content analysis. In *A Companion to Qualitative Research*; Flick, U., Kardorff, E., von Steinke, I., Eds.; SAGE: London, UK, 2004; pp. 266–269, ISBN 0761973745.
68. Krippendorff, K. *Content Analysis: An Introduction to Its Methodology*; Sage Publications: London, UK, 2018; ISBN 1506395678.
69. Pauer, E.; Wohner, B.; Heinrich, V.; Tacker, M. Assessing the Environmental Sustainability of Food Packaging: An Extended Life Cycle Assessment including Packaging-Related Food Losses and Waste and Circularity Assessment. *Sustainability* **2019**, *11*, 925. [CrossRef]
70. Otto, S.; Strenger, M.; Maier-Nöth, A.; Schmid, M. Food packaging and sustainability—Consumer perception vs. correlated scientific facts: A review. *J. Clean. Prod.* **2021**, *298*, 126733. [CrossRef]
71. Singh, P.; Wani, A.A.; Langowski, H.-C. (Eds.) *Food Packaging Materials: Testing & Quality Assurance*; CRC Press Taylor & Francis Group: Boca Raton, FL, USA; London, UK; New York, NY, USA, 2017; ISBN 9781466559943.
72. Robertson, G.L. *Food Packaging: Principles and Practice*, 3rd ed.; CRC Press: Boca Raton, FL, USA, 2013; ISBN 9781439862414.
73. Robertson, G. (Ed.) *Food Packaging and Shelf Life: A Practical Guide*; CRC Press: Boca Raton, FL, USA; London, UK; New York, NY, USA, 2009; ISBN 9781420078459.
74. Soroka, W. *Fundamentals of Packaging Technology*, 5th ed.; Institute of Packaging Professional: Herndon, VI, USA, 2014; ISBN 0615709346.
75. Wolf, B. *Confectionery and Sugar-Based Foods. Reference Module in Food Science*; Elsevier: Amsterdam, The Netherlands, 2016; ISBN 978-0-08-100596-5.
76. Caruana, R.; Glozer, S.; Eckhardt, G.M. 'Alternative Hedonism': Exploring the Role of Pleasure in Moral Markets. *J. Bus. Ethics* **2019**, *166*, 143–158. [CrossRef]
77. Cervellon, M.-C.; Shammas, L. The Value of Sustainable Luxury in Mature Markets. *J. Corp. Citizsh.* **2013**, *2013*, 90–101. [CrossRef]
78. Soper, K. Alternative hedonism, cultural theory and the role of aesthetic revisioning. *Cult. Stud.* **2008**, *22*, 567–587. [CrossRef]
79. Soper, K. Towards a sustainable flourishing: Ethical consumption and the politics of prosperity. In *Ethics and Morality in Consumption*; Routledge: London, UK, 2016; pp. 43–59, ISBN 1315764326.
80. Bansal, P.; Roth, K. Why companies go green: A model of ecological responsiveness. *Acad. Manag. J.* **2000**, *43*, 717–736.
81. Maziriri, E.T. Green packaging and green advertising as precursors of competitive advantage and business performance among manufacturing small and medium enterprises in South Africa. *Cogent Bus. Manag.* **2020**, *7*, 1719586. [CrossRef]
82. Jiménez-Guerrero, J.F.; Gázquez-Abad, J.C.; Ceballos-Santamaría, G. Innovation in eco-packaging in private labels. *Innovation* **2015**, *17*, 81–90. [CrossRef]
83. European Commission. Initiative on Substantiating Green Claims. Available online: https://ec.europa.eu/environment/eussd/smgp/initiative_on_green_claims.htm (accessed on 19 February 2022).
84. Talati, Z.; Egnell, M.; Hercberg, S.; Julia, C.; Pettigrew, S. Food Choice Under Five Front-of-Package Nutrition Label Conditions: An Experimental Study Across 12 Countries. *Am. J. Public Health* **2019**, *109*, 1770–1775. [CrossRef]
85. Julia, C.; Hercberg, S. Big Food's Opposition to the French Nutri-Score Front-of-Pack Labeling Warrants a Global Reaction. *Am. J. Public Health* **2018**, *108*, 318–320. [CrossRef]
86. Vandevijvere, S. Uptake of Nutri-Score during the first year of implementation in Belgium. *Arch. Public Health* **2020**, *78*, 107. [CrossRef]

87. Schorb, F. Ernährung als Gegenstand politischer Kommunikation. In *Ernährungskommunikation: Interdisziplinäre Perspektiven—Theorien—Methoden*; Godemann, J., Bartelmeß, T., Eds.; Springer VS: Wiesbaden/Heidelberg, Germany, 2021; pp. 345–359, ISBN 978-3-658-27313-2.
88. Orset, C.; Monnier, M. How do lobbies and NGOs try to influence dietary behaviour? *Rev. Agric. Food Environ. Stud.* **2020**, *101*, 47–66. [CrossRef]
89. European Commission. Communication from the Commission to the European Parliament, the Council, the European Economic and Social Committee and the Committee of the Regions a European Strategy for Plastics in a Circular Economy. Available online: https://eur-lex.europa.eu/legal-content/EN/TXT/?qid=1516265440535&uri=COM:2018:28:FIN (accessed on 2 April 2022).

 *foods*

MDPI

Review

Cereal and Confectionary Packaging: Assessment of Sustainability and Environmental Impact with a Special Focus on Greenhouse Gas Emissions

Victoria Krauter [1,*], Anna-Sophia Bauer [1], Maria Milousi [2], Krisztina Rita Dörnyei [3], Greg Ganczewski [4], Kärt Leppik [5,6], Jan Krepil [1] and Theodoros Varzakas [7]

1 Packaging and Resource Management, Department Applied Life Sciences, FH Campus Wien, University of Applied Sciences, 1030 Vienna, Austria; anna-sophia.bauer@fh-campuswien.ac.at (A.-S.B.); jan.krepil@fh-campuswien.ac.at (J.K.)
2 Department of Chemical Engineering, University of Western Macedonia, 50100 Kozani, Greece; mmilousi@uowm.gr
3 Institute of Marketing, Corvinus University of Budapest, 1093 Budapest, Hungary; krisztina.dornyei@uni-corvinus.hu
4 Management in Networked and Digital Societies (MINDS) Department, Kozminski University, 03-301 Warsaw, Poland; ganczewski@gmail.com
5 Center of Food and Fermentation Technologies, 12618 Tallinn, Estonia; kart@tftak.eu
6 Department of Chemistry and Biotechnology, School of Science, Tallinn University of Technology, 19086 Tallinn, Estonia
7 Department of Food Science and Technology, University of Peloponnese, 24100 Kalamata, Greece; theovarzakas@yahoo.gr
* Correspondence: victoria.krauter@fh-campuswien.ac.at; Tel.: +43-(0)-1-606-6877-3592

Citation: Krauter, V.; Bauer, A.-S.; Milousi, M.; Dörnyei, K.R.; Ganczewski, G.; Leppik, K.; Krepil, J.; Varzakas, T. Cereal and Confectionary Packaging: Assessment of Sustainability and Environmental Impact with a Special Focus on Greenhouse Gas Emissions. *Foods* **2022**, *11*, 1347. https://doi.org/10.3390/foods11091347

Academic Editor: Ana Teresa Sanches-Silva

Received: 13 April 2022
Accepted: 3 May 2022
Published: 6 May 2022

Publisher's Note: MDPI stays neutral with regard to jurisdictional claims in published maps and institutional affiliations.

Abstract: The usefulness of food packaging is often questioned in the public debate about (ecological) sustainability. While worldwide packaging-related CO_2 emissions are accountable for approximately 5% of emissions, specific packaging solutions can reach significantly higher values depending on use case and product group. Unlike other groups, greenhouse gas (GHG) emissions and life cycle assessment (LCA) of cereal and confectionary products have not been the focus of comprehensive reviews so far. Consequently, the present review first contextualizes packaging, sustainability and related LCA methods and then depicts how cereal and confectionary packaging has been presented in different LCA studies. The results reveal that only a few studies sufficiently include (primary, secondary and tertiary) packaging in LCAs and when they do, the focus is mainly on the direct (e.g., material used) rather than indirect environmental impacts (e.g., food losses and waste) of the like. In addition, it is shown that the packaging of cereals and confectionary contributes on average 9.18% to GHG emissions of the entire food packaging system. Finally, recommendations on how to improve packaging sustainability, how to better include packaging in LCAs and how to reflect this in management-related activities are displayed.

Keywords: food; packaging; cereals; confectionary; snacks; life cycle assessment; LCA; environmental impact; CO_2 footprint; food losses and food waste

1. Introduction

The sustainability of food and, in particular, its packaging continues to be at the center of public and political debate. In order to make objective and knowledge-based decisions, it is of utmost importance to understand the requirements of a food product on its packaging on the one hand and to be able to select the optimal packaging solution for the respective purpose on the other hand. While the former has already been covered in the review paper "Cereal and Confectionary Packaging: Background, Application and Shelf-Life Extension" [1], the present review aims to address the important issue of sustainability and assessment thereof.

Recently, it has been shown and further substantiated by Crippa et al. that food systems are accountable for a major share, namely 34%, of global anthropogenic greenhouse gas (GHG) emissions (data representing 2015). The authors also showed that this percentage predominantly originates from agriculture and land-use and land-use change activities (71%). The remaining fraction (29%) represents activities along the food supply chain such as processing, distribution (e.g., packaging, retail, transport), consumption and corresponding end-of-life scenarios. Being of increased importance and use, packaging resulted in a 5.4% share, which was calculated considering relevant materials and industries (e.g., pulp and paper, aluminum, metal, glass). This value is slightly above the shares for transportation (4.8%) and the cold chain (5%) [2].

The seemingly relatively small contribution of packaging to total GHG emissions in relation to food products against the background of current discussions about packaging and sustainability has also been shown by Poore and Nemecek [3]. The authors likewise calculated a 5% share of packaging but also showed that the results for product groups differed greatly from one another. For instance, alcoholic beverages, such as beer and wine, exhibited packaging-related emissions of around 40% (with glass packaging as the main driving impact factor), while fruit and vegetables showed packaging-related emissions of around 10 to 20% [3]. This difference in the impact ratio between packaging and food for different products has also been shown by other authors and studies [4–7]. For example, Verghese et al. stated that packaging of meat, fish and eggs accounts for 2% of GHG emissions, while packaging for dairy as well as fruits, vegetables and nuts account for 10 and 12%, respectively [6]. Heller et al. underlined this by visualizing that resource- and emission-intensive food products, such as meat or milk, tend to have a high food-to-packaging ratio, while less resource- and emission-intensive food products, such as leafy greens, show a small ratio [7].

Especially for food products with a (very) high impact, these results point out the importance of the protective function of packaging [6–10]. Optimizing and sometimes increasing packaging can reduce food losses and waste along the food supply chain while at the same time reducing the overall environmental impact [11]. For food products with a low impact, on the other hand, more precise consideration must be given to which packaging (e.g., material) should be used and which trade-offs must be considered [10–14]. Therefore, the sustainability (including ecological, economic and social dimensions) of product packaging systems is the subject of current research and finds more and more attention in policies and legislation [15–17].

Due to the great importance of high-impact foods (e.g., products of animal origin such as meat and milk [18]) and foods with high food losses and waste (e.g., fruits and vegetables), publications on these topics are a priority in the scientific literature. This is reflected by different studies and reviews [3,18–22]. However, to the author's best knowledge, no comprehensive work taking into account the important group of cereal and confectionary products [23–25], their packaging and related GHG emissions exists. This shortcoming is also underlined by different authors [26–32]. Against this background, the aim of the present review is to:

- Contextualize packaging and sustainability as well as sustainability assessment methods;
- Display and discuss how and to what extent food packaging is included in existing life cycle assessments (LCAs) in the cereals and confectionary sector;
- Point out the environmental impact of cereal and confectionary packaging in relation to the food product with a special focus on GHG emissions;
- Highlight improvement strategies to optimize (cereal and confectionary) packaging systems as well as LCA of the same.

This provides a valuable basis for decision makers as well as practitioners in research, development and innovation to take further steps towards sustainable food packaging.

2. Packaging and Sustainability

2.1. Sustainable Packaging

2.1.1. Definition

Despite its common usage, the term "sustainable packaging" is defined and utilized in different ways by various stakeholders along the food supply chain and beyond [33]. Accordingly, several approaches, frameworks and methodologies with differing foci, principles, criteria and connected indicators can be found in the relevant literature [34]. These, amongst others, encompass legal texts on packaging and packaging waste [35,36], guidelines for producers and retail focusing on specific topics such as design for recycling [37–41], as well as more holistic packaging sustainability frameworks [42–45].

A condensed but comprehensive framework is that of the Sustainable Packaging Alliance (Australia) [42]. This so-called Packaging Sustainability Framework defines a total of four principles, namely that sustainable packaging must be (i) effective, (ii) efficient, (iii) cyclic and (iv) safe. In this context, "effective" means that the respective packaging is fit for purpose and fulfils its essential functions (e.g., containment, protection, communication, convenience [46–48]) with as little effort as possible. "Efficient", on the other hand, refers to packaging that minimizes resource consumption (e.g., materials) as well as emissions (e.g., CO_2) along its life cycle and "cyclic" emphasizes that it is necessary to keep resources in the biological (e.g., bio-based or biodegradable materials) or technical (e.g., recycling, use of recycled materials) cycle. Furthermore, "safe" focuses on packaging that does not pose a risk to people (e.g., migration of harmful substances from the packaging material to the food product) or the environment (e.g., pollution) along its life cycle [42,43,45,49].

It is important to point out that the above four principles are closely interrelated and that (increased) efforts in one area can lead to positive or negative changes in another [43]. The latter case and corresponding trade-offs are represented, for example, by the use of multilayer flexible food packaging. While this often offers a high level of product protection (e.g., barrier) with low material input and correspondingly low emissions (e.g., CO_2), the combination of different materials (e.g., different plastics, aluminium, paper) makes it difficult to recycle them [50]. Another possible trade-off is the reduction or minimization of packaging. While this is desirable in principle, underpackaging can lead to undermining the effectiveness of a packaging system, resulting in increased food losses and/or waste and corresponding environmental impacts. Overpackaging, on the other hand, also leads to elevated environmental impacts due to the excess material used [43].

2.1.2. Development

Taking this into account, finding the optimum point (as little as possible, as much as necessary) with balancing the above-mentioned principles is of the utmost interest in a packaging (re)-design process. Since "THE" sustainable packaging is not a specific, existing product that can be applied to any given (food) product, but rather a system that must be constantly adapted to the changing needs of, for example, the (food) product, the value chain, consumers and legal requirements, the resulting "sustainable" packaging solutions can be as diverse as the initial factors [43].

Consequently, developing a successful packaging solution not only at the primary but also at the secondary and tertiary packaging level [51] is a complex and critical undertaking that requires dedication, investment and, most importantly, a holistic and collaborative approach [43,48,52]. While holistic refers to life cycle thinking and assessment, collaborative refers to pro-active and dedicated action of not only single actors but connected and communicating companies, supply chains, science and research as well as stakeholders such as governments or consumers. This allows the development of (eco)efficient and effective solutions that enable the transition from a linear to a circular economy and show benefits in multiple dimensions (ecologic, economic, social) [43,52–56].

To evaluate or compare different developed packaging solutions with regard to ecological, economic and social aspects, different criteria, indicators, metrics and evaluation methods can be used. While economic and social effects can be assessed using, for instance,

Life Cycle Costing (LCC) [57–61] and Social Life Cycle Assessment [62–65], ecological effects are usually assessed using a (full) Life Cycle Assessment (LCA) ([66–69], simplified (or streamlined) LCA, non-LCA tools or scorecards (see also Figure 1) [70–72].

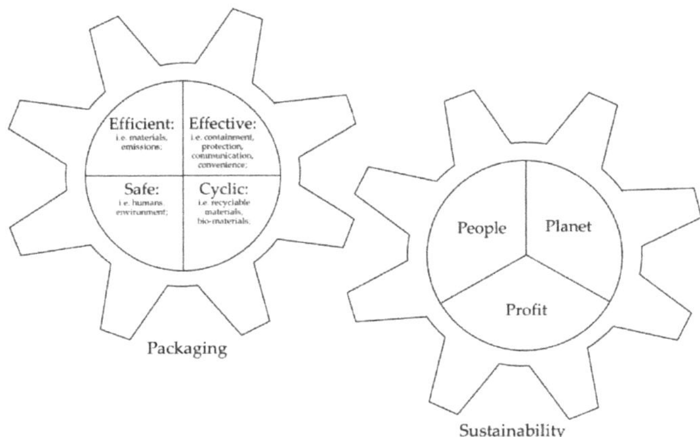

Figure 1. Principles of sustainable packaging and their impact on ecological, economic and social sustainability. Graphic based on [42,43,46].

2.1.3. Challenges

Sustainable packaging development frequently involves high production costs, long development time and technical difficulties [43,54]. Therefore, many sustainable packaging solutions are not implemented without significant sales increase or cost reduction. Findings also show that sustainable packaging ambitions often stay on the firm's strategic level because companies might prioritize a product's market potential and a limitation of commercial risks over sustainability considerations on an operational level. As a result, sustainable advances in packaging development frequently remain limited [73].

Companies' sustainability commitment is also reduced if such packaging solutions' commercial success is questionable or if it does not positively influence consumer behavior [53]. Unfortunately, from the consumer perspective, sustainable packaging does not always refer to a truly sustainable solution but to a specific design, which evokes explicitly or implicitly the perception of sustainability via its structure and its visual and informational cues [74,75]. Moreover, consumer perception of sustainable packaging is controversial: some consumers have a generally positive attitude toward sustainable packaging [76,77], and others regard such packaging as an environmental villain due to the way the media have recently communicated about packages. However, in general, they have limited awareness, recognition and knowledge of the different sustainable functions (such as labels, materials, disposal processes, and manufacturing technologies) of such packaging solutions [78–80] and often focus their environmental concerns solely on the packaging's end-of-life [56]. They also associate sustainable packages with certain risks (lower perceived quality, lower functionality, less attractiveness, perceived contamination), which leads to lower perceived functionality and lower willingness to purchase [76,81]. Consumers can also be easily deceived by packaging communication [82], and some even perceive sustainable claims as greenwashing, especially when these claims are not in line with their subjective sustainable packaging expectations [80,83]. It is, therefore, important to study and include consumer insights in sustainability packaging analysis and also include other necessary steps to avoid failures [43].

2.2. Life Cycle Assessment

One of the first LCAs focusing on food packaging was initiated by the Midwest Research Institute (MRI) for the Coca-Cola Company in 1969 [70,84–86]. In 1974, the same institute conducted a follow up of this study for the United States Environmental Protection Agency [87]. Similarly, Unilever has performed several LCA studies for various product groups such as margarine and ice cream in the late 1980s. Since then, and in the context of the need for more sustainable products and processes, numerous further studies have been conducted in this research field [85–95]. Building on this, LCA has also increasingly found its way into more than just industrial decision-making [96]. For instance, a comparative LCA study on different beverage packaging formed the basis of the political decision of the German Federal Ministry for the Environment with regard to the German deposit system on disposable packaging (single-use deposit) in the early 2000s. However, since conditions (e.g., legal framework, economy, inventory data) are not static but constantly adapting, the study was repeated recently and is again influencing policy-making [97,98]. Being just one example, it is expected that LCA will be more and more applied to improve policy- and decision-making in the future (e.g., waste management policies) since it offers transparent and valuable information about the actual sustainability of a product or process. However, a sound methodology and expert knowledge in conducting such analyses is a prerequisite to achieving meaningful output [99–101].

A full LCA should consider the following life cycle stages: raw material extraction and preprocessing (cradle), transportation of processed materials to the manufacturing site, production of components, assembly of the system, transportation to market (gate), use phase and end-of-life with transportations of the used equipment to the intended waste treatment plant, e.g., landfill (grave) or recycling/material recovery (back to cradle). An LCA study can be: (i) partial, referring to some phases of the product's lifecycle, i.e., cradle-to-gate, (ii) semi-complete, including landfilling or partial recycling, i.e., cradle-to-grave or (iii) complete, employing all life time phases and including material upscaling aspects as described in the circular economy principles, i.e., cradle-to-cradle [34]. The Product Environmental Footprint (PEF) is a multi-criteria method for modelling the potential environmental performance of a product, and it can easily be inferred through the LCA results, especially in cradle-grave or cradle-cradle approaches [102,103].

According to the guidance provided by the International Standardization Organization (ISO) in ISO 14040 and ISO 14044, an LCA study is generally carried out by iterating four distinct phases [66,67]:

In the first step, i.e., Goal and Scope, the objectives of the study are defined to clarify the intended application and the reasons for the study, including the target audience. Scope, on the other hand, describes the product system, as well as the functional unit (FU) and the system's boundaries. The selection of the FU is a basis for comparing similar products. Thus, a typical FU relates to the overall product function rather than focusing on a particular physical property, while it is normally time-bounded and can correlate the expected duration of use and desired quality under certain circumstances. The meaningful selection and definition of system boundaries is a crucial task as it determines the overall type of the LCA, i.e., whether it is a cradle-to-gate, a cradle-to-grave or a cradle-to-cradle approach [104].

During the second step, i.e., Life Cycle Inventory analysis (LCI), a comprehensive inventory of energy, materials and environmental inputs-outputs is created, identifying and quantifying all related data at every stage of the life cycle. The collection of data and determination of total emissions and resource use take place alongside a detailed definition of entailed production processes. All collected data are scaled based on the preset functional unit for the studied system. Lack of data availability and quality is a typical drawback and can usually refer to studies related to non-standardized procedures. Other inhibiting factors are geographic variations regarding the quality of raw materials and energy sources, production methods and relevant environmental impacts [105].

The next and third step, i.e., Life Cycle Impact Assessment (LCIA), is the phase of an LCA with particular respect to sustainability assessment. During the impact assessment, the potential environmental impacts associated with identified inputs and outputs are categorized into different categories. During LCIA, emissions and resource extractions are translated into a limited number of environmental impact scores by means of so-called characterization factors. There are two mainstream ways to derive these factors, i.e., at the midpoint and at the endpoint level. Midpoint indicators focus on single environmental problems, for example, climate change or acidification, while endpoint indicators present environmental impacts on three higher aggregation levels, i.e., (i) effect on human health, (ii) biodiversity and (iii) resource scarcity [106].

In the fourth step, i.e., Interpretation, the results of the inventory analysis and the impact assessment are interpreted and combined in order to make a more informed decision. During this phase, a comparison of the results with previous studies is made in order to determine whether they are aligned with the literature. Furthermore, a sensitivity analysis can be performed to validate the consistency of the findings. ISO standards provide a general framework of an iterative nature. Thus, if the outcomes of the impact assessment are incomplete for drawing conclusions, then the previous LCA steps must be repeated until the final results support the initial goals of the study [107].

As LCA is by default a holistic method that accounts for multiple environmental impact categories, carbon footprint analysis evaluates the GHG emissions generated by a product, activity, or process that contributes to global warming, and it is a subset of a complete LCA. Thus, it is always based on international standards such as ISO 14040/14044, ISO 14067, PAS 2050, and the GHG Product Life Cycle Standard [66,67,108,109].

One important aspect of applying LCA in food packaging is to quantify the inherent direct and indirect effects in order to assess the environmental sustainability of the sector. Direct effects of packaging include impacts from the production and end-of-life of the related materials. Additionally, indirect effects derive from life cycle losses and waste that occur in different phases of the food supply chain [110].

3. Sustainability of Cereal and Confectionary Packaging

3.1. Literature Analysis

To display and discuss how and to which extent packaging is present in existing LCA studies in the cereal and confectionary sector and to point out the environmental impact (focus on GHG emissions) of the packaging in relation to the respective food product, a literature search in different databases was conducted, similar to Molina-Besch et al. [111]. Firstly, and for the identification of relevant LCA studies, the keywords "Life Cycle Assessment" and "Carbon Footprint" were used. Secondly, to identify relevant food products, keywords given in the guidance document in Part E of Annex II of the regulation (EC) No 1333/2008 on food additives were used. (Sub)categories considered were: confectionary products (cocoa and chocolate products, other confectionaries including breath-freshening micro-sweets), cereals and cereal products (whole, broken or flaked grain, flours and other milled products, breakfast cereals, pasta, noodles, batters, pre-cooked or processed cereals), bakery wares (bread and rolls, fine bakery wares) as well as ready-to-eat savories and snacks (potato-, cereal-, flour- or starch-based snacks, processed nuts) [112]. The first keywords were combined with "or". The second keywords were individually added using "and". Articles written in English and published since 2009 were considered for review. Of these, relevant studies including food, packaging and related LCA results were analyzed in detail. Where results (on packaging) were included in graphics (e.g., bar chart) but not in numeric form, the online tool Web-Plot Digitizer was used to extract the data [113]. Further, for each study, the percentage of packaging-related GHG emissions was taken from the results or extracted (calculated) where necessary.

Based on the available data set, commonalities and differences between the studies were investigated in a multi-step approach based on ISO 14040 and 14044: (i) goal and scope, (ii) life cycle inventory, (iii) life cycle impact assessment and (iv) interpretation [66,67]. This stands in contrast to Molina-Besch et al., who focused primarily on (i) and (iv) [111]. Since the present review not only aims to highlight how packaging is included in the studies but also to point out improvement opportunities for packaging and assessment, the authors also focused on LCA methodology, represented by (ii) and (iii).

As it is well known that the direct comparison of results from different LCA studies (e.g., due to different goals and scope, data used, cut-offs) is difficult [111,114,115], the present study aims at rather comparing approaches, magnitudes and ranges than exact values.

3.2. Results

3.2.1. Goal and Scope

Focus

In total, 28 LCA studies covering 108 products in the categories of confectionary, cereals and cereal products, bakery wares and ready-to-eat savories and snacks fulfilled the above-given criteria (see also Table 1). Within these studies, products from the confectionary category (total 42%) and especially the sub-category of cocoa and chocolate products were assessed most frequently (38%). On the contrary, the sub-category of other confectionaries, including breath-freshening micro-sweets, only resulted in a low number of entries (4%). Products covered were, for example, jelly and foam sweets as well as sugar and milk-based confectionary. This focus on cocoa and chocolate products may be due to the high economic relevance of cocoa [23,24] and is well in line with, for example, the findings of Miah et al. [26], who stated that diverse confectionary products are underrepresented in LCA studies and that chocolate products dominate the literature body.

Table 1. Reviewed cereal and confectionary life cycle assessment (LCA) studies (*n* = 28).

Category	Sub-Category	LCAs * *n* = 28		Products *n* = 108		Greenhouse Gas Emissions		
		n	%	*n*	%	Food-Packaging System [kg CO_2eq]	Packaging [kg CO_2eq]	Packaging (%)
Confectionary	Cocoa and chocolate products	9	32	41	38	3.28	0.25	9.86
	Other confectionary including breath-freshening micro-sweets	2	7	4	4	2.80	0.16	4.68
Cereals and cereal products	Whole, broken or flaked grain	2	7	9	8	12.53	0.14	1.25
	Flours and other milled products and starches	2	7	3	3	0.65	0.04	5.30
	Breakfast cereals	2	7	4	4	0.87	0.15	19.68
	Pasta	4	14	10	9	1.33	0.10	7.24
Bakery wares	Bread and rolls	5	18	20	19	1.03	0.04	4.37
	Fine bakery wares	3	11	12	11	1.93	0.04	11.22
Ready-to-eat savories and snacks	Potato-, cereal-, flour- or starch-based snacks	1	4	1	1	0.43	0.04	8.14
	Processed nuts	1	4	4	4	1.87	0.33	20.10
	Overall (average)					2.67	0.13	9.18

* Some LCA studies covered more than one (sub)category. Therefore, given numbers do not sum up to *n* = 28 or 100%.

A total of 24% of the products were located in the area of cereal and cereal products. On the forefront in the sub-category of whole, broken or flaked grain (8%) was rice. For the sub-category of flours and other milled products and starches (3%), oat, potato and wheat were represented. Further, the sub-category of breakfast cereals (4%) was covered by one known brand's products as well as porridge. The sub-category of pasta (9%) included different products made from different raw materials. Interestingly, the category of bakery wares (30%) showed an elevated number of packaged products in the sub-categories of bread and rolls (e.g., (sliced) bread) (19%) as well as fine bakery wares (e.g., biscuits, cakes) (11%).

Last but not least, the category ready-to-eat savories and snacks only displayed one product example (5%), namely crisps, for the sub-category of potato-, cereal-, flour- or starch-based snacks (1%) and some examples for the sub-category of processed nuts (e.g., pistachio) (4%).

Aim

Analyzing the studies with regard to packaging, it quickly becomes clear that the focus (overall goal and scope) is mainly on the food products themselves. Molina-Besch et al. [111] name these types of studies *food LCAs*, whereas studies with a focus on the impact of the packaging system are called *packaging LCAs*. In total, 7 out of 28 studies explicitly mentioned packaging in one form or another in their aim. While some studies seem to mention packaging in passing, others go more into detail. For example, Boakye-Yiadom et al. [116] mentioned "environmental impacts associated with the production of a packaged chocolate", Cimini et al. [117] included "pasta in 0.5 kg polypropylene (PP) bags" in their aim, and Volpe et al. [118] focused on "bags of" nuts. Büsser and Jungbluth [119], on the other hand, aimed at analyzing "the environmental performance of packaging with respect to its function within the life cycle of chocolate" and Espinoza-Orias et al. [120] included " . . . the influence on the carbon footprint of several parameters . . . including . . . type of packaging (plastic and paper bags) . . . ". Further, with an explicit focus not only on the direct but also indirect effects of packaging, Svanes et al. [121] aimed to " . . . establish environmental hotspots; to examine the role of . . . packaging . . . and to identify potential measures to reduce this wastage", and Williams and Wikström [11] aimed to " . . . analyze the potential of decreasing environmental impact of five food items . . . through the development of packaging that reduces food losses in the consumer phase". These studies are, however, exceptions and mirror the findings of Molina-Besch et al. [111], who likewise, but for a wider product range, found that packaging is currently insufficiently considered in LCAs.

Functional Unit

The strong focus on the food product itself is also reflected by the functional units given; slightly more than half of the authors do not even name packaging in this regard [27,30,118,120–131]. Those who do [11,26,28,29,31,32,116,117,119,132–134] almost exclusively (with the exception of (Nilsson et al. [132]) give the functional unit as "one kilogram of product in the respective packaging". This corresponds to a formulation as laid down in the Product Category Rules (PCR) rules of the International Environmental Product Declaration (EPD) system [31,135,136], as well as other sources [104,137].

In this context, EPDs, as such, which are based on LCAs, should also be discussed in a short excurse. According to the definition of ISO 14025, these are so-called Type III environmental declarations. Specifically, they are independently verified and registered documents that make the environmental impact of products transparent and comparable over their entire life cycle. Type I and II stand for third-party and self-declared eco-labels, respectively [138,139]. Interestingly, the EPD Library (search criteria: product category food & beverages; PCR bakery products) already contains more than 100 EPDs [140]. These are highly relevant for the present review with regard to the categories of cereals and cereal products as well as bakery wares, but outside the scope (e.g., scientific literature) defined in chapter 3.1. Moreover, the EPDs are structured very similarly to each other. Accordingly, these will not be analyzed in detail in the coming chapters but will be used for comparison and discussion where appropriate.

System/Scope

While a considerable amount of the studies reviewed followed a cradle-to-gate or a gate-to-gate approach [116,118,119,122,123,125,127,131–133,141], the majority considered the product life cycle in a cradle-to-grave approach [11,26–32,117,120,121,124,126,128–130,134]. The latter is a prerequisite for assessing not only the direct environmental effects of packaging (impacts caused by production and end-of-life) but also the indirect environmental effects of the same (influence on, e.g., food waste and transport efficiency), a research field gaining more and more importance due to the high environmental impacts of food systems and the valuable role of packaging in avoiding or reducing food losses and waste [19,43,111,142,143]. The packaging-relevant direct and indirect effects in this context are: primary packaging (direct), secondary and tertiary packaging (direct), transport from producer to retail (indirect), food waste in transport, distribution and retail (indirect), food transport, storage and preparation by households (indirect), food waste in households (indirect), packaging end-of-life (direct) and food waste end-of-life (indirect) [111].

On closer examination of the studies with a cradle-to-grave approach, it becomes apparent that some did not include all key LCA steps necessary to evaluate the indirect effects of packaging at the point of sale or consumption. Transport (from producer to retail as well as to households), however, was covered in almost all the studies in the form of distance travelled. Factors influenced by the packaging, such as transport efficiency due to efficient and/or lighter packaging, on the other hand, were not in the foreground [11,26–32,117,120,121,124,126,128–130,134]. Regarding food losses and waste during transport, distribution and retail, Miah et al. [26], for example, gave information on the percentage of waste generated at the different life cycle stages for confectionary. Likewise, Sieti et al. [130] did the same for breakfast cereals. Cimini et al. [117] even named package breakage as a reason for waste during distribution. Additionally, Svanes et al. [121] explicitly calculated the direct and indirect effects of waste at the production, retail and household level for bread and rolls. Further, information on food waste was included by Espinoza-Orias et al. [120] for bread and rolls, Konstantas [29] for cakes, Miah et al. [26] for confectionary, Cimini et al. [117] for pasta and Sieti et al. [130] for breakfast cereals, making this the most-noticed form of indirect effects. Direct connection to the (packaging-related) cause was again not in focus. Data were rather derived from reports instead of actual conducted studies for the respective food product under consideration [120,144].

In the reviewed studies, considerations of end-of-life (e.g., recycling, landfill, incineration) were varied. Some studies excluded the end-of-life phase altogether [116,122,123,125, 127,128,131,133]. Some cited similar studies that excluded end-of-life due to many different scenarios that needed to be considered, making it difficult for standardization and comparison [116]. The remaining studies included end-of-life in some respect, either as end-of-life of packed food and/or end-of-life of the actual packaging solutions (often referenced as simply post-consumer waste, but also as the full packaging system, including primary, secondary and transport packaging). Though the end-of-life of packaging solutions was not often regarded as very significant in the results (as compared to other life cycle phases), commendably, some studies took a long and detailed look at the issue [117,120,121,129,130,132]. The inclusion and study of end-of-life scenarios are currently important, as with novel emerging products and materials, established waste management systems are continuously presented with new challenges to protect humans and the environment [145].

In terms of system boundaries, the picture is similar for EPDs. In principle, an attempt is made to cover the entire life cycle in three successive steps, namely upstream (e.g., raw material production, packaging and auxiliary material production), core (e.g., food production) and downstream (e.g., distribution up to shelf, primary packaging end-of-life). While most EPDs are limited to the named examples (e.g., EPD on crispbread [146]), others go beyond and include, for instance, domestic food losses or food preparation (e.g., cooking) (e.g., EPD on pasta [147]).

3.2.2. Life Cycle Inventory

Table 2 lists the LCA studies reviewed and gives a comprehensive overview of the product (sub)categories, product names, the given packaging-related information, as well as the percentage of packaging-related GHG emissions.

Table 2. Reviewed cereal and confectionary life cycle assessment (LCA) studies: information on packaging and its percentage share of total greenhouse gas (GHG) emissions.

Category	Sub-Category	Product	Primary Packaging Level	Secondary Packaging Level	Tertiary Packaging Level	GHG [%]	Ref.
Confectionery	Cocoa and chocolate products	Chocolate-covered hazelnut	Modified atmosphere in LDPE bag, label	Box	-	17.80	[118]
		Chocolate-covered almond	Modified atmosphere in LDPE bag, label	Box	-	6.00	
		Dark chocolate	Aluminum foil, cardboard	-	-	13.02	[32]
		Chocolate (100%)	Aluminum foil, paper	-	-	8.56	[122]
		Malty chocolates (in bags)	Aluminum foil	Corrugated cardboard boxes	LDPE stretch-film, LDPE consumer plastic bags	13.00	[28]
		Chocolate-coated wafers (contlines)	Aluminum foil	Corrugated cardboard boxes	LDPE stretch-film, LDPE consumer plastic bags	8.00	
		Milk chocolate (molded)	Aluminum foil	Corrugated cardboard boxes	LDPE stretch-film, LDPE consumer plastic bags	6.00	
		Milk chocolate	Aluminum foil, paper	-	-	6.94	[119]
		Dark chocolate				11.90	
		White chocolate				6.10	
		Chocolate with sultanas				10.42	

Table 2. *Cont.*

Category	Sub-Category	Product	Primary Packaging Level	Secondary Packaging Level	Tertiary Packaging Level	GHG [%]	Ref.
		Milk chocolate confectionary	Aluminum foil	Corrugated board box		2.27	
		Dark chocolate confectionary	PET tray, corrugated cardboard component	Corrugated board box	Not considered	5.18	[26]
		Milk chocolate biscuit confectionary	PP film	Corrugated board box		3.00	
		Dark chocolate	PP			4.71	
		Dark chocolate	Aluminum foil, fiber-based layer (cardboard)	-	-	24.87	
		Dark chocolate	Aluminum foil, fiber-based layer (Kraft paper)			18.82	
		Milk chocolate	PP			2.20	
		Milk chocolate	Aluminum foil, fiber-based layer (cardboard)	-	-	11.65	[129]
		Milk chocolate	Aluminum foil, fiber-based layer (Kraft paper)			8.82	
		White chocolate	PP			2.26	
		White chocolate	Aluminum foil, fiber-based layer (cardboard)	-	-	11.94	
		White chocolate	Aluminum foil, fiber-based layer (Kraft paper)			9.04	
		Extra dark chocolate, 65 g strip				23.64	
		Dark chocolate, 65 g strip	Paper covered Aluminum foil, paper sticker	Paper box	Cardboard/ carton box	23.35	[116]
		Milk chocolate, 65 g strip				9.31	
		Flavored milk chocolate, 65 g strip				9.26	

Table 2. *Cont.*

Category	Sub-Category	Product	Primary Packaging Level	Secondary Packaging Level	Tertiary Packaging Level	GHG [%]	Ref.
		Extra dark chocolate, 100 g bar	Aluminum foil)	Printed paper wrapper	Cardboard/ carton box	12.12	
		Dark chocolate, 100 g bar				11.98	
		Milk chocolate, 100 g bar				4.77	
		Flavored milk chocolate, 100 g bar				4.75	
		Extra dark chocolate, 300 g pouch	Paper covered aluminum foil, paper sticker	Paper box	Cardboard/ carton box	13.94	
		Dark chocolate, 300 g pouch				13.77	
		Milk chocolate, 300 g pouch				5.49	
		Flavored milk chocolate, 300 g pouch				5.46	
		Conventional monoculture chocolate (min. transport)	Aluminum foil, paper	-	-	8.71	[123] based on [32,122]
		Conventional agroforestry chocolate, (min. transport)				11.84	
		Organic agroforestry chocolate, (min. transport)				13.24	
		Conventional monoculture chocolate, (max. transport)				5.79	
		Conventional agroforestry chocolate, (max. transport)				7.03	
		Organic agroforestry chocolate, (max. transport)				7.50	
	Other confectionaries, including breath-freshening micro-sweets	Jelly sweets	PP bags	Not included	Not included	8.75	[132]
		Foam sweets	PP container			1.88	
		Sugar confectionary	Aluminum foil, paper	Corrugated board box	Not considered	5.26	[26]
		Milk-based confectionary	PP film	Corrugated board box		2.85	

Table 2. *Cont.*

Category	Sub-Category	Product	Primary Packaging Level	Secondary Packaging Level	Tertiary Packaging Level	GHG [%]	Ref.
Cereals and cereal products	Whole, broken or flaked grain	Rice (IT)	Plastic bag	-	-	1,95	[124]
		Rice organic (IT)				1.33	
		Rice (US)	Cardboard box			0.36	
		Rice parboiled (US)				0.91	
		Rice upland (CH)				1.82	
		Minimal tillage white rice	LDPE bags	-	-	1.46	[125]
		Minimal tillage brown rice				1.82	
		Organic cultivation white rice				0.62	
		Organic cultivation brown rice				1.02	
	Flours and other milled products and starches	Oatmeal	-	-	-	6.02	[126]
		Potato flour				7.69	
		Wheat flour	-	-	-	2.17	[141] based on [148]
	Breakfast cereals	Breakfast cereals	Printed board folding-box, HDPE bag/liner	Corrugated-board box, HDPE stretch film/wrap	Corrugated pallet layer pads, Wooden pallet	15.00	[27]
		Dry ready-made porridge	LDPE bag, cardboard box ("bag in box")	Not considered	Not considered	9.93	[130]
		Wet ready-made porridge	Glass jar, cab (aluminum and plastics)			38.02	
		Wet ready-made porridge (scenario)	Pouch, cap			15.77	
	Pasta	Dried short pasta 0.5 kg	Re-closeable PP bag	Carton, adhesive label, scotch tape	Stretch and shrink film, label, EPAL wood pallet, different layers of cartons	5.90	[117]
		Dried long pasta 0.5 kg	Re-closeable PP bag			3.40	
		Dried short pasta 0.5 kg	Paperboard box			13.90	
		Dried long pasta 0.5 kg	Paperboard box			9.40	
		Dried short pasta 3 kg	PE bag			8.20	

Table 2. *Cont.*

Category	Sub-Category	Product	Primary Packaging Level	Secondary Packaging Level	Tertiary Packaging Level	GHG [%]	Ref.
		Dried long pasta 3 kg	PE bag			3.10	
		Pasta	Paper	Cardboard paper, plastic film	Corrugated board	1.00	[133]
		Pasta (wheat, 0% straw)	Low-density PET film, cardboard box, printing	Corrugated board, PP film	Pallet	10.00	[127]
		Pasta (wheat, 80% straw)				10.20	
		Pasta (egg)	-	-	Pallet	7.26	[128] based on [149]
Bakery wares	Bread and rolls	White bread (medium slices, 40 g)	PE bag	-	-	1.61	[120]
		Wholemeal bread (medium slices, 40 g)				1.73	
		White bread (thick slices, 57.5 g)				1.67	
		Whole meal bread (thick slices, 57.5 g)				1.80	
		White bread, medium slices (generic study)				2.73	
		Wholemeal bread, medium slices (generic study)				2.91	
		Brown bread, medium slices				2.84	
		White bread, thick slices (generic study)				2.86	
		Wholemeal bread, thick slices (generic study)				3.07	
		Brown bread, thick slices (generic study)				2.99	
		White bread (medium slices, 40 g) (generic study)				5.31	

Table 2. *Cont.*

Category	Sub-Category	Product	Primary Packaging Level	Secondary Packaging Level	Tertiary Packaging Level	GHG [%]	Ref.
		Wholemeal bread (medium slices, 40 g) (generic study)	Wax coated paper bag			5.66	
		Brown bread, medium slices (generic study)				5.51	
		White bread (thick slices, 57.5 g) (generic study)				5.56	
		Whole meal bread (thick slices, 57.5 g) (generic study)				5.95	
		Brown bread, thick slices (generic study)				5.80	
		Bread (wheat)	Paper bag (paper and polylactide)	-	-	11.58	[131]
		Rye bread	LDPE bag, plastic clip	Returnable plastic box	-	6.10	[134] based on [11]
		Bread	PET and paper	HDPE box	HDPE trolley, extra packaging used by consumers	7.07	[121]
		Bread	LDPE bag, PS clip	Returnable plastic box	-	4.59	[11]
		Biscuits	Tray, wrap, cardboard case, plastic film	-	-	17.62	[31]
	Fine bakery wares	Crackers	PP film	Cardboard box	LDPE film, LDPE shopping bag	7.00	[30]
		Low fat/sugar biscuits	PP film		LDPE film, LDPE shopping bag	6.00	

Table 2. *Cont.*

Category	Sub-Category	Product	Primary Packaging Level	Secondary Packaging Level	Tertiary Packaging Level	GHG [%]	Ref.
		Semi-sweet biscuits	PP film		LDPE film, LDPE shopping bag	6.00	
		Chocolate-coated biscuits	PP film		LDPE film, LDPE shopping bag	4.00	
		Sandwich (Chocolate cream) biscuits	Metallized (aluminum) PP film	Cardboard box	LDPE film, LDPE shopping bag	8.00	
		Sandwich (vanilla cream) biscuits				7.00	
		Whole cakes	PP, cardboard folding box	Cardboard	LDPE wrap, consumer shopping bags	7.00	
		Cake slices	Cardboard folding box, LDPE	Cardboard	LDPE wrap, consumer shopping bags	19.00	
		Apple pie	Cardboard folding box, LDPE, aluminum foil	Cardboard	LDPE wrap, consumer shopping bags	24.00	[29]
		Cupcakes	Cardboard folding box, LDPE, paper	Cardboard	LDPE wrap, consumer shopping bags	24.00	
		Cheesecake	PP, cardboard folding box, LDPE	Cardboard	LDPE wrap, consumer shopping bags	5.00	
Ready-to-eat savories and snacks	Potato-, cereal-, flour- or starch-based snacks	Crisps	OPP and (aluminum) metallized OPP	Not included	Not included	8.14	[132]
	Processed nuts	Pistachio	Modified atmosphere in LDPE bag, label	Box	-	12.80	[118]
		Almond				12.90	
		Hazelnut				29.80	
		Peanut				24.90	

Packaging

Focusing solely on packaging, in the category of confectionaries and the sub-category of cocoa and chocolate products, the primary level of packaging was in most cases aluminum foil [26,28,32,116,119,122,123,129] or combinations of aluminum foil with fiber-based packaging materials like paper [26,116,119,122,123,129] and board [26,32,129]. In some packages, additional packaging aids such as paper stickers were used [116], and information on finishing (e.g., print) [116] was given. Plastic packaging was less prominently represented. Found examples included chocolate-covered products (nuts) packaged in labelled plastic (low-density polyethylene (LDPE)) bags containing a modified atmosphere based on N_2 [118], dark chocolate confectionary in a polyethylene terephthalate (PET) tray including a (corrugated) cardboard component, milk chocolate biscuit confectionary [26], as well as different chocolates [129] packaged in polypropylene (PP). Regarding the primary packaging concepts presented, product-typical solutions aimed at maintaining the product quality were given throughout. For example, the necessary barrier functions against light, oxygen, water vapor as well as aroma were met in almost all cases. In the cases where only plastic packaging (e.g., milk chocolate biscuit confectionary [26]; dark chocolate [129]) was mentioned and not further specified if a light barrier [150] in the form of a colored material or a secondary packaging level made of, e.g., cardboard was present, product quality and thus shelf-life may be potentially impaired [46]. The secondary packaging level of other products was exclusively fiber-based packaging, namely (corrugated) cardboard boxes [26,28,118], paper wrappers or boxes [116].

In the sub-category of other confectionaries, including breath-freshening micro-sweets, primary packaging concepts were similar to those given above and met product requirements which mainly covered protection from moisture uptake or loss [46]. Jelly and foam sweets [132], as well as milk-based confectionaries, were packaged in PP, while sugar confectionaries were packaged in aluminum foil and paper [26]. Secondary levels, where mentioned, were paper [26].

Cereals and cereal products, including the four sub-categories of whole, broken or flaked grain, flours and other milled products and starches, breakfast cereals as well as pasta, frequently used [46] plastic [117,124,125] and fiber-based [124,133] primary packaging concepts or a combination thereof [27,127,128,130]. All packaging concepts given aim to protect low-moisture or dried products (especially, e.g., breakfast cereals [27]) with low fat content from mainly water vapor, aroma, mechanical damage or oxidation [47]. In the case of ready-made wet porridge, a glass jar with an aluminum-plastic lid and alternatively a multilayer pouch with a cap was mentioned [130]. Secondary packaging levels were not thoroughly described, but if mentioned, they were mainly corrugated cardboard boxes [27,127,133] or cartons [117]. Additionally, high-density polyethylene (HDPE) [27], PP [127] or other unspecified plastic films [133] and labels [117] were named. One study even listed scotch tape used for closing cartons [117].

Comparing this with the EPDs found for this product group, one can see a strong overlap of packaging concepts. Flours and other milled products, for example, are likewise packaged in fiber-based solutions (paper bags) [151,152]. Additionally, bulk packaging (paper sacks, big plastic bags) is mentioned [153]. Breakfast cereals are packaged in plastic bags in paper box solutions [154], and pasta is packaged in either plastic [155–167], cardboard [156,157,168] or a combination thereof [147,157,158,169,170]. Additional packaging levels, where given, frequently included cardboard boxes, interlayers, pallets and plastic (stretch) films [147,154,155,158–162,165–170].

The shelf-life of bakery wares is significantly influenced by water exchange processes as well as interlinked structural changes, aroma uptake and (microbial) spoilage [46,47]. To limit this and prolong shelf-life, products in the sub-category of bread and rolls were primarily packaged in polyethylene (PE) bags [120], LDPE bags with (polystyrene (PS)) clips [11,134] or (wax-coated) paper bags [120]. Further, material combinations such as paper and polylactide (PLA) [131] or paper and PET [121] were used. Secondary packaging was (HDPE [121]) plastic boxes. In two sequential studies, it was stated that these were returnable [11,134].

The EPDs belonging to this product category, on the other hand, show only one packaging concept, namely that of a plastic bag with an associated clip. Additional packaging levels again include cardboard boxes and plastic films [171–183].

The sub-group of fine bakery wares showed a more diverse and elaborated packaging spectrum. While primary packaging for some biscuits was solely PP or a metallized PP film [30], others were packaged in multiple levels [29,31]. The latter may be due to higher product requirements in terms of quality. For example, cream fillings of biscuits as well as cakes [29,30] exhibit higher moisture and fat content and thus spoil more easily [46,47]. Additionally, elevated packaging [29,31] may be due to the fact that these products are more hedonistic than, e.g., cereal products such as breakfast cereals [184]. Secondary packaging in all given cases was cardboard/cardboard boxes [29,30].

The more diverse and elaborated packaging spectrum is also reflected in the EPDs. Here, different multilayer materials with or without paper are described. Additionally, different combinations of plastic or paper board trays, films, banderoles and/or boxes are given. Additional packaging layers are comparable to the above-mentioned ones [146,185–217].

Last but not least, the category of ready-to-eat savories and snacks, including potato-, cereal-, flour- or starch-based snacks using the example of crisps, were primarily packaged in a multilayer film made of oriented polypropylene (OPP) and metallized OPP [132], a common solution found in this category due to the superior gas and light barrier allowing stable product quality in terms of, e.g., crispness and lipid oxidation (rancidity) [46,47]. Processed nuts were packaged in LDPE bags with a label. Additionally, a modified atmosphere was applied [118] to protect the oxidation-sensitive products [46,47]. Secondary packaging (box, unspecified) was only given for the last-mentioned product [118].

Insofar as stated, tertiary packaging of all considered product (sub)categories was mainly represented by plastic materials such as (LDPE) (stretch-)films [28–30,117] and shrink-films [117] as well as (wooden) pallets [27,127,128]. Further materials described were cardboard/carton boxes [116], corrugated pallet layer pads [27] and labels [117]. In one case, an HDPE trolley was given [121]. Besides this, some authors even calculated consumer (plastic) bags in [28,30,121]. However, for the majority of products, no information on tertiary packaging levels was available.

Summing up, it can be seen from the reviewed studies taken together in Tables 1 and 2 that predominantly plastic and aluminum packaging solutions were used in direct product contact. Further, it can be observed that packaging-specific information is not always given and that the detail of the same varies remarkably. Regarding the packaging levels, most authors give information on the primary packaging level, whereas secondary and especially tertiary levels are less frequently given [31,32,119,120,122–126,128–132,141]. In some cases, secondary and/or tertiary levels are even intentionally excluded [26,130,132]. Miah et al. [26], for example, justify not considering tertiary packaging (cut-off), for example, by the low weight percentage that comes from the tertiary packaging. Similarly, so do Sieti et al. [130]. Consequently, in many cases, only the primary packaging, and not the whole packaging system, is analyzed. This fact is also shown by Molina-Besch et al. [111]. Interestingly, different authors also seem to delineate packaging levels differently. For example, some authors include stretch films, which are often used to secure pallets [48], in secondary packaging [27,127,133], whereas others include them in tertiary packaging levels [28]. Additionally and interestingly, the EPDs under consideration distinguish between primary packaging and packaging for transport and do not go into detail about secondary/tertiary packaging levels (e.g., EPD on American sandwich [175]).

Furthermore, the level of detail of the information is deviating strongly. While some authors only mention the material, others include further information on, for instance, packaging containers (e.g., bag, tray, foil) [11,26–32,116–120,122–125,127–129,131–134], packaging aids (e.g., labels, adhesive tape, clips) [11,27,116–118,134], packaging weight [26–30,32,116,122,123,127,129,132,133], or dimensions [27,116], material composition (e.g., recycled content) [27,28,32,131], multilayer structure [27,30,132], usage of modified atmosphere packaging [118] or finishing processes such as printing [27,127]. EPDs usually reduce the information to the material used (e.g., EPD on crispbread [187]).

In some cases, information is directly included in the scientific paper, while in other cases, it is given as the supplementary material of the studies [26,28–30,32,117,118,123,127,129,130,134]. In addition, it is noticeable that packaging-specific information is often not given condensed at the beginning of the paper (e.g., materials and methods section, life cycle inventory) but spread over the text. Moreover, differences were also notable with regard to the data source. While some authors used primary data (e.g., specifications, information from companies), others used secondary data or based their calculations on assumptions. The most detailed information on packaging was found in the study by Cimini et al. [117].

Packaging End-of-Life

Regarding the packaging end-of-life, particularly waste management, country-specific scenarios are most frequently considered in studies where packaging (material) is mentioned and a cradle-to-grave approach is followed. This applies to, for example, rates of recycling, incineration or landfilling. For instance, Konstantas et al. [28] focused on chocolate production and consumption in the United Kingdom and included post-consumer waste management activities for the corrugated cardboard (recycling > incineration with energy recovery), aluminum (recycling > landfill) and plastic packaging (landfill > incineration with/without energy recovery) components. Additionally, efficiencies of the corrugated board and aluminum recycling processes were counted in. Further, authors who include disposal routes are, inter alia, Miah et al. [26] (United Kingdom), Bianchi et al. [129] and Cimini et al. [117] (Italy). Further, EPDs usually include primary packaging end-of-life (e.g., EPD on durum wheat semolina [151]).

Interestingly, most of the statements in the studies under review, as well as EPDs, are made based on, for example, reports on the national recycling rates of (packaging) materials (e.g., Cimini et al. [117,218]). The actual recyclability of the specific packaging solutions is, however, hardly addressed or analyzed in the reviewed studies [130,132]. This, however, is a knowledge field gaining importance and momentum in recent years [50], which is accompanied by different (e.g., design for recycling) guidelines [41], instruments and certificates (e.g., cyclos-HTP [219]). This becomes interesting, for example, in the case of very small packaging components or multilayer materials, for which the necessary sorting and recycling facilities often are not applied or even do not exist to date [52]. Accordingly, it is necessary to discuss whether the specified end-of-life scenarios are actually realistic and to what extent the results change.

Data Quality

It is well known that an LCA is only as reliable as the sources and dataset base it is built upon. Multiple sources and handbooks on LCA even state that data quality may largely determine LCA results [220]. In LCA, there are two main categories of data: primary and secondary. While primary data refers to actual data collected from sources of the investigated life cycle step (farmer, manufacturer, distributor etc.), secondary data refers to information from literature and databases. Quality thereof is, among other factors, determined by the recentness of the data and the model, geographical coverage, variability, representativeness and reproducibility [43,144]. The investigated studies took varied approaches to data quality issues. The sources for packaging LCA data were secondary in the majority of studies [11,26–30,32,116,118,120,122,125,128–130,134,141], whereas the remaining studies used primary and a mixture of primary and secondary data for packaging [31,117,121,123,126,127,131–133]. The actual sources of primary data were in-depth interviews and questionnaires with packaging producers, and for secondary data, the sources were the Ecoinvent and GaBi databases. Two of the studies were reviews that used published reports and results of other studies (published in journals), including their supplementary materials [11,141].

Espinoza-Orias et al. [120] and Jensen and Arlbjorn [134] took up the topic of data quality and usability of the like for sustainability assessment in the product category of bakery wares, specifically in the sub-category bread and rolls. The former authors even compared calculations between mainly primary and secondary sourced data (generic study). Other studies worth commenting on from the perspective of their attention to data quality are Usva et al. [126], who created a whole set of criteria for data quality and development and explained them fully in the text, as well as Cimini et al. [117], who used PAS2050 requirements for data quality, including geographic and time scope as well as technology references. This is in line with the CEN/TR 13910:2010 report on criteria and methodologies for LCA of packaging, which mentions the importance of giving special attention to time, geography and technology aspects within the data collection phase of LCAs [221].

3.2.3. Life Cycle Impact Assessment

Impact Assessment Method and Impact Categories Used

As selected for, all of the examined studies assessed at least CO_2 emissions/global warming potential (GWP)/carbon footprint of the food packaging systems [118,120,124,125,128,133, 134,141]. In most cases, several other impact categories were also included. Examples are ozone depletion, fossil fuel depletion, terrestrial acidification, freshwater eutrophication, marine eutrophication and human toxicity [11,26–32,116,117,119,121–123,126,127,129–132]. The chosen impact categories depended on the used assessment method (e.g., ISO 14044 [67]) and the focus of the study in general. Using the above example of Espinoza-Orias et al. [120], two methodological approaches, namely PAS 2050 and ISO 14044 [67,108], were used. The former was used because it lays a focus on primary data, and the latter was used because the use of secondary data is allowed more. The aim was to compare the approaches and identify their influence on LCA results. It can be seen from this concrete example that the comparability of the studies is neither consistently given nor envisaged in this paper due to different scopes and applied assessment methods.

While carbon footprint is also covered by EPDs, other impact descriptive categories are, for instance, ecological footprint as well as water footprint (e.g., EPD on breakfast cereals [154]).

Sensitivity/Scenario Analysis

Of the present studies, only a few authors did not conduct a sensitivity/scenario analysis [122,124–126,128,132,141]. The others used this analysis to check for the robustness/generalizability of their results by alternating input data such as country of production [11,30,32,116,117,119,120,123,127,129,131,133,134]. Contrary to expectations, only a handful of studies included packaging in one or the other way in their sensitivity analysis [26–29,31,118,130]. For example, Volpe et al. [118] conducted an uncertainty and sensitivity analysis and concluded that abroad consumer markets and thus the final destination of (glass) packaging affect the LCA output (carbon footprint) significantly. However, the data for glass refers to nut spread cream packaged in a glass jar, which was excluded from the present review due to the product group exclusion reasons. Details for plastic bags used for the other products included in the present review were not given. Furthermore, Miah et al. [26] alternated packaging materials in an improvement analysis. Here, aluminum and PP were substituted with recycled material, paper with unbleached paper, and corrugated board with white lined board, while PET stayed unchanged. This led to " ... a mix change in total environmental impact across all five confectionary products ... " and, on average (across all confectionary products analyzed), an increase in GWP. Jeswani et al. [27], in the other case, exchanged some of the carton boxes with standalone HDPE bags in a hypothetical scenario, which resulted in a lowering of GWP. Additionally, Noya et al. [31] analyzed alternative waste management practices for packaging materials (increased recycling rates) with the result that the environmental burdens for the global process decreased (including climate change). Significance was, however, shown only for products with higher packaging requirements (plastic and cardboard). Last but not least, Konstantas et al. [29] focused on packaging losses (2 to 10%) in the manufacturing process and concluded that the results are not sensitive to packaging losses. Next to packaging, it can be mentioned that Miah et al. [26] and Noya et al. [31] also included food waste (reduction) in their analysis but did not interlink this with packaging (re)design. Surprisingly, although Williams and Wikström [11] had packaging embedded in their target, they did not conduct a corresponding sensitivity/scenario analysis.

3.2.4. Interpretation

Environmental Impacts and Mitigation Measures

While Table 2 exhibits values of packaging-related CO_2 emissions of different cereal and confectionary products on a single food item level, Table 1 provides an overview of product (sub) category-related emissions. As can be seen, single values range from 0.36 to 38.02% and in total, average packaging-related CO_2 emissions account for 9.18%. Despite the fact that different studies are hardly comparable due to, for example, different aims, scope, system boundaries and input data, it becomes apparent that the average value lies clearly above the estimated general global values of about 5% by Crippa et al. [2] and Poore and Nemecek [3]. However, the values well reflect the wide possible variation previously found by, among others, Poore and Nemecek [3], Verghese et al. [6] and Heller et al. [7]. When going into detail about the different (sub)categories, interesting tendencies and hotspots can be found. These are discussed in the following paragraphs.

In the category of confectionary and, further, in the sub-categories of cocoa and chocolate products as well as other confectionaries, including breath-freshening microsweets, where average CO_2 emissions (see Table 1) are 9.86 and 4.68%, respectively, the authors uni sono indicate that (raw)material sourcing is the main environmental impact driver. The provision and, in particular, the agricultural production of cocoa derivates, milk powder and sugar can be highlighted. This is also reflected by the environmental impacts of the respective products (Table 1). Boakye-Yiadom et al. [116] offer an illustrative example, where milk chocolate yielded significantly higher than dark or extra dark chocolate due to the high impact of the animal-derived food ingredients. Further, associated manufacturing processes and (fossil) energy consumption as well as (international) transport are ranked particularly high in the studies under review [26,28,32,116,118,119,123,129,132]. Further, reduction of (food)waste is mentioned as one way to cut carbon emissions [26,132]. In relation to packaging, behind the above-mentioned factors, significance has also been reported by different authors [26,28,116,118,119,129]. In this context, the main focus is on material choice [116,118,129]. In their work, Bianchi et al. [129] were able to show that a single PP layer is better than a combination of commonly used aluminum/fiber-based packaging solutions. Material (aluminum) substitution, if possible, is also on the agenda of Boakye-Yiadom et al. [116], who alternatively recommend using recycled or weight-reduced packaging solutions. Due to a lack of data, especially regarding thematic coverage, the studies [26,28,116,119] as well as Pérez-Neira et al. [123] do not go into detail about packaging but mention the importance of packaging optimization. Last but not least, collaboration with science and industry to develop packaging materials and solutions with lower impact were discussed by Miah et al. [26] and Boakye-Yiadom et al. [116].

Turning to cereals and cereal products, one can see that the average packaging-related CO_2 emissions from whole, broken or flaked grain, flours and other milled products and starches, breakfast cereals as well as pasta are 1.25, 5.30, 19.68 and 7.24% (see Table 1), respectively. The significantly higher value for breakfast cereals is justified by the fact that wet porridge in a single-use glass jar was included in one study [130]. This is a packaging solution known for its high environmental impact, mainly due to very high process temperatures and, thus, energy needed in the production of the same [43]. Accordingly, the authors suggest replacing this with a lightweight plastic packaging solution (pouch), which exhibits 15.77 instead of 38.02% with regard to CO_2 on a single product level [130]. A further change in material in the sub-category of breakfast cereals was proposed by Jeswani et al. [27], who found that replacing the well-known plastic bag and carton box combination for breakfast cereals with (standalone) plastic packaging (bags or pouches) could reduce carbon emissions. A possible preference for plastic packaging (PE bags) instead of paperboard boxes was also communicated by Cimini et al. [117] for dried pasta. The same authors also highlighted the correlation between high packaging density and the reduced packaging and transportation need for long pasta (e.g., spaghetti) in comparison with short pasta (e.g., spiral-shaped) due to the different shape and thus volume of pasta per functional unit. Furthermore, in the broader sense, relevant findings of packaging included the necessity to find the right trade-off between packaging function and environmental impact [141], to combine and prioritize actions [27,117], to engage relevant stakeholders (industry and government) to find best-practices and standards (e.g., packaging, types, mass reduction, recyclability) [130] and to intensify LCA applications and transparently communicate the results thereof (e.g., labelling) [124,141]. All in all, the packaging focus in this product category was less distinct than in the previous one, and the emphasis was mainly on the optimization of agricultural production and the provision of products [27,117,124–127,141], reformulation of recipes [128,130] and changing consumer habits. Here, for instance, the cooking of pasta [117,127], the consumption of cereal products with (cow's) milk [27] or the use of ingredients of animal origin (egg, milk) [128,130] were related to higher impacts.

Since no EPDs for whole, broken or flaked grain are available to date [140], only comparisons of flours and other milled products and starches [151–153], breakfast cereals [154] and pasta [147,155–170] can be made at this point. Here, the average values are found to be 3.22, 12.37 and 8.56%, respectively. Although, as stated above, direct comparison is difficult, interestingly, a similar ranking can be identified. Therefore, flours and other milled products and starches score the lowest, while pasta and breakfast cereals, in ascending order, score higher. A possible explanation for this is the level of complexity of the packaging solutions. While milled, powdery products are densely packaged in simple bags, more volume-taking pasta is packaged in more stable and elaborately designed packaging solutions partly combining different materials. Breakfast cereals, in the present case, exhibit even higher packaging effort with a plastic bag and an additional cardboard box.

In the case of bakery wares, such as bread and rolls, as well as fine bakery wares, an average contribution of packaging to the CO_2 emissions of 4.37 and 11.22% was found (Table 1). As expected, raw material (e.g., wheat, milk, palm oil, sugar) sourcing is the main environmental impact driver [29–31,120,121,131,134]. This is (not in strict chronological order) most often followed by processing and correlated energy use [29,30,131,134] as well as consumption (e.g., refrigeration, toasting) [120,134], although Svanes et al. [121] achieved a different result here. Further, waste at retail [121] and consumption level [120,121] as well as transport [30,31,120,131,134] and packaging are mentioned. The latter again played a less important role in other selected studies [29,30,120,121,131]. Of the packaging-related impacts, Konstantas et al. [30] named primary packaging as the most contributing factor. Several mitigation measures similar to the above product categories (e.g., efficient raw material sourcing) are given in the reviewed studies [11,29–31,120,121,131,134]. Regarding packaging, four main points were discussed by the authors, namely, portion size [120,121], packaging re-design [11,121] and light-weighting [29] as well as proper end-of-life management [31,134]. In the case of right-sizing portions, Espinoza-Orias et al. [120] as well as Svanes et al. [121] proposed that smaller sizes of bread (e.g., loafs) would reduce the amount of wasted bread (due to, e.g., spoilage) at the consumption stage but at the same time increase the need for packaging which, in the case of reduced food waste, still could lead to an environmental benefit–a finding that has already been shown in other contexts. Packaging re-design, on the other hand, included the substitution of a PET/paper packaging material with a material based on cellulose fibers and a perforated paper bag coated with PE on the inner side. While the former alteration allowed the bread to be kept fresher for one day, the latter solution allowed the product to be perceived as fresh even four days after production, which could lead to an environmental benefit since the impacts of producing the packaging alternatives are almost the same as with the packaging in comparison. The authors, who laid a strong focus on indirect packaging effects in their work, pointed out that further (large-scale) tests and the inclusion thereof in LCAs would be necessary to validate the results [121]. Studies on shelf-life extension strategies and waste prevention were also asked for by Williams and Wikström [11], who additionally highlighted that good product packaging should not encourage consumers to re-pack their products at home. This is a measure that could avoid unneeded extra packaging material. The latter also represents a recent research field where the understanding of consumer habits and social norms are focused, and food and packaging researchers are asked to more closely collaborate with social sciences and humanities [222]. Turning to the light-weighting of packaging, Konstantas et al. [29] calculated in their study on different cakes that a material reduction of 30% could lead to a significant drop in the GWP of cakes (except for whole cakes and cheesecakes). Food safety and shelf-life, however, must not be jeopardized as a result. The topic of end-of-life (improved waste management strategies and recycling rates [31,134]) was discussed by Jensen and Arlbjorn [134], who pointed out explicitly that hotspots should not only be identified on the basis of their impacts but also on the basis of their potential for change and that the awareness for possible burden shifting from one life cycle stage or impact category to another by just focusing on, for example, GWP values, should be kept at a high level.

Comparing the values found for the category of bakery wares and the sub-categories bread and rolls [171–183] as well as fine bakery wares [146,185–217] with the EPDs, values of 17.03 and 14.86% were found. In both cases, the values are higher than the ones from the studies under review. Possible causes for this may be, amongst others, the packaging material or the database used. The latter is frequently given to be mainly based on primary data. In the case of Italian bread (pagnotta), for example, it is stated that generic data contributes less than 10% to the calculation of environmental performance [182].

Lastly, in the category of ready-to-eat savories and snacks, which include potato-, cereal-, flour-, or starch-based snacks as well as processed nuts, the average contributions of packaging to the CO_2 emissions were 8.14 and 20.10% (Table 1). Since these products were also covered by the already discussed research from Nilsson et al. [132] and Volpe et al. [118] in the product category of confectionary products, no further detail on packaging can be named at this point.

Significance of the Results

In their parallel (mainly primary/secondary data) studies on bakery wares (loaves of sliced bread), Espinoza-Orias et al. [120] conclude that data quality is key for not only the accurateness of the LCA results but also for honest sustainability communication. While secondary LCI data may be useful for rather uncomplicated (company) internal detection of hotspots or projections at the (inter)national level, high-quality primary data is needed for communication to consumers via, e.g., carbon labelling [138]. Similarly, Jensen and Arlbjorn [134] conclude that high-quality data is needed to achieve robust results.

In relation to impact assessment, Williams and Wikström [11] address food losses and food waste as well as packaging optimization in their conclusion. Here, they call for the inclusion of these indirect packaging impacts in food and packaging LCAs to examine how waste and, in consequence, negative environmental impacts can be diminished. Further, they highlight that legal texts should more strongly include the topic of food losses and food waste prevention by appropriate packaging solutions.

When talking not only about one impact category (e.g., GWP), a multi-criteria decision analysis (MCDA) as used, for example, by Miah et al. [26] can be helpful. This allows to compare different environmental impact categories together and to ease decision-making and benchmarking. Accordingly, MCDA is increasingly being used in LCA [223].

4. Improvement Strategies

As described at the outset, food systems are responsible for a large proportion of environmental impacts, especially GHG emissions, worldwide [2]. Increasing efficiency in food production and, above all, reducing food losses and waste can, therefore, directly contribute to lowering the global footprint [19,224]. In the last decade, the focus has therefore been on targeting, measuring and reducing GHG emissions. Along with that, efforts by different stakeholders have been conducted or started, and respective policies have been outlined [52,225]. Packaging is playing an increasingly important role in this context. While efforts initially focused on the reduction of the direct environmental impacts of packaging (e.g., material use), today, the focus is increasingly on the indirect impact (e.g., reduction of food waste), as it has been recognized that this has a potential leverage effect [13,34,52,110,226,227]. However, the actual inclusion of the indirect impact in research, development and innovation activities lags behind [111], as has also been shown by the present review. Accordingly, strategies for the acceleration of the implementation are needed. In this context, Wikström et al. [52] elaborated a research agenda including 5 packaging-related issues. These include: (i) quantitatively understanding packaging's diverse functions and the influence on food losses and waste in the context of the (inter)national food system, (ii) more thoroughly understanding trade-offs between packaging and food losses and food waste, (iii) further improving representation thereof in LCA and (iv) designing processes and related methods as well as (v) setting stakeholder incentives such as profitable business models. To support this transition, the following text aims at aggregating possible points of action in the area of packaging, LCA and management beyond the topic of cereal and confectionary packaging.

4.1. Packaging

Starting with packaging, recommendations or suggestions found in this and other studies and texts can be very well set in the context of the existing Packaging Sustainability Framework with its four principles (effective, efficient, cyclic, safe) [42,43] (see also Table 3). This may act as a basis for future improvement regarding the reduction of the direct and indirect environmental impacts of food packaging. However, it must be clearly pointed out that there may be trade-offs and that verification of the respective product packaging system is essential [42,43].

Table 3. Recommendations for improving the sustainability of food packaging based on the structure given by [36,46].

Sustainable Packaging Principle	Recommendation	Reference
Effective	Usage of packaging fit for purpose Provision of appropriate shelf-life Employment of shelf-life extension strategies Avoidance of over-engineering Holistically integrate primary, secondary and tertiary packaging levels Provide packaging with high consumer value Target-group oriented packaging with consumer value Right-sized portions Provide clear and understandable communication	[43,44,46] [43,111] based on [228–230] [11,231] [43] [43] [10,11,43,111] based on [229] [10,11,43,111] based on [229] [111,120,121] based on [120,228,229] [11,37,43]
Efficient	Optimize packaging with regard to function and environmental impact Rethink material choice and packaging design Increase transport efficiency Decrease energy demand along the supply chain (e.g., process and transport) Focus on renewable resources (materials and energy)	[26,28,29,37,43,111,116,119,123,141] based on [27,232–245] [10,27,43,111,116–118,121,129,130] based on [27,120,233,235,236,238,240,244,246–251] [43,111,141] based on [232,237,244] [43,111] based on [243]
Cyclic	Avoid unneeded packaging Prevent and reduce food and packaging waste along the supply chain Use reusable, returnable or refillable (primary, secondary, tertiary) packaging solutions Design packaging *for* recycling	[111] based on [252] [26,43,111,132] based on [242]; [43,111] based on [240,246,252,253] [35,37,39,41,43]
	Design packaging *from* recycling Use bio-based and/or bio-degradable materials Assure proper end-of-life management Promote a circular economy	[37,43,111,116] based on [230,231,244,248,249] [37,43,44,111] [31,43,134] [35,36],
Safe	Focus clean production Install ecological stewardship Reduce possibility for litter formation	[35,37,43,44] [37,43] [43]

Going into detail about the effectiveness of food packaging and analyzing the findings with regard to packaging that is fit for its purpose and, thus, is satisfactorily fulfilling its containment, protection, communication and convenience function [43,44,46,47], one can see that authors currently lay a focus on protection and convenience. Regarding protection, which is enabled by the often-overseen basis function of containment [46,47], the provision of an appropriate or prolonged shelf life is frequently mentioned [43,111,228–230]. In this context, the application of well-established and modern shelf-life extension practices [11], such as modified atmosphere packaging (MAP) [46,254] or active and intelligent packaging solutions (AIP) [46,47,255–257], can be named. Attention, however, should be paid to the possible over-engineering of packaging and not losing a holistic view of the packaging system. With regard to over-engineering, it may be reasonable to re-assess the actual product requirements and avoid unneeded packaging, as well as reduce packaging complexity or components, where possible. This can be supported by, for example, market research or research on consumption patterns [43]. With regard to a holistic view, the interlinkage between primary, secondary and tertiary packaging must be considered, since changes on one level may also necessitate changes on other levels. For instance, a reduced or less mechanically stable primary packaging (material) may induce the need to design the secondary or tertiary packaging to be more stable [43,111]. With respect to the convenience aspect of packaging, several authors take up the topic of developing packaging with a high consumer value or target group orientation. This includes, inter alia, packaging that is easy to open, reclosable or easy to empty and, in general, does not frustrate or even encourage consumers to re-pack products at home [10,11,43,46,111,223,258]. A point emphasized several times is also the right-sizing of portions to avoid food waste at the consumer level. This is a measure that, despite the increased packaging effort, can lead to a lower total environmental impact [111,120,121,228]. Next, the communication function of packaging, which has been somewhat overlooked by studies, could additionally play a significant role in food waste prevention in the future, as it can have a considerable influence on consumer behavior [12,33,259,260]. Examples of implementation would be easy to read and understand directions on how to store, prepare and use products or information on how to interpret best-before or consume-by dates, as well as how to dispose of the packaging [11,37,38,43].

Turning to the cluster of recommendations on efficiency, it can be seen that in the past, an emphasis was placed on this topic by many authors and that three hotspots are reoccurring. These are packaging itself, transport and energy. In the case of packaging, the majority of authors are looking for a sweet spot, a point where minimal packaging is used, but at the same time, the quality of the product is not affected. The same applies to product waste. In this context, however, it is necessary to mention that the impetus should come from the area of optimization rather than the pure minimization or elimination of packaging. This is reported to be a target-oriented approach to find a satisfactory balance between effort and impact [28,29,37,43,111,116,119,123,141,144,261]. Further emphasis in the scientific literature is laid on material choice or substitution as well as the (re)design of product-packaging systems. For example, some authors change traditional packaging concepts such as a bag in a box to a free-standing plastic bag or a glass jar to a plastic pouch. (Re)design examples, on the other hand, are packages exhibiting a perforation, a wide neck or that stand upside-down. All are attempts to increase the efficiency of product emptying and thus product waste, which may also be achieved by altering the product itself (e.g., rheology) [11,27,43,111,116–118,120,121,129,130]. Further, the use of, for example, concentrated products is discussed. This can also lead to reduced packaging effort. The latter is also of interest for transport efficiency. Here, packaging weight, avoidance of void volume and stack-ability stand in direct correlation to transport efforts (e.g., frequency) and thus impacts. The measures applied are, next to packaging weight, the packaging-to-product ratio, cube utilization (volume) and pallet utilization. Alternatively, and where possible, bulk shipping could also be a way to increase efficiency [43,111,141,261]. With respect to energy, choosing materials with low embodied energy and further increasing the efficiency of production processes and transport as well as detachment from fossil energy sources can be named. In addition to this, the consumer stage should not be underestimated. Here, a product-packaging system that does not need to be, for example, stored under refrigerated conditions or long-life packaging (e.g., aseptic packaging) may have advantages compared to other solutions [43,111].

As for the other areas, for cyclic packaging, different recommendations are given in the scientific literature. Clustering and (potentially) ranking them could be a valuable approach to link them with the well-established waste hierarchy, which is laid down by the EU Waste Framework Directive. Here, waste prevention as well as (preparing for) reuse are the most favored options. Behind this, recycling (including the technical and biological cycle) and energy recovery are mentioned. The least preferred option should be waste disposal through a landfill [36]. Through clustering, it becomes clear that most of the points discussed by different authors already focus on the upper part of the waste hierarchy. While the prevention of waste has already been discussed in the paragraphs above, reuse strategies given include reusable, returnable and refillable solutions not only at the primary packaging levels but also at the secondary or tertiary levels. Examples are (plastic) trays and crates, molded plastic containers for specialty products, (beer) kegs, intermediate bulk containers, roll cages or (wooden or plastic) pallets. It is important to consider that strategies may work in one case but not in another. Therefore, it is necessary to identify if the respective business-to-business or business-to-consumer case allows for such solutions. Situations where this often works well are those where short distribution distances, frequent deliveries, a small number of parties or company-owned vehicles are present. Therefore, a (custom) closed-loop system can be maintained [43,111]. Where reuse is not possible but waste is still generated, the collection, sorting, and forwarding of the respective waste fractions for recycling should be the main target [36,262]. To support this, the past years have shown a steep increase in guidelines focusing on design for recycling [37,39,41,43,261,263]. While these today focus mainly on mechanical recycling, chemical recycling may also be in focus in the upcoming years. A constant point of discussion is, however, the trade-off between lightweight multilayer materials exhibiting a small environmental footprint and their recyclability [50,264]. Next to designs for recycling, designs from recycling are increasingly the focus of science and industry since they are often associated with reduced primary material and energy consumption. The use includes materials of all categories, such as glass, metal, paper and board, as well as plastic. In the latter case, it must be, however, highlighted that at the moment, mainly recycled PET is used as primary food packaging material. Most approval processes for, e.g., PE and PP are still pending due to safety concerns [50,265]. Another trend in the past years is the increased production and use of bio-based and/or bio-degradable materials (e.g., polymers) [266]. The latter may be used in scenarios where entry into the environment is foreseeable. This could be either in the form of controlled (home or industrial) composting or in the form of uncontrolled littering. This could, in certain circumstances, reduce the amount of food waste going to landfill. While there is still a debate about the actual advantages (e.g., lower carbon footprint, material properties, bio-degradability) and disadvantages (e.g., agricultural impacts, competition with food production, end-of-life management, costs) of bio-plastics in different fields of applications [267], it is well agreed that all materials, regardless the material type, should be kept in the circle as long as possible and that proper end-of-life management is needed to reduce environmental impacts. Therefore, the transformation from a linear to a recycling and ultimately to a circular economy can be accelerated [35,36,262,268,269].

Last but not least, the area of safe packaging seems not to be in the forefront focus of the reviewed literature since the effects are mainly noticeable in other impact categories than GHG emissions. What can be said is, however, that the avoidance of hazardous substances (including GHG active substances) as well as cleaner production (e.g., avoidance of volatile organic components) can, next to ecological stewardship and litter reduction (e.g., small parts of packaging), support the transition towards a more sustainable future [35,37,43,44,261,268].

4.2. Life Cycle Assessment

In the past, a large number of LCAs were carried out in the food sector. It is clear that not every issue requires the inclusion of packaging. However, where packaging has been included in LCAs in one way or another, this often has not been sufficiently addressed [13,111]. The following paragraphs, therefore, aim to provide suggestions that show the potential to improve the quality of future studies and the validity of packaging-related conclusions drawn from them. To structure this, the multi-step approach based on ISO 14040 and 14044, (i) goal and scope, (ii) life cycle inventory, (iii) life cycle impact assessment and (iv) interpretation, is used again for this purpose [66,67] (see also Table 4).

Table 4. Recommendations for improving food packaging life cycle assessments (LCAs) based on the structure given by [66,67].

Life Cycle Assessment Stage	Recommendation	Reference
Goal and scope	Holistic representation of the food packaging system	[43,111]
	Inclusion of all packaging levels	[43,111]
	Inclusion of direct and indirect packaging effects	[43,52,111]
	Awareness of interrelation	[43,111]
	Integration of Circular Economy principles within the goal and scope of food packaging LCAs	[270–272]
	Special attention to time, geography and technology aspects	[130,221,273]
Life cycle inventory	Focus on appropriate and reasonable high-quality data and software	[43,52,120,134,144]
	Provision of data transparency and consistency	[274]
	Usage of common language (definitions)	[51]
	Inclusion of details on packaging	[41]
	Inclusion of actual packaging recyclability and recycling quotas	[39,41]
	Inclusion of food and packaging waste	[111]
	Inclusion of consumer attitudes and behavior	[111]
Life cycle impact assessment	Use and build upon standards	[66,67,102]
	Include sensitivity or scenario analyses	[52,66,67,111] based on [12,13,275]
Interpretation	Discuss limitations	[43,52,111]
	Address trade-offs and burden-shifting	[31,134]
	Use multi-criteria decision analysis (MCDA)	[31,134]
	Only give sufficiently substantiated recommendations	[52,138]

Starting with the goal and scope of a packaging-related LCA, it has to be stressed that the holistic representation of the entire food packaging system is a prerequisite for all further steps. This means that packaging relevant points beyond production and waste management have to be included. These are, for example, indirect effects such as food waste or transport efficiency along the supply chain. Further, all packaging levels, from primary to tertiary packaging, should be considered, and awareness of their interrelationship should be given. This is relevant, for example, in comparative studies where different packaging variants are included [43,111,221].

Another issue that is worth addressing is the increasingly important concept of the Circular Economy. A new legislative initiative undertaken by the European Commission in adopting the Circular Economy Action Plan in 2015 had a significant impact on the field of packaging. This initiative led to changes in existing directives and the imposition of stricter rules as well as the introduction of the Product Environmental Footprint (PEF) circularity formula [270].

Further, the CEN/TR 13910:2010 report on criteria and methodologies for LCAs of packaging also mentions the importance of time, geography and technology aspects within the goal and scope definition as well as data collection phases of LCA. These time and technology aspects are important due to the characteristically short life cycle of packaging (e.g., design changes). The geographical aspect considers different supply chains across several countries and continents [221].

Building upon this sharpened approach, it is further necessary to increase efforts in the area of life cycle inventory to achieve meaningful results. First and foremost, data quality can be mentioned here [43,120,134]. Although it is well-known that data gathering can be quite resource-intensive (e.g., time, budget), ideally, primary data (e.g., directly (on-site) collected data) should be used. However, if not otherwise possible, secondary data (e.g., database, reports, statistics) may also be taken. Furthermore, in some cases, assumptions may be necessary [43,52,120,134]. With secondary data selection, there is also another issue. LCA software very often comes bundled with specific databases, and there is evidence that the choice of software used for environmental analysis can affect the relative comparisons between differing package system options and, therefore, the decisions that will be made. This effect is magnified by the natural inclination of the user to employ data sets that are "convenient" when using specific software packages [276]. Regardless of the source, however, it is helpful to present the information in the studies themselves or in the appendix in a transparent and bundled manner in order to promote the progress of the research field as well as comparability. This is a point that is increasingly requested by different stakeholders and encouraged by scientific journals on LCA such as The International Journal of Life Cycle Assessment and Environmental Impact Assessment Review [220,277]. Moreover, care should be taken to use widely accepted definitions (e.g., ISO standards) to avoid the misinterpretation of, for example, packaging levels [51].

In relation to primary, secondary and tertiary packaging, it is advisable to collect information that exceeds the one on the base material used. This refers to information on the packaging material (e.g., exact material, size, additives, barrier, color, print), packaging aids (e.g., closure, liner, gasket, valve) and decorations (e.g., labels, adhesives, decoration, size) [41] as well as any other relevant points such as modified atmosphere packaging (MAP) [46,254] or active and intelligent packaging (AIP) [46,47,255–257]. Although, at first glance, it may seem a bit far-fetched, addressing these points helps to assess the actual recyclability of a packaging solution in a target market or region (e.g., by using (inter)national guidelines) and potentially point out improvement possibilities [39,263]. Looking at the markets in more detail, it should be noted that some (federal) states have different collection, sorting and recycling practices, which means that recovery rates may differ in some cases from the average values for a country [278]. Accordingly, more focus should be placed on these currently rather underrepresented points to further increase the validity of LCA results.

Further, more attention should be paid to food and packaging waste generated at different supply chain stages (e.g., production waste, loss during transport and retail) and where the remainder of this waste is. Especially in efficiency-driven countries, data up to retail is often available. At the consumer level, however, the data situation is often less satisfactory. Therefore, more attention should be paid to better understanding consumer behavior and attitudes in the future. Points of interest could be consumers' preference for food/packaging, un/re-packing habits, storage and use of products, food waste as well as engagement in separation and disposal of packaging and preference for, e.g., bio-based and biodegradable/compostable packaging materials [56,111].

Turning to the LCIA, it can be reiterated that existing (e.g., ISO) and recently developed standards (e.g., PEF) provide a solid basis for the calculation of environmental impacts [66,67,102,103]. In the context of these, sensitivity or scenario analyses are mentioned, as they are a method to check for the validity of results or to describe possible variations/situations [66,67]. Applying this supports the authors if, for instance, different assumptions have to be made or the importance of different packaging attributes is to be tested [52,111]. A possible approach in relation to, for example, food waste originating from different packaging solutions would be the following: (i) examination of the situation (e.g., amount, reason) and gathering of supporting primary (e.g., experiments) or secondary data (e.g., literature), (ii) identification, definition and evaluation (e.g., experiments) of influencing packaging attributes, (iii) scenario development (e.g., alteration of packaging size) and evaluation as well as (iv) calculation and interpretation of results [52] based on [12,13,275].

Last but not least, interpretation of results has the potential to be improved in future LCAs. Depending on whether the respective study has a packaging focus (*packaging* LCA) or not (*food* LCA), different recommendations can be found in the literature. For *packaging* LCAs, awareness about limitations (even implicit ones) of the conducted study as well as transparent reflection thereof in the corresponding discussion can be highlighted [43,52,111]. This should include, once more, currently underrepresented points such as interdependencies of packaging levels, consumers or waste-related issues [52,111,221]. Furthermore, trade-offs and possible burden-shifting can be addressed using, for example, single-score values or multi-criteria decision analysis (MCDA) [31,134]. Where such critical discourse is, e.g., due to space limitation, not possible, giving recommendations or directions for packaging (re)design should therefore be refrained from. On the contrary, it would be more beneficial to underline the need for further research. The latter also applies to *food* LCAs [111].

4.3. Management

When it comes to promoting sustainable food packaging systems, different challenges and opportunities exist. The challenges include, for example, established economic systems that are traditionally strongly oriented toward growth and profit and are slow to implement necessary changes. In addition, there is often a need for improved holistic sustainability awareness, networking and exchange with the economic environment. This finds reflection until the single company and department level [43,52].

In order to more easily overcome the activation energy required for a change, various catalytic measures can be adopted on different levels (see also Table 5). At a meta or policy level, which rather reflects a top-down approach, incentives [52,111] such as corresponding legal frameworks, facilitation for exemplary companies [15,268,279], as well as support or funding for research, development and innovation can be named [222,280]. This motivates companies along the food supply chain to develop new business models in which saving resources and reducing or avoiding food losses and food waste are valued and gains and risks are shared equally [52]. Further impetus provides strong engagement and the cross-linking of relevant stakeholders (e.g., industry, government [130]) to promote best practices (e.g., recyclable packaging), standards, as well as an open (science) approach [274,281,282]. Education offensives at different levels are also seen as helpful. Therefore, for example, more and more schools and universities include packaging in their curricula [283].

Table 5. Recommendations for management-related activities to promote sustainable packaging.

Recommendation	Reference
Give incentives	[52]
Develop new business models	[52]
Engage and connect stakeholders	[130]
Follow an open (science) approach and promote best practices and standards	[274,284]
Promote education	[283]
Develop companies to sustaining corporations	[43,285]
Strengthen collaboration and communication	[26,116,130]
Avoid double efforts	[26,116,130]
Identification of environmental hotspots and potentials for change	[27,117]
Combine and prioritize actions	[27,117]
Extensively test (re)designed packaging solutions	[43,46–48]
Communicate sustainability aspects transparently and provide evidence	[121,138]
Avoid misleading or greenwashing	[124,141,286]

Next to this, the bottom-up approach also bears huge innovation potential. In particular, a lot can be expected from companies that, with reference to the sustainability phase model, have already left the phases of rejection, non-responsiveness, compliance and efficiency behind them and are already operating at the levels of strategic proactivity and a sustaining corporation [43,285,287]. As above, the cooperative approach should be emphasized here. For instance, science and industry can collaborate to develop improved food and packaging solutions, or communication along the supply chain can promote overall sustainability and avoid double efforts [26,43,116,130].

At the company level, the management of sustainable packaging development should target the identification of environmental hotspots and potentials for change (see also Section 4.2) as well as combining and prioritizing actions (see also Section 4.1) [27,117]. Here, it is especially important that supposedly more sustainable packaging approaches or solutions are also tested extensively (e.g., packaging performance, product quality, shelf life and waste, consumer attitudes and handling, environmental impact) in order to ultimately bring a product onto the market that is successful in all dimensions [43,46–48,70]. In times like these, when different consumers and other stakeholders are becoming increasingly aware of the sustainability of food packaging [74], it is vital to communicate the developments made in a transparent manner and provide factual information about the sustainability aspects of packaging. Explicit (e.g., text, labels, certificates) and implicit (e.g., pictures and graphics, colors, haptics, font, shape) communication thereby can take place through a variety of channels [56]. This can include, for example, on the packaging itself, but also on websites or various other advertising channels [121,138,140]. Whichever way is used to communicate, it is particularly important that there is no misleading or greenwashing [124,138,141,259,286] in this context, which is picked up in a recent initiative on substantiating green claims by the European Union [255,288,289].

5. Conclusions

In the past, it has been shown that packaging can have positive environmental effects, especially when it protects resource-intensive food products and thus prevents losses and waste of the same. This is an essential point when it comes to reducing GHG emissions associated with the global food supply chain. In the present review with a focus on LCA studies, it was shown that the average contribution of packaging to the overall footprint of the product packaging system is 9.18% for the product group of cereals and confectionery, which has not been the explicit focus of scientific literature to date. This value is approximately twice as high as the estimated value for global GHG emissions for packaging but fits in well with previous dimensions for packaging of various food groups, which range from a few percent to more than one-third. In this context, however, it must be emphatically pointed out that direct comparisons in this area are not permissible or are difficult to carry out, as the studies differ greatly in some cases. The results can therefore be seen more as a size estimate.

In addition, the present review provided valuable information about the type and quality with which packaging has been included in analyses so far. In particular, it showed that packaging was often not in focus, and if it was, it was often not sufficiently included at all levels (primary, secondary and tertiary). It also showed that mainly direct (e.g., material) and not indirect impacts (e.g., food waste, transport efficiency) were considered and that data quality and presentation could be improved.

Based on these evaluations and including further literature, recommendations for the sustainable design of food packaging, its analysis by means of LCA and innovation-supporting management could be given. In the area of packaging, it can be particularly emphasized that packaging must be designed to be effective, efficient, recyclable and safe, and that interrelationships between the individual packaging levels must always be considered. With LCA, on the other hand, it is necessary not to lose sight of packaging from the beginning, including the definition of the goal and the scope, through the LCI process over LCIA to the interpretation and issue of recommendations. In addition, to obtain accurate results, primary data should be used whenever possible, while secondary data are recommended for a rough estimate of influences. LCA practitioners should also refrain from issuing packaging-related recommendations if these have not previously been sufficiently included in the studies. In this case, the reference to the need for further studies is more appropriate. Last but not least, the management-related part dealt with how innovation can be fueled at different levels and showed that collaboration as well as transparent and honest communication of sustainability aspects within the supply chain and towards the consumer is a key instrument for realizing sustainability at all levels.

Against this background, the authors see considerable research and development potential in the areas of better coverage of the cereal and confectionary product group, optimization of packaging and evaluation of the actual influence of the same, the meaningful design of LCAs, the demonstration of indirect packaging effects along the supply chain, new business models and models for cooperation as well as communication of sustainability aspects.

Author Contributions: Conceptualization, V.K. and A.-S.B.; methodology, V.K.; validation, M.M., K.R.D., G.G. and T.V.; investigation, J.K., K.L., V.K.; writing—original draft preparation, V.K.; writing—review and editing, V.K., A.-S.B., M.M., K.R.D., G.G., K.L., T.V.; visualization, A.-S.B.; supervision, T.V.; project administration, V.K. and A.-S.B.; funding acquisition, V.K. All authors have read and agreed to the published version of the manuscript.

Funding: This article/publication is based upon work from COST Action Circul-a-bility, supported by COST (European Cooperation in Science and Technology), www.cost.eu (accessed on 28 March 2022).

Acknowledgments: The authors want to thank Bernd Brand for providing valuable feedback on the manuscript.

Conflicts of Interest: The authors declare no conflict of interest. The funders had no role in the design of the study; in the collection, analyses, or interpretation of data; in the writing of the manuscript, or in the decision to publish the results.

References

1. Bauer, A.-S.; Leppik, K.; Galić, K.; Anestopoulos, I.; Panayiotidis, M.I.; Agriopoulou, S.; Milousi, M.; Uysal-Unalan, I.; Varzakas, T.; Krauter, V. Cereal and Confectionary Packaging: Background, Application and Shelf-Life Extension. *Foods* **2022**, *11*, 697. [CrossRef] [PubMed]
2. Crippa, M.; Solazzo, E.; Guizzardi, D.; Monforti-Ferrario, F.; Tubiello, F.N.; Leip, A. Food systems are responsible for a third of global anthropogenic GHG emissions. *Nat. Food* **2021**, *2*, 198–209. [CrossRef]
3. Poore, J.; Nemecek, T. Reducing food's environmental impacts through producers and consumers. *Science* **2018**, *360*, 987–992. [CrossRef]
4. Vermeulen, S.J.; Campbell, B.M.; Ingram, J.S. Climate Change and Food Systems. *Annu. Rev. Environ. Resour.* **2012**, *37*, 195–222. [CrossRef]

5. Jungbluth, N.; Tietje, O.; Scholz, R.W. Food purchases: Impacts from the consumers' point of view investigated with a modular LCA. *Int. J. Life Cycle Assess.* **2000**, *5*, 134–142. [CrossRef]

6. Verghese, K.; Crossin, E.; Clune, S.; Lockrey, S.; Williams, H.; Rio, M.; Wikström, F. The greenhouse gas profile of a "Hungry Planet"; quantifying the impacts of the weekly food purchases including associated packaging and food waste of three families. In Proceedings of the 19th IAPRI World Conference on Packaging, Melbourne, Australia, 15–18 June 2014.

7. Heller, M.C.; Selke, S.E.M.; Keoleian, G.A. Mapping the Influence of Food Waste in Food Packaging Environmental Performance Assessments. *J. Ind. Ecol.* **2019**, *23*, 480–495. [CrossRef]

8. Olsson, A.; Hellström, D. (Eds.) *Managing Packaging Design for Sustainable Development: A Compass for Strategic Directions*; John Wiley & Sons: Chichester, UK; Hoboken, NJ, USA, 2017; ISBN 9781119151036.

9. Licciardello, F. Packaging, blessing in disguise. Review on its diverse contribution to food sustainability. *Trends Food Sci. Technol.* **2017**, *65*, 32–39. [CrossRef]

10. Wikström, F.; Williams, H. Potential environmental gains from reducing food losses through development of new packaging-a life-cycle model. *Packag. Technol. Sci.* **2010**, *23*, 403–411. [CrossRef]

11. Williams, H.; Wikström, F. Environmental impact of packaging and food losses in a life cycle perspective: A comparative analysis of five food items. *J. Clean. Prod.* **2011**, *19*, 43–48. [CrossRef]

12. Wikström, F.; Williams, H.; Venkatesh, G. The influence of packaging attributes on recycling and food waste behaviour–An environmental comparison of two packaging alternatives. *J. Clean. Prod.* **2016**, *137*, 895–902. [CrossRef]

13. Wikström, F.; Williams, H.; Verghese, K.; Clune, S. The influence of packaging attributes on consumer behaviour in food-packaging life cycle assessment studies-a neglected topic. *J. Clean. Prod.* **2014**, *73*, 100–108. [CrossRef]

14. Hopewell, J.; Dvorak, R.; Kosior, E. Plastics recycling: Challenges and opportunities. *Philos. Trans. R. Soc. Lond. B Biol. Sci.* **2009**, *364*, 2115–2126. [CrossRef] [PubMed]

15. European Commission. Communication from the Commission to the European Parliament, the European Council, the Council, the European Economic and Social Committee and the Committee of the Regions: The European Green Deal, Brussels. 2019. Available online: https://eur-lex.europa.eu/resource.html?uri=cellar:b828d165-1c22-11ea-8c1f-01aa75ed71a1.0002.02/DOC_1&format=PDF (accessed on 2 February 2022).

16. European Commission. Communication from the Commission to the European Parliament, the Council, the European Economic and Social Committee and the Committee of the Regions: A Farm to Fork Strategy for a Fair, Healthy and Environmentally-Friendly Food System, Brussels. 2020. Available online: https://eur-lex.europa.eu/legal-content/EN/TXT/?uri=CELEX:52020DC0381 (accessed on 2 February 2022).

17. United Nations. Resolution Adopted by the General Assembly on 25 September 2015: Transforming Our World: The 2030 Agenda for Sustainable Development. 2015. Available online: https://www.un.org/ga/search/view_doc.asp?symbol=A/RES/70/1&Lang=E (accessed on 2 February 2022).

18. Clune, S.; Crossin, E.; Verghese, K. Systematic review of greenhouse gas emissions for different fresh food categories. *J. Clean. Prod.* **2017**, *140*, 766–783. [CrossRef]

19. HLPE. Food Losses and Waste in the Context of Sustainable Food Systems: A Report by the High Level Panel of Experts on Food Security and Nutrition of the Committee on World Food Security, Rome. 2014. Available online: http://www.fao.org/3/i3901e/i3901e.pdf (accessed on 2 February 2022).

20. Food and Agriculture Organization of the United Nations. Global Food Losses and Food Waste: Extent, Causes and Prevention. Study Conducted for the International Congress SAVE FOOD! At Interpack2011 Düsseldorf, Germany, Rome. 2011. Available online: http://www.fao.org/3/mb060e/mb060e.pdf (accessed on 2 February 2022).

21. Notarnicola, B.; Tassielli, G.; Renzulli, P.A.; Castellani, V.; Sala, S. Environmental impacts of food consumption in Europe. *J. Clean. Prod.* **2017**, *140*, 753–765. [CrossRef]

22. Garnett, T. Where are the best opportunities for reducing greenhouse gas emissions in the food system (including the food chain)? *Food Policy* **2011**, *36*, S23–S32. [CrossRef]

23. Caobisco. Facts and Figures: Key Data of the European Sector (EU27 + Switzerland and Norway). Available online: https://caobisco.eu/facts/ (accessed on 17 January 2022).

24. EUROSTAT. EU Production of Chocolate. Available online: https://ec.europa.eu/eurostat/web/products-eurostat-news/-/ddn-20200831-1 (accessed on 9 February 2022).

25. EUROSTAT. Main Producers of Chocolate in the EU. Available online: https://ec.europa.eu/eurostat/en/web/products-eurostat-news/-/edn-20190417-1 (accessed on 9 February 2022).

26. Miah, J.H.; Griffiths, A.; McNeill, R.; Halvorson, S.; Schenker, U.; Espinoza-Orias, N.D.; Morse, S.; Yang, A.; Sadhukhan, J. Environmental management of confectionery products: Life cycle impacts and improvement strategies. *J. Clean. Prod.* **2018**, *177*, 732–751. [CrossRef]

27. Jeswani, H.K.; Burkinshaw, R.; Azapagic, A. Environmental sustainability issues in the food–energy–water nexus: Breakfast cereals and snacks. *Sustain. Prod. Consum.* **2015**, *2*, 17–28. [CrossRef]

28. Konstantas, A.; Jeswani, H.K.; Stamford, L.; Azapagic, A. Environmental impacts of chocolate production and consumption in the UK. *Food Res. Int.* **2018**, *106*, 1012–1025. [CrossRef]

29. Konstantas, A.; Stamford, L.; Azapagic, A. Evaluating the environmental sustainability of cakes. *Sustain. Prod. Consum.* **2019**, *19*, 169–180. [CrossRef]

30. Konstantas, A.; Stamford, L.; Azapagic, A. Evaluation of environmental sustainability of biscuits at the product and sectoral levels. *J. Clean. Prod.* **2019**, *230*, 1217–1228. [CrossRef]
31. Noya, L.I.; Vasilaki, V.; Stojceska, V.; González-García, S.; Kleynhans, C.; Tassou, S.; Moreira, M.T.; Katsou, E. An environmental evaluation of food supply chain using life cycle assessment: A case study on gluten free biscuit products. *J. Clean. Prod.* **2018**, *170*, 451–461. [CrossRef]
32. Recanati, F.; Marveggio, D.; Dotelli, G. From beans to bar: A life cycle assessment towards sustainable chocolate supply chain. *Sci. Total Environ.* **2018**, *613–614*, 1013–1023. [CrossRef] [PubMed]
33. Zeng, T.; Durif, F. The Impact of Eco-Design Packaging on Food Waste Avoidance: A Conceptual Framework. *J. Promot. Manag.* **2020**, *26*, 768–790. [CrossRef]
34. Pauer, E.; Wohner, B.; Heinrich, V.; Tacker, M. Assessing the Environmental Sustainability of Food Packaging: An Extended Life Cycle Assessment including Packaging-Related Food Losses and Waste and Circularity Assessment. *Sustainability* **2019**, *11*, 925. [CrossRef]
35. *European Parliament and Council Directive 94/62/EC of 20 December 1994 on Packaging and Packaging Waste*; European Council: Brussels, Belgium, 1994.
36. *Directive 2008/98/EC of the European Parliament and of the Council of 19 November 2008 on Waste and Repealing Certain Directives (Text with EEA Relevance)*; European Council: Brussels, Belgium, 2008.
37. Walmart. Sustainable Packaging Playbook: A Guidebook for Suppliers to Improve Packaging Sustainability. Available online: https://s4rbimagestore.blob.core.windows.net/images/rightnow/walmartsustainability.custhelp.com/for_answers/packagingplaybook.pdf (accessed on 9 February 2022).
38. Australian Packaging Covenant Organization. Sustainable Packaging Guidelines. 2021. Available online: https://documents.packagingcovenant.org.au/public-documents/Principles%20of%20the%20SPGs%20-%20Content%20For%20Translation (accessed on 9 February 2022).
39. GS1 Austria GmbH; ECR Austria; FH Campus Wien; Circular Analytics TK GmbH. Packaging Design for Recycling: A Global Recommendation for 'Circular Packaging Design'. Available online: https://www.ecr-community.org/global-recyclable-packaging-guide/ (accessed on 9 February 2022).
40. The Consumer Goods Forum. Global Protocol on Packaging Sustainability 2.0. 2011. Available online: https://www.theconsumergoodsforum.com/wp-content/uploads/2017/11/CGF-Global-Protocol-on-Packaging.pdf (accessed on 9 February 2022).
41. FH Campus Wien; Circular Analytics TK GmbH. Circular Packaging Design Guideline: Empfehlungen für die Gestaltung Recyclinggerechter Verpackungen, Vienna. 2021. Available online: https://www.fh-campuswien.ac.at/fileadmin/redakteure/Forschung/FH-Campus-Wien_Circular-Packaging-Design-Guideline_V04_DE.pdf (accessed on 9 February 2022).
42. Sustainable Packaging Alliance. Sustainable Packaging Alliance. Available online: https://www.sustainablepack.org/ (accessed on 9 February 2022).
43. Verghese, K.; Lewis, H.; Fitzpatrick, L. (Eds.) *Packaging for Sustainability*; Springer: London, UK, 2012; ISBN 9780857299871.
44. Sustainable Packaging Coalition. Definition of Sustainable Packaging. 2011. Available online: https://sustainablepackaging.org/wp-content/uploads/2017/09/Definition-of-Sustainable-Packaging.pdf (accessed on 9 February 2022).
45. Lewis, H.; Sonneveld, K.; Fitzpatrick, L.; Nicol, R. Towards Sustainable Packaging: Discussion Paper. 2002. Available online: http://www.sustainablepack.org/database/files/filestorage/Towards%20Sustainable%20 (accessed on 14 May 2009).
46. Robertson, G.L. *Food Packaging: Principles and Practice*, 3rd ed.; CRC Press: Boca Raton, FL, USA, 2013; ISBN 9781439862414.
47. Singh, P.; Wani, A.A.; Langowski, H.-C. (Eds.) *Food Packaging Materials: Testing & Quality Assurance*; CRC Press: Boca Raton, FL, USA; Taylor & Francis Group: London, UK; New York, NY, USA, 2017; ISBN 9781466559943.
48. Soroka, W. *Fundamentals of Packaging Technology*, 5th ed.; Institute of Packaging Professional: Herndon, VA, USA, 2014; ISBN 0615709346.
49. Lewis, H.; Fitzpatrick, L.; Verghese, K.; Sonneveld, K.; Jordon, R. Sustainable Packaging Redefined: DRAFT. 2007. Available online: http://www.helenlewisresearch.com.au/wp-content/uploads/2012/03/Sustainable-Packaging-Redefined-Nov-2007.pdf (accessed on 9 February 2022).
50. Bauer, A.-S.; Tacker, M.; Uysal-Unalan, I.; Cruz, R.M.S.; Varzakas, T.; Krauter, V. Recyclability and Redesign Challenges in Multilayer Flexible Food Packaging-A Review. *Foods* **2021**, *10*, 2702. [CrossRef]
51. *ISO. 21067:2007(en)*; Packaging—Vocabulary. Available online: https://www.iso.org/obp/ui/#iso:std:iso:21067:ed-1:v1:en (accessed on 2 February 2022).
52. Wikström, F.; Verghese, K.; Auras, R.; Olsson, A.; Williams, H.; Wever, R.; Grönman, K.; Kvalvåg Pettersen, M.; Møller, H.; Soukka, R. Packaging Strategies That Save Food: A Research Agenda for 2030. *J. Ind. Ecol.* **2019**, *23*, 532–540. [CrossRef]
53. Jiménez-Guerrero, J.F.; Gázquez-Abad, J.C.; Ceballos-Santamaría, G. Innovation in eco-packaging in private labels. *Innovation* **2015**, *17*, 81–90. [CrossRef]
54. Wandosell, G.; Parra-Meroño, M.C.; Alcayde, A.; Baños, R. Green Packaging from Consumer and Business Perspectives. *Sustainability* **2021**, *13*, 1356. [CrossRef]
55. Lindh, H.; Olsson, A.; Williams, H. Consumer Perceptions of Food Packaging: Contributing to or Counteracting Environmentally Sustainable Development? *Packag. Technol. Sci.* **2016**, *29*, 3–23. [CrossRef]
56. Herbes, C.; Beuthner, C.; Ramme, I. Consumer attitudes towards biobased packaging–A cross-cultural comparative study. *J. Clean. Prod.* **2018**, *194*, 203–218. [CrossRef]

57. Woodward, D.G. Life cycle costing—Theory, information acquisition and application. *Int. J. Proj. Manag.* **1997**, *15*, 335–344. [CrossRef]
58. Laso, J.; García-Herrero, I.; Margallo, M.; Vázquez-Rowe, I.; Fullana, P.; Bala, A.; Gazulla, C.; Irabien, Á.; Aldaco, R. Finding an economic and environmental balance in value chains based on circular economy thinking: An eco-efficiency methodology applied to the fish canning industry. *Resour. Conserv. Recycl.* **2018**, *133*, 428–437. [CrossRef]
59. Hunkeler, D.; Lichtenvort, K.; Rebitzer, G. *Environmental Life Cycle Costing*, 1st ed.; CRC Press: Boca Raton, FL, USA, 2008; ISBN 9780429140440.
60. Konstantas, A.; Stamford, L.; Azapagic, A. Economic sustainability of food supply chains: Life cycle costs and value added in the confectionary and frozen desserts sectors. *Sci. Total Environ.* **2019**, *670*, 902–914. [CrossRef] [PubMed]
61. Martinez-Sanchez, V.; Tonini, D.; Møller, F.; Astrup, T.F. Life-Cycle Costing of Food Waste Management in Denmark: Importance of Indirect Effects. *Environ. Sci. Technol.* **2016**, *50*, 4513–4523. [CrossRef] [PubMed]
62. Jørgensen, A.; Le Bocq, A.; Nazarkina, L.; Hauschild, M. Methodologies for social life cycle assessment. *Int. J. Life Cycle Assess.* **2008**, *13*, 96–103. [CrossRef]
63. Vinyes, E.; Oliver-Solà, J.; Ugaya, C.; Rieradevall, J.; Gasol, C.M. Application of LCSA to used cooking oil waste management. *Int. J. Life Cycle Assess.* **2013**, *18*, 445–455. [CrossRef]
64. Benoît, C.; Norris, G.A.; Valdivia, S.; Ciroth, A.; Moberg, A.; Bos, U.; Prakash, S.; Ugaya, C.; Beck, T. The guidelines for social life cycle assessment of products: Just in time! *Int. J. Life Cycle Assess.* **2010**, *15*, 156–163. [CrossRef]
65. Chhipi-Shrestha, G.K.; Hewage, K.; Sadiq, R. 'Socializing' sustainability: A critical review on current development status of social life cycle impact assessment method. *Clean Technol. Environ. Policy* **2015**, *17*, 579–596. [CrossRef]
66. *ISO. 14040:2006*. Available online: https://www.iso.org/standard/37456.html (accessed on 9 February 2022).
67. *ISO. 14044:2006*. Available online: https://www.iso.org/standard/38498.html (accessed on 9 February 2022).
68. European Commission. Single Market for Green Products-The Product Environmental Footprint Pilots-Environment-European Commission. Available online: https://ec.europa.eu/environment/eussd/smgp/ef_pilots.htm (accessed on 9 February 2022).
69. Klöpffer, W.; Grahl, B. *Life Cycle Assessment (LCA): A Guide to Best Practice*; Wiley-VCH: Weinheim an der Bergstrasse, Germany, 2014; ISBN 1306550475.
70. Pauer, E.; Heinrich, V.; Tacker, M. Methods for the Assessment of Environmental Sustainability of Packaging: A review. *IJRDO-J. Agric. Res.* **2018**, *3*, 33–62.
71. United Nations Environment Programme. Guidelines for Social Life Cycle Assessment of Products and Organizations. 2020. Available online: https://www.lifecycleinitiative.org/wp-content/uploads/2021/01/Guidelines-for-Social-Life-Cycle-Assessment-of-Products-and-Organizations-2020-22.1.21sml.pdf (accessed on 9 February 2022).
72. Ramos Huarachi, D.A.; Piekarski, C.M.; Puglieri, F.N.; de Francisco, A.C. Past and future of Social Life Cycle Assessment: Historical evolution and research trends. *J. Clean. Prod.* **2020**, *264*, 121506. [CrossRef]
73. De Koeijer, B.; de Lange, J.; Wever, R. Desired, Perceived, and Achieved Sustainability: Trade-Offs in Strategic and Operational Packaging Development. *Sustainability* **2017**, *9*, 1923. [CrossRef]
74. Magnier, L.; Crié, D. Communicating packaging eco-friendliness. *Int. J. Retail Distrib. Manag.* **2015**, *43*, 350–366. [CrossRef]
75. Steenis, N.D.; van Herpen, E.; van der Lans, I.A.; Ligthart, T.N.; van Trijp, H.C. Consumer response to packaging design: The role of packaging materials and graphics in sustainability perceptions and product evaluations. *J. Clean. Prod.* **2017**, *162*, 286–298. [CrossRef]
76. Steenis, N.D.; van der Lans, I.A.; van Herpen, E.; van Trijp, H.C. Effects of sustainable design strategies on consumer preferences for redesigned packaging. *J. Clean. Prod.* **2018**, *205*, 854–865. [CrossRef]
77. Magnier, L.; Schoormans, J.; Mugge, R. Judging a product by its cover: Packaging sustainability and perceptions of quality in food products. *Food Qual. Prefer.* **2016**, *53*, 132–142. [CrossRef]
78. Herbes, C.; Beuthner, C.; Ramme, I. How green is your packaging—A comparative international study of cues consumers use to recognize environmentally friendly packaging. *Int. J. Consum. Stud.* **2020**, *44*, 258–271. [CrossRef]
79. Taufik, D.; Reinders, M.J.; Molenveld, K.; Onwezen, M.C. The paradox between the environmental appeal of bio-based plastic packaging for consumers and their disposal behaviour. *Sci. Total Environ.* **2020**, *705*, 135820. [CrossRef]
80. Nguyen, A.T.; Parker, L.; Brennan, L.; Lockrey, S. A consumer definition of eco-friendly packaging. *J. Clean. Prod.* **2020**, *252*, 119792. [CrossRef]
81. Magnier, L.; Mugge, R.; Schoormans, J. Turning ocean garbage into products–Consumers' evaluations of products made of recycled ocean plastic. *J. Clean. Prod.* **2019**, *215*, 84–98. [CrossRef]
82. Ketelsen, M.; Janssen, M.; Hamm, U. Consumers' response to environmentally-friendly food packaging-A systematic review. *J. Clean. Prod.* **2020**, *254*, 120123. [CrossRef]
83. Magnier, L.; Schoormans, J. Consumer reactions to sustainable packaging: The interplay of visual appearance, verbal claim and environmental concern. *J. Environ. Psychol.* **2015**, *44*, 53–62. [CrossRef]
84. Hunt, R.G.; Franklin, W.E. LCA—How it came about. *Int. J. Life Cycle Assess.* **1996**, *1*, 4–7. [CrossRef]
85. Guinée, J.B.; Lindeijer, E. (Eds.) *Handbook on Life Cycle Assessment: Operational Guide to the ISO Standards*; Kluwer: Dordrecht, The Netherlands, 2002; ISBN 978-1-4020-0557-2.
86. Hunt, R.G.; Sellers, J.D.; Franklin, W.E. Resource and environmental profile analysis: A life cycle environmental assessment for products and procedures. *Environ. Impact Assess. Rev.* **1992**, *12*, 245–269. [CrossRef]

87. Hunt, R.G.; Franklin, W.E.; Welch, R.O.; Cross, J.A.; Woodall, A.E. *Resource and Environmental Profile Analysis of Nine Beverage Container Alternatives*; EPA/530/SW-91c 1974; United States Environmental Protection Agency (US EPA), Office of Solid Waste Management Programs: Atlanta, GA, USA, 1974.

88. Detzel, A.; Mönckert, J. Environmental evaluation of aluminium cans for beverages in the German context. *Int. J. Life Cycle Assess.* **2009**, *14*, 70–79. [CrossRef]

89. Gasol, C.M.; Farreny, R.; Gabarrell, X.; Rieradevall, J. Life cycle assessment comparison among different reuse intensities for industrial wooden containers. *Int. J. Life Cycle Assess.* **2008**, *13*, 421–431. [CrossRef]

90. Belboom, S.; Renzoni, R.; Verjans, B.; Léonard, A.; Germain, A. A life cycle assessment of injectable drug primary packaging: Comparing the traditional process in glass vials with the closed vial technology (polymer vials). *Int. J. Life Cycle Assess.* **2011**, *16*, 159–167. [CrossRef]

91. Ustun Odabasi, S.; Buyukgungor, H. Comparison of Life Cycle Assessment of PET Bottle and Glass Bottle. Available online: https://www.researchgate.net/publication/314100348_Comparison_of_Life_Cycle_Assessment_of_PET_Bottle_and_Glass_Bottle (accessed on 12 April 2022).

92. Shi, S.; Yin, J. Global research on carbon footprint: A scientometric review. *Environ. Impact Assess. Rev.* **2021**, *89*, 106571. [CrossRef]

93. Von Falkenstein, E.; Wellenreuther, F.; Detzel, A. LCA studies comparing beverage cartons and alternative packaging: Can overall conclusions be drawn? *Int. J. Life Cycle Assess.* **2010**, *15*, 938–945. [CrossRef]

94. Ayres, R.U. Life cycle analysis: A critique. *Resour. Conserv. Recycl.* **1995**, *14*, 199–223. [CrossRef]

95. Verghese, K.L.; Horne, R.; Carre, A. PIQET: The design and development of an online 'streamlined' LCA tool for sustainable packaging design decision support. *Int. J. Life Cycle Assess.* **2010**, *15*, 608–620. [CrossRef]

96. Dorn, C.; Behrend, R.; Giannopoulos, D.; Napolano, L.; James, V.; Herrmann, A.; Uhlig, V.; Krause, H.; Founti, M.; Trimis, D. A Systematic LCA-enhanced KPI Evaluation towards Sustainable Manufacturing in Industrial Decision-making Processes. A Case Study in Glass and Ceramic Frits Production. *Procedia CIRP* **2016**, *48*, 158–163. [CrossRef]

97. Schonert, M.; Motz, G.; Meckel, H.; Detzel, A.; Giegrich, J.; Ostermayr, A.; Schorb, A.; Schmitz, S. Ökobilanz für Getränkeverpackungen II/Phase 2: Berichtsnummer UBA-FB 000363. 2002. Available online: https://www.umweltbundesamt.de/sites/default/files/medien/publikation/long/2180.pdf (accessed on 10 February 2022).

98. Detzel, A.; Kauertz, B.; Grahl, B.; Heinisch, J. Prüfung und Aktualisierung der Ökobilanzen für Getränkeverpackungen. 2016. Available online: https://www.umweltbundesamt.de/publikationen/pruefung-aktualisierung-der-oekobilanzen-fuer (accessed on 10 February 2022).

99. Säynäjoki, A.; Heinonen, J.; Junnila, S.; Horvath, A. Can life-cycle assessment produce reliable policy guidelines in the building sector? *Environ. Res. Lett.* **2017**, *12*, 13001. [CrossRef]

100. Sonneveld, K. The role of life cycle assessment as a decision support tool for packaging. *Packag. Technol. Sci.* **2000**, *13*, 55–61. [CrossRef]

101. Fullana i Palmer, P.; Puig, R.; Bala, A.; Baquero, G.; Riba, J.; Raugei, M. From Life Cycle Assessment to Life Cycle Management. *J. Ind. Ecol.* **2011**, *15*, 458–475. [CrossRef]

102. Manfredi, S.; Allacker, K.; Pelletier, N.; Chomkhamsri, K.; de Souza, D.M. Product Environmental Footprint (PEF) Guide. Available online: https://ec.europa.eu/environment/eussd/pdf/footprint/PEF%20methodology%20final%20draft.pdf (accessed on 12 April 2022).

103. Lehmann, A.; Bach, V.; Finkbeiner, M. Product environmental footprint in policy and market decisions: Applicability and impact assessment. *Integr. Environ. Assess. Manag.* **2015**, *11*, 417–424. [CrossRef] [PubMed]

104. Weidema, B.; Wenzel, H.; Peterson, C.; Hansen, K. The Product, Functional Unit and Reference Flows in LCA: Environmental News No. 70. 2004. Available online: https://lca-center.dk/wp-content/uploads/2015/08/The-product-functional-unit-and-reference-flows-in-LCA.pdf (accessed on 25 March 2022).

105. European Commission-Joint Research Centre-Institute for Environment and Sustainability: International Reference Life Cycle Data System (ILCD) Handbook-General Guide for Life Cycle Assessment-Detailed Guidance, Luxembourg. 2010. Available online: https://eplca.jrc.ec.europa.eu/uploads/ILCD-Handbook-General-guide-for-LCA-DETAILED-GUIDANCE-12March2010-ISBN-fin-v1.0-EN.pdf (accessed on 28 March 2022).

106. Weidema, B.P. Comparing Three Life Cycle Impact Assessment Methods from an Endpoint Perspective. *J. Ind. Ecol.* **2015**, *19*, 20–26. [CrossRef]

107. Zampori, L.; Saouter, E.; Schau, E.; Cristobal, J.; Castellani, V.; Sala, S. Guide for Interpreting Life Cycle Assessment Result, Luxembourg. 2016. Available online: https://publications.jrc.ec.europa.eu/repository/bitstream/jrc104415/lb-na-28266-en-n.pdf (accessed on 12 April 2022).

108. BSI. PAS 2050:2011: Specification for the Assessment of the Life Cycle Greenhouse Gas Emissions of Goods and Services. London, UK, 2011. 13.020.40. Available online: https://middleware.accord.bsigroup.com/pdf-preview?path=Preview%2F000000000030227173.pdf&inline=true (accessed on 25 March 2022).

109. *ISO. 14067:2018.* Available online: https://www.iso.org/standard/71206.html (accessed on 12 April 2022).

110. Wohner, B.; Pauer, E.; Heinrich, V.; Tacker, M. Packaging-Related Food Losses and Waste: An Overview of Drivers and Issues. *Sustainability* **2019**, *11*, 264. [CrossRef]

111. Molina-Besch, K.; Wikström, F.; Williams, H. The environmental impact of packaging in food supply chains—Does life cycle assessment of food provide the full picture? *Int. J. Life Cycle Assess.* **2019**, *24*, 37–50. [CrossRef]

112. European Commission. Guidance Document Describing the Food Categories in Part E of Annex II to Regulation (EC) No 1333/2008 on Food Additives. 2017. Available online: https://ec.europa.eu/food/system/files/2017-09/fs_food-improvement-agents_guidance_1333-2008_annex-2.pdf (accessed on 4 February 2022).
113. Rohatgi, A. WebPlotDigitizer-Extract Data from Plots, Images, and Maps: Version 4.5. Available online: https://automeris.io/WebPlotDigitizer/ (accessed on 10 February 2022).
114. van den Berg, N.; Huppes, G.; Lindeijer, E.; van der Ven, B.L.; Wrisberg, N.M. Quality Assessment for LCA: CML Report 152, Leiden, Netherlands. 1999. Available online: https://www.leidenuniv.nl/cml/ssp/publications/quality.pdf (accessed on 10 February 2022).
115. Curran, M.A. Strengths and Limitations of Life Cycle Assessment. In *Background and Future Prospects in Life Cycle Assessment*; Springer: Dordrecht, The Netherlands, 2014; pp. 189–206.
116. Boakye-Yiadom, K.A.; Duca, D.; Foppa Pedretti, E.; Ilari, A. Environmental Performance of Chocolate Produced in Ghana Using Life Cycle Assessment. *Sustainability* **2021**, *13*, 6155. [CrossRef]
117. Cimini, A.; Cibelli, M.; Moresi, M. Cradle-to-grave carbon footprint of dried organic pasta: Assessment and potential mitigation measures. *J. Sci. Food Agric.* **2019**, *99*, 5303–5318. [CrossRef]
118. Volpe, R.; Messineo, S.; Volpe, M.; Messineo, A. Carbon Footprint of Tree Nuts Based Consumer Products. *Sustainability* **2015**, *7*, 14917–14934. [CrossRef]
119. Büsser, S.; Jungbluth, N. LCA of Chocolate Packed in Aluminium Foil Based Packaging, Switzerland. 2009. Available online: http://www.alufoil.org/files/alufoil/sustainability/ESU_-_Chocolate_2009_-_Exec_Sum.pdf (accessed on 4 February 2022).
120. Espinoza-Orias, N.; Stichnothe, H.; Azapagic, A. The carbon footprint of bread. *Int. J. Life Cycle Assess.* **2011**, *16*, 351–365. [CrossRef]
121. Svanes, E.; Oestergaard, S.; Hanssen, O. Effects of Packaging and Food Waste Prevention by Consumers on the Environmental Impact of Production and Consumption of Bread in Norway. *Sustainability* **2019**, *11*, 43. [CrossRef]
122. Pérez-Neira, D. Energy sustainability of Ecuadorian cacao export and its contribution to climate change. A case study through product life cycle assessment. *J. Clean. Prod.* **2016**, *112*, 2560–2568. [CrossRef]
123. Pérez-Neira, D.; Copena, D.; Armengot, L.; Simón, X. Transportation can cancel out the ecological advantages of producing organic cacao: The carbon footprint of the globalized agrifood system of ecuadorian chocolate. *J. Environ. Manag.* **2020**, *276*, 111306. [CrossRef] [PubMed]
124. Kägi, T.; Wettstein, D.; Dinkel, F. Comparing rice products: Confidence intervals as a solution to avoid wrong conclusions in communicating carbon footprints. *Proc. LCA Food* **2010**, *1*, 229–233.
125. Nunes, F.A.; Seferin, M.; Maciel, V.G.; Flôres, S.H.; Ayub, M.A.Z. Life cycle greenhouse gas emissions from rice production systems in Brazil: A comparison between minimal tillage and organic farming. *J. Clean. Prod.* **2016**, *139*, 799–809. [CrossRef]
126. Usva, K.; Saarinen, M.; Katajajuuri, J.-M. Supply chain integrated LCA approach to assess environmental impacts of food production in Finland. *Agric. Food Sci.* **2009**, *18*, 460–476. [CrossRef]
127. Saget, S.; Costa, M.; Barilli, E.; Wilton de Vasconcelos, M.; Santos, C.S.; Styles, D.; Williams, M. Substituting wheat with chickpea flour in pasta production delivers more nutrition at a lower environmental cost. *Sustain. Prod. Consum.* **2020**, *24*, 26–38. [CrossRef]
128. Nette, A.; Wolf, P.; Schlüter, O.; Meyer-Aurich, A. A Comparison of Carbon Footprint and Production Cost of Different Pasta Products Based on Whole Egg and Pea Flour. *Foods* **2016**, *5*, 17. [CrossRef]
129. Bianchi, F.R.; Moreschi, L.; Gallo, M.; Vesce, E.; Del Borghi, A. Environmental analysis along the supply chain of dark, milk and white chocolate: A life cycle comparison. *Int. J. Life Cycle Assess.* **2021**, *26*, 807–821. [CrossRef]
130. Sieti, N.; Rivera, X.C.S.; Stamford, L.; Azapagic, A. Environmental impacts of baby food: Ready-made porridge products. *J. Clean. Prod.* **2019**, *212*, 1554–1567. [CrossRef]
131. Korsaeth, A.; Jacobsen, A.Z.; Roer, A.-G.; Henriksen, T.M.; Sonesson, U.; Bonesmo, H.; Skjelvåg, A.O.; Strømman, A.H. Environmental life cycle assessment of cereal and bread production in Norway. *Acta Agric. Scand. Sect. A–Anim. Sci.* **2012**, *62*, 242–253. [CrossRef]
132. Florén, B.; Sund, V.; Nilsson, K. Environmental Impact of the Consumption of Sweets, Crisps and Soft Drinks, Copenhagen. 2011. Available online: http://www.diva-portal.org/smash/get/diva2:702819/FULLTEXT01.pdf (accessed on 17 February 2022).
133. Röös, E.; Sundberg, C.; Hansson, P.-A. Uncertainties in the carbon footprint of refined wheat products: A case study on Swedish pasta. *Int. J. Life Cycle Assess.* **2011**, *16*, 338–350. [CrossRef]
134. Jensen, J.K.; Arlbjørn, J.S. Product carbon footprint of rye bread. *J. Clean. Prod.* **2014**, *82*, 45–57. [CrossRef]
135. EPD International AB. Product Category Rules. Available online: https://www.environdec.com/product-category-rules-pcr/the-pcr (accessed on 10 February 2022).
136. EPD International AB. PCR Library. Available online: https://environdec.com/pcr-library (accessed on 10 February 2022).
137. Weidema, B. Short Procedural Guideline to Identify the Functional Unit for a Product Environmental Footprint and to Delimit the Scope of Product Categories, 2.-0 LCA … 2017. Available online: https://lca-net.com/files/granularity-guideline-final_201703 31.pdf (accessed on 25 March 2022).
138. *ISO. 14025:2006*. Available online: https://www.iso.org/standard/38131.html (accessed on 10 February 2022).
139. EPD International AB. Environmental Product Declarations. Available online: https://www.environdec.com/all-about-epds/the-epd (accessed on 10 February 2022).
140. EPD International AB. EPD Library. Available online: https://www.environdec.com/library (accessed on 10 February 2022).

141. Sonesson, U.; Davis, J.; Ziegler, F. *Food Production and Emissions of Greenhouse Gases: An Overview of the Climate Impact of Different Product Groups*; Goteborg.se: Gothenburg, Swedish, 2010.
142. Lillford, P.; Hermansson, A.-M. Global missions and the critical needs of food science and technology. *Trends Food Sci. Technol.* **2021**, *111*, 800–811. [CrossRef]
143. HLPE. Nutrition and Food Systems: A Report by the High Level Panel of Experts on Food Security and Nutrition of the Committee on World Food Security, Rome. 2017. Available online: http://www.fao.org/3/i7846e/i7846e.pdf (accessed on 2 February 2022).
144. Miah, J.H.; Griffiths, A.; McNeill, R.; Halvorson, S.; Schenker, U.; Espinoza-Orias, N.; Morse, S.; Yang, A.; Sadhukhan, J. A framework for increasing the availability of life cycle inventory data based on the role of multinational companies. *Int. J. Life Cycle Assess.* **2018**, *23*, 1744–1760. [CrossRef]
145. Saner, D.; Walser, T.; Vadenbo, C.O. End-of-life and waste management in life cycle assessment—Zurich, 6 December 2011. *Int. J. Life Cycle Assess.* **2012**, *17*, 504–510. [CrossRef]
146. EPD International AB. Wasa Sandwich Cheese & Chives: Environmental Product Declaration. 2021. Available online: https://portal.environdec.com/api/api/v1/EPDLibrary/Files/b43098fc-14bf-48f7-a9a6-08d9c4927501/Data (accessed on 10 February 2022).
147. EPD International AB. Dichiarazione Ambientale di Prodotto: Pasta di Semola di Grano Duro 100% Italiano Confezionata in Astuccio di Cartoncino. 2021. Available online: https://portal.environdec.com/api/api/v1/EPDLibrary/Files/e43a8cfd-b9f6-4fc0-aa28-08d9c4927501/Data (accessed on 10 February 2022).
148. Cederberg, C.; Berlin, J.; Henriksson, M.; Davis, J. Utsläpp av växthusgaser i ett Livscykelperspektiv för Verksamheten vid Livsmedelsföretaget Berte Qvarn (Emissions of Greenhouse Gases in a Life Cycle Perspective from the Food Company Berte Quarn, in Swedih): SIK-Report 777. RISE Research Institutes of Sweden: Göteborg, Sweden, 2008.
149. Ruini, L.; Marino, M. LCA of semolina dry pasta produced by Barilla. In Proceedings of the Sustainable Development: A Challenge for European Research, Brussels, Belgium, 26 May 2009.
150. Morris, B.A. *The Science and Technology of Flexible Packaging*; Elsevier: Amsterdam, The Netherlands, 2017; pp. 259–308. ISBN 9780323242738.
151. EPD International AB. La Semola Bio: Emvironmental Product Declaration of Organic Durum Wheat Semolina. 2021. Available online: https://portal.environdec.com/api/api/v1/EPDLibrary/Files/8822ab25-5883-4fe6-279b-08d98899db1f/Data (accessed on 10 February 2022).
152. EPD International AB. La Farina Bio: Environmental Product Declaration of Soft Wheat Organic Flour Type 00. 2021. Available online: https://portal.environdec.com/api/api/v1/EPDLibrary/Files/d9903b95-632f-4715-279d-08d98899db1f/Data (accessed on 10 February 2022).
153. EPD International AB. La Semola Kronos: Environmental Product Declaration of Kronos Durum Wheat Semolina. 2021. Available online: https://portal.environdec.com/api/api/v1/EPDLibrary/Files/62db622e-4912-49ce-279a-08d98899db1f/Data (accessed on 10 February 2022).
154. EPD International AB. Gran Cereale Mix di Cereali Croccanti Classico, Con Mela e Succhi di Frutta, Con Cioccolato: Dichiarazione Ambientale di Prodotto. 2021. Available online: https://portal.environdec.com/api/api/v1/EPDLibrary/Files/cb2a9ca1-013b-4e26-4e9e-08d900a54cf5/Data (accessed on 10 February 2022).
155. EPD International AB. Product Environmental Statement: Dried Durum Wheat Semolina Pasta–Patrimoni D'italia. 2020. Available online: https://portal.environdec.com/api/api/v1/EPDLibrary/Files/ab6d6b61-3eb2-4313-b9b2-08d8cda02dc5/Data (accessed on 21 February 2022).
156. EPD International AB. De Cecco Durum Wheat Semolina Pasta: Environmental Product Declaration. 2017. Available online: https://portal.environdec.com/api/api/v1/EPDLibrary/Files/64aa83be-3e76-41c1-abdd-d82e16ec9f78/Data (accessed on 21 February 2022).
157. EPD International AB. De Cecco Durum Wheat Semolina Egg Pasta: Environmental Product Declaration. 2017. Available online: https://portal.environdec.com/api/api/v1/EPDLibrary/Files/0909bd0f-3d65-4aa6-86d1-27e551ac6dbd/Data (accessed on 21 February 2022).
158. EPD International AB. Environmental Product Declaration: Yellow Label Sgambaro Pasta. 2020. Available online: https://portal.environdec.com/api/api/v1/EPDLibrary/Files/19073946-0ccb-42b3-b778-1936b12e662c/Data (accessed on 21 February 2022).
159. EPD International AB. Environmental Product Declaration: Pasta la Marca del Consumatore. 2020. Available online: https://portal.environdec.com/api/api/v1/EPDLibrary/Files/7f928e88-b8d5-4be8-86fe-b81138e24f31/Data (accessed on 21 February 2022).
160. EPD International AB. Durum Wheat Semolina Pasta 5kg for FoodService: Environmental Product Declaration. 2021. Available online: https://portal.environdec.com/api/api/v1/EPDLibrary/Files/b193b95b-c170-43ac-a9e4-08d9c4927501/Data (accessed on 21 February 2022).
161. EPD International AB. Dry Semolina Pasta Selezione Oro Chef: Environmental Product Declaration. 2021. Available online: https://portal.environdec.com/api/api/v1/EPDLibrary/Files/f9981fe1-6b66-4178-a9d6-08d9c4927501/Data (accessed on 21 February 2022).

162. EPD International AB. Whole Durum Wheat Semolina Pasta 1 kg for Food Service: Environmental Product Declaration. 2021. Available online: https://portal.environdec.com/api/api/v1/EPDLibrary/Files/fd2b635a-2ce8-4d36-a9c4-08d9c4927501/Data (accessed on 21 February 2022).
163. EPD International AB. Climate Declaration: For the Pasta Sgambaro Food Service Bio. Available online: https://portal.environdec.com/api/api/v1/EPDLibrary/Files/280e8f6b-7b9a-4b28-80fc-51e1ce523b01/Data (accessed on 21 February 2022).
164. EPD International AB. Climate Declaration: For the Pasta Sgambaro Food Service. Available online: https://portal.environdec.com/api/api/v1/EPDLibrary/Files/54c03318-54c0-4419-9d93-d8688ce8e2d5/Data (accessed on 21 February 2022).
165. EPD International AB. Misko Dry Semolina Pasta: Environmental Product Declaration. 2020. Available online: https://portal.environdec.com/api/api/v1/EPDLibrary/Files/537a6cc4-1ed7-4782-ba2d-093cff8734e8/Data (accessed on 21 February 2022).
166. EPD International AB. Filiz Dry Semolina Pasta: Environmental Product Declaration. 2020. Available online: https://portal.environdec.com/api/api/v1/EPDLibrary/Files/ebf474c9-6a8c-4a34-aa98-ad0a1f93a7b6/Data (accessed on 21 February 2022).
167. EPD International AB. Pasta di Semola di Grano Duro Prodotta Nello Stabilimento di Marcianise: Environmental Product Declaration. 2020. Available online: https://portal.environdec.com/api/api/v1/EPDLibrary/Files/d0b3003a-1fb7-40cf-b875-c65a80c7133a/Data (accessed on 21 February 2022).
168. EPD International AB. Dichiarazione Ambientale di Prodotto: Emiliane Chef. 2021. Available online: https://portal.environdec.com/api/api/v1/EPDLibrary/Files/9d9688d8-91fa-407f-a9ce-08d9c4927501/Data (accessed on 21 February 2022).
169. EPD International AB. Durum Wheat Semolina Pasta in Paperboard Box: Environmental Product Declaration. 2021. Available online: https://portal.environdec.com/api/api/v1/EPDLibrary/Files/106e48ea-59a2-4f53-aa01-08d9c4927501/Data (accessed on 21 February 2022).
170. EPD International AB. Dichiarazione Ambientale di Prodotto: Pasta All'uovo. 2021. Available online: https://portal.environdec.com/api/api/v1/EPDLibrary/Files/190f73a6-6b4f-459d-8af5-08d8c43682c8/Data (accessed on 21 February 2022).
171. EPD International AB. 100% Mie Nature: Environmental Product Declaration. 2020. Available online: https://portal.environdec.com/api/api/v1/EPDLibrary/Files/afd47b97-0538-4fe3-be49-74b7b67d53bf/Data (accessed on 21 February 2022).
172. EPD International AB. Extra Moelleux Nature: Environmental Product Declaration. 2020. Available online: https://portal.environdec.com/api/api/v1/EPDLibrary/Files/033ee9c2-597e-4c11-a0bb-d60914474eba/Data (accessed on 21 February 2022).
173. EPD International AB. American Sandwich Complet: Environmental Product Declaration. 2020. Available online: https://portal.environdec.com/api/api/v1/EPDLibrary/Files/78e7ac3b-8a10-408c-b455-2ed9ba31a943/Data (accessed on 21 February 2022).
174. EPD International AB. Harry's Beau&Bon Semi-Complet: Environmental Product Declaration. 2021. Available online: https://portal.environdec.com/api/api/v1/EPDLibrary/Files/d4a1ec1c-a775-4616-aa1d-08d9c4927501/Data (accessed on 21 February 2022).
175. EPD International AB. American Sandwich Nature: Environmental Product Declaration. 2020. Available online: https://portal.environdec.com/api/api/v1/EPDLibrary/Files/6bd63f5b-eafd-4a20-8bd1-da48e8563b59/Data (accessed on 19 February 2022).
176. EPD International AB. Pan Goccioli: Dichiarazione Ambientale di Prodotto. 2020. Available online: https://portal.environdec.com/api/api/v1/EPDLibrary/Files/848a5614-8dd6-418d-afc0-b0b5ebeaf5ec/Data (accessed on 21 February 2022).
177. EPD International AB. Brioches Tranchée Nature: Environmental Product Declaration. 2020. Available online: https://portal.environdec.com/api/api/v1/EPDLibrary/Files/262969ee-db8f-468c-90df-00568cabe234/Data (accessed on 21 February 2022).
178. EPD International AB. Pan Brioscè: Dichiarazione Ambientale di Prodotto. 2021. Available online: https://portal.environdec.com/api/api/v1/EPDLibrary/Files/7639ac50-e766-436b-39d8-08d99c9745fc/Data (accessed on 21 February 2022).
179. EPD International AB. Cuor di Lino: Dichiarazione Ambientale di Prodotto. 2021. Available online: https://portal.environdec.com/api/api/v1/EPDLibrary/Files/9e6e5863-e8fc-462d-39d4-08d99c9745fc/Data (accessed on 21 February 2022).
180. EPD International AB. Pan Bauletto Bianco, Grano duro, Cereali e soia, Integrale: Dichiarazione Ambientale di Prodotto. 2021. Available online: https://portal.environdec.com/api/api/v1/EPDLibrary/Files/662dd665-ab91-4eea-39d0-08d99c9745fc/Data (accessed on 21 February 2022).
181. EPD International AB. Gran Bauletto Grano Tenero e farro, Rustico, Erbe Aromatiche e Integrale Con Semi E Noci: Dichiarazione Ambientale di Prodotto. 2020. Available online: https://portal.environdec.com/api/api/v1/EPDLibrary/Files/a5c26146-b4a4-400d-a32e-c52f2253c511/Data (accessed on 21 February 2022).
182. EPD International AB. Pagnotta di Grano Duro e Integrale: Dichiarazione Ambientale di Prodotto. 2021. Available online: https://portal.environdec.com/api/api/v1/EPDLibrary/Files/7287bf42-716a-4f57-1942-08d972a96257/Data (accessed on 19 February 2022).
183. EPD International AB. PanCarrè: Dichiarazione Ambientale di Prodotto. 2020. Available online: https://portal.environdec.com/api/api/v1/EPDLibrary/Files/d1a7c8f1-c51f-4e4d-b839-b06e92fcf624/Data (accessed on 21 February 2022).
184. Wolf, B. Confectionery and Sugar-Based Foods. In *Reference Module in Food Science*; Elsevier: Amsterdam, The Netherlands, 2016; ISBN 978-0-08-100596-5.
185. EPD International AB. Wasa Ragi, Original: Environmental Product Declaration. 2021. Available online: https://portal.environdec.com/api/api/v1/EPDLibrary/Files/2d1fdfb7-2ee4-4e0c-a9fc-08d9c4927501/Data (accessed on 21 February 2022).
186. EPD International AB. Wasa Light Rye, Integrale & Delikatess: Environmental Product Declaration. 2021. Available online: https://portal.environdec.com/api/api/v1/EPDLibrary/Files/62f7828d-abe6-4da5-a9f9-08d9c4927501/Data (accessed on 21 February 2022).

187. EPD International AB. Wasa Havre and Vitalité: Environmental Product Declaration. 2021. Available online: https://portal. environdec.com/api/api/v1/EPDLibrary/Files/556e2d20-ef90-4702-a9f4-08d9c4927501/Data (accessed on 19 February 2022).
188. EPD International AB. Wasa Celebrating 100: Environmental Product Declaration. 2021. Available online: https://portal. environdec.com/api/api/v1/EPDLibrary/Files/be88b6e7-7852-490b-a9ee-08d9c4927501/Data (accessed on 21 February 2022).
189. EPD International AB. Wasa Frukost: Environmental Product Declaration. 2021. Available online: https://portal.environdec. com/api/api/v1/EPDLibrary/Files/cf56e76d-c4c6-4611-a9eb-08d9c4927501/Data (accessed on 21 February 2022).
190. EPD International AB. Sfoglia di Grano: Dichiarazione Ambientale di Prodotto. 2020. Available online: https://portal.environdec. com/api/api/v1/EPDLibrary/Files/ac02fb10-73e7-46de-91a5-77d4c7d3b634/Data (accessed on 21 February 2022).
191. EPD International AB. Wasa Rounds Sesame & Sea Salt: Environmental Product Declaration. 2021. Available online: https://portal.environdec.com/api/api/v1/EPDLibrary/Files/5d142443-04cc-4426-a9b1-08d9c4927501/Data (accessed on 21 February 2022).
192. EPD International AB. Wasa Multigrain, Surdeg Flerkorn: Environmental Product Declaration. 2021. Available online: https://portal.environdec.com/api/api/v1/EPDLibrary/Files/7f3e9948-c324-4ccb-a99b-08d9c4927501/Data (accessed on 21 February 2022).
193. EPD International AB. Wasa Husman: Environmental Product Declaration. 2021. Available online: https://portal.environdec. com/api/api/v1/EPDLibrary/Files/dc6b5576-d135-4b33-a994-08d9c4927501/Data (accessed on 21 February 2022).
194. EPD International AB. Wasa Crisp Rosemary & Seasalt: Environmental Product Declaration. 2021. Available online: https://portal. environdec.com/api/api/v1/EPDLibrary/Files/88860279-1a5f-413c-a9a2-08d9c4927501/Data (accessed on 21 February 2022).
195. EPD International AB. Gran Pavesi: Dichiarazione Ambientale di Prodotto. 2020. Available online: https://portal.environdec. com/api/api/v1/EPDLibrary/Files/57212d22-3896-4351-8f93-81848397d396/Data (accessed on 23 February 2022).
196. EPD International AB. Fiori d'acqua: Dichiarazione Ambientale di Prodotto. 2020. Available online: https://portal.environdec. com/api/api/v1/EPDLibrary/Files/44eca5f8-c6c8-4370-b3da-d6a039f75adf/Data (accessed on 23 February 2022).
197. EPD International AB. Granetti Classici e Integrali: Dichiarazione Ambientale di Prodotto. 2020. Available online: https://portal. environdec.com/api/api/v1/EPDLibrary/Files/f2ff3ed9-dc5b-4cd1-b945-2cc56fda0835/Data (accessed on 23 February 2022).
198. EPD International AB. Michetti: Dichiarazione Ambientale di Prodotto. 2020. Available online: https://portal.environdec.com/ api/api/v1/EPDLibrary/Files/ecf89ab9-abb0-4115-82b6-a983e50d58ae/Data (accessed on 23 February 2022).
199. EPD International AB. Biscotto Pan di Stelle: Dichiarazione Ambientale di Prodotto. 2021. Available online: https://portal. environdec.com/api/api/v1/EPDLibrary/Files/9ed5c329-2f4f-4b8a-aa10-08d9c4927501/Data (accessed on 23 February 2022).
200. EPD International AB. Macine: Dichiarazione Ambientale di Prodotto. 2021. Available online: https://portal.environdec.com/ api/api/v1/EPDLibrary/Files/cb2e0333-ec45-45ec-aa08-08d9c4927501/Data (accessed on 23 February 2022).
201. EPD International AB. Abbracci: Dichiarazione Ambientale di Prodotto. 2021. Available online: https://portal.environdec.com/ api/api/v1/EPDLibrary/Files/c08f30f7-28d6-4ccf-d9a1-08d9b3162149/Data (accessed on 23 February 2022).
202. EPD International AB. Batticuori: Dichiarazione Ambientale di Prodotto. 2021. Available online: https://portal.environdec.com/ api/api/v1/EPDLibrary/Files/e7996471-de4f-4009-d99b-08d9b3162149/Data (accessed on 23 February 2022).
203. EPD International AB. Buongrano: Dichiarazione Ambientale di Prodotto. 2020. Available online: https://portal.environdec. com/api/api/v1/EPDLibrary/Files/f87dad27-d3a6-41a0-9dec-aa92e7efa9ea/Data (accessed on 23 February 2022).
204. EPD International AB. Campagnole: Dichiarazione Ambientale di Prodotto. 2020. Available online: https://portal.environdec. com/api/api/v1/EPDLibrary/Files/f5a121b8-3f6e-4a3d-8043-efa659a3710f/Data (accessed on 23 February 2022).
205. EPD International AB. Gran Cereale Biscotto Classico, Frutta, Cioccolato, Croccante, Digestive, Legumi Croccanti: Dichiarazione Ambientale di Prodotto. 2021. Available online: https://portal.environdec.com/api/api/v1/EPDLibrary/Files/1b7751a2-3d35-4b30-4e9f-08d900a54cf5/Data (accessed on 23 February 2022).
206. EPD International AB. Pavesini Classico, al Caffè, al Cacao: Dichiarazione Ambientale di Prodotto. 2020. Available online: https://portal.environdec.com/api/api/v1/EPDLibrary/Files/70c7b37a-cbc3-453f-ad85-2aafa75aa436/Data (accessed on 23 February 2022).
207. EPD International AB. Petit Pavesi: Dichiarazione Ambientale di Prodotto. 2020. Available online: https://portal.environdec. com/api/api/v1/EPDLibrary/Files/be18f17e-9a6d-4260-a5f3-ea0920bb8a96/Data (accessed on 23 February 2022).
208. EPD International AB. Ringo cacao, vaniglia, nocciola: Dichiarazione Ambientale di Prodotto. 2020. Available online: https://portal.environdec.com/api/api/v1/EPDLibrary/Files/b9794225-dd22-4bf6-9125-54849ea7f9f0/Data (accessed on 23 February 2022).
209. EPD International AB. Tarallucci: Dichiarazione Ambientale di Prodotto. 2020. Available online: https://portal.environdec.com/ api/api/v1/EPDLibrary/Files/8c9aafe6-67c5-48ac-8b16-08d8c43682c8/Data (accessed on 23 February 2022).
210. EPD International AB. Galletti: Dichiarazione Ambientale di Prodotto. 2020. Available online: https://portal.environdec.com/ api/api/v1/EPDLibrary/Files/ee75691d-ebee-42c6-be7f-59881ba2e854/Data (accessed on 23 February 2022).
211. EPD International AB. Girotondi: Dichiarazione Ambientale di Prodotto. 2020. Available online: https://portal.environdec.com/ api/api/v1/EPDLibrary/Files/4f13f759-963b-42f0-96dd-bfbfbcca0038/Data (accessed on 23 February 2022).
212. EPD International AB. Camille: Dichiarazione Ambientale di Prodotto. 2021. Available online: https://portal.environdec.com/ api/api/v1/EPDLibrary/Files/6b5afcb6-5256-486a-aa07-08d9c4927501/Data (accessed on 23 February 2022).
213. EPD International AB. Merendina Pan di Stelle: Dichiarazione Ambientale di Prodotto. 2021. Available online: https://portal. environdec.com/api/api/v1/EPDLibrary/Files/e4c2c88b-cd36-42e0-d9a8-08d9b3162149/Data (accessed on 23 February 2022).

214. EPD International AB. Torta Pan di Stelle: Dichiarazione Ambientale di Prodotto. 2021. Available online: https://portal. environdec.com/api/api/v1/EPDLibrary/Files/328195e6-bdbc-42a2-d9ab-08d9b3162149/Data (accessed on 23 February 2022).
215. EPD International AB. Torta Limone: Dichiarazione Ambientale di Prodotto. 2020. Available online: https://portal.environdec. com/api/api/v1/EPDLibrary/Files/e1ae7a16-3621-4547-8e65-7c9de212c5b6/Data (accessed on 23 February 2022).
216. EPD International AB. Mooncake Pan di Stelle: Dichiarazione Ambientale di Prodotto. 2021. Available online: https://portal. environdec.com/api/api/v1/EPDLibrary/Files/dcf7cd43-ab7f-4524-d9a4-08d9b3162149/Data (accessed on 23 February 2022).
217. EPD International AB. Plumcake Classico, Integrale, Con Gocce di Cioccolato, Senza Zuccheri Aggiunti: Dichiarazione Ambientale di Prodotto. 2020. Available online: https://portal.environdec.com/api/api/v1/EPDLibrary/Files/4968396f-dfa9-4919-9d49-7a625efa6e48/Data (accessed on 23 February 2022).
218. Ronchi, E.; Nepi, M.L. L'Italia del Riciclo 2017, Rome. 2017. Available online: https://www.fondazionesvilupposostenibile.org/wp-content/uploads/dlm_uploads/2017/12/Rapporto_Italia_del_riciclo_2017.pdf (accessed on 19 February 2022).
219. Institut Cyclos-HTP GmbH. Cyclos-HTP Institute for Recyclability and Product Responsibilty. Available online: https://www.cyclos-htp.de/cyclos-htp/ (accessed on 10 February 2022).
220. Bicalho, T.; Sauer, I.; Rambaud, A.; Altukhova, Y. LCA data quality: A management science perspective. *J. Clean. Prod.* **2017**, *156*, 888–898. [CrossRef]
221. BSI. PD CEN/TR 13910:2010 Packaging. *Report on Criteria and Methodologies for Life Cycle Analysis of Packaging.* 2010. Available online: https://www.en-standard.eu/pd-cen-tr-13910-2010-packaging-report-on-criteria-and-methodologies-for-life-cycle-analysis-of-packaging/ (accessed on 25 March 2022).
222. European Commission. Horizon Europe. Available online: https://ec.europa.eu/info/research-and-innovation/funding/funding-opportunities/funding-programmes-and-open-calls/horizon-europe_en (accessed on 23 February 2022).
223. Wohner, B.; Gabriel, V.H.; Krenn, B.; Krauter, V.; Tacker, M. Environmental and economic assessment of food-packaging systems with a focus on food waste. Case study on tomato ketchup. *Sci. Total Environ.* **2020**, *738*, 139846. [CrossRef]
224. Food and Agriculture Organization of the United Nations. Sustainable Development Goals: 12.3.1 Global Food Losses. Available online: https://www.fao.org/sustainable-development-goals/indicators/1231/en/ (accessed on 23 February 2022).
225. EPD International AB. EPD Applications. Available online: https://www.environdec.com/all-about-epds/epd-applications (accessed on 12 April 2022).
226. Kooijman, J.M. Environmental assessment of packaging: Sense and sensibility. *Environ. Manag.* **1993**, *17*, 575–586. [CrossRef]
227. Silvenius, F.; Grönman, K.; Katajajuuri, J.-M.; Soukka, R.; Koivupuro, H.-K.; Virtanen, Y. The Role of Household Food Waste in Comparing Environmental Impacts of Packaging Alternatives. *Packag. Technol. Sci.* **2014**, *27*, 277–292. [CrossRef]
228. Davis, J.; Sonesson, U. Life cycle assessment of integrated food chains—A Swedish case study of two chicken meals. *Int. J. Life Cycle Assess.* **2008**, *13*, 574–584. [CrossRef]
229. Flysjö, A. Potential for improving the carbon footprint of butter and blend products. *J. Dairy Sci.* **2011**, *94*, 5833–5841. [CrossRef]
230. Girgenti, V.; Peano, C.; Baudino, C.; Tecco, N. From "farm to fork" strawberry system: Current realities and potential innovative scenarios from life cycle assessment of non-renewable energy use and green house gas emissions. *Sci. Total Environ.* **2014**, *473–474*, 48–53. [CrossRef]
231. Bacenetti, J.; Cavaliere, A.; Falcone, G.; Giovenzana, V.; Banterle, A.; Guidetti, R. Shelf life extension as solution for environmental impact mitigation: A case study for bakery products. *Sci. Total Environ.* **2018**, *627*, 997–1007. [CrossRef]
232. Amienyo, D.; Camilleri, C.; Azapagic, A. Environmental impacts of consumption of Australian red wine in the UK. *J. Clean. Prod.* **2014**, *72*, 110–119. [CrossRef]
233. Amienyo, D.; Azapagic, A. Life cycle environmental impacts and costs of beer production and consumption in the UK. *Int. J. Life Cycle Assess.* **2016**, *21*, 492–509. [CrossRef]
234. Bonamente, E.; Scrucca, F.; Rinaldi, S.; Merico, M.C.; Asdrubali, F.; Lamastra, L. Environmental impact of an Italian wine bottle: Carbon and water footprint assessment. *Sci. Total Environ.* **2016**, *560–561*, 274–283. [CrossRef]
235. Dalla Riva, A.; Burek, J.; Kim, D.; Thoma, G.; Cassandro, M.; de Marchi, M. Environmental life cycle assessment of Italian mozzarella cheese: Hotspots and improvement opportunities. *J. Dairy Sci.* **2017**, *100*, 7933–7952. [CrossRef]
236. Fusi, A.; Guidetti, R.; Benedetto, G. Delving into the environmental aspect of a Sardinian white wine: From partial to total life cycle assessment. *Sci. Total Environ.* **2014**, *472*, 989–1000. [CrossRef] [PubMed]
237. Hanssen, O.J.; Vold, M.; Schakenda, V.; Tufte, P.-A.; Møller, H.; Olsen, N.V.; Skaret, J. Environmental profile, packaging intensity and food waste generation for three types of dinner meals. *J. Clean. Prod.* **2017**, *142*, 395–402. [CrossRef]
238. Humbert, S.; Loerincik, Y.; Rossi, V.; Margni, M.; Jolliet, O. Life cycle assessment of spray dried soluble coffee and comparison with alternatives (drip filter and capsule espresso). *J. Clean. Prod.* **2009**, *17*, 1351–1358. [CrossRef]
239. Manfredi, M.; Vignali, G. Life cycle assessment of a packaged tomato puree: A comparison of environmental impacts produced by different life cycle phases. *J. Clean. Prod.* **2014**, *73*, 275–284. [CrossRef]
240. Point, E.; Tyedmers, P.; Naugler, C. Life cycle environmental impacts of wine production and consumption in Nova Scotia, Canada. *J. Clean. Prod.* **2012**, *27*, 11–20. [CrossRef]
241. Rinaldi, S.; Barbanera, M.; Lascaro, E. Assessment of carbon footprint and energy performance of the extra virgin olive oil chain in Umbria, Italy. *Sci. Total Environ.* **2014**, *482–483*, 71–79. [CrossRef] [PubMed]
242. Schmidt Rivera, X.C.; Espinoza Orias, N.; Azapagic, A. Life cycle environmental impacts of convenience food: Comparison of ready and home-made meals. *J. Clean. Prod.* **2014**, *73*, 294–309. [CrossRef]

243. Thoma, G.; Popp, J.; Nutter, D.; Shonnard, D.; Ulrich, R.; Matlock, M.; Kim, D.S.; Neiderman, Z.; Kemper, N.; East, C.; et al. Greenhouse gas emissions from milk production and consumption in the United States: A cradle-to-grave life cycle assessment circa 2008. *Int. Dairy J.* **2013**, *31*, S3–S14. [CrossRef]

244. Zufia, J.; Arana, L. Life cycle assessment to eco-design food products: Industrial cooked dish case study. *J. Clean. Prod.* **2008**, *16*, 1915–1921. [CrossRef]

245. Hassard, H.A.; Couch, M.H.; Techa-erawan, T.; McLellan, B.C. Product carbon footprint and energy analysis of alternative coffee products in Japan. *J. Clean. Prod.* **2014**, *73*, 310–321. [CrossRef]

246. Amienyo, D.; Gujba, H.; Stichnothe, H.; Azapagic, A. Life cycle environmental impacts of carbonated soft drinks. *Int. J. Life Cycle Assess.* **2013**, *18*, 77–92. [CrossRef]

247. Bevilacqua, M.; Braglia, M.; Carmignani, G.; Zammori, F.A. Life cycle assessment of pasta production in italy. *J. Food Qual.* **2007**, *30*, 932–952. [CrossRef]

248. Calderón, L.A.; Iglesias, L.; Laca, A.; Herrero, M.; Díaz, M. The utility of Life Cycle Assessment in the ready meal food industry. *Resour. Conserv. Recycl.* **2010**, *54*, 1196–1207. [CrossRef]

249. Cellura, M.; Longo, S.; Mistretta, M. Life Cycle Assessment (LCA) of protected crops: An Italian case study. *J. Clean. Prod.* **2012**, *28*, 56–62. [CrossRef]

250. Garofalo, P.; D'Andrea, L.; Tomaiuolo, M.; Venezia, A.; Castrignanò, A. Environmental sustainability of agri-food supply chains in Italy: The case of the whole-peeled tomato production under life cycle assessment methodology. *J. Food Eng.* **2017**, *200*, 1–12. [CrossRef]

251. Laso, J.; Margallo, M.; Fullana, P.; Bala, A.; Gazulla, C.; Irabien, Á.; Aldaco, R. When product diversification influences life cycle impact assessment: A case study of canned anchovy. *Sci. Total Environ.* **2017**, *581–582*, 629–639. [CrossRef]

252. Tasca, A.L.; Nessi, S.; Rigamonti, L. Environmental sustainability of agri-food supply chains: An LCA comparison between two alternative forms of production and distribution of endive in northern Italy. *J. Clean. Prod.* **2017**, *140*, 725–741. [CrossRef]

253. Cordella, M.; Tugnoli, A.; Spadoni, G.; Santarelli, F.; Zangrando, T. LCA of an Italian lager beer. *Int. J. Life Cycle Assess.* **2008**, *13*, 133–139. [CrossRef]

254. Lee, D.S. *Modified Atmosphere Packaging of Foods: Principles and Applications*; John Wiley & Sons Inc; Institute of Food Technologists: Hoboken, NJ, USA; Chichester, UK, 2021; ISBN 9781119530770.

255. European Parliament, Council of the European Union. *Regulation (EC) No 1935/2004 of the European Parliament and of the Council of 27 October 2004 on Materials and Articles Intended to Come into Contact with Food and Repealing Directives 80/590/EEC and 89/109/EEC*; European Council: Brussels, Belgium, 2004.

256. *Commission Regulation (EC) No 450/2009 of 29 May 2009 on Active and Intelligent Materials and Articles Intended to Come into Contact with Food (Text with EEA Relevance)*; European Council: Brussels, Belgium, 2009.

257. Han, J.H. (Ed.) *Innovations in Food Packaging*; Elsevier Ltd.: Amsterdam, The Netherlands, 2005; ISBN 978-0-12-311632-1.

258. Wohner, B.; Schwarzinger, N.; Gürlich, U.; Heinrich, V.; Tacker, M. Technical emptiability of dairy product packaging and its environmental implications in Austria. *PeerJ* **2019**, *7*, e7578. [CrossRef]

259. Boz, Z.; Korhonen, V.; Koelsch Sand, C. Consumer Considerations for the Implementation of Sustainable Packaging: A Review. *Sustainability* **2020**, *12*, 2192. [CrossRef]

260. Zeng, T.; Durif, F.; Robinot, E. Can eco-design packaging reduce consumer food waste? an experimental study. *Technol. Forecast. Soc. Change* **2021**, *162*, 120342. [CrossRef]

261. Australian Packaging Covenant Organization. Sustainable Packaging Guidelines (SPGs). 2020. Available online: https://documents.packagingcovenant.org.au/public-documents/Sustainable%20Packaging%20Guidelines%20(SPGs) (accessed on 28 March 2022).

262. Kirchherr, J.; Reike, D.; Hekkert, M. Conceptualizing the circular economy: An analysis of 114 definitions. *Resour. Conserv. Recycl.* **2017**, *127*, 221–232. [CrossRef]

263. RecyClass. Recyclass Recyclability Methodology. 2021. Available online: https://recyclass.eu/wp-content/uploads/2022/01/Recyclass_methodology_UPDATED_-JANUARY-2022.pdf (accessed on 27 March 2022).

264. Marrone, M.; Tamarindo, S. Paving the sustainability journey: Flexible packaging between circular economy and resource efficiency. *J. Appl. Packag. Res.* **2018**, *10*, 53–60.

265. European Food Safety Authority. Food Ingredients and Packaging. Available online: https://www.efsa.europa.eu/en/topics/topic/food-ingredients-and-packaging (accessed on 28 March 2022).

266. European Bioplastics. Bioplastics Market Development Update 2021. Available online: https://docs.european-bioplastics.org/publications/market_data/Report_Bioplastics_Market_Data_2021_short_version.pdf (accessed on 27 March 2022).

267. Rosenboom, J.-G.; Langer, R.; Traverso, G. Bioplastics for a circular economy. *Nat. Rev. Mater.* **2022**, *7*, 117–137. [CrossRef] [PubMed]

268. European Commission. Communication from the Commission to the European Parliament, the Council, the European Economic and Social Committee and the Committee of the Regions: A New Circular Economy Action Plan. *For a Cleaner and More Competitive Europe, Brussels.* Available online: https://eur-lex.europa.eu/legal-content/EN/TXT/?qid=1583933814386&uri=COM:2020:98:FIN (accessed on 2 February 2022).

269. Circularity Gap Report 2022: Five Years of Analysis by Circle Economy I European Circular Economy Stakeholder Platform. Available online: https://circulareconomy.europa.eu/platform/en/knowledge/circularity-gap-report-2022-five-years-analysis-circle-economy (accessed on 27 March 2022).

270. Sazdovski, I.; Bala, A.; Fullana-i-Palmer, P. Linking LCA literature with circular economy value creation: A review on beverage packaging. *Sci. Total Environ.* **2021**, *771*, 145322. [CrossRef] [PubMed]

271. Slorach, P.C.; Jeswani, H.K.; Cuéllar-Franca, R.; Azapagic, A. Environmental and economic implications of recovering resources from food waste in a circular economy. *Sci. Total Environ.* **2019**, *693*, 133516. [CrossRef] [PubMed]

272. Schmidt Rivera, X.C.; Leadley, C.; Potter, L.; Azapagic, A. Aiding the Design of Innovative and Sustainable Food Packaging: Integrating Techno-Environmental and Circular Economy Criteria. *Energy Procedia* **2019**, *161*, 190–197. [CrossRef]

273. Rosa, D.; Figueiredo, F.; Castanheira, É.G.; Freire, F. Life-cycle assessment of fresh and frozen chestnut. *J. Clean. Prod.* **2017**, *140*, 742–752. [CrossRef]

274. European Commission. Open Science: An Approach to the Scientific Process That Focuses on Spreading Knowledge as Soon as It Is Available Using Digital and Collaborative Technology. *Expert Groups, Publications, News and Events.* Available online: https://ec.europa.eu/info/research-and-innovation/strategy/strategy-2020-2024/our-digital-future/open-science_en (accessed on 23 February 2022).

275. Di Polizzi Sorrentino, E.; Woelbert, E.; Sala, S. Consumers and their behavior: State of the art in behavioral science supporting use phase modeling in LCA and ecodesign. *Int. J. Life Cycle Assess.* **2016**, *21*, 237–251. [CrossRef]

276. Speck, R.; Selke, S.; Auras, R.; Fitzsimmons, J. Choice of Life Cycle Assessment Software Can Impact Packaging System Decisions. *Packag. Technol. Sci.* **2015**, *28*, 579–588. [CrossRef]

277. Kennedy, D.J.; Montgomery, D.C.; Quay, B.H. Data quality. *Int. J. Life Cycle Assess.* **1996**, *1*, 199–207. [CrossRef]

278. Antonioli, B.; Massarutto, A. The municipal waste management sector in Europe: Shifting boundaries between public service and the market: Snythesis Report. *Ann. Public Coop. Econ.* **2012**, *83*, 505–532. [CrossRef]

279. European Commission. Communication from the Commission to the European Parliament, the Council, the European Economic and Social Committee and the Committee of the Regions: A European Strategy for Plastics in a Circular Economy, Brussels. 2018. Available online: https://eur-lex.europa.eu/resource.html?uri=cellar:2df5d1d2-fac7-11e7-b8f5-01aa75ed71a1.0001.02/DOC_1&format=PDF (accessed on 2 February 2022).

280. COST. European Cooperation in Science and Technology. Available online: https://www.cost.eu/ (accessed on 28 March 2022).

281. EUROSTAT. Data Explorer. Available online: http://appsso.eurostat.ec.europa.eu/nui/show.do?dataset=env_waspac&lang=en (accessed on 12 April 2022).

282. EUROPEN. European and National Legislation on Packaging and the Environment, Brussels. 2016. Available online: https://www.europen-packaging.eu/wp-content/uploads/2012/03/European-and-National-Legislation-on-Packaging-and-the-Environment.pdf (accessed on 12 April 2022).

283. Ellen MacArthur Foundation. Education and Learning: Learning to Apply Circular Economy Thinking. Available online: https://ellenmacarthurfoundation.org/resources/education-and-learning/overview (accessed on 19 February 2022).

284. ISO. ISO/TC 122/SC 4-Packaging and the Environment. 2022. Available online: https://www.iso.org/committee/52082.html (accessed on 12 April 2022).

285. Benn, S.; Dunphy, D.; Griffiths, A. Enabling Change for Corporate Sustainability: An Integrated Perspective. *Australas. J. Environ. Manag.* **2006**, *13*, 156–165. [CrossRef]

286. TerraChoice Environmental Marketing Inc. The "Six Sins of GreenwashingTM": A Study of Environmental Claims in North American Consumer Markets. 2007. Available online: https://sustainability.usask.ca/documents/Six_Sins_of_Greenwashing_nov2007.pdf (accessed on 19 February 2022).

287. Escursell, S.; Llorach-Massana, P.; Roncero, M.B. Sustainability in e-commerce packaging: A review. *J. Clean. Prod.* **2021**, *280*, 124314. [CrossRef] [PubMed]

288. European Commission. Screening of Websites for 'Greenwashing': Half of Green Claims Lack Evidence. Available online: https://ec.europa.eu/commission/presscorner/detail/en/ip_21_269 (accessed on 19 February 2022).

289. European Commission. Initiative on Substantiating Green Claims. Available online: https://ec.europa.eu/environment/eussd/smgp/initiative_on_green_claims.htm (accessed on 19 February 2022).

Article

Design of a Blockchain-Enabled Traceability System Framework for Food Supply Chains

Lixing Wang [1,2,*], Yulin He [1] and Zhenning Wu [3]

1   School of Computers and Engineering, Northeastern University, Shenyang 110000, China; hyl1026789105@163.com
2   Laboratory for Soft Machines and Electronics, School of Packaging, Michigan State University, East Lansing, MI 48824, USA
3   College of Information Science and Engineering, Northeastern University, Shenyang 110000, China; wuzhenning@ise.neu.edu.cn
*   Correspondence: wanglixing@mail.neu.edu.cn

Abstract: Tracing food products along the entire supply chain is important for achieving better management of food products. Traditionally, centralized traceability systems have been developed for such purposes. One major drawback of this approach is that different users of the supply chain have their own systems with their own complexities and distinct features; thus, the interaction among them creates challenges when implementing a single centralized system. Therefore, a decentralized traceability system is favorable for tracing food products along the supply chain. In this study, we develop a supply chain traceability system framework based on blockchain and radio frequency identification (RFID) technology. The system consists of a decentralized blockchain-enabled data storage platform for data management and an RFID system at the packaging level for data collection and storage. We applied a consortium blockchain to the application. Fabric 2.0 in Hyperledger was chosen as the development platform. The proposed blockchain-enabled platform can provide decentralized data management and its underlying algorithm can guarantee data security. The system includes a creatively designed blockchain-enabled data structure in the RFID tag. When people scan the tag, the relevant information is written in the tag as a block linked to the previous blocks; simultaneously, the information is transmitted to the blockchain platform and recorded on the platform. No battery is required and the system works when there is an RFID reader nearby. The usage conditions included shipment, stocking, and storage. The RFID tag can be directly attached to paper packaging. This approach embeds the blockchain technique into the RFID tag and develops a corresponding system. The new traceability system has the potential to simplify the tracking of products and can be scaled for industrial use.

Keywords: blockchain; RFID; food package; food supply chain

Citation: Wang, L.; He, Y.; Wu, Z. Design of a Blockchain-Enabled Traceability System Framework for Food Supply Chains. *Foods* **2022**, *11*, 744. https://doi.org/10.3390/foods11050744

Academic Editors: Theodoros Varzakas and Rui M.S. Cruz

Received: 8 February 2022
Accepted: 1 March 2022
Published: 3 March 2022

1. Introduction

People often have contact with food products in their daily lives and food quality is a universal concern. Food quality problems also lead to serious social issues. Some examples of the most serious incidents include milk powder contaminated with melamine, pork contaminated with clenbuterol, Sudan dyes found in duck eggs, and recycled gutter oil used in cooking [1]. Most fresh food products are perishable. Their quality is sensitive to temperature and other environmental factors. Incorrect storage or transportation usually results in considerable economic loss to the corresponding companies. For example, 25–30% of food losses and damages are caused by inadequate transportation and distribution facilities such as cold storage units, dedicated fleets, and cold trucks [2]. In a report published by the United Nations, the National Development and Reform Commission [3] estimated that every year approximately one-third of all food items are lost globally with a value of approximately $8.3 billion. Consequently, tracing and tracking food products along the supply chain is an important topic for researchers.

The Internet of Things (IoT) technologies are immensely helpful in the food supply chain [4,5]. Radio frequency identification (RFID) is one of the most popular techniques. RFID has been increasingly used in logistics and supply chain management, because it can increase the efficiency of data input [6]. Between 2005 and 2006, an electronic pedigree (ePedigree) was proposed. It is an electronic document that provides data on the history of a particular product item. Therefore, RFID technology allows the tracking of items in real-time across the supply chain. This makes it easier to identify and trace products [7,8]. Jedermann, R. et al., developed a real-time autonomous sensor system for monitoring products when they were being transported [9]. Woo, S. H. et al., proposed an activity product state tracking system architecture for tracking products even when they were in a box or container [10].

With the widespread use of RFID, many researchers have designed systems to enhance food safety management. Abad, E. et al., developed an RFID-based system for the tracking and cold-chain monitoring of food [11]. The ePedigree was advocated to gain full supply chain visibility with detailed trace and track information. In another study, Wang, L. et al., proposed a system called the "supply-chain pedigree" [12].

Recently, the quick response (QR) code has also become a popular technology in traceability systems. Compared with barcodes, QR codes have a larger storage capacity without increasing costs. Compared with RFID, the cost of implementation of a QR code system is lower. People can easily obtain product information by scanning the code using a smartphone. There is no need to invest in dedicated reading devices such as RFID readers. For this application, Dong, Y. et al., applied QR codes to analyze China's leafy vegetable supply chain [13]. Xu, Z. and Gao, H. M. used QR code technology to build a white guard traceability system for SHUNZI Vegetable Cooperatives in order to achieve farm-to-table whole process recordability and traceability [14]. Peng, Y. et al., presented a QR code based tracing method for a meat quality tracing system [15]. Li, Z. et al., combined RFID and QR code techniques together to realize real-time tracking and tracing for a prepackaged food supply chain [16].

Many approaches provide interactive and dynamic explorations for the public and governments to explore the information in each part of the food chain and provide easy implementation of hazard analysis and critical control points in the food industry. One major drawback of these approaches is the complexity and distinct features in the individual systems from different users/customers owing to the different requirements, facilities, and conditions among them, resulting in difficulty in adopting a centralized system for all customers. The emerging blockchain technique can help improve this situation.

Blockchain was first proposed by Nakamoto, S. in 2008 [17]. The well-known cryptocurrency Bitcoin was created based on this technique. A blockchain is a distributed database. It comprises a series of orderly blocks that are chained together. Data are stored in blocks such as the ledger [18]. One defining characteristic of blockchain is decentralization. Many researchers have noticed this advantage and have applied it in supply chain management [19–21], including the food supply chain [22–25]. However, most studies have ignored data collection. Different companies still need to upload information to a centralized platform. Thus, a practical model incorporating decentralization has not been demonstrated.

In summary, it is necessary to trace and track food products along the supply chain. In recent years, several types of approaches to achieve this goal have been developed. However, each company has its own system, which makes it difficult to implement a traceability system for all the stakeholders along the supply chain. Any missing data in the links in the supply chain makes product tracing inaccurate. The integration of RFID and blockchain techniques in food packaging can help the implementation of a traceability system with different stakeholders; consequently, the objective of this study is to develop a packaging system framework for tracing and tracking food supply chains based on blockchain and RFID technologies. The system consists of a decentralized blockchain-enabled data storage platform for data management and an RFID system for data collection and storage. It works when there is an RFID reader nearby. The usage conditions include shipment, stocking, and storage. The new traceability system makes tracing and tracking products much easier and cheaper.

The remainder of this paper is organized as follows: Section 2 briefly reviews previous research on blockchain and RFID techniques in supply chains and food supply chains, in Section 3, the entire framework of the proposed system is introduced, Section 4 presents a case study to demonstrate the workflow of the proposed system, Section 5 presents the research conclusions and provides an outlook for future work.

2. Literature Review

The extensive research interest in blockchain began in 2018. We looked at research records covering the last five years (2017–September 2021) using keyword searches such as "blockchain", "blockchain + supply chain", and "blockchain + food" to explore the Web of Science database. The results are listed in Table 1a. We also calculated the percentages of the publications with the latter two keywords in the publications with the keyword "blockchain". The results are listed in Table 1b. The publication trends are shown in Figure 1. We can see that there is an increasing interest in blockchain research. Studies applying this technique to supply chains and food areas are increasing yearly. This means that researchers have noticed that this emerging technique has the potential to bring benefits to these areas. However, the limited number of publications suggests that the current studies remain in the early stages.

Table 1. (**a**) Absolute numbers of recent publications regarding blockchain and relevant applications, (**b**) percentages of recent publications about blockchain computing related to food or supply chains (based on keyword search).

(a)					
Keywords	2017	2018	2019	2020	January–September 2021
Blockchain	743	2933	8144	13,685	8811
Blockchain + Supply Chain	31	125	379	620	508
Blockchain + Food	6	33	91	173	128
(b)					
Keywords	2017	2018	2019	2020	January–September 2021
Blockchain + Supply Chain	4.17%	4.26%	4.65%	4.53%	5.77%
Blockchain + Food	0.81%	1.13%	1.12%	1.26%	1.45%

Decentralization makes blockchain technology attractive to researchers aiming to apply it to the supply chain, but there is little research to apply blockchain in supply chain management [26–29]. Smart contracts and consensus algorithms need to be redesigned for blockchain applications in supply chains [30–32]. Based on these studies, there are also some successful cases of blockchain applications in the supply chain. Wang et al., built a novel blockchain-based information management framework for a precast supply chain, thereby extending the applications of blockchain in the domain of construction supply chains [33]. Ho, G. T. et al., proposed a blockchain-based system for the accurate recording of aircraft spare parts traceability data with organizational consensus and validation using Hyperledger Fabric [34]. Omar, I. A. et al., presented a blockchain-based approach using Ethereum smart contracts and decentralized storage systems to transform vendor-managed inventory supply chain operations [35].

Insofar as studies targeting the food supply chain are concerned [36–39], among these studies how to apply blockchain for food traceability in the supply chain is a hot topic. Lin, Q. et al., designed a decentralized traceability system for food safety problems based on blockchain and an electronic product code information services network [40]. Tsang, Y. P. et al., proposed a blockchain–IoT based food traceability system that integrates a blockchain, IoT technology, and fuzzy logic for managing perishable food. A new consensus mechanism for considering the shipment transit time, stakeholder assessment, and shipment volume was developed to address the needs for food traceability, weight minimization, and vaporization characteristics [41]. Powell, W. et al., analyzed an ongoing beef supply chain project integrating the blockchain and IoT for supply chain event tracking and beef provenance assurance and proposed two solutions for increasing data integrity and trust in the blockchain and IoT-enabled food supply chain [42].

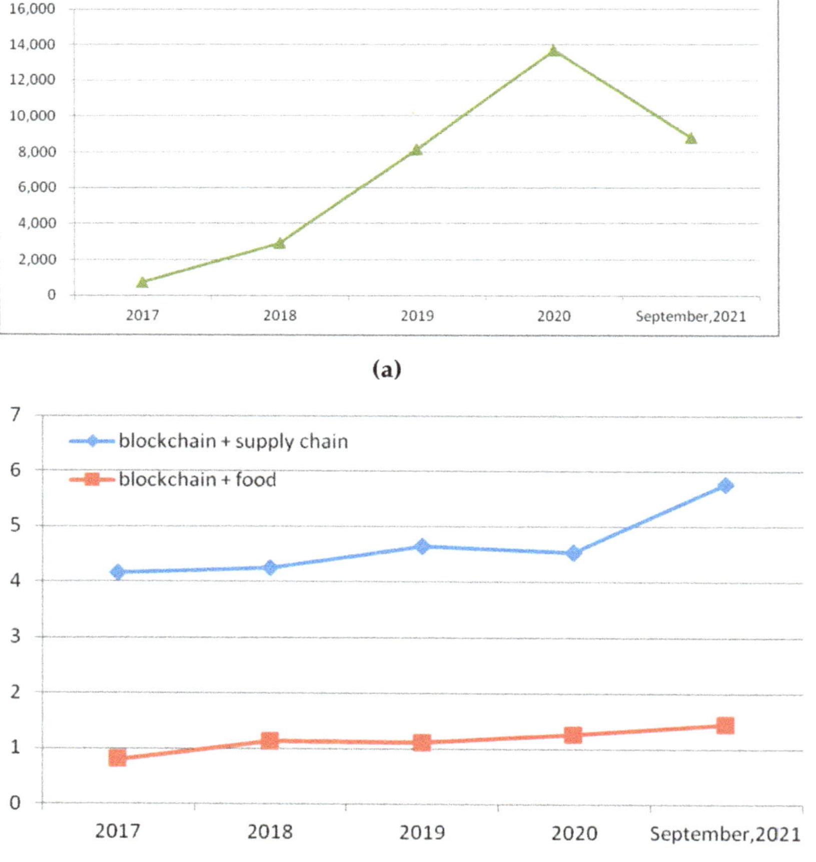

Figure 1. Publication trends: (**a**) blockchain, (**b**) percentages in publications about blockchain.

The aforementioned studies primarily focused on data management. However, data collection is also important for supply chain management. As stated in Section 1, RFID and QR codes are currently two popular techniques used in tracing and tracking systems to increase the efficiency of data input. Therefore, some studies have considered hybrid RFID approaches, QR codes, IOT techniques, and blockchains. Lakshmi, G. V. et al., used QR codes and a blockchain for inventory management [43]. Baralla, G. et al., developed a generic agri-food supply chain traceability system based on blockchain technology using

Hyperledger Sawtooth. The system acquired data simply through a QR code scan [44]. Dey, S. et al., proposed a blockchain-technology-based framework, which made food product information easily accessible, traceable, and verifiable by consumers and producers by using QR codes to embed the information [45]. Bencic, F. M. et al., proposed a smart tag, called a DL-tag, which is used to track products during their lifecycle. The tags contain QR codes and the system implements a blockchain [46]. Compared with RFID, the cost of deploying a QR code system is much lower. However, the main drawback of QR codes is that they are easy to physically duplicate or destroy. There are safety risks in food supply chain management, where food safety is the main concern. In view of these different concerns, and in light of the current literature, we focused this research on a combination of blockchain and RFID technologies.

Feng, T. first utilized RFID and blockchain techniques to develop an agri-food supply chain traceability system framework. The system could realize traceability with trusted information across the entire agri-food supply chain, thereby effectively guaranteeing food safety by gathering, transferring, and sharing the relevant data in the production, processing, warehousing, distribution, and selling stages [47]. Meanwhile, the same author proposed the concept of "BigchainDB" [48], aiming to improve performance by applying blockchain for high-volume data storage. Helo, P. and Shamsuzzoha, H. M. considered the challenges of tracing and tracking in multi-company project environments. They proposed a pilot system formed by the combination of RFID, IoT, and blockchain technologies for real-time tracking and tracing of logistics and supply chains [49]. Mondal, S. et al., proposed a blockchain-inspired IoT architecture for creating a transparent food supply chain. The architecture was realized by integrating an RFID-based sensor at the physical layer and a blockchain at the cyber layer [50]. Mazzei, D. et al., described the implementation of a portable, platform-agnostic, secured blockchain tokenizer for industrial IoT trustless applications. The system was designed, implemented, and tested in two supply chain scenarios [51]. Sfa, A. et al., proposed two blockchain-based decentralized mutual authentication protocols for IoT systems. Moreover, combining blockchain with RFID can also help anti-counterfeiting efforts in the supply chain as these tags can be easily cloned in the post supply chain [52]. Toyoda, K. et al., proposed a novel product ownership management system for anti-counterfeiting RFID-attached products [53].

From the review, it can be found that most studies have considered the use of RFID for data collection, and blockchain for uploaded data processing and storage. In fact, these two techniques are generally used separately and only in the supply chain. Based on previous studies, this research focused on improving food packaging. We propose a traceability system framework for supply chain management. The system integrates blockchain with RFID much more deeply at the package level. The detailed design of the system is described in the following sections.

3. Traceability System Design

Figure 2 illustrates the application scenarios of the system. The system has three main parts. One part comprises blockchain-enabled RFID tags attached to the product packaging. The blockchain-enabled RFID tags follow the products along the entire supply chain. The relevant product information, transactions, storage, and delivery status are all recorded. Some tags can be coupled with sensors to measure and record environmental conditions. When the RFID tags are read by the RFID readers at each link of the supply chain, the information is transmitted to the blockchain light nodes. These comprise the second part of the system. The light nodes are integrated with the RFID readers. They are responsible for the communication and verification of RFID tags. Owing to the limited capacity of the RFID reader, the light node does not store all the ledgers. In addition, cross-validation is performed to guarantee safety. The approach is described in Section 3.3. The third part of the system comprises a blockchain platform. This platform stores all the ledgers. It also contains a designed consensus mechanism for making decisions on how to record ledgers.

Users are also provided with an interface for product information inquiry. Figure 3 shows the relationship between these three parts, along with the system architecture.

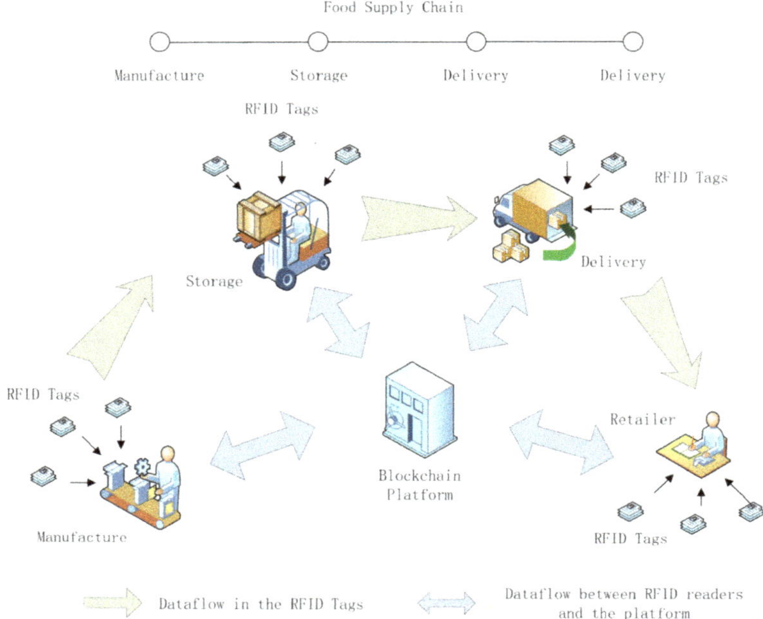

Figure 2. Application scenarios of the traceability system.

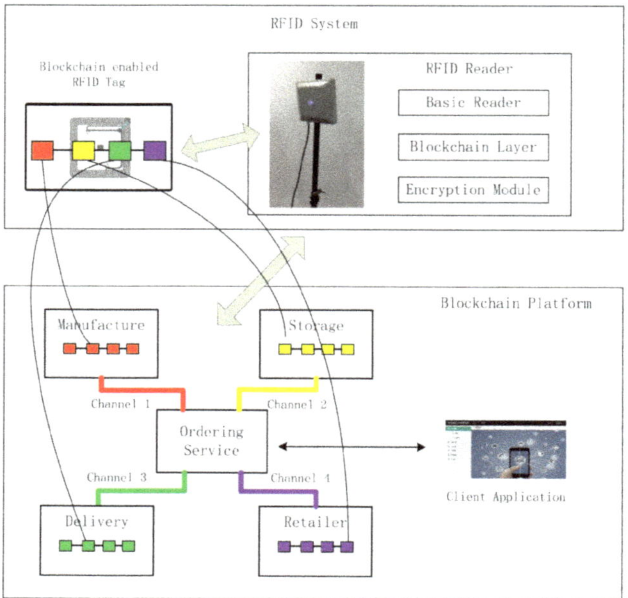

Figure 3. Traceability system architecture (The color is used to distinguish the blocks from different sessions in the supply chain.).

3.1. System Architecture

As presented in Figure 3, the system comprises three parts: the RFID tags, RFID reader, and backend blockchain platform. A simple blockchain is contained in a blockchain-enabled RFID tag. Each time the tag passes a reader, a block is generated. The information written in the tag and the hash value of the previous block are calculated using the SHA-256 algorithm. The result is also written in the tag as the hash value of the current block. In this way, the blocks are individually linked. The second part is an RFID reader. Similar to the role of the normal RFID system, the RFID reader builds communication between the RFID tags and the backend system. In addition to the basic functions of a normal RFID reader, the reader in this blockchain-based traceability system contains a blockchain layer to justify whether the tag is a blockchain-enabled tag and for deciding what information should be written on the tag after encryption. An encryption module is also included in the RFID reader and is used to calculate the hash value of the input information. Finally, the RFID reader sends the information that the reader writes to the backend blockchain platform. The reader also stores parts of the ledgers of the blockchain platform as a light blockchain node. The third part is the blockchain platform. All the ledgers are stored on it. The main functions of the platform are data management and verification. A user application linked to the blockchain platform was also designed. Its main purpose is to provide a convenient user interface for inquiries regarding food product supply chain information.

3.2. Blockchain-Enabled Radio Frequency Identification (RFID) Tag Design

This is the base of the system. As stated in Section 2, in most previous studies integrating RFID and blockchain, RFID tags were only used to collect data. In this study, the RFID tags contain a blockchain data structure. The tags must be attached to all the products in the entire supply chain. When the reader reads a tag, the information is recorded and a block is formed. The blocks form a chain and are stored in the tags; consequently, people only need to scan the tags. Then, they can obtain all the information regarding the product and what it has experienced from source to destination. There is no need for a central database or system for management because all the information users might need is stored in the RFID tags. Moreover, the blockchain structure guarantees data safety. If anyone wanted to modify it, it would have to be modified from the beginning.

With the explosion of chip technologies, the storage space in RFID tags has experienced exponential growth; consequently, it is possible for RFID tags to store all product information and transaction records, that is, much more than a mere ID number. Furthermore, the safety mechanism of an RFID system uses a hash chain scheme, as shown in Figure 4, which is the same as the basic principle of blockchain. In the hash chain, the tag and reader share two hash functions: $G(\ )$ and $H(\ )$. $G(\ )$ is used to calculate response messages. $H(\ )$ is used for updating. They also share an initial random identifier (s). When the RFID reader investigates the tag, the tag responds to the hash value $a_i = G(s_i)$ of the current identifier s_i; simultaneously, the tag updates the current identifier s_i to $s_{i+1} = H(s_i)$. These two functions make it feasible to integrate blockchain with RFID.

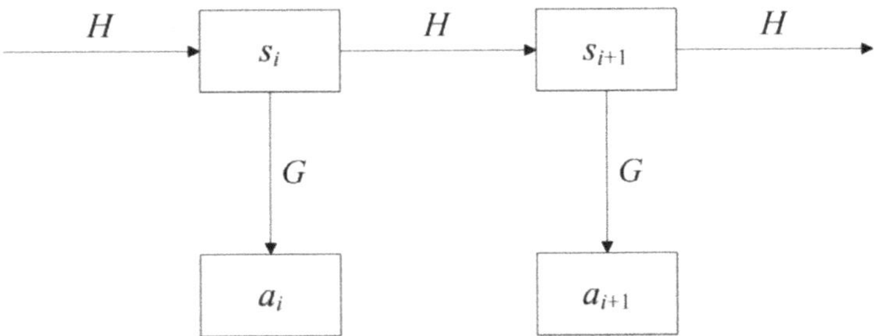

Figure 4. Safety mechanism of radio frequency identification (RFID).

The blockchain design in the RFID tags and the corresponding mechanisms are discussed below. When a food product is packaged, all the relevant information is coded by the hash algorithm and then recorded in the block; this block is called the "genesis block". The block is stored in an RFID tag attached to the food package. When the product enters the next session, for example, storage in the warehouse, the relevant information is coded by the hash algorithm (SHA-256) and then recorded in a second block linked to the genesis block. When the product proceeds to the next session, the same work is repeated until the product reaches the consumer. Figure 5 illustrates the procedure. The data structure design of the blocks is presented in Table 2. It contains an RFID tag ID, reader IP address, product information, and relevant sensor data. The table also provides an example of the data (the only example). In fact, there is no need to fill all the columns or follow the data format. In addition to the necessary reader IP address, location, timestamp, and hash value the reader in each link of the supply chain only needs to input the information it can provide. In this way, it effectively solves the data-sharing problem of different companies with different types of information systems in a complex supply chain and at a low cost. The only thing the manufacturer needs to do is update the RFID reader at the software level.

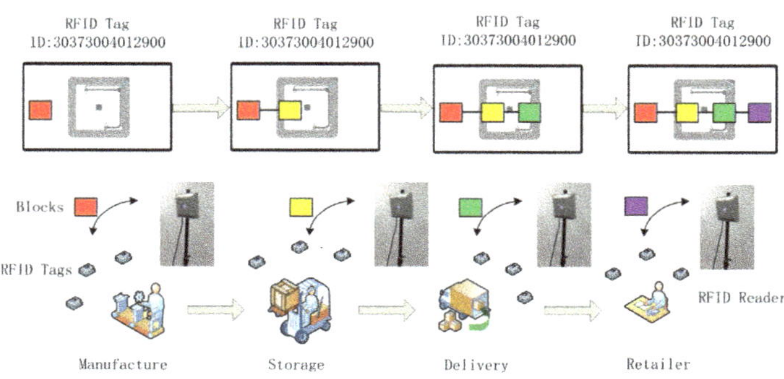

Figure 5. Work procedure of the RFID system (The color is used to distinguish the blocks from different sessions in the supply chain.).

Table 2. Sample record for a single product data structure design.

Data Structure	Manufacture	Storage	Delivery	Retailer
Tag ID	30373004012900	30373004012900	30373004012900	30373004012900
Reader IP	175.160.137.100	127.155.155.100	165.129.69.105	168.122.53.107
Timestamp	2110231409	2110231611	2110250632	2110261017
Product Name	Milk	Milk	Milk	Milk
Product ID	30373004740000	30373004740000	30373004740000	30373004740000
Location	39.773, 119.253	38.476, 117.245	36.279, 118.366	34.773, 113.727
Company	A	B	C	D
Job	Manufacture	Storage	Delivery	Retailer
Sensor	Not Applicable	Temperature	Temperature	Temperature
Sensor Data	Not Applicable	5	7	3
Previous Hash	Not Applicable	3a6fed5fc11392b3ee9f81 caf017b48640d74587 66a8eb0382899a60 5b41f2b9	d807aa9812835b0 1243185be550c 7dc372be5d7480 deb1fe9bdc06a 7c19bf174	748f82ee78a563 6f84c878148cc 7020890befffaa4 506cebbef9a3f 7c67178f2
Hash Value	3a6fed5fc1139 2b3ee9f81caf017b 48640d7458766 a8eb0382899 a605b41f2b9	d807aa981283 5b01243185 be550c7dc372 be5d7480deb 1fe9bdc06 a7c19bf174	748f82ee78 a5636f84c878 148cc70208 90befffaa4 506cebbef9 a3f7c67178f2	

3.3. Blockchain Light Node Design

The RFID reader takes the role of a bridge between the RFID tag and the blockchain platform. The main function of this system is communication. For the RFID tag, the RFID reader is responsible for reading the data of the tags, decrypting the data, and recording it in the database of the corresponding system. This system is also responsible for writing data in the RFID tags and for encryption. Meanwhile, ledgers in a period relevant to when the tags were read are also stored in the reader. If there are any data safety problems, the reader can also provide cross-verification. For a given (limited) storage capacity, the ledgers stored in the reader have a maximum size requirement. An RFID reader is a blockchain light node. Therefore, we need to develop a blockchain platform with full functionality. For the blockchain platform, the RFID reader includes a communication module for data processing and uploading.

For the communication with RFID tags, the work procedure is as follows:

Step 1: The RFID reader reads the RFID tag and determines whether it is a blockchain-enabled tag. If it is, proceed to Step 2; otherwise, the normal RFID reader procedure is completed.

Step 2: The reader determines whether the tag has been read before. If it has, proceed to Step 3. If it has not, proceed to Step 5.

Step 3: The reader determines whether the time interval between reads is longer than a user-defined value, t. If it is, proceed to Step 4. If it is not, proceed to Step 5.

Step 4: The reader reads the hash value from the second to the last block, then proceeds to Step 6.

Step 5: The reader reads the hash value of the last block.

Step 6: The reader writes the data into a tag with the hash value of the last step.

Step 7: The reader calculates the hash value of the input information using the SHA-256 algorithm.

Step 8: The reader writes the hash value into the tag.

Step 9: After data processing, the reader transforms the data into a blockchain platform. The communication module is described in the next section. Figure 6 shows a flowchart of how the RFID reader writes the data in the RFID tags.

The communication module in the RFID reader comprises four parts. The first part includes a radio signal transmitting and processing module, command parser, and data integration and processing module, which realize the basic RFID reader function. If the reader finds that the tag it has read is a blockchain-enabled RFID tag, the data are transmitted to the blockchain layer in the RFID reader. This is the second part of the communication module and represents its core. The blockchain layer reads the hash value of the last block in the tag and sends it, together with the new information it wants to write to the tag, to the third part of the communication module: the encryption module. The encryption module uses the SHA-256 algorithm to calculate a new hash value derived from the previous hash value and the input data. After encryption, this new (now current) hash value and input data are transmitted back from the blockchain layer, via the RFID reader, to the RFID tag as a new block. The blockchain layer also stores recent ledgers relevant to the tags it has read (within the limits of its storage capacity). When there is any verification requirement, these recorded ledgers can be used for cross-validation. The fourth part is the network module. It connects the reader to the Internet. Then, the hash value, input data, tag ID, and product ID can be transmitted to the blockchain platform. The recent ledgers of this RFID system will also be downloaded and updated from the blockchain platform. Figure 7 shows the architecture of the communication module between the light node and blockchain platform. This module also solves the problems of data integration, processing, and bulk data upload.

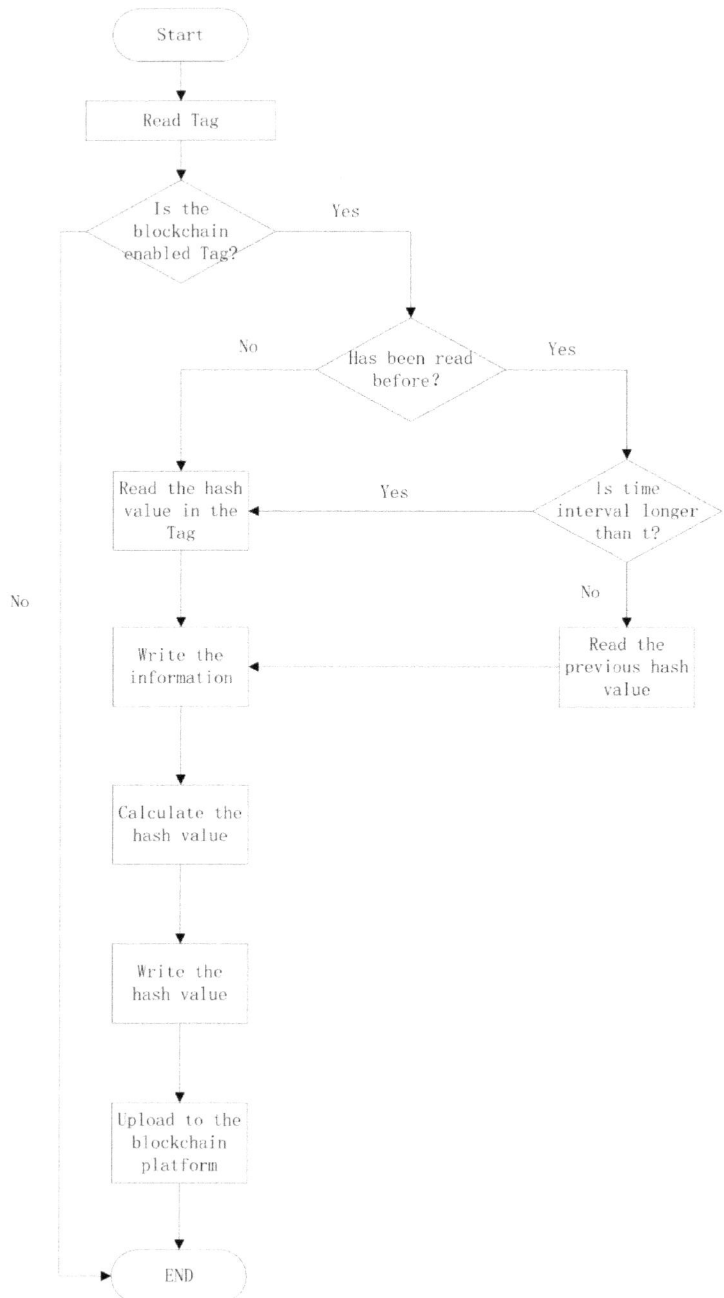

Figure 6. Workflow of the RFID reader.

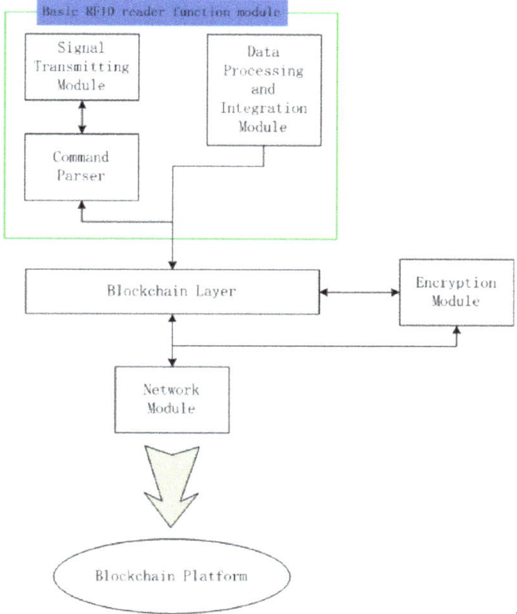

Figure 7. Data operational process to the blockchain platform.

3.4. Blockchain Platform Design

A blockchain platform was developed on Hyperledger Fabric. The system architecture is illustrated in Figure 8. The central part of the architecture is an ordering service. It organizes different parties in the supply chain actions for verification. Consensus algorithms and chain codes are both located here. The ordering service node also connects to a client application, as introduced in Section 3.5.

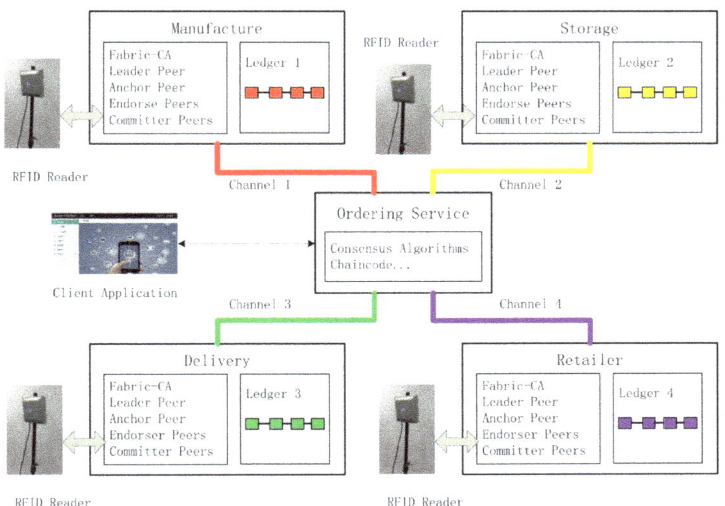

Figure 8. System architecture (The color is used to distinguish the blocks from different sessions in the supply chain.).

According to the characteristics of the supply chain, the parties can be divided into four groups: manufacturers, warehouses, logistics, and retailers. Each group is the responsibility of a different company. Channels are built between the group and ordering service; consequently, there are four channels in this platform, thus there are four ledgers. Each group contains the fabric certification authority, leader peer, anchor peer, endorser peers, and committer peers. The endorser peers and committer peers are connected to the responding RFID readers. RFID readers are the light nodes of their peers, as they only store parts of the ledgers. All the peer nodes, regardless of the company they belong to, store the entire ledger for their group. The records can also be the reader data, transmitted back by the RFID reader, and set as blocks linked together by a hash value. They are saved in their respective ledgers. In contrast to the client application, the tag ID in the ledger is set as the inquiry number. In fact, the RFID tag in the platform can also be considered as a peer node, however it contains parts of all four ledgers along with their relevant smart contracts.

The workflow for the data transmitted from the RFID reader to the platform is illustrated in Figure 9. First, the platform verifies the validity of the reader. Second, the RFID tag ID is extracted for the platform to identify the block of the RFID tag from the previous session. Then, the platform compares the hash value recorded in the last block of this RFID tag with the hash value of the previous block in the tag transmitted back by the RFID reader. If they are identical, all the information transmitted back is recorded as a block of that tag and linked with other blocks in the same group. If they are different, a warning is issued.

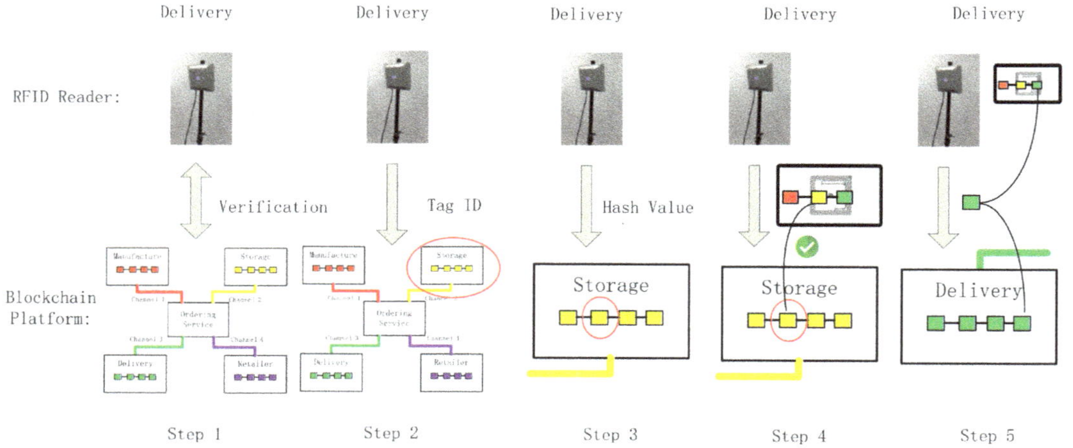

Figure 9. Workflow of the communication between the RFID reader and blockchain platform (The color is used to distinguish the blocks from different sessions in the supply chain).

3.5. Client Application Design

We also designed a client application for the traceability system (the platform was developed in Chinese; we translated the words in the following figures for clarity). This application only provides an interface for users for inquiries. There is no need for companies in the product supply chain to use it; they only need to write relevant information on the tag through the RFID reader.

As shown in Figure 10, there are six modules in the application: product information, manufacturing management, storage management, delivery management, retail management, and food safety analysis. The product information module includes basic product information such as shelf life and name. The manufacturing management module processes the information from the manufacturer. The details can be decided by the company itself, for example, details about storage management, delivery management, and retail management. The relevant parties in the supply chain can only provide their company names or more detailed information such as the storage environment, delivery truck, and arrival time. The final part is the food safety analysis module. In this module, the food pedigree is analyzed to locate the key node in the supply chain. If a product has any problems, the module can also help to recall the problem product and analyze which chain has the problem. Detailed information can be found in our previous study [12,54].

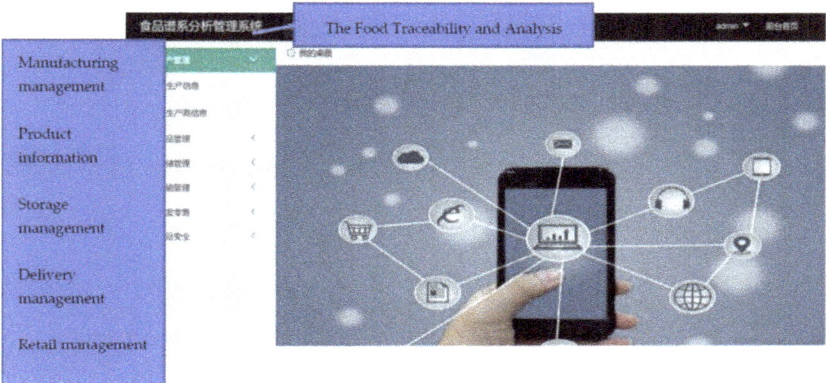

Figure 10. Client Application Interface.

In this user application, all modules contained the product ID. The product ID is the index number for this system. The tag ID and hash value are not shown to the users owing to safety considerations.

4. Case Study

In this section, a case study is presented to demonstrate the operation of the system. The case is based on Tiandi Yuan Ltd., a famous beverage company. Their main products are white wines.

When a product is manufactured, a serial number is generated. We use the example "6926582705239LW-2013-0300001" for demonstration. Furthermore, a UHF RFID tag, as shown in Figure 11, is also attached to the product package. An RFID reader scans the tag and the tag ID is linked to the product ID. The system uploads relevant product information to the system, as shown in Figure 12. A hash value is calculated using SHA-256 with the input of the tag ID and the product information. The hash value is written on the tag by the RFID reader as the genesis block (block 0). The RFID reader also connects to the blockchain platform and transmits the tag ID, product information, and hash value to the platform. These inputs and the previous hash value of the block in the platform are calculated using SHA-256. Then, a new block is generated in the platform. Parts of the ledger of the manufacturing node are also downloaded onto the RFID reader. When the product enters the next session, the same steps are repeated. In the storage session, the reader reads the tag, links the tag ID with the storage information, and uploads the information to the system. Meanwhile, the reader uses the information and the hash value recorded in block 0 to generate a new hash value associated with block 1 to be recorded in the RFID tag. The rest of the procedure is the same as in the manufacturing session. The procedure for the remaining sessions is also the same for each session.

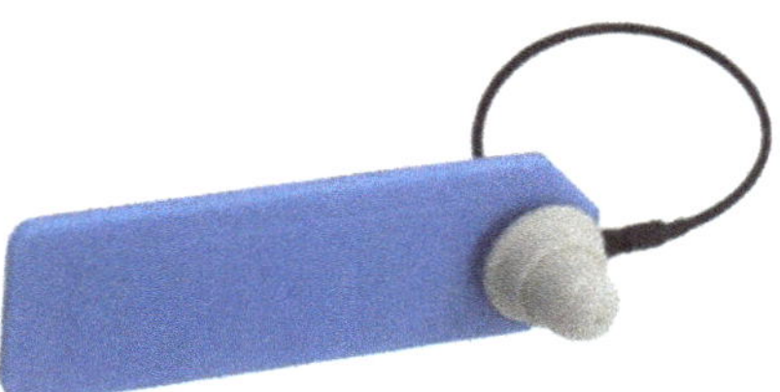

Figure 11. The attached UHF RFID Tag.

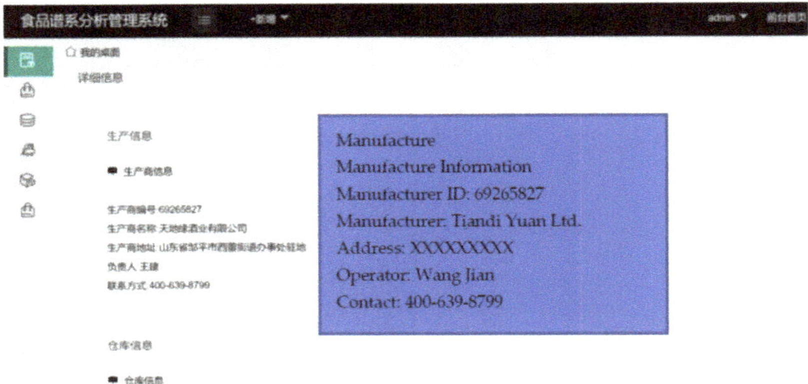

Figure 12. Detailed Information in Client Application.

Finally, as shown in Figure 13, when the consumer inputs the product ID at the interface, the application shows the entire supply chain of the product. The user can also find detailed information by clicking the nodes in the supply chain or clicking the buttons in the left column.

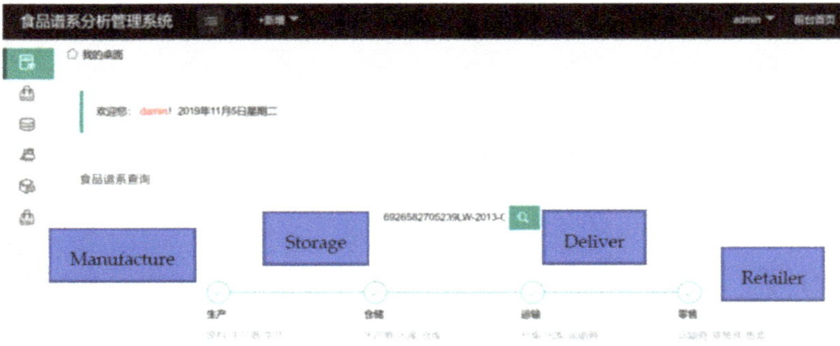

Figure 13. Food Supply Chain in Client Application.

This company has different partners in its product supply chain, especially in logistics and retailer sessions. In the past, it was difficult for companies to follow their products along the entire supply chain, because not all companies are willing to upload the information to the company's information platform. With the help of this system, the company has better information about its products in the supply chain, such as final consumers. The company merely needs to update its partners' RFID readers. The reader can automatically identify the blockchain-enabled RFID tag in order to perform its functions, as stated above.

Although some companies in the supply chain are resistant to sharing information, basic information such as time and location will also be recorded automatically through the RFID reader. With a more transparent supply chain, the company can better manage its products. The data also helps the company analyze its product supply chain. If there are any problem products, they can be easily recalled. Moreover, the application of blockchain techniques in both the platform and the RFID tags significantly increases product security and facilitates anti-counterfeiting measures.

The case study proves that the proposed system can effectively help different companies with different information system collect and upload data to the blockchain plat-form. With the information, all the traces of a product along the supply chain can be clearly presented to the consumers. Furthermore, the system also has the potential to enable companies to manage their products better with some data analysis modules.

5. Conclusions

This research proposed a blockchain and RFID-enabled traceability system for the food industry. The system consists of three parts. The first is a blockchain–RFID tag. We deeply integrated blockchain and RFID techniques. A simple blockchain storage structure was designed for data storage in an RFID tag based on the original safety scheme of the RFID system. All information regarding the product as it follows the supply chain is recorded in the tags in a blockchain format. The second part is an RFID reader with a blockchain light node. Its main function is to build communication between the RFID and backend blockchain platforms in a normal RFID system. In addition to these basic functions, the reader also contains blockchain and encryption modules. These two modules change the reader to a blockchain light node. The blockchain layer can determine whether a tag is a blockchain-enabled RFID tag. If it is, the data will be dealt with by the encryption module and written on the tag with a calculated hash value. Moreover, the recent ledgers of the tags that the RFID reader has read previously are also stored in it. If there is any uncertainty about the data, the reader can also help in cross-validation. The third part is the blockchain platform. The platform was developed using Hyperledger Fabric. It stores all the ledgers. The reader transmits the data written on the tags to the platform. The platform manages and verifies the data and the reader. Finally, we designed a user application which has the look and feel of a webpage. It provides a user interface for convenient inquiries regarding the pedigrees of food products. Detailed information on each chain, as provided by the responding company, can also be obtained using this application.

The main contribution of this research is the creative design of a blockchain-based RFID tag. The tag can be attached to a food package with low costs for the application of this new emerging blockchain technique. Blockchain technology allows supply chain monitoring to be decentralized. For the various companies referred to in the complex supply chain, there is no need to use different types of information systems or to follow different types of data formats for different companies. They need to (and can simply) update their RFID readers at the software level. Anyone requiring the product supply chain information can download it from the blockchain platform. With the help of blockchain, the food supply chain can be more transparent and can be easily traced and tracked.

Comparisons between our proposed system and other popular blockchain-based methods for tracing and tracking systems are shown in Table 3. From the comparison, it can be seen that our proposed system has a high security level and relatively low cost, allowing convenient data input, and it needs to be equipped with RFID readers without simultaneously demanding a uniform information system.

Table 3. Comparisons between different blockchain based methods for tracing and tracking systems.

Approaches	Security Level	Convenient for Data Input	Cost
Blockchain	High	No (data need to be input manually)	Low
Blockchain + QR code	Low	Yes (But data need to be scanned individually)	Low
Blockchain + RFID	Medium	Yes (Each stakeholder needs to connect their RFID system to a uniform information system)	Medium
Blockchain + IoT	High	Yes (Each stakeholder should guarantee the device can be connected to the Internet)	High
Our method	High	Yes (Each stakeholder merely needs an RFID reader)	Medium

We did not redesign the consensus mechanism for blockchain platforms. The system is only in the research phase, therefore there is no problem with the current volume of data. However, a traditional consensus mechanism, such as practical Byzantine fault tolerance, may not satisfy the data throughput in the supply chain. The data in the supply chain have their own characteristics. In future studies, we will consider developing a new consensus mechanism that is more suitable for the food supply chain. We will also consider printing QR codes on the RFID tags. Companies without RFID readers can also be involved in this system. Moreover, with the help of the blockchain technique, the food supply chain becomes much more transparent and easier to safeguard. The analysis of these data will also be our research focus in the next phase.

Author Contributions: L.W. and Z.W. conceived the ideas and designed the framework of the system; L.W. contributed to the blockchain platform design and the client application design; Z.W. contributed to the RFID system design; L.W. prepared the draft; Y.H. contributed to the manuscript writing and editing. All authors commented on the manuscript. All authors have read and agreed to the published version of the manuscript.

Funding: This research was funded by the National Natural Science Foundation of China under Grant 71502029 and Grant 61703087.

Acknowledgments: The authors would like to acknowledge the financial support from the China Scholarship Council (CSC) and technical support from Michigan State University.

Conflicts of Interest: The authors declare no conflict of interest.

References

1. Liu, S.; Xie, Z.; Zhang, W.; Cao, X.; Pei, X. Risk assessment in Chinese food safety. *Food Control* **2013**, *30*, 162–167. [CrossRef]
2. Raut, R.D. Improvement in the food losses in fruits and vegetable supply chain—A perspective of cold third-party logistics approach. *Oper. Res. Perspect.* **2019**, *6*, 100–117. [CrossRef]
3. NRDC. Wasted: How America Is Losing up to 40 Percent of Its Food from Farm to Fork to Landfill. 2012. Available online: https://www.nrdc.org/sites/default/files/wasted-food-IP.pdf (accessed on 2 August 2018).
4. Jagtap, S.; Duong, L.; Trollman, H.; Bader, F.; Garcia-Garcia, G.; Skouteris, G.; Li, J.; Pathare, P.; Martindale, W.; Swainson, M.; et al. Iot Technologies in The Food Supply Chain. In *Food Technology Disruptions*; Academic Press: Cambridge, MA, USA, 2021; pp. 175–211.
5. Bader, F.; Jagtap, S. Internet of Things-Linked Wearable Devices for Managing Food Safety in The Healthcare Sector. In *Wearable and Implantable Medical Devices*; Academic Press: Cambridge, MA, USA, 2019.
6. Jedermann, R.; Ruiz-Garcia, L.; Lang, W. Spatial temperature profiling by semi-passive RFID loggers for perishable food transportation. *Comput. Electron. Agric.* **2009**, *65*, 145–154. [CrossRef]
7. Kwok, S.K.; Ting, J.S.L.; Tsang, A.H.C.; Lee, W.B.; Cheung, B.C.F. Design and development of a mobile epc-rfid-based self-validation system (mess) for product authentication. *Comput. Ind.* **2010**, *61*, 624–635. [CrossRef]
8. Kwok, S.K.; Ting, S.L.; Tsang, A.; Cheung, C.F. A counterfeit network analyzer based on RFID and EPC. *Ind. Manag. Data Syst.* **2010**, *110*, 1018–1037. [CrossRef]

9.   Jedermann, R.; Behrens, C.; Westphal, D.; Lang, W. Applying autonomous sensor systems in logistics-combining sensor networks, RFIDs and software agents. *Sens. Actuators* **2006**, *132*, 370–375. [CrossRef]

10. Woo, S.H.; Choi, J.Y.; Kwak, C.J.; Kim, C.O. An active product state tracking architecture in logistics sensor networks. *Comput. Ind.* **2009**, *60*, 149–160. [CrossRef]

11. Abad, E.; Palacio, F.; Nuin, M.; AGDZárate Marco, S. RFID smart tag for traceability and cold chain monitoring of foods: Demonstration in an intercontinental fresh fish logistic chain. *J. Food Eng.* **2009**, *93*, 394–399. [CrossRef]

12. Wang, L.; Ting, S.L.; Ip, W.H. Design of supply-chain pedigree interactive dynamic explore (SPIDER) for food safety and implementation of hazard analysis and critical control points. *Comput. Electron. Agric.* **2013**, *90*, 14–23. [CrossRef]

13. Dong, Y.; Fu, Z.; Stankovski, S.; Wang, S.; Li, X. Nutritional quality and safety traceability system for China's leafy vegetable supply chain based on fault tree analysis and QR code. *IEEE Access* **2020**, *8*, 161261–161275. [CrossRef]

14. Xu, Z.; Gao, H.M. The construction of melon traceability system based on QR code for SHUNZI vegetable cooperative. In Proceedings of the International Conference on Management Science and Management Innovation, Guilin, China, 15–16 August 2015; Atlantis Press: Paris, France, 2015; pp. 120–126.

15. Peng, Y.; Zhang, L.; Song, Z.; Yan, J.; Li, X.; Li, Z. A QR code based tracing method for fresh pork quality in cold chain. *J. Food Process Eng.* **2018**, *41*, 12685. [CrossRef]

16. Li, Z.; Liu, G.; Liu, L.; Lai, X.; Xu, G. Iot-based tracking and tracing platform for prepackaged food supply chain. *Ind. Manag. Data Syst.* **2017**, *117*, 1906–1916. [CrossRef]

17. Nakamoto, S. Bitcoin: A Peer-to-Peer Electronic Cash System. Social Science Electronic Publishing. 2018. Available online: https://bitcoin.org/bitcoin.pdf (accessed on 1 February 2022).

18. Lee, D.; Chuen, K. *Handbook of Digital Currency*; Elsevier: Amsterdam, The Netherlands, 2015.

19. Wang, Y.; Han, J.H.; Beynon-Davies, P. Understanding blockchain technology for future supply chains: A systematic literature review and research agenda. *Supply Chain. Manag.* **2018**, *24*, 62–84. [CrossRef]

20. Kouhizadeh, M.; Saberi, S.; Sarkis, J. Blockchain technology and the sustainable supply chain: Theoretically exploring adoption barriers. *Int. J. Prod. Econ.* **2021**, *231*, 1078311. [CrossRef]

21. Chang, S.E.; Chen, Y.C.; Lu, M.F. Supply chain re-engineering using blockchain technology: A case of smart contract based tracking process. *Technol. Forecast. Soc. Change* **2019**, *144*, 1–11. [CrossRef]

22. Feng, H.; Wang, X.; Duan, Y.; Zhang, J.; Zhang, X. Applying blockchain technology to improve agri-food traceability: A review of development methods, benefits and challenges. *J. Clean. Prod.* **2020**, *260*, 121031. [CrossRef]

23. Kamilaris, A.; Fonts, A.; Prenafeta-Boldu, F.X. The rise of blockchain technology in agriculture and food supply chains. *Trends Food Sci. Technol.* **2019**, *91*, 640–652. [CrossRef]

24. Galvez, J.F.; Mejuto, J.C.; Simal-Gandara, J. Future challenges on the use of blockchain for food traceability analysis. *TrAC Trends Anal. Chem.* **2018**, *107*, 222–232. [CrossRef]

25. Kamble, S.S.; Gunasekaran, A.; Sharma, R. Modeling the blockchain enabled traceability in agriculture supply chain. *Int. J. Inf. Manag.* **2019**, *52*, 101967. [CrossRef]

26. Azzi, R.; Kilany, R.; Sokhn, M. The power of a blockchain-based supply chain. *Comput. Ind. Eng.* **2019**, *135*, 582–592. [CrossRef]

27. Abeyratne, S.; Monfared, R. Blockchain ready manufacturing supply chain using distributed ledger. *Int. J. Res. Eng. Technol.* **2016**, *5*, 1–10.

28. Bai, C.; Sarkis, J. A supply chain transparency and sustainability technology appraisal model for blockchain technology. *Int. J. Prod. Res.* **2019**, *58*, 2142–2162. [CrossRef]

29. Cole, R.; Stevenson, M.; Aitken, J. Blockchain technology: Implications for operations and supply chain management. *Supply Chain. Manag.* **2019**, *24*, 469–483. [CrossRef]

30. Huertas. Eximchain: Supply Chain Finance Solutions on a Secured Public, Permissioned Blockchain Hybrid. 2018. Available online: https://eximchain.com/Whitepaper%20-%20Eximchain.pdf (accessed on 1 March 2018).

31. Dolgui, A.; Ivanov, D.; Potryasaev, S.A.; Sokolov, B.V.; Werner, F. Blockchain-oriented dynamic modelling of smart contract design and execution in the supply chain. *Int. J. Prod. Res.* **2020**, *58*, 2184–2199. [CrossRef]

32. Xu, X.; Zhu, D.; Yang, X.; Wang, S.; Qi, L.; Dou, W. Concurrent practical byzantine fault tolerance for integration of blockchain and supply chain. *ACM Trans. Internet Technol. (TOIT)* **2020**, *21*, 1–17. [CrossRef]

33. Wang, Z.; Wang, T.; Hu, H.; Gong, J.; Ren, X.; Xiao, Q. Blockchain-based framework for improving supply chain traceability and information sharing in precast construction. *Autom. Constr.* **2020**, *111*, 103063. [CrossRef]

34. Ho GT, S.; Tang, Y.M.; Tsang, K.Y.; Tang, V.; Ka, Y.C. A blockchain-based system to enhance aircraft parts traceability and trackability for inventory management. *Expert Syst. Appl.* **2021**, *179*, 115101.

35. Omar, I.A.; Jayaraman, R.; Salah, K.; Debe, M.; Omar, M.A. Enhancing vendor managed inventory supply chain operations using blockchain smart contracts. *IEEE Access* **2020**, *8*, 182704–182719. [CrossRef]

36. Bumblauskas, D.; Mann, A.; Dugan, B.; Rittmer, J. A blockchain use case in food distribution: Do you know where your food has been? *Int. J. Inf. Manag.* **2019**, *52*, 102008. [CrossRef]

37. Biswas, K.; Muthukkumarasamy, V.; Lum, W. Blockchain based wine supply chain traceability system. In Proceedings of the Future Technologies Conference (FTC) 2017, Vancouver, BC, Canada, 29–30 November 2017; pp. 56–62.

38. Caro, M.P.; Ali, M.S.; Vecchio, M.; Gia Ff Reda, R. Blockchain-based traceability in Agri-Food supply chain management: A practical implementation. In Proceedings of the 2018 IoT Vertical and Topical Summit on Agriculture—Tuscany (IOT Tuscany), Tuscany, Italy, 8–9 May 2018; pp. 1–4.

39. Mhaa, B.; Lc, C.; Ak, D.; Sz, E.; Kht, F. A sustainable blockchain framework for the halal food supply chain: Lessons from malaysia. *Technol. Forecast. Soc. Change* **2021**, *170*, 120870.

40. Lin, Q.; Wang, H.; Pei, X.; Wang, J. Food safety traceability system based on blockchain and EPCIS. *IEEE Access* **2019**, *7*, 20698–20707. [CrossRef]

41. Tsang, Y.P.; Choy, K.L.; Wu, C.H.; Ho GT, S.; Lam, H.Y. Blockchain-driven iot for food traceability with an integrated consensus mechanism. *IEEE Access* **2019**, *7*, 129000–129017. [CrossRef]

42. Powell, W.; Foth, M.; Cao, S.; Natanelov, V. Garbage in garbage out: The precarious link between IoT and blockchain in food supply chains. *J. Ind. Inf. Integr.* **2021**, *25*, 100261. [CrossRef]

43. Lakshmi, G.V.; Gogulamudi, S.; Nagaeswari, B.; Reehana, S. BlockChain based inventory management by QR code using open CV. In Proceedings of the International Conference on Computer Communication and Informatics (ICCCI-2021), Coimbatore, India, 27–29 January 2021.

44. Baralla, G.; Pinna, A.; Corrias, G. Ensure traceability in european food supply chain by using a blockchain system. In Proceedings of the 2019 IEEE/ACM 2nd International Workshop on Emerging Trends in Software Engineering for Blockchain (WETSEB), Montreal, QC, Canada, 27–27 May 2019.

45. Dey, S.; Saha, S.; Singh, A.K.; Mcdonald-Maier, K. Foodsqrblock: Digitizing food production and the supply chain with blockchain and qr code in the cloud. *Sustainability* **2021**, *13*, 3486. [CrossRef]

46. Benčić, F.M.; Skočir, P.; Žarko, I.P. Dl-tags: DLT and smart tags for decentralized, privacy-preserving and verifiable supply chain management. *IEEE Access* **2019**, *7*, 46198–46209. [CrossRef]

47. Feng, T. An agri-food supply chain traceability system for China based on RFID & blockchain technology. In Proceedings of the 2016 13th International Conference on Service Systems and Service Management (ICSSSM), Kunming, China, 24–26 June 2016; IEEE Press: Piscataway, NJ, USA, 2016; pp. 1–6.

48. Feng, T. A supply Chain Traceability System for Food Safety Based on HACCP, Blockchain & Internet of things. In Proceedings of the 2017 International Conference on Service Systems and Service Management, Dalian, China, 6–18 June 2017; IEEE Press: Piscataway, NJ, USA, 2017; pp. 1–6.

49. Helo, P.; Shamsuzzoha, A.H.M. Real-time supply chain—A blockchain architecture for project deliveries. *Robot. Comput. Integr. Manuf.* **2020**, *63*, 101909. [CrossRef]

50. Mondal, S.; Wijewardena, K.; Karuppuswami, S.; Kriti, N.; Kumar, D.; Chahal, P. Blockchain inspired RFID-based information architecture for food supply chain. *IEEE Internet Things J.* **2019**, *6*, 5803–5813. [CrossRef]

51. Mazzei, D.; Baldi, G.; Fantoni, G.; Montelisciani, G.; Rizzello, L. A blockchain tokenizer for industrial IOT trustless applications. *Future Gener. Comput. Syst.* **2019**, *105*, 432–445. [CrossRef]

52. Sfa, A.; Hm, B.; Cs, C.; Ms, D.; Rt, D. Closed-loop and open-loop authentication protocols for blockchain-based IOT systems. *Inf. Process. Manag.* **2021**, *58*, 102568.

53. Toyoda, K.; Mathiopoulos, P.T.; Sasase, I.; Ohtsuki, T. A novel blockchain-based product ownership management system (poms) for anti-counterfeits in the post supply chain. *IEEE Access* **2017**, *5*, 17465–17477. [CrossRef]

54. Wang, L. Design optimization of food safety monitoring system with social network analysis. *IEEE Trans. Comput. Soc. Syst.* **2018**, *5*, 676–686. [CrossRef]

Review

Cereal and Confectionary Packaging: Background, Application and Shelf-Life Extension

Anna-Sophia Bauer [1,†], Kärt Leppik [2,3,†], Kata Galić [4], Ioannis Anestopoulos [5,6], Mihalis I. Panayiotidis [5,6], Sofia Agriopoulou [7], Maria Milousi [8], Ilke Uysal-Unalan [9,10], Theodoros Varzakas [7,*] and Victoria Krauter [1,*]

[1] Packaging and Resource Management, Department Applied Life Sciences, FH Campus Wien, 1030 Vienna, Austria; anna-sophia.bauer@fh-campuswien.ac.at

[2] Center of Food and Fermentation Technologies, Akadeemia tee 15a, 12618 Tallinn, Estonia; kart@tftak.eu

[3] Department of Chemistry and Biotechnology, School of Science, Tallinn University of Technology, Ehitajate tee 5, 19086 Tallinn, Estonia

[4] Faculty of Food Technology and Biotechnology, University of Zagreb, HR10000 Zagreb, Croatia; kata.galic@pbf.unizg.hr

[5] Department of Cancer Genetics, Therapeutics & Ultrastructural Pathology, The Cyprus Institute of Neurology & Genetics, AyiosDometios, Nicosia 2371, Cyprus; ioannisa@cing.ac.cy (I.A.); mihalisp@cing.ac.cy (M.I.P.)

[6] The Cyprus School of Molecular Medicine, AyiosDometios, Nicosia 2371, Cyprus

[7] Department of Food Science and Technology, University of the Peloponnese, Antikalamos, 24100 Kalamata, Greece; s.agriopoulou@uop.gr

[8] Department of Chemical Engineering, University of Western Macedonia, 50100 Kozani, Greece; mmilousi@uowm.gr

[9] Department of Food Science, Aarhus University, Agro Food Park 48, 8200 Aarhus, Denmark; iuu@food.au.dk

[10] CiFOOD—Center for Innovative Food Research, Aarhus University, Agro Food Park 48, 8200 Aarhus, Denmark

* Correspondence: t.varzakas@uop.gr (T.V.); victoria.krauter@fh-campuswien.ac.at (V.K.); Tel.: +43-1-606-68-77-3592 (V.K.)

† These authors contributed equally to this work.

Citation: Bauer, A.-S.; Leppik, K.; Galić, K.; Anestopoulos, I.; Panayiotidis, M.I.; Agriopoulou, S.; Milousi, M.; Uysal-Unalan, I.; Varzakas, T.; Krauter, V. Cereal and Confectionary Packaging: Background, Application and Shelf-Life Extension. *Foods* 2022, 11, 697. https://doi.org/10.3390/foods11050697

Academic Editor: Haiying Cui

Received: 5 February 2022
Accepted: 15 February 2022
Published: 26 February 2022

Publisher's Note: MDPI stays neutral with regard to jurisdictional claims in published maps and institutional affiliations.

Abstract: In both public and private sectors, one can notice a strong interest in the topic of sustainable food and packaging. For a long time, the spotlight for optimization was placed on well-known examples of high environmental impacts, whether regarding indirect resource use (e.g., meat, dairy) or problems in waste management. Staple and hedonistic foods such as cereals and confectionary have gained less attention. However, these products and their packaging solutions are likewise of worldwide ecologic and economic relevance, accounting for high resource input, production amounts, as well as food losses and waste. This review provides a profound elaboration of the status quo in cereal and confectionary packaging, essential for practitioners to improve sustainability in the sector. Here, we present packaging functions and properties along with related product characteristics and decay mechanisms in the subcategories of cereals and cereal products, confectionary and bakery wares alongside ready-to-eat savories and snacks. Moreover, we offer an overview to formerly and recently used packaging concepts as well as established and modern shelf-life extending technologies, expanding upon our knowledge to thoroughly understand the packaging's purpose; we conclude that a comparison of the environmental burden share between product and packaging is necessary to properly derive the need for action(s), such as packaging redesign.

Keywords: food packaging; cereals; confectionary; bakery; food quality; shelf-life; sustainable packaging; active and intelligent packaging; modified atmosphere packaging; vacuum packaging

1. Introduction

Over the past decades, global awareness about environmental, social and economic sustainability challenges, as well as the need for immediate action to limit their negative short- and long-term impacts, has risen considerably. With regard to environmental sustainability, challenges encompass, but are not limited to, the use of resources, land, water, energy, and generation of associated emissions and waste. In order to facilitate the transition towards a sustainable future, several (inter)national goals, commitments, and legal bases have already been initiated or applied. These include, for instance, the Paris Agreement on climate change and the United Nations Sustainable Development Goals (SDGs) on a global scale, the European Green Deal including the New Circular Economy Action Plan, as well as the Farm to Fork Strategy on European level and numerous implementations into national law systems [1–6].

Regarding food, it is well-agreed in the scientific community and beyond, that a great share of negative environmental impacts such as global anthropogenic greenhouse gas emissions or waste originate from food systems [7–9]. These systems are defined as the whole of actors and activities involved, from production to the disposal of food products of different origins, as well as herewith associated natural, social, and economic environments [10]. Moreover, they are composed of subsystems (e.g., farming) and connected to other systems (e.g., energy). A complex network in which changes (e.g., policies) made in one sector may also affect others. Against this background, different international efforts have been taken to achieve sustainable food systems, which will provide present and future generations with a secure supply of safe food [11].

Packaging is strongly associated with food, allowing, amongst other functions, containment, protection, and transportation of contents, and thus can be seen as an integral part of food systems [12,13]. Nevertheless, nowadays it is the subject of intense debates and even stricter legal requirements, mainly due to massive circularity gaps including, for example, unsatisfactory end-of-life scenarios such as limited recyclability or (marine) litter [14,15]. However, the simple omission of packaging is hardly possible, since a well-chosen packaging system frequently shows positive (indirect) effects on the total environmental sustainability of a food system by, for example, reducing food losses and food waste or increasing transport efficiency [16]. Therefore, when aiming at developing sustainable packaging solutions, it is important to apply a holistic and interdisciplinary approach over the whole life cycle of both food and its corresponding packaging [17].

Since packaging offers a service to the food product and does not fulfil an end in itself, it is often worth starting a packaging development or a redesign process from the food perspective. By gaining profound knowledge of the food product itself, together with the intrinsic and extrinsic factors that affect quality along the food supply chain, further packaging requirements can be defined and considered in the innovation process [12,13,17].

Due to their high environmental impact, the focus of research and development activities is often on (animal protein-rich) foods such as meat or milk [18–20]. Despite their high nutritional value that shouldn't be underestimated, cereal and confectionary products are rather underrepresented, regarding their impact in health but also in economic and environmental sustainability [21–27]. For instance, about 50% of daily required carbohydrates are consumed through bread in industrialized countries. Further, cereals are also an important source of proteins, minerals, and trace elements [28]. Expressed in figures, retail sales of bread alone were expected to reach about 92 billion euros in Europe in 2021 [29]. On the other hand, confectionary products reached a production volume of 14.7 million tons with an annual turnover of 60 billion euros along with an export value of 9.2 euros and an import value of two billion euros in Europe (EU28) in 2019 [30].

In more detail, the present review aims at building a comprehensive basis for future sustainable packaging development activities in the area of cereal and confectionary products by:

- Presenting relevant information on packaging functions and properties of packaging materials,
- detailing product group specific decay mechanisms and frequently used packaging solutions,
- and highlighting packaging-related shelf-life extension technologies.

The text is therefore structured as follows: After the introduction, a general background on food packaging is discussed, followed by product group specific decay mechanisms and packaging solutions. Finally, packaging measures that can extend the shelf-life are presented (see also Figure 1).

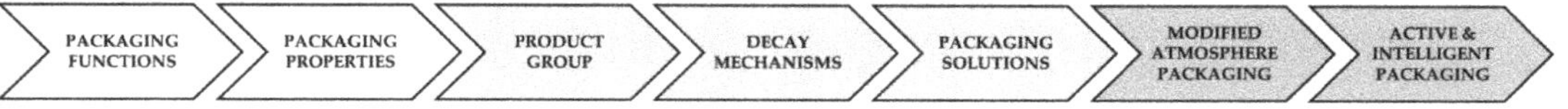

Figure 1. Outline of discussed topics, based on the review's aims.

2. Packaging

2.1. Packaging Functions

No matter how diverse individual products and packaging solutions may be on the market, it is well-agreed in relevant literature that the main functions of packaging can be broken down into a few. Next to the concept of primary and secondary functions, where the former describes in particular the technical functions like storage and transport, and the latter describes functions related to e.g., communication, a more holistic concept is frequently mentioned in the packaging literature. This concept describes the four basic functions of food packaging as (i) containment, (ii) protection, (iii) convenience, and (iv) communication [12,13,31–33].

Although the containment function is often overlooked, it can be considered one of the most essential, since it prevents product loss and contamination and facilitates storage, transportation, and distribution. There are only a few exceptions, where containment and thus packaging is not needed. Such examples are relatively large, chunky products that are often regionally produced and consumed within a short period of time or that show long shelf-life [12,13,31].

The protection function is often recognised as well as highlighted and can be indeed considered as the most important aspect of packaging. It limits or excludes intrinsic as well as extrinsic physical, chemical, and biological factors that may have negative influences on the quality of the respective food product. In the best case, the packaging is even capable of extending the shelf-life of the product. Therefore, it is of upmost importance to match the food product's properties and requirements along the supply chain with packaging to achieve optimal results. Both under- and over-packaging should be avoided since this may result, on one hand, in food losses or waste and, on the other hand, in excessive packaging [12,13,31].

Further, the convenience function refers to the practical aspects or user-friendliness of packaging. As an example, easy-to-open or -empty, microwave- or heat-able, resealable, or portion packaging can be named. These features are more and more implemented in package designs, since they allow to specifically address target groups (e.g., children, elderly, single-households, on-the-go lifestyle) and therefore frequently influence the market success of a product [12,13,31].

Last but not least, the communication function allows for information transfer and marketing. While the former allows to display legally required (e.g., product name, ingredients, shelf-life), necessary (e.g., barcodes), or voluntary (e.g., certificates, cooking recipe) information, the latter enables to transfer an often unique brand image (e.g., form, colour, shape), which may be of great recognition value [12,13,31].

It is worth mentioning that a successful package on the market does not only need a strong product in terms of quality but also an effective packaging, which in a clever way combines the above described four functions of containment, protection, convenience and communication. Otherwise, it may result in a short-term success (weak product and effective packaging), a situation where the potential is not achieved (strong product and ineffective packaging), or even failure (weak product and ineffective packaging) [31].

2.2. Packaging Properties

From a technical point of view, the functions containment and protection are closely linked to the right selection of packaging materials which consequently poses a key decision in the development process. The available material classes cover mainly glass, metal, paper/board, (bio)plastic, as well as composite materials (laminated, coextruded, blended). Composites can consist of two or more components combined to form, for example, multi-layer materials (e.g., plastic-coated cardboard) which frequently show superior functional properties (e.g., barrier) and reduced weight [31], but on the downside also reduced recyclability [34,35]. Touching upon the topic of recyclability, many packaging solutions face obstacles, if it is at the stage of collection, sorting, or in general limited technical recyclability. Not even the use of mono-materials guarantees actual recycling, as it is the case for PET trays versus PET bottles (bottles are highly likely to be recycled). On the other hand, specific combinations of compatible materials, even high barrier films, for example, metallized polyolefins, might be considered recyclable in the appropriate infrastructure [36,37]. Summing up, it can be stated that each of the named materials show advantages and disadvantages (see Table 1) and the decision for a specific material must be based on the prevailing requirements (e.g., product, supply chain, use, end-of-life). Support is often provided by material specifications and declaration of compliance documents. However, it is recommended to test the materials in question under respective conditions by means of a field or laboratory test. This ensures that deviations from the target value can be recognized at an early stage in the development process [12,13,31,38,39].

Table 1. Overview of the properties and applications of most widely used materials for packaging.

Packaging Material		Barrier			Heat Seal-Ability	Mechanical, Physical and Chemical Properties	Application	Reference
		Oxygen	Moisture	Light				
Plastic	Low-density polyethylene (LDPE)		High			Toughness, flexibility, resistance to grease and chemicals, temperature range −50 – +80 °C	Bags, flexible lids and bottles	[12]
	Linear low-density polyethylene (LLDPE)	Very low	High	Low	Yes	Toughness, extensibility, resistant to grease, temperature range −30 – +100 °C	(Strech) wrap	
	High-density polyethylene (HDPE)		Extremely high			Toughness, stiffness, resistance to grease and chemicals, easy processing and forming, temperature range −40 – +120 °C	Bottles, cardboard liners, tubs, bags	
	Polypropylene (PP)	Low	High	Low	Yes	Moderate stiffness, strong, resistant to grease and chemicals, temperature range −40 – +120 °C	Bottles, cardboard liners, tubs, microwavable packaging, bags	
	Polyethylene terephthalate (PET)	Good	Good	Low	Yes	Stiffness, strong, resistance to grease and oil, temperature range −60 – +200 °C	Bottles, jars, tubs, trays, blisters, films (bags and wrappers)	[12,40]
Glass	Transparent	Absolute		Low	No	High temperature and pressure stability, brittle, chemical resistance, microwave-able	Bottles, jars	[12,40–42]
	Green			Good				
	Brown			High				
Metal (aluminium, tinplate, tin-free steel)		Absolute			No	High temperature stability	Bottles, cans, tubs, caps	[12,40]
Paper and board		Extremely low	High – extremely high		No	Mechanical stability	Boxes, liners	[12,40,41]

The key properties of packaging materials of interest are physical and mechanical strength, barrier, migration, as well as hygiene. Regarding the physical and mechanical strength, it can be noted that static as well as dynamic stress challenges the packages along the supply chain from packing, storage, and transport to consumer use. Examples for static stress are stacking and increased pressure (vacuum or modified atmosphere packaging—MAP), as well as pointed or angular products. Dynamic stress on the other hand may be caused by the production process (e.g., printing, forming, filling) or transport (e.g., vibration). The right selection of the material, but also the shape of the packaging, therefore plays a vital role in the success of a primary, secondary or tertiary package (see also Figure 2) [12,13,38,43].

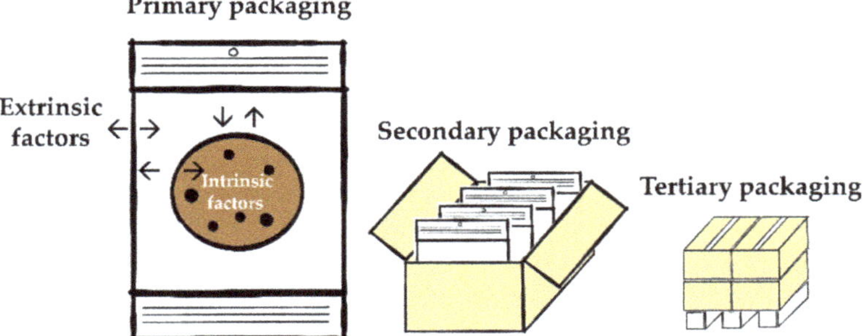

Figure 2. Schematic packaging levels of fine bakery ware (example: chocolate chip cookie), adapted from [12,13,31].

Another key characteristic of materials to be considered is the barrier property. Especially, the barriers against oxygen (O_2) and water vapour (H_2O) transmission are determinant since these can exhibit significant influences on product quality and safety. The former for example can promote oxidation reactions, loss of quality-determining ingredients (e.g., vitamins), and growth of spoilage and pathogenic microorganisms. The latter can influence structural changes such as hardening, agglomeration, or softening of products and promote microbial growth (see Section 3.2). Additionally, barriers against carbon dioxide (CO_2) and nitrogen (N_2), which are the often-used gases in MAP, as well as aroma components, are decisive. Depending on the use case and product requirement, material with an appropriate barrier, i.e., permeation characteristics, should be chosen. Complementary to the above described, the barrier against other substances like fat may be considered [12,13,38,44]. Furthermore, electromagnetic radiation (light) has to be taken into consideration, since oxidative or other chemical reactions as well as structural changes may be induced or accelerated, thus impairing product quality [12,41,45–47].

What is important regarding chemical safety is the migration of compounds from packaging materials into the food. Migration describes the mass transfer of substances from a packaging material into the food product or vice versa. As for the permeation, the driving force behind this phenomenon is the concentration gradient. Additionally, factors such as material, storage temperature, relative humidity, and time play an influencing role [38,39,48].

Against common perception, possible migration of, for example, additives, are not only present in plastic packaging materials. Migration can also be found in other (primary or secondary (recycled)) materials such as glass (e.g., silicates), metal (e.g., corrosion of the metal, additive migration from organic coatings), paper and board (e.g., fillers, contaminations like mineral oils) and may, next to the packaging material itself, find its origin in packaging aids (e.g., labels, closures, coatings) or even set-off processes (e.g., printed and role-to-role processed or stapled materials) [12,13,38]. To ensure safety of food contact materials (including packaging), several legal requirements are in place in the European Union and beyond [39,48–53]. It should be noted that in addition to the migration from the packaging material to the food, migration processes from the food to the packaging can also be observed. This process is also called sorption or scalping and may cause alteration of the product (e.g., flavour loss) as well as reduced reusability of packaging containers due to the re-release of previously migrated substances [12,13].

In addition to chemical safety, packaging materials also play a role in the hygiene and biological safety of food products. Depending on the material used, a barrier against contamination, microorganisms and animals (e.g., food pests) can be given. To achieve a high standard of hygiene, it is crucial to utilize materials that pose a sufficient barrier and that are free from contamination. Further, it is important to use materials that do not support microbial growth. Lastly, it is important to recognise, that most packaging materials carry a low microbial count when freshly produced due to often high process temperatures (e.g., melting of glass). So, the microbial burden is often a result of recontamination during finishing processes, storage, and application, which can sometimes make it necessary to implement decontamination measures prior to the filling process [38,54].

3. Cereal and Confectionary Products

Against the above-summarized background, food packaging can be seen as a mediator between product and the environment, capable of significantly influencing food quality, safety, and shelf-life [12]. Regarding cereal and confectionary products, the following text aims at summarizing and categorizing the product group, presenting an overview of category specific decay mechanisms, as well as respective packaging solutions.

3.1. Categorization of Cereal and Confectionary Products

As shown by Belitz et al. [28], cereal and confectionary products cover a wide and diverse range of food products. They summarized different products in two groups, namely cereals and cereal products. The first group is mainly made from important staple foods such as wheat, rye, rice, barley, millet, oats and corn. These are used to produce different kinds of products. For example, Smith et al. [55] made the following division: "...unsweetened goods (bread, rolls, buns, crumpets, muffins and bagels), sweet goods (pancakes, doughnuts, waffles and cookies) and filled goods (fruit and meat pies, sausage rolls, pastries, sandwiches, cream cakes, pizza and quiche)".

The group of confectionery products are mainly sugar-based products that, in contrast to cereal products, are predominantly consumed as a "treat" rather than a full meal. These include products such as chocolate, hard candy, and pralines [56,57]. In addition to sweet confectionery, savory snacks can also be found on the market. According to Robertson [13], these include "...a very wide range of products, including potato and corn chips, alkali-cooked corn tortilla chips, pretzels, popcorn, extruded puffed and baked/fried products, half-products, meat snacks and rice-based snacks" [13,58]. In addition to that, there are combinations of sweet and savory snacks like chocolate covered pretzels or sweet popcorn [59].

In the available literature and other sources including statistics, codices and regulations, different approaches to properly (sub)categorize cereal and confectionary products can be found [59–61]. Taking a food and shelf-life perspective, it is reasonable to cluster products that exhibit similar characteristics or spoilage mechanisms. In the European Union, where there is a strong food law [62] in place, a comprehensive list can be, for example, found in the guidance document to Annex II of regulation (EC) No 1333/2008 on food additives [59,63]. For the field of cereals and confectionary, the four groups of confectionary, cereals and cereal products, bakery wares, and ready-to-eat savories and snacks are of special interest. While confectionary is further subdivided into cocoa and chocolate products, other confectionery products including breath freshening micro-sweets, chewing gum as well as decorations, coatings and fillings, cereals and cereal products are divided into whole, broken or flaked grain, flours, milled products and starches, breakfast cereals as well as pasta, noodles, batters and pre-cooked or processed cereals. For bakery wares, a classification into bread and rolls and fine bakery wares is given. Last but not least, savories and snacks are broken down into potato-, cereal-, flour- or starch-based snacks as well as processed nuts. For each of the above-mentioned subgroups, a comprehensive list of product examples is given in the mentioned document [59]. The present review adopts this categorization approach and structures relevant information on cereal and confectionary shelf-life, packaging, and shelf-life extension strategies accordingly (Figure 3).

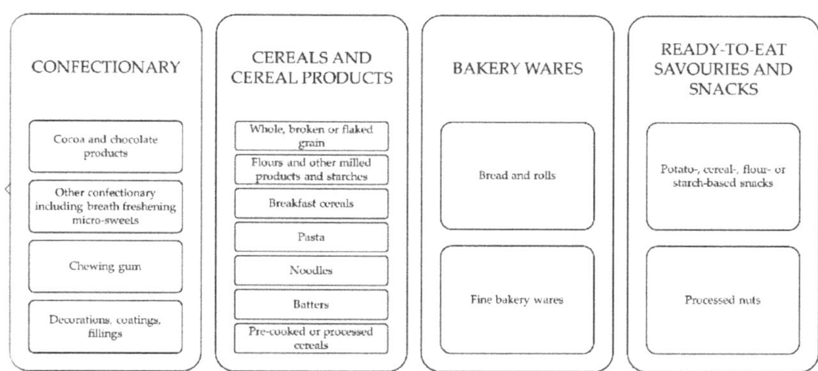

Figure 3. Representation of the followed product categorization. Adapted from [59].

3.2. Decay Mechanisms and Shelf-Life

It is well-established that intrinsic as well as extrinsic factors influence the quality of food and thus its shelf-life [13], which can be defined as the period of time a food maintains its safety and/or quality under reasonably foreseeable conditions of distribution, storage, and use [12,64–66]. Intrinsic factors include, amongst others, pH, water activity (a_w), initial microbial population, redox potential value (Eh), and nutrient content and therefore determine the nature of decay mechanisms of a food product. On the other hand, extrinsic factors determine how fast decay mechanisms proceed. Typical examples are atmosphere, climatic conditions, and illumination. Packaging itself acts as mediator or separator between intrinsic and extrinsic systems [13,67]. The following paragraphs highlight the main challenges of quality maintenance of cereal and confectionary products but do not go into detail about the physical, chemical, or biological bases of these mechanisms (e.g., oxidation). This information can be found in the relevant scientific literature [13,67,68].

Focusing on cereal and confectionary products (see Table 2), moisture content (MC) and water activity (a_w) are some of the most important quality-affecting parameters. Kong and Singh [69] define, that the a_w value is "…the vapour pressure of water above a sample (p) divided by that of pure water at the same temperature (p0); i.e, $a_w = \frac{p}{p0}$. It describes the degree to which water is free or bound to other components". They state that this is related to "…the composition, temperature, and physical state of the compounds" [69,70]. This is of importance regarding the potential growth of microorganisms as they depend on free water presence [71].

Table 2. Water activity and moisture content of confectionery products, breakfast cereals, snacks, and bakery products.

Product category	Subcategory	Product	Water Activity [a_w]	Moisture Content [%]	Reference
Confectionery	Cocoa and chocolate products	Chocolate	0.42–0.60	1.2	[72]
	Other confectionery including breath freshening micro-sweets	Hard candy	0.25–0.40	2.0–5.0	[73,74]
		Fudge, toffee	0.45–0.60	6.0–18.0	
		Nougat (white, dark)	0.55	8.00–10.0	[13,75]
		Jelly, liquorice	0.50–0.75	8.0–22.0	[73,74]
		Marshmallow	0.60–0.75	12.0–22.0	
		Marzipan	0.75–0.80	–	[13]
	Chewing gum	Chewing gum	0.40–0.65	3.0–6.0	[73,74]

Table 2. *Cont.*

Product category	Subcategory	Product	Water Activity [a_w]	Moisture Content [%]	Reference
Cereals and cereal products	Whole, broken, or flaked grain	Oats, grains, cereals	0.34–0.70	8.8–9.2	[13,72]
	Breakfast cereals	Cornflakes	0.25–0.38	1.7–3.5	
		Puffs	0.17–0.20	0.48–1.70	
	Fresh pasta	Fresh pasta	0.91–0.98	≥24	
	Dry pasta	Dry pasta	0.33–0.57	5.4–8.3	
Bakery wares	Fine bakery wares	Sponge cake, muffins	0.84–0.95	21.0–40.0	[76,77]
		Croissant crust	0.59–0.61	8.0–10–0	
		Croissant crumb	0.92–0.94	30.0–33.0	
		Biscuits	0.60–0.63	1.5–3.0	[72,78]
		Wafers	0.13–0.15	2.1	[72]
		Cookies	0.18–0.64	1.4–11.7	
	Bread and rolls	Flat bread (no yeast)	-	33.0–35.0	[79]
		Sourdough bread, yeast bread crumb	0.91–0.95	29.0–40.0	[72]
		Sourdough bread, yeast bread crust	0.88–0.94	26.0–32.0	
		Bagel crust	0.96	38.5	
		Bagel crumb	0.92	31.0	
Ready-to-eat savouries and snacks	Potato-, cereal-, flour- or starch-based snacks	Popcorn	0.07	0.28	
		Chips	0.09–0.27	0.3–1.3	
		Crackers, grissini, sticks, pretzels	0.05–0.54	1.1–5.4	
	Processed nuts	Nuts, seeds, nibs	0.15–0.75	0.5–3.1	

With an a_w lower than 0.75, a large proportion of the products listed in Table 2 falls into the group of low-moisture or dried foods that additionally exhibit low (e.g., cornflakes) or high (e.g., crisps) fat content. In this group, water uptake and thus loss of, e.g., crispness, which occurs, e.g., in potato chips and breakfast cereals after gaining moisture at a range of 0.35 to 0.5 a_w, is the main decay mechanism [12,13,69,80]. Other mechanisms include loss of aroma (e.g., flavoured products) or aroma uptake from the products' surrounding due to the often porous structure of the food products. Further, structural changes such as loss of integrity due to e.g., mechanical damage (e.g., breakage), softening, or caking may occur. While microbial growth is the basis for both, low and high fat types, oxidative mechanisms, which may lead to off-odours and -tastes and subsequently to quality loss in terms of overall acceptance, are often linked to the fat content and thus tend to increase with the same [12]. Examples that can be named are nuts, chips, biscuits, and cookies. All in all, this product group can, however, be described as rather stable and therefore storage under dry and ambient conditions is recommended and possible. For example, breakfast cereals and dry pasta stay stable under temperate conditions for 6–18 months and 48 months, respectively [72,81]. Confectionary products like pulled sugar are stable for 6–9 months under temperate conditions (e.g., ~20 °C) [68].

Other products, including chocolate for example, can be allocated to compact foods with high fat content, a group mainly susceptible to the uptake of unwanted flavours and some (often minor) water exchange (uptake or loss) processes [12]. The latter can induce so-called blooming effects [13]. Sugar bloom on the one hand is often provoked by humid storage or rapid temperature changes and leads to the loss of surface gloss. Fat bloom on the other side is also known to cause quality related issues visible as a fine whitish layer [82]. Growth of microorganisms is, however, of minor importance in this product group. Storage under temperate or chilled conditions is therefore possible for up to 12–24 months [57].

Microbial growth is of major concern in the group of ready-to-eat and ready-to-cook convenience food products (e.g., fresh pasta). At this point, in addition to spoilage microorganisms, pathogenic microorganisms play an essential role [65,83]. Further, water loss and structural changes can be named. Additionally, oxidation can significantly gain importance regarding shelf-life. Accordingly, chilled storage is often preferred [13,67].

The area of bakery products can be divided into fresh bakery wares and ready-to-bake products. The first group (e.g., bread) shows high a_w values (>0.8) and thus short shelf-life, which is heavily influenced by water exchange processes that are often interlinked with structural changes (softening of the crust and drying of the crumb). Connected to this, starch retrogradation, which is the main mechanism of staling, can be highlighted [69]. Further, loss of moisture and hardening with a_w values below 0.5–0.7 [13,69,80] quickly result in low sensory acceptance of the products. While oxidation and rancidity play a minor role in this food category, uptake of flavours as well as microbial spoilage play a more elaborated role in this product group. The latter point is mainly driven by the often visible growth of moulds and yeasts on the food surface. Characteristic microorganisms are *Penicillium roqueforti, Hansenula anomala, Pichia anomala, Candida guilliermondii, C. parapsilosis, Saccharomyces cerevisiae, S. exiguus, S. unisporus, S. bayanus, S. pastorianus*. Additionally, *Clostridium* and *Bacillus* genera are known bacteria potentially affecting bakery wares (spore-forming), with e.g. *Bacillus* spp. causing "rope" or "ropy spoilage" (*Bacillus amyloliquefaciens, Bacillus subtilis, Bacillus pumilus, Bacillus cereus*) [71,84,85]. Oxidation and rancidity play a minor role in this product category. Accordingly, the average shelf-life of fresh bread and cake under ambient conditions is often less than one week [86]. In some cases, chilled or frozen storage is advisable. The group of ready-to-bake rolls show very similar decay mechanisms. However, due to the higher water content, drying and spoilage is even more pronounced. In the case of frozen products, these mechanisms are delayed. A special focus has to be laid on water exchange (freezer burn) and structural damage [87].

3.3. Product Group Specific Packaging

Responding to the above-mentioned predominant decay mechanisms of cereal and confectionary products, the following section aims at highlighting common packaging concepts and material choices (compare also Table 1).

Chocolate packaging has to provide a good barrier against aroma, gas (especially O_2 and H_2O) as well as light. This is conventionally achieved by using aluminium foil of different thickness to wrap the product. Since aluminium alone cannot be heat sealed, the per se excellent barrier of the material is, however, interrupted at, e.g., overlapping areas or gaps. Hence, diffusion (mass transfer) of aroma, gas and other molecules (e.g., mineral oil components) to the product cannot be excluded. Additionally, the originality of the product, an important factor of food safety, may not be ensured [13,67]. For this and other reasons (e.g., communication), many described packaging concepts (still) include an additional packaging layer, namely paper or paperboard [13,27,88–92].

Today, more and more multilayer materials can be found on the market. For example, laminates of LDPE (low density polyethylene) and aluminium allow for heat sealing of the aluminium by at the same time keeping the superior barrier and dead-fold properties of aluminium. Further, multilayer materials including paper or other aluminium replacing barrier materials (e.g., polyvinylidene dichloride (PVdC)) are available. Possible build-ups may include LDPE/aluminium/paper or LDPE/PVdC, respectively [13]. Nowadays, a shift towards packaging made (solely) from (oriented) PP, which exhibits, due to a stretching process, inter alia, improved mechanical and barrier properties, is notable [21,92]. Additionally, cold sealing, is more and more adopted, since it avoids exposing sensitive products, such as chocolate, to elevated temperatures during heat sealing. This alternative is made possible by applying cold-seal adhesives on the intended sealing areas of the packaging film and pressing of two of the sealing areas together [31].

Individually packed chocolate products, such as chocolate coated bars or pralines, are often bought for hedonistic reasons (e.g., treats, gift function) and thus the communication function (design) of these packages is frequently at the forefront [13,56]. While the functions of containment and protection are already met, these packages often use excess packaging materials and/or layers and for example consist of a (e.g., polyethylene terephthalate (PET)) tray with individual cavities, (e.g., aluminium) wrapping of the individual pieces, a (e.g., paperboard) box, (e.g., polyethylene (PE) or polypropylene (PP)) overwrapping and packaging aids (e.g., labels, stickers). Glass or metal is also used in some cases [13].

Many confections, such as hard candies, gums, toffees and caramels are likewise (twist) wrapped individually. This is either for technical reasons such as provision of an adequate (H_2O) barrier and thus avoidance of moisture loss or uptake, resulting in e.g., drying or agglutination of the product pieces, hygienic reasons or distinction from other products. As for chocolate, tightness of the package should be in the ideal case assured [73]. Due to their in general good barrier properties and sealability, the market dominating polyolefins (PE and PP) as well as PET [93] are also frequently used in this product category (e.g., multipacks) [21,94]. If elevated barriers are needed, different multilayer materials are also adopted. Further, glass and metal packaging can be found on the market and traditional materials include waxed paper, waxed glassine and waterproof, plasticized cellulose fibre [57]. Plain paper and board are, however, hardly used as a primary packaging material, since products tend to stick to the material. The packaging types in this product category are manifold and include, for example, trays, flow packs, boxes (for example cardboard and metal) and jars [13].

Other products such as biscuits, (processed) nuts and fruits are traditionally packaged in regenerated cellulose (trade name Cellophane) fibres (RCF). Therefore, RCF is usually coated with either LDPE or PVdC copolymer and often with a layer of glassine in direct contact with the product if it contains fat. Currently, this combination of materials is replaced by PP, either as plain or pearlized OPP film, coextruded OPP (OPPcoex) film, or acrylic-coated (Ac) on both sides. Plain OPP films require a heat seal coating to improve sealability while coextruded OPP provides superior seal strength. If a high O_2 barrier is required, then acrylic-coated OPP (AcOPP) is used. One side is sometimes coated with PVdC copolymer rather than Ac. In addition, Ac and PVdC copolymer-coated OPP films provide a superior flavour and aroma barrier compared with that of uncoated OPP. Biscuits are often packed in PP and additionally a cardboard box, acting as secondary packaging [13,25].

In comparison to other products, the dry and low in fat group of cereals and cereal products, (such as whole, broken, flaked or milled) grains (e.g., wheat and rice) show rather low packaging demands. Mostly used are paper bags, flexible plastic bags (e.g., PE [95]), as well as cardboard boxes [96,97]. There are also variations of these packages, for example inner flexible plastic bag and a secondary cardboard box. If paper is used and high barriers are needed, LDPE liners for example can be applied [13], also to avoid mineral oil migration [98]. Rigid laminates with paper content and plastic lids usually known in snack product packaging, are also available. Flours for example are commercially packaged in bags or bulk bins [13]. In addition to that, woven PP bags are commonly used in developing countries. However, Forsido et al. [99] discussed that the low moisture barrier led to chemical, physical, sensorial, and microbial changes of flour. Another successful approach for flour packaging that was used for decades, was bags made from cotton twill [13].

The barrier requirements for breakfast cereals packaging are set higher than in the above-mentioned group since crispness, formation of off-flavours, loss of aroma and vitamins or breakage are more critical for consumer acceptance [13]. Consequently, the inner packaging/primary packaging level of these products is a plastic bag, mostly HDPE (high density polyethylene), giving a sufficient water vapour barrier since moisture vapour transmission rates less than or equal to 15 g/m^2-day-atm are often required. Sealant polymers such as EVA (ethylene vinyl acetate), ionomer, mPE (metallocene polyethylene), or blends are used for low temperature seals, form-fill-seal packaging, and easy opening seals [95]. In order to increase barrier characteristics, HDPE is also coextruded with a thin layer of EVA or PA (polyamide) and EVOH (ethylene vinyl alcohol) polymers [95,100]. Other O$_2$ barrier materials for breakfast cereals are PVdC and coated polypropylene-low density polyethylene [101]. In addition, PP-bags are common liners. The secondary packaging/outer packaging is most frequently a fibreboard box [13,22]. Alternative packaging concepts include coated paperboard, plastic cups, as well as metal boxes and glass jars [13,102].

Dried pasta is often packaged in paperboard carton, containing a plastic window. At the moment, most pasta products are packaged in plastic films, such as PE or oriented polypropylene [13,103–107]. For fresh pasta/noodle products, packaging solutions might be different, as appropriate barriers (gas and/or water vapour) and/or MAP (e.g. CO$_2$:N$_2$ 20:80% MAP for pasta) is needed [107,108]. The selection of packaging materials for fresh pasta products can also depend on whether or not the product is pasteurized (thus, the package must be able to withstand the pasteurization conditions) and whether or not the product is to be heated in its package (the package must be able to withstand either heating in boiling water or microwave conditions) by the consumer. For products which are not pasteurized nor intended to be heated in their package, a rigid tray of PVC-LDPE sealed with PA-LDPE film is common. When microwave heating is used, the rigid tray is usually made from crystalline polyethylene terephthalate (PET-C), or polystyrene-ethylene vinyl alcohol copolymer-LDPE (PS-EVOH-LDPE) laminate, and the film may be based on PVdC copolymer-coated PET, OPET-EVOH-LDPE, or PP [109].

Packaging of fresh bakery products such as bread is a moisture balancing act. On one hand, moisture needs to be contained to prevent drying of the product and on the other hand, moisture has to be released from the product to avoid softening of the crust and microbial spoilage. Since there is a wide range of products and product characteristics, also a wide range of packaging solutions can be found. Frequently, paper-based materials, LDPE, LLDPE, HDPE bags as well as OPP, either as plain, pearlized, OPPcoex, or Ac/OPP/Ac films are used [13,95,110–114]. The bags are usually closed either with a strip of adhesive tape or a (plastic) clip in order to reduce moisture loss [111,113,115]. EVA polymers are also used for sealability and optics [95]. Perforated LDPE bags are used (for crusty products) in order to prevent the formation of a leathery consistency of the crust due to moisture migration from the crumb [115]. If aroma and taste barriers are needed, PA is used [95]. Vacuum packaging including the use of respective barrier packaging materials is only used in some exceptions (e.g., flat breads) in this product category due to mechanical impairment of the often soft products. MAP rich in CO$_2$ is whereas more frequently used (e.g., sliced bread, convenience applications). For example, CO$_2$:N$_2$ 60:40% MAP for bread, cakes, crumpets, crepes, fruit pies and pita bread. This is also the case for ready-to-bake products, which are intended to have a longer shelf-life [13].

Packaging for fried snack foods such as potato or tortilla chips, which exhibit, due to their production process, low moisture and high fat contents, preliminarily aims at providing a barrier against gases (H$_2$O and O$_2$) and light to avoid loss of crispness and increased oxidation/rancidity levels of the product [95]. Hence, these products are mainly packaged in high barrier multilayer films containing aluminium foil or metallisation (e.g., PET/Alu/LDPE; PETmet/LDPE; BOPP/BOPPmet) [31,94,116]. In addition, barrier polymers such EVOH or PVDC can be found in these materials. Further, rigid multilayer paper solutions with aluminium (for example spiral wound paper-board cans) or metal cans are also used. Since extruded and puffed snack foods exhibit lower fat levels and thus primarily rely on a package that provides a barrier against water vapour; these products are less often packaged in metallized materials. An example is OPP/LDPE/OPP [95]. In both scenarios, and whether flexible or rigid packaging is adopted, modified atmosphere packaging is frequently used. For example, the package is usually flushed with an inert gas (N$_2$) before closing [116]. Additional mechanical protection of the often fragile products and dry storage is recommended. This might lead to the use of secondary packaging, such as cardboard boxes [31].

4. Shelf-Life Extension

As can be seen from the above text, choosing the right packaging material concept can have a positive effect on quality maintenance and therefore shelf-life of cereal and confectionary products and food in general. Where particularly sensitive products (e.g., high a_w value, high fat content or oxidation potential) are present (e.g., fresh pasta, fried snacks) or an elevated shelf-life has to be achieved (e.g., ready-to-bake rolls, fine bakery wares), modern packaging concepts such as modified atmospheric packaging or active (AP) and intelligent packaging (IP) are used (combined abbreviation: AIP). Manifold different approaches can be found regarding MAP, AP, and IP, each with different relevance for the discussed product subgroups, cereals and cereal products, confectionary, bakery wares and ready-to-eat savouries and snacks. However, for an impression of these, Figure 4 depicts selected examples.

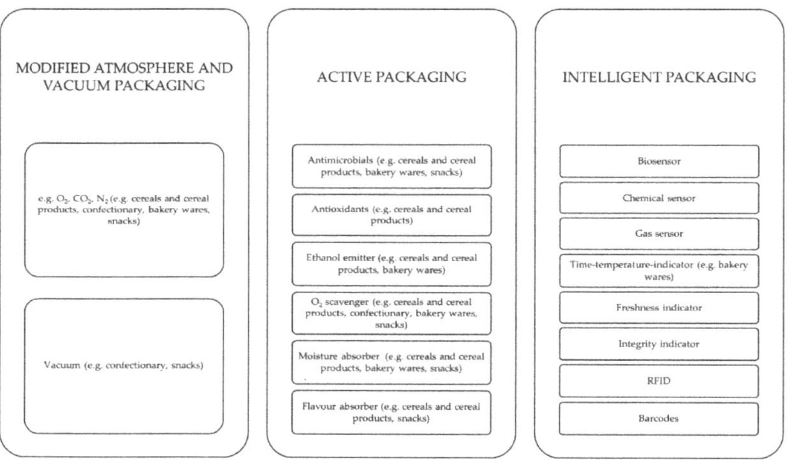

Figure 4. Selected examples of modified atmosphere, vacuum, as well as active and intelligent packaging approaches with certain use cases for cereal and confectionary packaging. Adapted from [13,108,117–140].

Using these approaches, other product preservation actions (e.g., heating, use of preservatives) may be reduced, which supports attempts to reach a healthier diet (e.g., reduction of salt) or a clean label (e.g., avoidance of excess additives) [141] These allow specifically addressing other remaining challenges in the chemical, biological, mechanical, and physical fields [12,13]. Thus, they are also often implemented in the hurdle technology, a concept of combining diverse adverse factors or treatments to control microbial growth in food products [13,142]. According to studies found, also biobased and/or biodegradable packaging material is experimentally combined with AIP approaches. These materials offer new opportunities, for example in making use of different barrier properties, that allow a certain shelf-life extension [134,135]. Examples for MAP and AP with traditional as well as biobased/biodegradable packaging materials can be found in Table 3.

Table 3. Effects of packaging material selection, active packaging (AP) and modified atmosphere packaging (MAP) on shelf-life extension of cereal and confectionary products. Abbreviations: m = month; d = day; RH = relative humidity; RT = room temperature.

Category	Product	Packaging Material	AIP/MAP Applied	Storage	Shelf-Life	Reference
Confectionary	Dark chocolate with hazelnuts	Alu (commercial)	Air	20 °C in dark	8 m	[119]
		PET/LDPE	Vacuum or N_2		8–9 m	
		PET-SiOx/LDPE			11 m	
		PET/LDPE or PET-SiOx/LDPE	Oxygen absorber		$\geq$ 12 m	
Cereals and cereal products	Muesli with chocolate and apricots	Paper bag: PAP + PP window	Air	20 °C, RH 55 %	2 m	[143]
		Pouch: PAP/Alu/PE			9 m	
		Can:PAP/Alu + LDPE lid				
	Fresh pasta	PS tray + PVC film	Air	8 °C	20 d	[120]
		PA/EVOH/LLDPE	CO_2:N_2 22:78% MAP		40 d	
	Fresh pasta filled with cheese	Tray: EVOH/PS/PE wrapped in film: EVOH/OPET/PE	Air	4 °C	7–14 d	[108]
			CO_2:N_2 50:50% MAP		42 d	
	Gluten-free fresh filled pasta	Tray: PETFilm: antifog PET film	Air	4 °C	14 d	[121]
		Tray: EVOH/PS/PEFilm: EVOH/OPET/PE	CO_2:N_2 30:70% MAP		42 d	
Bakery wares	Sponge cake	PA/LLDPE	Combinations of oxygen scavengers with / without ethanol emitter	30 °C, RH 60%	$\leq$42 d	[139]
		PVDC/PA/cPP				
	Sliced wheat bread	PET-SiOx/LDPE	Bread	20 °C	4 d	[130]
			Bread + preservatives		6 d	
			Ethanol emitter		24 d	
			Ethanol emitter + oxygen absorber		30 d	
	Ciabatta bread	OPA/PE	Air (control)	21 °C	5 d	[122]
			Air + ethanol spray		11 d	
			CO_2:N_2 10:90% MAP		12 d	
			MAP + ethanol spray		13 d	
			Air + ethanol emitter		25 d	
			MAP + ethanol emitter		30 d	

Table 3. *Cont.*

Category	Product	Packaging Material	AIP/MAP Applied	Storage	Shelf-Life	Reference
	Wheat bread	HDPE/PE	-	25.8 °C, 275.5 lx, RH 31.2%	2 d	[144]
		Unpackaged bread	-		3 d	
		HDPE/Nanoparticles/PE	Ag-TiO$_2$		>6 d	
	Calcium-enriched wholemeal bread	PA/PE bag + cardboard box	CO$_2$:N$_2$ 60:40% MAP	20 °C	24 d	[145]
	Whole wheat bread	PA/PE	N$_2$	RT	2–3 w	[123]
	Part-baked flat bread (Sangak)	PA/PE	Air	25 °C	9 d	[124]
			CO$_2$:N$_2$ 20:80% MAP		18 d	
			CO$_2$ 100% MAP		21 d	
	Sliced wheat bread	Tray: APET/EVOH/PEAntifog-film: PA/PE	Air without potassium sorbate & with 0.15% potassium sorbate	20 °C, RH 60%	14 d	[125]
			N$_2$ 100% MAP, CO$_2$:N$_2$ 30:70% MAP, CO$_2$:N$_2$ 50:50% MAP, CO$_2$:N$_2$ 70:30% MAP, CO$_2$ 100 %MAP;with & without potassium sorbate		21 d	
			Air with 0.30% potassium sorbate		>21 d	
	Bread	Plastic bag	E-Poly-L-Lysine Biofilms1.6/3.2/6.5 mg of E-Poly-L-Lysine /cm^2	RT for 7 days inoculated with *A. parasitus*	+1 d	[131]
			E-Poly-L-Lysine Biofilms6.5 mg of E-Poly-L-Lysine /cm^2	RT for 7 days inoculated with *P. expansum*	+3 d	
	Sliced wheat bread	PP/PET/LDPE	Star anise oil, thymol	25 °C inoculated with *P. roqueforti*	14 d	[132]
	Bread	Starch-based bionanocomposite film	Chitosan, grapefruit seed extract	25 °C, RH 59%	20 d	[133]
	Sliced white pan bread	PP bag	–	30°C	3 d	[134]
		PBAT-PLA bag	Trans-cinnamaldehyde		≥21 d	
	Bread	BOPP	–	25 °C, RH 75%	3 d	[135]
		PLA			6 d	
		PLA-PBSA bag	Thymol		7–9 d	

4.1. Modified Atmosphere Packaging (MAP)

Leaving quality sensitive products exposed to atmospheric conditions (gas composition of N_2, O_2, Ar, CO_2, traces of other gases) can trigger undesirable changes such as quality-related oxidative decay or growth of (non)pathogenic aerobic microorganisms. On the contrary, modifying the atmosphere inside a packaging can help maintain the quality of a product over an elevated timeframe. Consequently, common mitigation strategies include the reduction of packaging headspace and, thus, total available atmosphere or even removal of the atmosphere (to a value below one percent), which in turn results in vacuum packaging. To maintain these conditions over time, it is necessary to assure an appropriate containment function of the packaging by choosing packaging materials with an appropriate gas barrier and proper sealing. Challenges in this case are often the structure of the products and the corresponding residual oxygen in the packaging in the case of e.g., pores and the collapse of the product in the case of e.g., a soft structure [13,125,146].

A more advanced modification can be found in a so-called modified atmosphere packaging, MAP [147]. Here, an active modification takes place in a two-step process, where first the initial atmosphere is removed (vacuum) and then replaced with a specific artificially composed atmosphere before closure of the barrier packaging. Commonly, in product-dependent concentrations used, colourless and odourless gases in MAP mainly encompass CO_2 and N_2. Due to its formation of hydrated carbonate species in aqueous phase CO_2 is valued for its bacteriostatic and fungistatic effect, which increases with increasing concentration. Due to the solubility in water and fat, formation of under-pressure in the package and, consequently, possible collapse of the latter is possible. To avoid this and to act as a filler gas, the inexpensive and inert N_2 is applied. Hence, passively, also this gas contributes to quality maintenance of the product. Furthermore, O_2 is a frequently used gas but of little relevance for the cereal and confectionary sector. Its field of application is mostly in meat (e.g., bright-red colour preservation via high-oxygen MAP) and fish products and to lower extent in plant products [145,148,149]. More recently, permitted noble gases such as argon are subject to research but not broadly applied on cereal and confectionary products [150,151]. Depending on the chosen MAP gas composition, food shelf-life can increase manifold (50–400%) and with this advantage along the supply chain can be recorded (e.g., less food waste, longer remaining shelf-life, less frequent production and transport). However, disadvantages linked to MAP, in general encompass the need for more sophisticated packaging materials and filling equipment, costs for gas and increased packaging volume [13].

Regarding the food categories at the centre of the present review, confectionary products are less frequently in the centre of research and application of MAP than cereals and cereal products, bakery wares or ready-to-eat savouries and snacks (see Table 3). One case of MAP use, however, is reported by Mexis et al. [119], for dark chocolate with hazelnuts. The authors found, that when conventionally used aluminium packaging together with storage under surrounding atmosphere was replaced with a PET/LDPE or PET-SiO$_x$ packaging and vacuum or N_2, the shelf-life (dark storage at 20 °C) was increased from 8 to 8–9 and 11 months, respectively. Also Kita et al. [152], investigated the effects of different packaging types and shelf-life extension strategies for chocolate coated products (fruits and nuts). They analysed air, vacuum and MAP ($N_2 \geq 98\%$) of coated cherries, figs, hazelnuts and almonds in long term storage conditions in three different types of packaging. PP film closed with a clip was chosen for air, PP film sealed for vacuum and metallized sealed film for MAP. They resumed that the best packaging solutions for the chosen chocolate coated products, ensuring quality (for example bioactive compounds, antioxidative activity) were, on one hand, air and vacuum packaging for fruits, vacuum packaging for hazelnuts and MAP for almonds.

In the category of cereals and cereal products, and in more detail in fresh pasta, MAP often contains elevated amounts of CO_2 (up to 80%) and corresponding low N_2 values (balance) [13,108,120,121]. For instance, Lee et al. [120] conducted a comparative study on fresh pasta packaged under air (PS tray with PVC film) and under CO_2:N_2 78:22% MAP

(PA/EVOH/LLDPE). As a result, the shelf-life was doubled from 20 to 40 days at a storage temperature of 8 °C. Even higher rates of shelf-life increase for fresh filled pasta were shown in two other studies [108,121]. In the first case, samples included fresh pasta filled with cheese in a sealed tray (EVOH/PS/PE) with a barrier film (EVOH/OPET/PE) and two different atmospheres (air; CO_2:N_2 50:50% MAP). Quality maintenance was increased from 7–10 days up to 42 days [108]. Similarly, in the second case, gluten-free fresh pasta was packaged in trays (control: PET; test: EVOH/PS/PE) sealed with films (control: PET; test: EVOH/OPET/PE). Shelf life under air was compared to CO_2:N_2 30:70% MAP. Here, an increase from 14 to 42 days was notable [121].

Turning to bakery wares such as (pita)bread, cakes, crumpets, crepes, (fruit)pies, Robertson [13] reports a frequent use of CO_2:N_2 60:40% MAP. However, in the scientific literature, a more diverse application of CO_2:N_2 MAP can be seen. For example, Rodriguez et al. [126] investigated extending the shelf-life of bread using MAP packaging in a combination with preservatives. The research referred to bread slices packaged in a 60 μm bag. The results showed that in the samples without added preservative, and CO_2:N_2 50:50% MAP, the increases in shelf-life were 117% and 158% (at 22–25 °C and 15–20 °C). For the samples with calcium propionate addition and in N_2 100% MAP, shelf-life was increased by 116%. Furthermore, calcium propionate addition and CO_2:N_2 20:80% MAP increased the shelf-life by 150% and 131% at 22–25 °C and 15–20 °C. When the CO_2 concentration was increased to 50%, the increased shelf-life of the samples with added preservative was 167% at 22–25 °C. For the same settings at 15–20 °C the increase was even 195%. Fernandez et al. [149], conducted a similar research with soy bread. They as well used different settings of MAP and preservative adding but expanded the question of packaging options. They used two multilayer packaging solutions, high and medium barrier. The high barrier was LLDPE/PA/EVOH/PA/LLDPE, whereas the medium barrier solution was LLDPE/PA/LLDPE. As controls, LDPE and air atmospheres were used. The combination of high barrier packaging in CO_2:N_2 50:50% or CO_2:N_2 20:80% MAP without calcium propionate addition extended the shelf-life of the samples by at least 200%.

Turning to ready-to-eat savouries and snacks (e.g., crisps) Sanches et al. [128] investigated inter alia the effects of different packaging atmospheres under 40 °C and room temperature on multiple crisp samples, linked to lipid oxidation. They included marketed products under unknown MAP concentrations, air, N_2, vacuum and oxygen scavengers in the analysis. Reflecting changes in the fatty acid profile of the crisps, it was resumed that changes in the package's atmospheres, mostly cutting out oxygen, was crucial for the shelf-life of the crisps. Vacuum packaging options would also allow stable lipid profiles, however, they are not suitable for easily breakable crisps. Del Nobile [129] was similarly questioning the optimal packaging for crisps, however, focused on finding the best headspace gas composition for two different multilayer film packages (metallized PP and PVdC coated PE) through simulated storage. He proposed that N_2 combined with water vapour would lead to a shelf-life extension up to 80%.

4.2. Active and Intelligent Packaging (AIP)

While MAP is firmly established in the market, active and intelligent packaging has not yet reached its full potential in food packaging applications but is at the threshold of more widespread use in the European market and subject to intense research and development activities [153–155]. Accordingly, the following paragraphs aim at outlining the concept of AIP and highlighting applications most relevant for cereal and confectionary packaging.

Just as conventional packaging applications, AIP define as food contact materials as given in Regulation (EC) No 1935/2004. While conventional packaging has to be sufficiently inert not to transfer substances to the food in quantities that endanger human health or bring an unacceptable change of the food product (composition, organoleptic properties), AIP are intentionally designed not to be inert. This allows them to actively maintain or even improve the quality or shelf-life of food products [39]. Hence, AIP deliberately includes "active" components that are either aimed to be released to the food or that aim at absorbing substances from it. This justifies the division of active packaging into so-called releaser and absorber systems. However, a clear distinction is made to traditional substance releasing materials (e.g., wooden barrels) in food contact. The use of active substances aimed to be released to the food must also comply with the Directive 1333/2008 on food additives and should be authorized accordingly by applicable community provisions [63]. Furthermore, specific requirements regarding labelling and information, avoidance of misleading consumers as well as safety assessment and authorisation is given [39]. In addition to Regulation (EC) No 1935/2004, Commission Regulation (EC) No 450/2009 gives specific rules for the use of AIP (e.g., community list of allowed substances for use and evaluation of these) [39,156].

In response to major challenges in food quality and safety [12,13], key technologies in the area of active packaging are emitters (e.g., CO_2, ethanol, antimicrobials, antioxidants) and scavengers (e.g., O_2, CO_2, ethylene), absorbers (e.g., H_2O, flavour and odour), self-venting packages, microwave susceptors, and temperature control packaging [13,40,157–165]. Intelligent packaging on the other hand refers to packaging that monitors the food product and provides information about its condition [39]. Related key technologies are mostly indicators and sensors (e.g., time, temperature) and linked processing and communication systems (e.g., (printed) electronics). Further, tamper evident packaging and anti-counterfeiting applications exist [163,166].

Due to their effectiveness, the growth forecasts for AIP in the coming years are high, but it must be emphasised that the sustainability of such sophisticated packaging solutions should be evaluated case by case [167]. In addition to the actual reduction of food losses and food waste, factors such as, e.g., the recyclability of AIP, which may include metal-based components, should be evaluated [153,163,168,169].

Going into detail about cereal and confectionary packaging (see also Table 3), an application example for oxygen absorbers is in sliced bread. Where O_2 concentration decreased below 0.1% within a few days of packaging, microbial shelf-life was shown to be extended. It was reported that there was no effect on sensory quality [170]. Oxygen absorber can also be used in combination with MAP. In 2003, Del Nobile et al. [127] showed that the application of CO_2:N_2 80:20% MAP in the packaging of durum wheat bread prolonged the shelf-life from 3 to about 18 days at 30 °C. However, if the packaging film itself possesses a high barrier against oxygen, neither the use of scavengers nor MAP are necessary to achieve the desired shelf-life of white bread [171]. Finally, an oxygen scavenger system, consisting of a multilayer coextruded bag associated with an oxygen scavenger, was tested in different storage conditions (accelerated storage, room temperature, refrigerator), for its effect on preservative-free tortillas shelf life. The results indicated a protective effect of the packages including the oxygen scavenger system. Specifically, the weight and thickness of flour tortillas under room temperature conditions could be maintained, opposed to respective decreases detected in control packages (consisting of LDPE/LLDPE). In parallel, yeast and mold growth were hold back in the packages containing the oxygen scavenger versus control (room temperature and accelerated storage). Under refrigerated conditions, a shelf-life up to 31 days was estimated, however, independed of the use of oxygen scavengers [172].

It has been also shown that the use of ethanol emitters extend shelf-life even without establishment of an additional modified atmosphere. For ciabatta, a shelf-life of 16 days, at 21 °C could be obtained, packaged in air atmosphere and ethanol emitter addition [122].

Antimicrobial, antifungal, and antioxidative agents as active food packaging include multiple research topics. Options include the applications of essential oils, edible films, and nanocomposites, which are often used with products susceptible to microbiological degradation, e.g., sliced bread. For example, oregano essential oil has been observed to be a successful application against yeasts and moulds in sliced bread. It was applied in the form of antimicrobial sachet at concentrations of 5, 10, and 15% (v/w) at room temperature [136]. In addition to that, methylcellulose edible films produced with clove and oregano essential oil have displayed antimicrobial activity against spoilage fungi in bakery products and have improved sliced bread shelf-life to 15 days, at 25 ± 2 °C [137].

Also, cinnamaldehyde was used as an active ingredient to increase the shelf-life of sliced bread. It was incorporated in gliadin films (5%), which allowed to keep the quality of the product for 27 days of storage at 23 °C [173]. Next to having antimicrobial effects, essential oils are also antioxidative agents that can be included in packaging material like HDPE, LDPE, EVA. Zhu et al. [138] for example tested this approach with sesame essential oils for the packaging of oat cereals. However, there are also biological threats that could shorten the shelf-life of cereal and confectionery products. Essential oils from garlic, black pepper, ginger, fennel, and onion already have been tested as insect repellents for grain packaging. All these tested essential oils were characterized by significant fumigant insecticidal properties. For example, allyl mercaptan deriving from allium plants added as a sachet with rice flour, was proven as potential protective active packaging against *S. oryzae* contamination leaving sensory properties unaffected [174]. In general, the incorporation of essential oils in packaging materials is a growing sector [175,176]. One background can be that they are waterproof, so it could be the ideal material for the incorporation into a film, which will turn it from a conventional packaging material to an active one, increasing both its value and its functionality [175].

One further option of active packaging is the targeted use of composites at the nanoscale, whether organic (oils/proteins/carbohydrates) and/or inorganic, e.g., clays. This topic is of interest as active agents might have different properties in smaller scales. Materials of which at least one of its external dimensions belongs to the nanoscale (1 to 100 nm) are considered nanomaterials [177,178]. They are characterized for their unique properties such us high surface-area-to-volume ratio, fine particle size, and high reactivity [179]. One common area of research interest is represented by publications including essential oils. For example, bio-nano-composite films prepared with corn starch incorporated with chitosan nano-clay, and further enriched with a variety of ratios of grapefruit seed extracts have been studied. It was shown that this solution was capable of inhibiting fungal proliferation for a period of 20 days, compared to that of 6 days in bread packaged samples with synthetic plastic, indicating a successful active packaging approach to extent the shelf-life of bakery products [133]. Furthermore, two different formulations mainly consisting of essential oils from several plants were evaluated for their potential antifungal properties in maize grains. Specifically, in a recent study, bioactive EVOH films including various essential oils have been characterized. Cinnamaldehyde, citral, linalool and isoeugenol were investigated to decrease the activity of *A. steynii* and *A. tubingensis* strains. It was shown that the ochratoxin A production by these strains in partly milled maize grains could be reduced significantly. The inhibitory effect was the highest in EVOH with cinnamaldehyde, followed by isoeugenol and citral [180]. In parallel, EVOH copolymer films incorporated with essential oils from *Origanum vulgare*, *Cinnamomum zeylanicum* and/or their major active constituents have been studied. The results showed that carvacrol and cinnamaldehyde were effective in decreasing *Aspergillus flavus* and *A. parasiticus*-induced aflatoxin production in maize, respectively. Overall, cinnamaldehyde showed the highest inhibitory effect, followed by combinations of EVOH with essential oils from *Origanum vulgare*, *Cinnamomum zeylanicum* and carvacrol [181].

Next to these highly discussed organic nanoparticles, inorganic particles like Ag (silver) and TiO_2 (titan dioxide) have also been applied to packaging solutions, for example cereal products, due to their antimicrobial effects [182–185]. However, there is a concern on potential risk of nanoparticles migrating into food, although limited data showed that obtained values were within the limits set by the legislation [185–189]. It was shown that Ag-TiO_2 nanocomposite incorporated in HDPE considerably extended shelf-life and microbiological safety of bread in comparison with control sample stored in an open atmosphere or in HDPE bags [144]. Not only the characteristics of plastic packaging can be optimized by the inclusion of nanoparticles. The modification of paper with Ag-TiO_2-SiO_2 (silicon dioxide) or Ag/N-TiO_2 composites can improve the papers material characteristics. It was shown that such paper was capable to extend the shelf-life of bread by 2 days in comparison to the control, in both ambient (18–20 °C) and refrigerated (0–4 °C) conditions [190].

Research in optimizing packaging with nanostructures goes even further to high-tech materials. An example is a packaging material with a montmorillonite layer. It was shown that montmorillonite composite polyamide 6 nano-fibres placed over PP films, increased the shelf-life of bread by 2 days at room temperature, due to inhibition of microbial growth [191].

Intelligent packaging, on the other hand, is a special packaging technique aiming to monitor the quality of the packaged food and to predict or measure the safe shelf-life better than a best before marking date [122,130,171,192–194]. It provides functions beneath the ones considered as conventional e.g., protection and containment and is used to monitor the condition and provide quality information of packed foods to the consumers [158]. Different indicators, such as time-temperature, microbial growth, product freshness, pack integrity etc., are used as intelligent packaging systems. High temperatures and/or temperature fluctuation are often correlated with food deterioration as result of detrimental biochemical reactions combined with microbial growth. Depending on the food sensitivity specific intelligent indicators can be applied to specific food products. The time-temperature indicator measures the change that imitates the targeted quality characteristics with the same behaviour under the same time-temperature exposure. The pH and enzymatic indicators can also give information about the quality of food [195]. Commercially available time-temperature indicators can be used to monitor quality changes of many perishable and semi-perishable foods. Among other products, these indicators have been applied to canned fruitcake for 10 days' storage at constant (12, 25 and 37 °C) temperatures. Sensory analysis, as quality characteristic of the product, was correlated with indicator response [140,196].

Reflecting the above chapters and findings, it can be summarized and confirmed that, if chosen correctly, cereal and confectionary packaging, as well as food packaging in general can make a valuable contribution to maintaining the quality and safety of food [12,13,17]. Accordingly, it can also help to prevent food losses and waste, an important point when it comes to making our food systems more sustainable [11,16]. This point is also taken up in the SDGs and influences current political efforts such as the European Union's Green Deal [2,3,6].

However, packaging redesign or optimizations should not simply be carried out without evaluating the effects on ecological, social, and economic sustainability as objectively as possible. This is the only way to avoid possible hidden trade-offs [17].

In addition, close cooperation between a wide range of disciplines is required. In this context, and among others, material science, sustainability science and social sciences, and humanities can be mentioned in addition to food science and technology. The latter in particular has, however, an important enabling function [197,198]. The future focus here could be on the points of promoting (i) diverse and sustainable primary produce, (ii) new processes and systems for sustainable manufacture, (iii) reduction of food and material waste along the supply chain, (iv) safety and traceability, (v) affordable and balanced nutrition, (vi) healthy diets as well as (vii) digitalization. MAP and AIP are important approaches in this context, which are particularly present in the points (ii), (iii) and (iv) [198].

5. Conclusions

The ongoing discussion about packaging optimization towards the enhancement of the sustainability of certain products, asks for a profound review of the status quo in specific food groups. Cereal and confectionary were found to be underrepresented in recent publications addressing this topic, despite their global economic and ecologic importance. To take the right steps aspiring more sustainable production and consumption of goods, it is essential for practitioners along the food supply chain to thoroughly understand packaging functions (containment, protection, convenience, communication), properties (physical and mechanical strength, barrier, migration, hygiene), product group specific decay mechanisms, used packaging solutions, and shelf-life extension strategies.

Commonly available packaging solutions vary in material selection (glass, metal, plastic, paper), as well as in shape (rigid, semi-rigid, flexible) and size. Therefore, each packaging solution offers unique benefits and limitations regarding its optimization potential. Important decay mechanisms mediated by packaging in cereal and confectionary products and snacks include inter alia oxidative mechanisms and changes in moisture content. Especially for products for which quality is easily impaired through such mechanisms, packaging solutions and technologies extending the shelf-life need to be considered as ways to improve the products´ sustainability. This, in combination with a proper material selection, includes the applications of MAP and AIP (e.g., scavengers, indicators, active ingredients) as well as novel approaches (e.g., nanotechnology).

However, sustainability improvement includes different other aspects. After the proper understanding of the packaging's purpose in these certain product categories and subcategories, the question of burden shares between the environmental impacts of the food product itself in comparison to its packaging must be considered along the whole life cycle. Thus, further research is deemed necessary to investigate data from related Life Cycle Assessment (LCA) studies and to combine the findings with the current status quo, in order to derive proper redesign steps for cereal and confectionary products. However, LCA is by default limited to environmental analysis and does not cover all sustainability dimensions. The inclusion of economic and social aspects would finally provide a holistic picture on how to attain more sustainable products.

Author Contributions: Conceptualization, V.K.; resources, A.-S.B., K.L., K.G., I.A., M.I.P., S.A., M.M., I.U.-U., T.V. and V.K.; writing—original draft preparation, A.-S.B., K.L., K.G., I.A., M.I.P., S.A., M.M., I.U.-U., T.V. and V.K.; writing—review and editing, A.-S.B., K.L., and V.K.; supervision, V.K.; project administration, V.K.; funding acquisition, V.K. All authors have read and agreed to the published version of the manuscript.

Funding: This article/publication is based upon work from COST Action Circul-a-bility, supported by COST (European Cooperation in Science and Technology). www.cost.eu (accessed on 17 February 2022).

Conflicts of Interest: The authors declare no conflict of interest.

References

1. United Nations Framework Convention on Climate Change. The Paris Agreement. 2016. Available online: https://unfccc.int/sites/default/files/resource/parisagreement_publication.pdf (accessed on 2 February 2022).
2. United Nations. Transforming Our World: The 2030 Agenda for Sustainable Development. *Resolution Adopted by the General Assembly on 25 September 2015.* Available online: http://www.un.org/ga/search/view_doc.asp?symbol=A/RES/70/1&Lang=E (accessed on 2 February 2022).
3. Communication from the Commission to the European Parliament, the European Council, the Council, the European Economic and Social Committee and the Committee of the Regions, Brussels, Belgium. 2019. Available online: https://eur-lex.europa.eu/resource.html?uri=cellar:b828d165-1c22-11ea-8c1f-01aa75ed71a1.0002.02/DOC_1&format=PDF (accessed on 2 February 2022).
4. European Commission. Annex to the Communication from the Commission to the European Parliament, the European Council, the Council, the European Economic and Social Committee and the Committee of the Regions: The European Green Deal, Brussels. Available online: https://eur-lex.europa.eu/resource.html?uri=cellar:b828d165-1c22-11ea-8c1f-01aa75ed71a1.0002.02/DOC2&format=PDF (accessed on 2 February 2022).
5. European Commission. Communication from the Commission to the European Parliament, the Council, the European Economic and Social Committee and the Committee of the Regions. A New Circular Economy Action Plan. *For a Cleaner and More Competitive Europe. COM/2020/98 Final, European Commission Brussels, Belgium, 2020, 20.* Available online: https://eur-lex.europa.eu/legal-content/EN/TXT/?qid=1583933814386&uri=COM:2020:98:FIN (accessed on 2 February 2022).
6. European Commission. Communication from the Commission to the European Parliament, the Council, the European Economic and Social Committee and the Committee of the Regions a Farm to Fork Strategy for a Fair, Healthy and Environmentally Friendly Food System. *COM (2020) 381 Final.* 2020. Available online: https://eur-lex.europa.eu/legal-content/EN/TXT/?uri=CELEX:52020DC0381 (accessed on 2 February 2022).

7. Food and Agriculture Organization of the United Nations. Global Food Losses and Food Waste: Extent, Causes and Prevention. *Study Conducted for the International Congress SAVE FOOD! at Interpack2011 Düsseldorf, Germany, Rome.* 2011. Available online: http://www.fao.org/3/mb060e/mb060e.pdf (accessed on 2 February 2022).
8. FAO—Food and Agriculture Organization of the United Nations. *The State of Food and Agriculture 2019. Moving forward on Food Loss and Waste Reduction*; Licence: CC BY-NC-SA 3.0 IGO; FAO: Rome, Italy, 2019; Available online: https://www.fao.org/3/ca6030en/ca6030en.pdf (accessed on 2 February 2022).
9. Crippa, M.; Solazzo, E.; Guizzardi, D.; Monforti-Ferrario, F.; Tubiello, F.N.; Leip, A. Food systems are responsible for a third of global anthropogenic GHG emissions. *Nat. Food* **2021**, *2*, 198–209. [CrossRef]
10. Hawkes, C.; Ruel, M. Value Chains for Nutrition: 2020 Conference Brief 4, Washington, DC, USA. 2011. Available online: https://a4nh.cgiar.org/files/2013/06/ValueChainsForNutrition.pdf (accessed on 2 February 2022).
11. HLPE. Nutrition and Food Systems: A Report by the High-Level Panel of Experts on Food Security and Nutrition of the Committee on World Food Security, Rome. 2017. Available online: http://www.fao.org/3/i7846e/i7846e.pdf (accessed on 2 February 2022).
12. Singh, P.; Wani, A.A.; Langowski, H.-C. (Eds.) *Food Packaging Materials: Testing & Quality Assurance*; CRC Press Taylor & Francis Group: Boca Raton, FL, USA, 2017; ISBN 9781466559943.
13. Robertson, G.L. *Food Packaging: Principles and Practice*, 3rd ed.; CRC Press: Boca Raton, FL, USA, 2013; ISBN 9781439862414.
14. European Commission. Communication from the Commission to the European Parliament, the Council, the European Economic and Social Committee and the Committee of the Regions: A European Strategy for Plastics in a Circular Economy, Brussels. 2018. Available online: https://eur-lex.europa.eu/resource.html?uri=cellar:2df5d1d2-fac7-11e7-b8f5-01aa75ed71a1.0001.02/DOC_1&format=PDF (accessed on 2 February 2022).
15. European Commission. ANNEXES to the Communication from the Commission to the European Parliament, the Council, the European Economic and Social Committee and the Committee of the Regions: A European Strategy for Plastics in a Circular Economy, Brussels. 2018. Available online: https://eur-lex.europa.eu/resource.html?uri=cellar:2df5d1d2-fac7-11e7-b8f5-01aa75ed71a1.0001.02/DOC_2&format=PDF (accessed on 2 February 2022).
16. HLPE. Food Losses and Waste in the Context of Sustainable Food Systems: A Report by the High-Level Panel of Experts on Food Security and Nutrition of the Committee on World Food Security, Rome. 2014. Available online: http://www.fao.org/3/i3901e/i3901e.pdf (accessed on 2 February 2022).
17. Verghese, K.; Lewis, H.; Fitzpatrick, L. (Eds.) *Packaging for Sustainability*; Springer: London, UK, 2012; ISBN 9780857299871.
18. Clune, S.; Crossin, E.; Verghese, K. Systematic review of greenhouse gas emissions for different fresh food categories. *J. Clean. Prod.* **2017**, *140*, 766–783. [CrossRef]
19. Garnett, T. Where are the best opportunities for reducing greenhouse gas emissions in the food system (including the food chain). *Food Policy* **2011**, *36*, S23–S32. [CrossRef]
20. Poore, J.; Nemecek, T. Reducing food's environmental impacts through producers and consumers. *Science* **2018**, *360*, 987–992. [CrossRef]
21. Miah, J.; Griffiths, A.; McNeill, R.; Halvorson, S.; Schenker, U.; Espinoza-Orias, N.; Morse, S.; Yang, A.; Sadhukhan, J. Environmental management of confectionery products: Life cycle impacts and improvement strategies. *J. Clean. Prod.* **2018**, *177*, 732–751. [CrossRef]
22. Jeswani, H.K.; Burkinshaw, R.; Azapagic, A. Environmental sustainability issues in the food–energy–water nexus: Breakfast cereals and snacks. *Sustain. Prod. Consum.* **2015**, *2*, 17–28. [CrossRef]
23. Konstantas, A.; Jeswani, H.K.; Stamford, L.; Azapagic, A. Environmental impacts of chocolate production and consumption in the UK. *Food Res. Int.* **2018**, *106*, 1012–1025. [CrossRef]
24. Konstantas, A.; Stamford, L.; Azapagic, A. Evaluating the environmental sustainability of cakes. *Sustain. Prod. Consum.* **2019**, *19*, 169–180. [CrossRef]
25. Konstantas, A.; Stamford, L.; Azapagic, A. Evaluation of environmental sustainability of biscuits at the product and sectoral levels. *J. Clean. Prod.* **2019**, *230*, 1217–1228. [CrossRef]
26. Noya, L.I.; Vasilaki, V.; Stojceska, V.; González-García, S.; Kleynhans, C.; Tassou, S.; Moreira, M.T.; Katsou, E. An environmental evaluation of food supply chain using life cycle assessment: A case study on gluten free biscuit products. *J. Clean. Prod.* **2018**, *170*, 451–461. [CrossRef]
27. Recanati, F.; Marveggio, D.; Dotelli, G. From beans to bar: A life cycle assessment towards sustainable chocolate supply chain. *Sci. Total Environ.* **2018**, *613–614*, 1013–1023. [CrossRef] [PubMed]
28. Belitz, H.-D.; Grosch, W.; Schieberle, P. Cereals and Cereal Products. In *Food Chemistry*, 3rd ed.; Belitz, H.-D., Grosch, W., Schieberle, P., Eds.; Springer: Berlin/Heidelberg, Germany, 2004; pp. 673–746. ISBN 978-3-540-40818-5.
29. Statista. Retail Sales of Bread Sold in Europe from 2012 to 2021: (in Million U.S. Dollars). Available online: https://www.statista.com/statistics/782120/bread-retail-sales-europe/ (accessed on 2 February 2022).
30. Caobisco. Facts and Figures: Key Data of the European Sector (EU27 + Switzerland and Norway). Available online: https://caobisco.eu/facts/ (accessed on 17 January 2022).
31. Soroka, W. *Fundamentals of Packaging Technology*, 5th ed.; Institute of Packaging Professional: Herndon, WV, USA, 2014; ISBN 0615709346.

32. Kaßmann, M. *Grundlagen der Verpackung: Leitfaden für die Fächerübergreifende Verpackungsausbildung*; DIN Deutsches Institut für Normung: Berlin, Germany, 2014; ISBN 3410241922.
33. Wani, A.A.; Singh, P.; Langowski, H.-C. Food Technologies: Packaging. In *Encyclopedia of Food Safety*; Motarjemi, Y., Ed.; Elsevier Science: Burlington, UK, 2014; ISBN 978-0-12-378613-5.
34. Bauer, A.-S.; Tacker, M.; Uysal-Unalan, I.; Cruz, R.M.S.; Varzakas, T.; Krauter, V. Recyclability and Redesign Challenges in Multilayer Flexible Food Packaging—A Review. *Foods* **2021**, *10*, 2702. [CrossRef] [PubMed]
35. Dahlbo, H.; Poliakova, V.; Mylläri, V.; Sahimaa, O.; Anderson, R. Recycling potential of post-consumer plastic packaging waste in Finland. *Waste Manag.* **2018**, *71*, 52–61. [CrossRef] [PubMed]
36. Ceflex. Designing for a Circular Economy: Recyclability of Polylefin-Based Flexible Packaging. 2020. Available online: https://guidelines.ceflex.eu (accessed on 16 February 2021).
37. FH Campus Wien; Circular Analytics TK GmbH. *Circular Packaging Design Guideline: Empfehlungen für Die Gestaltung Recyclinggerechter Verpackungen, Vienna*. 2021. Available online: https://www.fh-campuswien.ac.at/fileadmin/redakteure/Forschung/FH-Campus-Wien_Circular-Packaging-Design-Guideline_V04_DE.pdf (accessed on 9 February 2022).
38. Bergmair, J.; Washüttl, M.; Wepner, B. *Prüfpraxis für Kunststoffverpackungen: Lebensmittel-, Pharma- und Kosmetikverpackungen*; Behr: Hamburg, Germany, 2012; ISBN 9783899479072.
39. *Regulation (EC) No 1935/2004 of the European Parliament and of the Council of 27 October 2004 on Materials and Articles Intended to Come into Contact with Food and Repealing Directives 80/590/EEC and 89/109/EEC*; European Parliament, Council of the European Union: Brussels, Belgium, 2004.
40. Han, J.H. (Ed.) *Innovations in Food Packaging*; Elsevier Ltd.: Amsterdam, The Netherlands, 2005; ISBN 978-0-12-311632-1.
41. Campbell-Platt, G. (Ed.) *Food Science and Technology*, 2nd ed.; John Wiley & Sons, Ltd.: Hoboken, NJ, USA, 2017; ISBN 9780470673423.
42. Marsh, K.; Bugusu, B. Food Packaging? Roles, Materials, and Environmental Issues. *J. Food Sci.* **2007**, *72*, R39–R55. [CrossRef]
43. ISO 5801:2007 Industrial Fans Performance (Testing Using Standardized Airways). Available online: https://www.iso.org/obp/ui/#iso:std:iso:5801:ed-2:v1:en (accessed on 2 February 2022).
44. Detzel, A.; Bodrogi, F.; Kauertz, B.; Bick, C.; Welle, F.; Schmid, M.; Schmitz, K.; Müller, K.; Käb, H. *Biobasierte Kunststoffe als Verpackung von Lebensmitteln*; Bundesministerium für Ernährung und Landwirtschaft: Heidelberg, Germany, 2018; Available online: https://www.ifeu.de/fileadmin/uploads/Endbericht-Bio-LVp_20180612.pdf (accessed on 27 September 2021).
45. Carlsson, D.J.; Wiles, D.M. Composites, Fabrication to Die Design. In *Encyclopedia of Polymer Science and Engineering*, 2nd ed.; Mark, H., Bikales, N.M., Overberger, C.G., Menges, G., Eds.; John Wiley & Sons: New York, NY, USA, 1986; p. 665.
46. Fellows, P. *Food Processing Technology: Principles and Practice*, 3rd ed.; Woodhead Publishing Limited, CRC Press: Cambridge, UK, 2009; ISBN 1615830413.
47. Yam, K.L. (Ed.) *The Wiley Encyclopedia of Packaging Technology*, 3rd ed.; John Wiley & Sons: Hoboken, NJ, USA, 2009; ISBN 0470087048.
48. *Commission Regulation (EU) No 10/2011 of 14 January 2011 on Plastic Materials and Articles Intended to Come into Contact with Food Text with EEA Relevance*; European Commission: Brussels, Belgium, 2011.
49. *Commission Regulation (EC) No 2023/2006 of 22 December 2006 on Good Manufacturing Practice for Materials and Articles Intended to Come into Contact with Food (Text with EEA Relevance)*; European Commission: Brussels, Belgium, 2006.
50. BFR. Database BfR Recommendations on Food Contact Materials: Recommendations. Available online: https://bfr.ble.de/kse/faces/DBEmpfehlung_en.jsp (accessed on 3 February 2022).
51. EDQM Council of Europe. Food Contact Materials and Articles. Available online: https://www.edqm.eu/en/food-contact-materials-and-articles (accessed on 11 October 2021).
52. SR 817. 023.21—*Verordnung des EDI vom 16. Dezember 2016 über Materialien und Gegenstände, die Dazu Bestimmt Sind, mit Lebensmitteln in Berührung zu Kommen (Bedarfsgegenständeverordnung): Bedarfsgegenständeverordnung*; Eidgenössische Departement des Innern: Switzerland, 2016.
53. European Printing Ink Association EuPIA. *EuPIA Guideline on Printing Inks Applied to Food Contact Materials*. 2020. Available online: https://www.eupia.org/fileadmin/Documents/Food_contact_material/2020-12-22_EuPIA_Guideline_on_Printing_Inks_applied_to_Food_Contact_Materials.pdf (accessed on 3 February 2022).
54. Barone, C.; Bolzoni, L.; Caruso, G.; Salvatore, P. (Eds.) *Food Packaging Hygiene*; Springer: Berlin/Heidelberg, Germany, 2015; ISBN 9783319148267.
55. Smith, J.P.; Daifas, D.P.; El-Khoury, W.; Koukoutsis, J.; El-Khoury, A. Shelf Life and Safety Concerns of Bakery Products—A Review. *Crit. Rev. Food Sci. Nutr.* **2004**, *44*, 19–55. [CrossRef]
56. Wolf, B. *Confectionery and Sugar-Based Foods. Reference Module in Food Science*; Elsevier: Amsterdam, The Netherlands, 2016; ISBN 978-0-08-100596-5.
57. Subramaniam, P. The Stability and Shelf Life of Confectionery Products. In *Stability and Shelf Life of Food*, 2nd ed.; Subramaniam, P., Wareing, P., Eds.; Elsevier Science & Technology: Cambridge, UK, 2016; ISBN 9780081004364.
58. Lusas, E.W.; Rooney, L.W. (Eds.) *Snack Foods Processing*; CRC Press LLC: Boca Raton, FL, USA, 2001; ISBN 1566769329.
59. European Commission. *Guidance Document Describing the Food Categories in Part E of Annex II to Regulation (EC) No 1333/2008 on Food Additives*. 2017. Available online: https://ec.europa.eu/food/system/files/2017-09/fs_food-improvement-agents_guidance_1333-2008_annex-2.pdf (accessed on 4 February 2022).

60. FAO. Cereal Supply and Demand Brief. Available online: http://www.fao.org/worldfoodsituation/csdb/en/ (accessed on 4 October 2021).
61. EUROSTAT. Prodcom Annual Data. 2020. Available online: https://ec.europa.eu/eurostat/web/prodcom/data/excel-files-nace-rev.2 (accessed on 4 October 2021).
62. *Regulation (EC) No 178/2002 of the European Parliament and of the Council of 28 January 2002 Laying Down the General Principles and Requirements of Food Law, Establishing the European Food Safety Authority and Laying Down Procedures in Matters of Food Safety*; EC: Brussels, Belgium, 2002.
63. *Regulation (EC) No 1331/2008 of the European Parliament and of the Council of 16 December 2008 Establishing a Common Authorisation Procedure for Food Additives, Food Enzymes and Food Flavourings (Text with EEA Relevance)*; EC: Brussels, Belgium, 2008.
64. *Regulation (EU) No 1169/2011 of the European Parliament and of the Council of 25 October 2011 on the Provision of Food Information to Consumers, Amending Regulations (EC) No 1924/2006 and (EC) No 1925/2006 of the European Parliament and of the Council, and Repealing Commission Directive 87/250/EEC, Council Directive 90/496/EEC, Commission Directive 1999/10/EC, Directive 2000/13/EC of the European Parliament and of the Council, Commission Directives 2002/67/EC and 2008/5/EC and Commission Regulation (EC) No 608/2004 Text with EEA relevance*; EC: Brussels, Belgium, 2011.
65. *Commission Regulation (EC) No 2073/2005 of 15 November 2005 on Microbiological Criteria for Foodstuffs (Text with EEA Relevance)*; EC: Brussels, Belgium, 2005.
66. *Regulation (EC) No 852/2004 of the European Parliament and of the Council of 29 April 2004 on the Hygiene of Foodstuffs*; EC: Brussels, Belgium, 2004.
67. Robertson, G.L. (Ed.) *Food Packaging and Shelf Life: A Practical Guide*; CRC Press: Boca Raton, FL, USA, 2009; ISBN 9781420078459.
68. Subramaniam, P.; Wareing, P. (Eds.) *Stability and Shelf Life of Food*, 2nd ed.; Elsevier Science & Technology: Cambridge, UK, 2016; ISBN 9780081004364.
69. Kong, F.; Singh, R.P. Chemical Deterioration and Physical Instability of Foods and Beverages. In *Stability and Shelf Life of Food*, 2nd ed.; Subramaniam, P., Wareing, P., Eds.; Elsevier Science & Technology: Cambridge, UK, 2016; pp. 43–76. ISBN 9780081004364.
70. Fabra, M.J.; Talens, P.; Moraga, G.; Martínez-Navarrete, N. Sorption isotherm and state diagram of grapefruit as a tool to improve product processing and stability. *J. Food Eng.* **2009**, *93*, 52–58. [CrossRef]
71. Lianou, A.; Panagou, E.Z.; Nychas, G.-J.E. Microbiological spoilage of foods and beverages. In *The Stability and Shelf Life of Food*, 2nd ed.; Subramaniam, P., Ed.; Woodhead Publishing: Cambridge, UK, 2016; pp. 3–42. [CrossRef]
72. Schmidt, S.J.; Fontana, A.J. Appendix E: Water Activity Values of Select Food Ingredients and Products. In *Water Activity in Foods: Fundamentals and Applications*; John Wiley & Sons, Inc.: Hoboken, NJ, USA, 2007; pp. 407–420.
73. Ergun, R.; Lietha, R.; Hartel, R.W. Moisture and Shelf Life in Sugar Confections. *Crit. Rev. Food Sci. Nutr.* **2010**, *50*, 162–192. [CrossRef]
74. Bussiere, G.; Serpelloni, M. Confectionery and Water Activity Determination of AW by Calculation. In *Properties of Water in Foods: In Relation to Quality and Stability*; Simatos, D., Multon, J.L., Eds.; Springer: Dordrecht, The Netherlands, 1985; pp. 627–645. ISBN 978-94-010-8756-8.
75. Subramaniam, P.J. Shelf-Life Prediction and Testing. In *Science and Technology of Enrobed and Filled Chocolate, Confectionery and Bakery Products*; Talbot, G., Ed.; CRC Press: Boca Raton, FL, USA, 2009; pp. 233–254. ISBN 9781845693909.
76. Cauvain, S.P.; Young, L.S. *Bakery Food Manufacture and Quality: Water Control and Effects*, 2nd ed.; Wiley-Blackwell: Chichester, UK, 2008; ISBN 9781444301083.
77. Dury-Brun, C.; Jury, V.; Guillard, V.; Desobry, S.; Voilley, A.; Chalier, P. Water barrier properties of treated-papers and application to sponge cake storage. *Food Res. Int.* **2006**, *39*, 1002–1011. [CrossRef]
78. Davidson, I. *Biscuit, Cookie and Cracker Production: Process, Production and Packaging Equipment*, 2nd ed.; Academic Press, Elsevier: Amsterdam, The Netherlands, 2018; ISBN 0128155795.
79. Pekmez, H. Properties of Flour used in Flat Bread (Gaziantep pita) Production. *Turk. J. Agric. Food Sci. Technol.* **2019**, *7*, 209–213. [CrossRef]
80. Taoukis, P.; Labuza, T.; Sam Saguy, I. *Kinetics of Food Deterioration and Shelf-Life Prediction. Handbook of Food Engineering Practice*; CRC Press: New York, NY, USA, 1997.
81. Costa, A.L.C. Combination of Process Technology and Packaging Conditions to Improve the Shelf Life of Fresh Pasta. *J. Food Process. Technol.* **2014**, *5*. [CrossRef]
82. Machálková, L.; Hřivna, L.; Nedomová, Š.; Jůzl, M. The effect of storage temperature on the quality and formation of blooming defects in chocolate confectionery. *Potravinarstvo Slovak J. Food Sci.* **2015**, *9*. [CrossRef]
83. Jaroni, D.; Ravishankar, S.; Juneja, V. Microbiology of Ready-to-Eat Foods. In *Ready-to-Eat Foods*, 1st ed.; Hwang, A., Huang, L., Eds.; CRC Press: Boca Raton, FL, USA, 2010; pp. 1–60. [CrossRef]
84. Valerio, F.; de Bellis, P.; Di Biase, M.; Lonigro, S.L.; Giussani, B.; Visconti, A.; Lavermicocca, P.; Sisto, A. Diversity of spore-forming bacteria and identification of Bacillus amyloliquefaciens as a species frequently associated with the ropy spoilage of bread. *Int. J. Food Microbiol.* **2012**, *156*, 278–285. [CrossRef] [PubMed]
85. Pepe, O.; Blaiotta, G.; Moschetti, G.; Greco, T.; Villani, F. Rope-producing strains of Bacillus spp. from wheat bread and strategy for their control by lactic acid bacteria. *Appl. Environ. Microbiol.* **2003**, *69*, 2321–2329. [CrossRef] [PubMed]
86. Galić, K.; Gabrić, D.; Ćurić, D. *Packaging and the Shelf Life of Bread*; Reference Module in Food Science; Elsevier: Amsterdam, The Netherlands, 2019.

87. Omedi, J.O.; Huang, W.; Zhang, B.; Li, Z.; Zheng, J. Advances in present-day frozen dough technology and its improver and novel biotech ingredients development trends—A review. *Cereal. Chem.* **2018**, *96*, 34–56. [CrossRef]

88. Neira, D.P. Energy sustainability of Ecuadorian cacao export and its contribution to climate change. A case study through product life cycle assessment. *J. Clean. Prod.* **2016**, *112*, 2560–2568. [CrossRef]

89. Büsser, S.; Jungbluth, N. LCA of Chocolate Packed in Aluminium Foil Based Packaging, Switzerland. 2009. Available online: http://www.alufoil.org/files/alufoil/sustainability/ESU_-_Chocolate_2009_-_Exec_Sum.pdf (accessed on 4 February 2022).

90. Boakye-Yiadom, K.; Duca, D.; Pedretti, E.F.; Ilari, A. Environmental Performance of Chocolate Produced in Ghana Using Life Cycle Assessment. *Sustainability* **2021**, *13*, 6155. [CrossRef]

91. Pérez-Neira, D.; Copena, D.; Armengot, L.; Simón, X. Transportation can cancel out the ecological advantages of producing organic cacao: The carbon footprint of the globalized agrifood system of ecuadorian chocolate. *J. Environ. Manag.* **2020**, *276*, 111306. [CrossRef] [PubMed]

92. Bianchi, F.R.; Moreschi, L.; Gallo, M.; Vesce, E.; Del Borghi, A. Environmental analysis along the supply chain of dark, milk and white chocolate: A life cycle comparison. *Int. J. Life Cycle Assess.* **2020**, *26*, 807–821. [CrossRef]

93. PlascticsEurope. Plasctics—The Facts. 2020. Available online: https://plasticseurope.org/knowledge-hub/plastics-the-facts-2020/ (accessed on 19 January 2022).

94. Nilsson, K.; Sund, V.; Florén, B. Environmental Impact of the Consumption of Sweets, Crisps and Soft Drinks, Copenhagen. 2011. Available online: http://www.diva-portal.org/smash/get/diva2:702819/FULLTEXT01.pdf (accessed on 17 February 2022).

95. Morris, B. *Examples of Flexible Packaging Film Structures. The Science and Technology of Flexible Packaging*; Elsevier: Amsterdam, The Netherlands, 2017; pp. 697–709. ISBN 9780323242738.

96. Kägi, T.; Wettstein, D.; Dinkel, F. Comparing rice products: Confidence intervals as a solution to avoid wrong conclusions in communicating carbon footprints. In Proceedings of the 7th International Conference on Life Cycle Assessment in the Agrifood Sector, Bari, Italy, 24 September 2010; pp. 229–233.

97. Nunes, F.A.; Seferin, M.; Maciel, V.G.; Flôres, S.H.; Ayub, M.A.Z. Life cycle greenhouse gas emissions from rice production systems in Brazil: A comparison between minimal tillage and organic farming. *J. Clean. Prod.* **2016**, *139*, 799–809. [CrossRef]

98. Urbelis, J.H.; Cooper, J.R. Migration of food contact substances into dry foods: A review. *Food Addit. Contam. Part A* **2021**, *38*, 1044–1073. [CrossRef]

99. Forsido, S.F.; Welelaw, E.; Belachew, T.; Hensel, O. Effects of storage temperature and packaging material on physico-chemical, microbial and sensory properties and shelf life of extruded composite baby food flour. *Heliyon* **2021**, *7*, e06821. [CrossRef] [PubMed]

100. Monahan, E.J. Packaging of ready-to-eat breakfast cereals. *Cereal Foods World* **1988**, *33*, 215–221.

101. Sakamaki, C.; Gray, J.I.; Harte, B.R. The influence of selected barriers and oxygen absorbers on the stability of oat cereal during storage. *J. Packag. Technol.* **1988**, *2*, 98–103.

102. Sieti, N.; Rivera, X.C.S.; Stamford, L.; Azapagic, A. Environmental impacts of baby food: Ready-made porridge products. *J. Clean. Prod.* **2018**, *212*, 1554–1567. [CrossRef]

103. Cimini, A.; Cibelli, M.; Moresi, M. Cradle-to-grave carbon footprint of dried organic pasta: Assessment and potential mitigation measures. *J. Sci. Food Agric.* **2019**, *99*, 5303–5318. [CrossRef]

104. Röös, E.; Sundberg, C.; Hansson, P.-A. Uncertainties in the carbon footprint of refined wheat products: A case study on Swedish pasta. *Int. J. Life Cycle Assess.* **2011**, *16*, 338–350. [CrossRef]

105. Saget, S.; Costa, M.P.; Barilli, E.; de Vasconcelos, M.W.; Santos, C.S.; Styles, D.; Williams, M. Substituting wheat with chickpea flour in pasta production delivers more nutrition at a lower environmental cost. *Sustain. Prod. Consum.* **2020**, *24*, 26–38. [CrossRef]

106. Nette, A.; Wolf, P.; Schlüter, O.; Meyer-Aurich, A. A Comparison of Carbon Footprint and Production Cost of Different Pasta Products Based on Whole Egg and Pea Flour. *Foods* **2016**, *5*, 17. [CrossRef]

107. Park, C.; Szabo, R.; Jean, A. A Survey of Wet Pasta Packaged Under a C02:N2 (20:80) Mixture for Staphylococci and their Enterotoxins. *Can. Inst. Food Sci. Technol. J.* **1988**, *21*, 109–111. [CrossRef]

108. Sanguinetti, A.; Del Caro, A.; Mangia, N.; Secchi, N.; Catzeddu, P.; Piga, A. Quality Changes of Fresh Filled Pasta During Storage: Influence of Modified Atmosphere Packaging on Microbial Growth and Sensory Properties. *Food Sci. Technol. Int.* **2011**, *17*, 23–29. [CrossRef]

109. Rachtanapun, P.; Tangnonthaphat, T. Effects of packaging types and storage temperatures on the shelf life of fresh rice noodles under vacuum conditions. *Chiang Mai J. Sci.* **2011**, *38*, 579–589.

110. Espinoza-Orias, N.; Stichnothe, H.; Azapagic, A. The carbon footprint of bread. *Int. J. Life Cycle Assess.* **2011**, *16*, 351–365. [CrossRef]

111. Jensen, J.K.; Arlbjørn, J.S. Product carbon footprint of rye bread. *J. Clean. Prod.* **2014**, *82*, 45–57. [CrossRef]

112. Svanes, E.; Oestergaard, S.; Hanssen, O.J. Effects of Packaging and Food Waste Prevention by Consumers on the Environmental Impact of Production and Consumption of Bread in Norway. *Sustainability* **2018**, *11*, 43. [CrossRef]

113. Williams, H.; Wikström, F. Environmental impact of packaging and food losses in a life cycle perspective: A comparative analysis of five food items. *J. Clean. Prod.* **2011**, *19*, 43–48. [CrossRef]

114. Korsaeth, A.; Jacobsen, A.Z.; Roer, A.-G.; Henriksen, T.M.; Sonesson, U.; Bonesmo, H.; Skjelvåg, A.O.; Strømman, A.H. Environmental life cycle assessment of cereal and bread production in Norway. *Acta Agric. Scand. Sect. A Anim. Sci.* **2012**, *62*, 242–253. [CrossRef]

115. Cauvain, S.P.; Young, L.S. Chemical and physical deterioration of bakery products. In *Chemical Deterioration and Physical Instability of Food and Beverages*; Skibsted, L., Risbo, J., Andersen, M., Eds.; Woodhead Publishing: Washington, DC, USA, 2010; pp. 381–412. ISBN 9781845699260.

116. Chinnadurai, K.; Sequeira, V. *Packaging of Cereals, Snacks, and Confectionery. Reference Module in Food Science*; Elsevier: Amsterdam, The Netherlands, 2016; ISBN 978-0-08-100596-5.

117. Qian, M.; Liu, D.; Zhang, X.; Yin, Z.; Ismail, B.B.; Ye, X.; Guo, M. A review of active packaging in bakery products: Applications and future trends. *Trends Food Sci. Technol.* **2021**, *114*, 459–471. [CrossRef]

118. Kuswandi, B. Active and intelligent packaging, safety, and quality controls. In *Fresh-Cut Fruits and Vegetables: Technologies and Mechanisms for Safety Control*; Siddiqui, M.W., Ed.; Elsevier Science & Technology: San Diego, CA, USA, 2020; pp. 243–294. ISBN 9780128161845.

119. Mexis, S.; Badeka, A.; Riganakos, K.; Kontominas, M. Effect of active and modified atmosphere packaging on quality retention of dark chocolate with hazelnuts. *Innov. Food Sci. Emerg. Technol.* **2010**, *11*, 177–186. [CrossRef]

120. Lee, D.S.; Paik, H.D.; Im, G.H.; Yeo, I.H. Shelf life extension of Korean fresh pasta by modified atmosphere packaging. *J. Food Sci. Nutr.* **2001**, *6*, 240–243.

121. Sanguinetti, A.M.; Del Caro, A.; Scanu, A.; Fadda, C.; Milella, G.; Catzeddu, P.; Piga, A. Extending the shelf life of gluten-free fresh filled pasta by modified atmosphere packaging. *LWT* **2016**, *71*, 96–101. [CrossRef]

122. Hempel, A.W.; O'Sullivan, M.G.; Papkovsky, D.B.; Kerry, J.P. Use of smart packaging technologies for monitoring and extending the shelf-life quality of modified atmosphere packaged (MAP) bread: Application of intelligent oxygen sensors and active ethanol emitters. *Eur. Food Res. Technol.* **2013**, *237*, 117–124. [CrossRef]

123. Jensen, S.; Oestdal, H.; Clausen, M.R.; Andersen, M.L.; Skibsted, L.H. Oxidative stability of whole wheat bread during storage. *LWT* **2011**, *44*, 637–642. [CrossRef]

124. Khoshakhlagh, K.; Hamdami, N.; Shahedi, M.; Le-Bail, A. Quality and microbial characteristics of part-baked Sangak bread packaged in modified atmosphere during storage. *J. Cereal Sci.* **2014**, *60*, 42–47. [CrossRef]

125. Degirmencioglu, N.; Göcmen, D.; Inkaya, A.N.; Aydin, E.; Guldas, M.; Gonenc, S. Influence of modified atmosphere packaging and potassium sorbate on microbiological characteristics of sliced bread. *J. Food Sci. Technol.* **2010**, *48*, 236–241. [CrossRef] [PubMed]

126. Rodríguez, M.; Medina, L.M.; Jordano, R. Effect of modified atmosphere packaging on the shelf life of sliced wheat flour bread. *Food Nahr.* **2000**, *44*, 247–252. [CrossRef]

127. Del Nobile, M.A.; Martoriello, T.; Cavella, S.; Giudici, P.; Masi, P. Shelf life extension of durum wheat bread. *Ital. J. Food Sci.* **2003**, *15*, 383–394.

128. Silva, A.S.; Hernández, J.L.; Losada, P.P. Modified atmosphere packaging and temperature effect on potato crisps oxidation during storage. *Anal. Chim. Acta* **2004**, *524*, 185–189. [CrossRef]

129. Del Nobile, M. Packaging design for potato chips. *J. Food Eng.* **2001**, *47*, 211–215. [CrossRef]

130. Latou, E.; Mexis, S.; Badeka, A.; Kontominas, M. Shelf life extension of sliced wheat bread using either an ethanol emitter or an ethanol emitter combined with an oxygen absorber as alternatives to chemical preservatives. *J. Cereal Sci.* **2010**, *52*, 457–465. [CrossRef]

131. Luz, C.; Calpe, J.; Saladino, F.; Luciano, F.B.; Fernandez-Franzón, M.; Mañes, J.; Meca, G. Antimicrobial packaging based on ε-polylysine bioactive film for the control of mycotoxigenic fungi in vitro and in bread. *J. Food Process. Preserv.* **2017**, *42*, e13370. [CrossRef]

132. Lee, J.; Park, M.A.; Yoon, C.S.; Na, J.H.; Han, J. Characterization and Preservation Performance of Multilayer Film with Insect Repellent and Antimicrobial Activities for Sliced Wheat Bread Packaging. *J. Food Sci.* **2019**, *84*, 3194–3203. [CrossRef] [PubMed]

133. Jha, P. Effect of grapefruit seed extract ratios on functional properties of corn starch-chitosan bionanocomposite films for active packaging. *Int. J. Biol. Macromol.* **2020**, *163*, 1546–1556. [CrossRef] [PubMed]

134. Srisa, A.; Harnkarnsujarit, N. Antifungal films from trans-cinnamaldehyde incorporated poly(lactic acid) and poly(butylene adipate-co-terephthalate) for bread packaging. *Food Chem.* **2020**, *333*, 127537. [CrossRef]

135. Suwanamornlert, P.; Kerddonfag, N.; Sane, A.; Chinsirikul, W.; Zhou, W.; Chonhenchob, V. Poly(lactic acid)/poly(butylene-succinate-co-adipate) (PLA/PBSA) blend films containing thymol as alternative to synthetic preservatives for active packaging of bread. *Food Packag. Shelf Life* **2020**, *25*, 100515. [CrossRef]

136. Passarinho, A.T.P.; Dias, N.F.; Camilloto, G.P.; Cruz, R.S.; Otoni, C.; Moraes, A.R.F.; Soares, N.D.F.F. Sliced Bread Preservation through Oregano Essential Oil-Containing Sachet. *J. Food Process Eng.* **2014**, *37*, 53–62. [CrossRef]

137. Otoni, C.; Pontes, S.F.O.; Medeiros, E.A.A.; Soares, N.D.F.F. Edible Films from Methylcellulose and Nanoemulsions of Clove Bud (*Syzygium aromaticum*) and Oregano (*Origanum vulgare*) Essential Oils as Shelf Life Extenders for Sliced Bread. *J. Agric. Food Chem.* **2014**, *62*, 5214–5219. [CrossRef] [PubMed]

138. Zhu, X.; Schaich, K.; Chen, X.; Yam, K. Antioxidant Effects of Sesamol Released from Polymeric Films on Lipid Oxidation in Linoleic Acid and Oat Cereal. *Packag. Technol. Sci.* **2012**, *26*, 31–38. [CrossRef]

139. Janjarasskul, T.; Tananuwong, K.; Kongpensook, V.; Tantratian, S.; Kokpol, S. Shelf life extension of sponge cake by active packaging as an alternative to direct addition of chemical preservatives. *LWT* **2016**, *72*, 166–174. [CrossRef]

140. Wells, J.H.; Singh, R.P. Application of Time-Temperature Indicators in Monitoring Changes in Quality Attributes of Perishable and Semiperishable Foods. *J. Food Sci.* **1988**, *53*, 148–152. [CrossRef]

141. Vargas, M.C.A.; Simsek, S. Clean Label in Bread. *Foods* **2021**, *10*, 2054. [CrossRef]
142. Leistner, L.; Gorris, L.G. Food preservation by hurdle technology. *Trends Food Sci. Technol.* **1995**, *6*, 41–46. [CrossRef]
143. Senhofa, S.; Straumite, E.; Sabovics, M.; Klava, D.; Galoburda, R.; Rakcejeva, T. The effect of packaging type on quality of cereal muesli during storage. *Agron. Res.* **2015**, *13*, 1064–1073.
144. Cozmuta, A.M.; Peter, A.; Cozmuta, L.M.; Nicula, C.; Crisan, L.; Baia, L.; Turila, A. Active Packaging System Based on Ag/TiO2Nanocomposite Used for Extending the Shelf Life of Bread. Chemical and Microbiological Investigations. *Packag. Technol. Sci.* **2014**, *28*, 271–284. [CrossRef]
145. Fik, M.; Surówka, K.; Maciejaszek, I.; Macura, M.; Michalczyk, M. Quality and shelf life of calcium-enriched wholemeal bread stored in a modified atmosphere. *J. Cereal Sci.* **2012**, *56*, 418–424. [CrossRef]
146. Smith, J.; Ooraikul, B.; Koersen, W.; Jackson, E.; Lawrence, R. Novel approach to oxygen control in modified atmosphere packaging of bakery products. *Food Microbiol.* **1986**, *3*, 315–320. [CrossRef]
147. Lee, D.S. *Modified Atmosphere Packaging of Foods: Principles and Applications*; John Wiley & Sons Inc, Institute of Food Technologists: Hoboken, NJ, USA, 2021; ISBN 9781119530770.
148. Lucas, J. Integrating MAP with new germicidal techniques. In *Novel Food Packaging Techniques*; Ahvenainen, R., Ed.; CRC Press: Boca Raton, FL, USA, 2003; ISBN 128037294X.
149. Fernandez, U.; Vodovotz, Y.; Courtney, P.; Pascall, M.A. Extended Shelf Life of Soy Bread Using Modified Atmosphere Packaging. *J. Food Prot.* **2006**, *69*, 693–698. [CrossRef] [PubMed]
150. Heinrich, V.; Zunabovic, M.; Nehm, L.; Bergmair, J.; Kneifel, W. Influence of argon modified atmosphere packaging on the growth potential of strains of Listeria monocytogenes and Escherichia coli. *Food Control* **2016**, *59*, 513–523. [CrossRef]
151. *European Parliament and Council Directive No 95/2/EC of 20 February 1995 on Food Additives Other than Colours and Sweeteners*; EU Parliament: Brussels, Belgium, 1995.
152. Kita, A.; Lachowicz, S.; Filutowska, P. Effects of package type on the quality of fruits and nuts panned in chocolate during long-time storage. *LWT* **2020**, *125*, 109212. [CrossRef]
153. Tiekstra, S.; Dopico-Parada, A.; Koivula, H.; Lahti, J.; Buntinx, M. Holistic Approach to a Successful Market Implementation of Active and Intelligent Food Packaging. *Foods* **2021**, *10*, 465. [CrossRef]
154. Actinpak. Cost Action FP1405. Available online: http://www.actinpak.eu/ (accessed on 4 February 2022).
155. AIPIA. Active & Intelligent Packaging Industry Association. Available online: https://www.aipia.info/ (accessed on 4 February 2022).
156. *Commission Regulation (EC) No 450/2009 of 29 May 2009 on Active and Intelligent Materials and Articles Intended to Come into Contact with Food (Text with EEA Relevance)*; EC: Brussels, Belgium, 2009.
157. Topuza, F.; Uyarb, T. Antioxidant, antibacterial and antifungal electrospun nanofibers for food packaging applications. *Food Res. Int.* **2019**, *130*, 108927. [CrossRef]
158. Callaghan, K.A.O.; Kerry, J.P. Consumer attitudes towards the application of smart packaging technologies to cheese products. *Food Packag. Shelf Life* **2016**, *9*, 1–9. [CrossRef]
159. Agriopoulou, S. Active packaging for food applications. *EC Nutr.* **2016**, *6*, 86–87.
160. Conte, A.; Angiolillo, L.; Mastromatteo, M.; Del Nobile, M.A. Technological Options of Packaging to Control Food Quality. In *Food Industry*; Muzzalupo, I., Ed.; InTech: Houston, TX, USA, 2013. [CrossRef]
161. Kerry, J. *Smart Packaging Technologies for fast Moving Consumer Goods*; John Wiley: Hoboken, NJ, USA, 2008; ISBN 9780470753699.
162. Wilson, C.L. (Ed.) *Intelligent and Active Packaging for Fruits and Vegetables*; CRC Press: Boca Raton, FL, USA, 2007; ISBN 0849391660.
163. Smithers. Future of Active and Intelligent Packaging | Market Reports and Trends | Smithers. Available online: https://www.smithers.com/services/market-reports/packaging/the-future-of-active-and-intelligent-packaging (accessed on 19 January 2022).
164. Vilela, C.; Kurek, M.; Hayouka, Z.; Röcker, B.; Yildirim, S.; Antunes, M.D.C.; Nilsen-Nygaard, J.; Pettersen, M.K.; Freire, C.S.R. A concise guide to active agents for active food packaging. *Trends Food Sci. Technol.* **2018**, *80*, 212–222. [CrossRef]
165. Yildirim, S.; Röcker, B. Chapter 7—Active Packaging. In *Nanomaterials for Food Packaging: Properties, Processing and Regulation*; Cerqueira, M.A.P.R., Lagaron, J.M., Pastrana Castro, L.M., de Oliveira Soares Vicente, A.A.M., Eds.; Elsevier: Saint Louis, MO, USA, 2018; pp. 173–202. ISBN 978-0-323-51271-8.
166. Berryman, P. *Advances in Food and Beverage Labelling: Information and Regulations*; Woodhead Publishing: London, UK, 2014; ISBN 9781782420934.
167. Wikström, F.; Verghese, K.; Auras, R.; Olsson, A.; Williams, H.; Wever, R.; Grönman, K.; Pettersen, M.K.; Møller, H.; Soukka, R. Packaging Strategies That Save Food: A Research Agenda for 2030. *J. Ind. Ecol.* **2018**, *23*, 532–540. [CrossRef]
168. *Directive 2012/19/EU of the European Parliament and of the Council of 4 July 2012 on Waste Electrical and Electronic Equipment (WEEE) Text with EEA Relevance*; EU: Brussels, Belgium, 2012.
169. Licciardello, F. Packaging, blessing in disguise. Review on its diverse contribution to food sustainability. *Trends Food Sci. Technol.* **2017**, *65*, 32–39. [CrossRef]
170. Salminen, A.; Latva-Kala, K.; Randell, K.; Hurme, E.; Linko, P.; Ahvenainen, R. The effect of ethanol and oxygen absorption on the shelf-life of packed sliced rye bread. *Packag. Technol. Sci.* **1996**, *9*, 29–42. [CrossRef]
171. Upasen, S.; Wattanachai, P. Packaging to prolong shelf life of preservative-free white bread. *Heliyon* **2018**, *4*, e00802. [CrossRef]

172. Antunez, P.D.; Omary, M.B.; Rosentrater, K.; Pascall, M.; Winstone, L. Effect of an Oxygen Scavenger on the Stability of Preservative-Free Flour Tortillas. *J. Food Sci.* **2011**, *77*, S1–S9. [CrossRef]
173. Balaguer, M.P.; Lopez-Carballo, G.; Catala, R.; Gavara, R.; Hernandez-Munoz, P. Antifungal properties of gliadin films incorporating cinnamaldehyde and application in active food packaging of bread and cheese spread foodstuffs. *Int. J. Food Microbiol.* **2013**, *166*, 369–377. [CrossRef]
174. Chang, Y.; Lee, S.-H.; Na, J.H.; Chang, P.-S.; Han, J. Protection of Grain Products from *Sitophilus oryzae* (L.) Contamination by Anti-Insect Pest Repellent Sachet Containing Allyl Mercaptan Microcapsule. *J. Food Sci.* **2017**, *11*, 2634–2642. [CrossRef]
175. Carpena, M.; Nuñez-Estevez, B.; Soria-Lopez, A.; Garcia-Oliveira, P.; Prieto, M.A. Essential Oils and Their Application on Active Packaging Systems: A Review. *Resources* **2021**, *10*, 7. [CrossRef]
176. López-Gómez, A.; Navarro-Martínez, A.; Martínez-Hernández, G.B. Active Paper Sheets Including Nanoencapsulated Essential Oils: A Green Packaging Technique to Control Ethylene Production and Maintain Quality in Fresh Horticultural Products—A Case Study on Flat Peaches. *Foods* **2020**, *9*, 1904. [CrossRef] [PubMed]
177. Huang, Y.; Mei, L.; Chen, X.; Wang, Q. Recent Developments in Food Packaging Based on Nanomaterials. *Nanomaterials* **2018**, *8*, 830. [CrossRef]
178. Agriopoulou, S.; Stamatelopoulou, E.; Skiada, V.; Varzakas, T. Nanobiotechnology in Food Preservation and Molecular Perspective. In *Nanotechnology-Enhanced Food Packaging*; Parameswaranpillai, J., Krishnankutty, R.E., Jayakumar, A., Rangappa, S.M., Siengchin, S., Eds.; Wiley-VCH: Weinheim, Germany, 2022; pp. 327–359. ISBN 978-3-527-82770-1.
179. Ariyarathna, I.R.; Rajakaruna, R.; Karunaratne, D.N. The rise of inorganic nanomaterial implementation in food applications. *Food Control* **2017**, *77*, 251–259. [CrossRef]
180. Tarazona, A.; Gómez, J.V.; Gavara, R.; Mateo-Castro, R.; Gimeno-Adelantado, J.V.; Jiménez, M.; Mateo, E.M. Risk management of ochratoxigenic fungi and ochratoxin A in maize grains by bioactive EVOH films containing individual components of some essential oils. *Int. J. Food Microbiol.* **2018**, *269*, 107–119. [CrossRef] [PubMed]
181. Mateo, E.M.; Gómez, J.V.; Domínguez, I.; Gimeno-Adelantado, J.V.; Mateo-Castro, R.; Gavara, R.; Jiménez, M. Impact of bioactive packaging systems based on EVOH films and essential oils in the control of aflatoxigenic fungi and aflatoxin production in maize. *Int. J. Food Microbiol.* **2017**, *254*, 36–46. [CrossRef] [PubMed]
182. Alhendi, A.; Choudhary, R. Current practices in bread packaging and possibility of improving bread shelf life by nano-technology. *Int. J. Food Sci. Nutr. Eng.* **2013**, *3*, 55–60.
183. Sharma, C.; Dhiman, R.; Rokana, N.; Panwar, H. Nanotechnology: An Untapped Resource for Food Packaging. *Front. Microbiol.* **2017**, *8*, 1735. [CrossRef]
184. Metak, A.M.; Ajaal, T.T. Investigation on Polymer Based Nano-Silver as Food Packaging Materials. *Int. J. Chem. Mol. Eng.* **2013**, *7*, 1103–1109. [CrossRef]
185. Metak, A.M. Effects of nanocomposite based nano-silver and nano-titanium dioxide on food packaging materials. *Int. J. Appl. Sci. Technol.* **2015**, *5*, 26–40.
186. European Commission. *Commission Directive 2007/19/EC of 30 March 2007 amending Directive 2002/72/EC Relating to Plastic Materials and Articles Intended to Come into Contact with Food and Council Directive 85/572/EEC Laying Down the List of Simulants to be Used for Testing Migration of Constituents of Plastic Materials and Articles Intended to Come into Contact with Foodstuffs*; EC: Brussels, Belgium, 2007.
187. Avella, M.; De Vlieger, J.J.; Errico, M.; Fischer, S.; Vacca, P.; Volpe, M.G. Biodegradable starch/clay nanocomposite films for food packaging applications. *Food Chem.* **2005**, *93*, 467–474. [CrossRef]
188. Echegoyen, Y.; Nerin, C. Nanoparticle release from nano-silver antimicrobial food containers. *Food Chem. Toxicol.* **2013**, *62*, 16–22. [CrossRef] [PubMed]
189. Rešček, A.; Ščetar, M.; Hrnjak-Murgić, Z.; Dimitrov, N.; Galić, K. Polyethylene/Polycaprolactone Nanocomposite Films for Food Packaging Modified with Magnetite and Casein: Oxygen Barrier, Mechanical, and Thermal Properties. *Polym. Technol. Eng.* **2016**, *55*, 1450–1459. [CrossRef]
190. Peter, A.; Mihaly-Cozmuta, L.; Mihaly-Cozmuta, A.; Nicula, C.; Ziemkowska, W.; Basiak, D.; Danciu, V.; Vulpoi, A.; Baia, L.; Falup, A.; et al. Changes in the microbiological and chemical characteristics of white bread during storage in paper packages modified with Ag/TiO_2–SiO_2, Ag/N–TiO_2 or Au/TiO_2. *Food Chem.* **2016**, *197*, 790–798. [CrossRef]
191. Agarwal, A.; Raheja, A.; Natarajan, T.; Chandra, T. Effect of electrospun montmorillonite-nylon 6 nanofibrous membrane coated packaging on potato chips and bread. *Innov. Food Sci. Emerg. Technol.* **2014**, *26*, 424–430. [CrossRef]
192. Melini, V.; Melini, F. Strategies to Extend Bread and GF Bread Shelf-Life: From Sourdough to Antimicrobial Active Packaging and Nanotechnology. *Fermentation* **2018**, *4*, 9. [CrossRef]
193. Gutiérrez, L.; Batlle, R.; Andújar, S.; Sánchez, C.; Nerín, C. Evaluation of Antimicrobial Active Packaging to Increase Shelf Life of Gluten-Free Sliced Bread. *Packag. Technol. Sci.* **2011**, *24*, 485–494. [CrossRef]
194. Muizniece-Brasava, S.; Dukalska, L.; Murniece, I.; Dabina-Bicka, I.; Kozlinskis, E.; Sarvi, S.; Santars, R.; Silvjane, A. Active Packaging Influence on Shelf Life Extension of Sliced Wheat Bread. *Int. J. Nutr. Food Eng.* **2012**, *6*, 480–486. [CrossRef]
195. Opara, U.L. A review on the role of packaging in securing food system: Adding value to food products and reducing losses and waste. *AJAR* **2013**, *8*, 2621–2630. [CrossRef]
196. Wells, J.H.; Singh, R.P. Response characteristics of full-history time-temperature indicators suitable for perishable food handling. *J. Food Process. Preserv.* **1988**, *12*, 207–218. [CrossRef]

197. Floros, J.D.; Newsome, R.; Fisher, W.; Barbosa-Cánovas, G.V.; Chen, H.; Dunne, C.P.; German, J.B.; Hall, R.L.; Heldman, D.R.; Karwe, M.V.; et al. Feeding the World Today and Tomorrow: The Importance of Food Science and Technology. *Compr. Rev. Food Sci. Food Saf.* **2010**, *9*, 572–599. [CrossRef] [PubMed]
198. Lillford, P.; Hermansson, A.-M. Global missions and the critical needs of food science and technology. *Trends Food Sci. Technol.* **2020**, *111*, 800–811. [CrossRef]

 *foods*

Article

Effects of Chitosan and Duck Fat-Based Emulsion Coatings on the Quality Characteristics of Chicken Meat during Storage

Dong-Min Shin, Yea-Ji Kim, Jong-Hyeok Yune, Do-Hyun Kim, Hyuk-Cheol Kwon, Hyejin Sohn, Seo-Gu Han, Jong-Hyeon Han, Su-Jin Lim and Sung-Gu Han *

Department of Food Science and Biotechnology of Animal Resources, Konkuk University, Seoul 05029, Korea; jeff.shin90@gmail.com (D.-M.S.); dpwl961113@konkuk.ac.kr (Y.-J.K.); skyun0423@konkuk.ac.kr (J.-H.Y.); secret311@konkuk.ac.kr (D.-H.K.); rnjs1024@konkuk.ac.kr (H.-C.K.); sonhjin123@konkuk.ac.kr (H.S.); tjrn8854@konkuk.ac.kr (S.-G.H.); hyeon4970@konkuk.ac.kr (J.-H.H.); tnwlsdl1110@konkuk.ac.kr (S.-J.L.)
* Correspondence: hansg@konkuk.ac.kr

Abstract: Chicken meat is a popular food commodity that is widely consumed worldwide. However, the shelf-life or quality maintenance of chicken meat is a major concern for industries because of spoilage by microbial growth. The aim of this study was to evaluate the effects of chitosan and duck fat-based emulsion coatings on the quality characteristics and microbial stability of chicken meat during refrigerated storage. The coated chicken meat samples were as follows: control (non-coated), DFC0 (coated with duck fat), DFC0.5 (coated with duck fat and 0.5% chitosan), DFC1 (coated with duck fat and 1% chitosan), DFC2 (coated with duck fat and 2% chitosan), and SOC2 (coated with soybean oil and 2% chitosan). The results showed that the apparent viscosity and coating rate were higher in DFC2 than in other groups. Physicochemical parameters (pH, color, and Warner–Bratzler shear force) were better in DFC2 than those in other groups during 15 days of storage. Moreover, DFC2 delayed lipid oxidation, protein deterioration, and growth of microorganisms during storage. These data suggest that chitosan-supplemented duck fat-based emulsion coating could be used to maintain the quality of raw chicken meat during refrigerated storage.

Keywords: duck fat; chitosan; edible coating; chicken meat; shelf-life

Citation: Shin, D.-M.; Kim, Y.-J.; Yune, J.-H.; Kim, D.-H.; Kwon, H.-C.; Sohn, H.; Han, S.-G.; Han, J.-H.; Lim, S.-J.; Han, S.-G. Effects of Chitosan and Duck Fat-Based Emulsion Coatings on the Quality Characteristics of Chicken Meat during Storage. *Foods* **2022**, *11*, 245. https://doi.org/10.3390/foods11020245

Academic Editors: Theodoros Varzakas and Rui M. S. Cruz

Received: 22 December 2021
Accepted: 15 January 2022
Published: 17 January 2022

Publisher's Note: MDPI stays neutral with regard to jurisdictional claims in published maps and institutional affiliations.

1. Introduction

The consumption of chicken meat has increased over recent decades because of its low-cost, low-fat content, high nutritional value, and unique flavor [1]. However, chicken meat is a perishable product because it enables the growth of spoilage and pathogenic microorganisms [2]. This is because of its high moisture and protein contents and high pH value. The shelf-life of chicken meat is as short as 3–5 days in a refrigerator [3]. Hence, the chicken meat industry is interested in extending the shelf-life of raw chicken meat.

In recent years, edible coating technology has received attention for improving food quality and shelf-life. Edible emulsion coatings are described as a thin and continuous layer of edible biomaterial that may be formed or placed on or between foods [4]. These coating biomaterials are mainly derived from natural materials, including proteins (e.g., gelatin, whey, and zein), polysaccharides (e.g., chitosan and alginate), and lipids (e.g., soybean oil and sunflower oil) [5,6]. However, protein and polysaccharide-based coating materials are highly vulnerable to moisture and are not suitable for water-resistant coatings [7]. Among various coating materials, lipid-based edible coatings provide a better moisture barrier and protection for foods, and vegetable oils and waxes are the main components of edible coatings [6]. However, the use of vegetable oils causes lipid oxidation due to the high levels of unsaturated fatty acids, as the predominant fatty acid in vegetable oils is linoleic acid [8]. According to a previous study, sunflower oil–chitosan edible films for pork hamburgers were more vulnerable to oxidation than non-coated samples [9]. Moreover, for emulsion coatings, rheological properties, such as apparent viscosity and yield stress, are important for determining the coating quality [7]. Linoleic acid-rich lipid products have poorer rheological and textural properties than oleic acid-rich lipid products [10].

Duck meat is well-known for its unique flavor and aroma, and high nutritional values such as essential amino acids and unsaturated fatty acids [11]. Duck fat is usually obtained as a by-product during duck meat production [12]. Duck fat contains high levels of unsaturated fatty acids (64.51%), including oleic acid (48.7%) and linoleic acid (15.8%), as well as low levels of saturated fatty acids (28.53%) compared with other animal fats (e.g., beef fat and pork fat) [13]. Hence, duck fat intake has the potential to provide health benefits to humans by decreasing the risk of cardiovascular diseases [14,15]. In addition, unlike linoleic acid-rich oil, the presence of oleic acid in duck fat can delay lipid oxidation due to its resistance to oxidation [16]. In addition, the high oleic acid content of duck fat can provide strong physical and thermal resistance in lipid-based products [10]. Like duck fat, olive oil is also rich in oleic acid. However, olive oil is not a suitable material for manufacturing an edible coating solution due to its dark color and price, compared to other oils [17,18]. Considering the cost efficiency, duck fat is a cheaper source for an edible coating solution than olive oil [13].

Edible coatings are a new approach to controlling microbial growth, and thereby improve the shelf-life and safety of meat, fish, and poultry products [19]. In fact, lipid-based edible coatings are insufficient to control microorganisms, and, thus, the use of antimicrobial agents is required [9]. Chitosan is made by the deacetylation of chitin and is a versatile biopolymer. Chitosan is used as a natural preservative for edible coating manufacturing because of its strong antimicrobial properties against several foodborne microorganisms [3]. Many studies have reported that chitosan-added edible coatings can extend the shelf-life of fresh fruits and vegetables [20,21]. Additionally, chitosan-added edible coatings can be applied to fresh poultry, meat, and fish products. For example, the effects of chitosan coatings and gamma irradiation on chicken meat [22], chitosan–gelatin edible coatings with nisin and grape seed extract on fresh pork [23], chitosan coatings incorporated with lactoperoxidase on trout [24], and chitosan–soybean oil emulsion coatings on eggs [25] have been reported.

There are limited publications that provide practical and effective coating techniques for the chicken meat industry. Therefore, the objective of this study was to evaluate the effects of chitosan and duck fat-based edible coatings on the quality characteristics and microbial stability of chicken meat during refrigerated storage (at 4 °C for 15 days).

2. Materials and Methods

2.1. Materials

Chicken meat and soybean oil (SO; Beksul, Incheon, Korea) were purchased from a local market. Duck fat (DF) was kindly provided by Taekyung Nongsan (Seoul, Korea). Chitosan (molecular weight of 310–375 kDa, acid-soluble, and coarse ground flakes and powder from crustacean shells), lecithin, Tween® 80 (polyoxyethylene-20 sorbitan monooleate), thiobarbituric acid (TBA), chloroform, bromocresol green, methyl red, boric acid, sulfuric acid, and acetic acid were obtained from Sigma-Aldrich Co. (St. Louis, MO, USA).

2.2. Preparation of Coating Solution and Coating of Chicken Meat

The edible coating solution was prepared as previously described [25]. Briefly, chitosan (final pH of 4.52) was prepared by dissolving chitosan in 1% acetic acid (v/v) solution (i.e., 0.5, 1.0, and 2.0 g of chitosan/100 mL acetic acid (w/v)). The chitosan solution and DF were mixed at a ratio of 40:60 by adding the Tween® 80 emulsifier. This chitosan/DF mixture was blended for 3 min at low speed, followed by blending for 6 min at high-speed using a hand blender (Tefal Co., Ltd., Mayenne, France). The mixture was homogenized at 20,000 rpm for 3 min using a homogenizer (DAIHAN Scientific Co., Ltd., Gangwon, Korea).

To coat the chicken meat, the meat samples were immersed in the coating solution for 2 min under magnetic stirring at 800 rpm. Samples were then placed in a biological hood for 2 h at 25 ± 2 °C to form an edible coating. The coated sample was sealed in a polyethylene bag and stored at 4 ± 1 °C. Sample analyses were performed on days 0, 3, 5, 7, 10, and 15 of refrigerated storage. The total number of chicken breast meat slices used for physicochemical and microbiological analyses was 288. The control and treatment groups were prepared as follows: control (NC, non-coated), DFC0 (coated with duck fat), DFC0.5 (coated with duck fat and 0.5% chitosan), DFC1 (coated with duck fat and 1% chitosan), DFC2 (coated with duck fat and 2% chitosan), and SOC2 (coated with soybean oil and 2% chitosan).

2.3. Apparent Viscosity of Coating Solution

The apparent viscosity of the coating solution was measured using a rheometer (model MCR 92, Anton Paar, Graz, Austria) at 25 °C, and the data were collected between shear rates of 0.1 and 100 Hz ($n = 3$/group). The results are expressed in units of Pascal-seconds (Pa-s). The data were analyzed using an Anton Paar RheoCompass Ver. 1.25.

2.4. Coating Rate of Samples

The coating rate was determined as described previously [26]. The coating rate (%) of samples was calculated as (weight of coated chicken meat (g)—weight of raw chicken meat (g))/weight of coated chicken meat (g) $\times$ 100.

2.5. pH and Color Measurements of Chicken Meat

The pH of the coated chicken meat was determined using a pH meter (LAQUA, Horiba, Ltd., Kyoto, Japan). Briefly, 5 g of sample and 20 mL distilled water were homogenized at 10,000 rpm for 30 s using a homogenizer (DAIHAN Scientific Co., Ltd., Gangwon, Korea), and the pH of the homogenate was measured. Color was measured on the surface of the coated samples using a CR-210 colorimeter (Minolta Camera Co., Ltd., Osaka, Japan) with standard white calibration plates. The data were expressed as L* (lightness), a* (redness), and b* (yellowness) values.

2.6. Warner–Bratzler Shear Force (WBSF)

The WBSF of the chicken meat was measured using a TA-XT2i texture analyzer (Stable Micro Systems Ltd., Godalming, UK) equipped with a Warner–Bratzler shear attachment (V-type blade set). Samples were cut to sizes of 2.0×2.0 cm ($n = 8$). The WBSF was analyzed under the following conditions: a test speed of 2.0 mm/s, a post-test speed of 4.0 mm/s, and a distance of 25.0 mm. The maximum force required to shear through the samples was determined and analyzed as WBSF.

2.7. Lipid Oxidation

The lipid oxidation of coated chicken meat was evaluated by measuring the development of thiobarbituric acid reactive substances (TBARS) according to a previously described method [27]. Briefly, the coated sample (10 g) was blended with distilled water (50 mL), and then the mixture was homogenized for 2 min using a Model AM-7 homogenizer (Nissei Co., Ltd., Tokyo, Japan). The homogenate was transferred to a distillation flask and 47.5 mL of distilled water, 2.5 mL of 4 N HCl solution, and 1 mL of antifoam agent (KMK-73, Shin-Etsu Silicone Co., Ltd., Seoul, Korea) were added to it. The mixture was distilled, and 40 mL of the distillate was collected. Then, 5 mL of the collected sample and 5 mL of TBA reagent (0.02 M in 90% acetic acid) were mixed in a test tube and heated at 95 °C for 30 min. After cooling, the absorbance of the samples was measured at 538 nm for TBARS measurements using a UV/VIS spectrophotometer (Optizen 2120 UV Plus, Mecasys Co., Ltd., Daejeon, Korea).

2.8. Volatile Basic Nitrogen (VBN)

The volatile basic nitrogen (VBN, mg%) content was determined using the Conway microdiffusion method, as reported previously [28]. In brief, 5 g of the coated chicken meat sample was mixed with 20 mL of distilled water. The mixtures were homogenized at

10,000 rpm for 1 min using a homogenizer (Model AM-7, Nihonseiki Kaisha Ltd., Tokyo, Japan) and filtered using Whatman No. 1 filter paper (Whatman International, Maidstone, UK). After filtering, 30 mL of distilled water was added. Then, 1 mL of the filtered sample and 1 mL of 50% K_2CO_3 solution were added to the outer section, and 100 µL of indicator (1:1 = 0.066% bromocresol green in ethanol–0.066% methyl red in ethanol) and 1 mL of 0.01 N H_3BO_3 were added to the inner section of the Conway microdiffusion cells. The cells were incubated for 90 min at 37 °C, and the solution in the inner section was titrated with 0.02 N H_2SO_4 solution.

2.9. Microbiological Analysis

Microbiological evaluation was performed on days 0, 3, 5, 7, 10, and 15 of storage. Briefly, 25 g of coated chicken meat sample was mixed using a stomacher (Masticator Paddle Blender, IUL Instrument, Barcelona, Spain) with 225 mL of 0.1% peptone water for 2 min. The mixtures were serially diluted with 0.1% peptone water. The total viable count (TVC) and *Listeria* spp. were counted on plate count agar (Merck, Darmstadt, Germany) and Oxford agar (Oxoid Ltd., Hampshire, UK), and each agar was incubated at 37 °C for 24 h. *Escherichia coli*, coliforms, molds, and yeasts were counted using Petrifilm (3M, St. Paul, MN, USA). *E. coli* and coliforms were incubated at 36 °C for 24 h, and molds and yeasts were incubated at 25 °C for 5 days. The results are expressed as log CFU/g.

2.10. Statistical Analysis

Each experiment was performed in triplicate and the data are expressed as mean $\pm$ standard deviation (SD). A two-way analysis of variance (ANOVA) followed by Duncan's multiple range test ($p < 0.05$) was conducted using SPSS Ver. 24.0 (SPSS Inc., Chicago, IL, USA) for assessing significant differences.

3. Results and Discussion

3.1. Apparent Viscosity of Coating Solution and Coating Rate

The apparent viscosities of the coating solutions are shown in Figure 1A. The apparent viscosity of the coating solution was significantly affected by the type of lipid and the addition of chitosan, where DFC2 exhibited the highest apparent viscosity among the groups ($p < 0.05$). This can be explained by the melting point of duck fat. In a previous study, duck fat-added margarine had higher apparent viscosity than soybean oil-added margarine due to the higher melting point of duck fat (6.21 °C) than that of soybean oil (−22.59 °C) [10]. The chitosan content also affected the apparent viscosity of the coating solution. This can be explained by the degree of chain entanglement in the coating solution. As the polymer concentration increases, the freedom of movement of polymer chains is restricted because of the correspondingly greater entanglement [29].

The coating rate of the coating solution showed trends similar to those of apparent viscosity (Figure 1B). The coating rate increased as the amount of chitosan increased in the duck fat. Duck fat had a higher coating rate than soybean oil. More specifically, the coating rate of DFC2 was higher than that of other coating solution groups ($p < 0.05$). This may be due to an increase in the apparent viscosity of the coating solution. A high viscosity of the solution can lead to a more stable shape, which leads to a higher coating yield [30]. In addition, polysaccharides, such as chitosan and dietary fiber have a high water holding capacity, which can enhance the emulsifying capacity, thereby increasing viscosity [31,32].

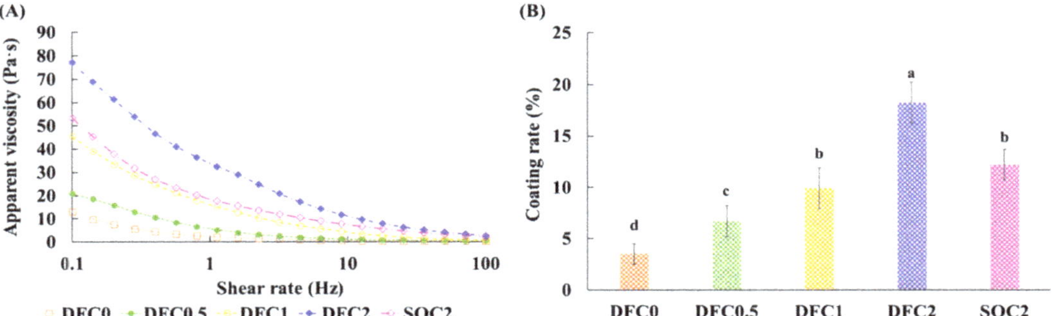

Figure 1. Apparent viscosity and coating rate of the coating solution. (**A**) Apparent viscosity ($n = 3$) and (**B**) coating rate ($n = 3$). DFC0, coated with duck fat, with no chitosan; DFC0.5, coated with duck fat and 0.5% chitosan; DFC1, coated with duck fat and 1% chitosan; DFC2, coated with duck fat and 2% chitosan; SOC2, coated with soybean oil and 2% chitosan. Bars with the different letter are significantly different ($p < 0.05$), and the error bars indicate SD.

3.2. pH and Color of Coated Chicken Meat

Chicken meat is more prone to rapid bacterial deterioration than pork and beef because raw chicken meat generally has a higher pH (0.2–0.4 higher than raw pork and beef) [33]. In addition, the changes in the pH values of chicken meat are highly related to microbial balance, which can lead to a low shelf-life [34]. The pH values of coated chicken meat during refrigerated storage are presented in Table 1. As expected, the pH of all samples tended to increase with storage until day 10 ($p < 0.05$). This increase was significantly higher in the NC group than in other groups ($p < 0.05$). This may be due to the antimicrobial activity of chitosan in the coating solution. A previous study reported that the antibacterial properties of chitosan in coated samples were associated with the lower pH values of samples [35]. Moreover, when chicken meat becomes spoilt, VBN values tend to increase due to the production of NH_3 along with other volatile amines [36]. Therefore, the higher pH of NC could be explained by faster spoilage than that of the coated groups.

Color is considered by consumers as the most important factor in the marketability of meat and poultry. Table 1 shows the L* (lightness), a* (redness), and b* (yellowness) values of the coated chicken meat during storage. Both the storage period and the presence of coating affected the color of the samples ($p < 0.05$). The L* values of all samples decreased during storage ($p < 0.05$). The coated samples showed higher L* values than the NC group ($p < 0.05$). Using chitosan-based emulsions for the coating solution could lead to an increase in the L* value of the samples. When the chitosan emulsion forms, the turbidity and opacity of the solution can increase, resulting in increased lightness [37]. Similar results were reported where a chitosan–essential oil solution increased the L* value of coated chicken meat [34]. The changes in a* and b* values are highly associated with the formation of metmyoglobin, which forms by the oxygenation of myoglobin [38]. As the storage period increased, the a* values of all samples decreased, and b* values increased. ($p < 0.05$). DFC1, DFC2, and SOC2 showed higher a* values and lower b* values than the NC group during the storage period ($p < 0.05$). This was probably due to the inhibition of myoglobin oxidation by the antioxidant activity of chitosan. Cooked pork chops coated with chitosan and bamboo vinegar effectively maintain their initial a* and b* values during storage because chitosan has high antioxidant properties and can maintain meat color because of its ability to act as a chelator of transition metal ions [39]. Overall, the chitosan and duck fat-based emulsion coating may be a good option for inhibiting the discoloration of chicken meat during refrigerated storage.

Table 1. pH and color of coated chicken meat during storage at 4 ± 1 °C for up to 15 days.

Parameter	Treatment [1]	Storage Period (Days)					
		0	3	5	7	10	15
pH	NC	5.99 ± 0.01 [Ad]	5.92 ± 0.01 [Ae]	5.98 ± 0.02 [Ad]	6.20 ± 0.01 [Aa]	6.14 ± 0.01 [Ab]	6.03 ± 0.01 [Ac]
	DFC0	5.86 ± 0.01 [Cd]	5.86 ± 0.03 [Bd]	5.92 ± 0.01 [Bc]	5.85 ± 0.01 [Cd]	6.08 ± 0.01 [Ba]	6.00 ± 0.01 [Bb]
	DFC0.5	5.88 ± 0.01 [Bb]	5.90 ± 0.01 [Ab]	5.84 ± 0.01 [Dc]	5.92 ± 0.01 [Ba]	5.94 ± 0.01 [Da]	5.89 ± 0.02 [Cb]
	DFC1	5.81 ± 0.01 [Ee]	5.85 ± 0.01 [Bc]	5.87 ± 0.03 [Cb]	5.86 ± 0.01 [Cbc]	5.90 ± 0.01 [Ea]	5.81 ± 0.01 [Dd]
	DFC2	5.87 ± 0.01 [Ccd]	5.90 ± 0.01 [Abc]	5.94 ± 0.01 [Bab]	5.83 ± 0.06 [Cd]	5.97 ± 0.03 [Ca]	5.79 ± 0.01 [Ee]
	SOC2	5.84 ± 0.01 [De]	5.87 ± 0.01 [Bcd]	5.97 ± 0.01 [Ab]	5.85 ± 0.02 [Cde]	5.98 ± 0.01 [Ca]	5.88 ± 0.01 [Cc]
L*	NC	58.68 ± 2.90 [Ba]	57.51 ± 5.34 [Bab]	57.46 ± 2.38 [Bab]	56.26 ± 2.91 [Bab]	56.17 ± 4.85 [Bab]	54.79 ± 2.61 [Bb]
	DFC0	62.00 ± 3.37 [Aa]	61.17 ± 2.48 [Aa]	60.32 ± 3.75 [Aab]	58.39 ± 2.84 [ABbc]	58.33 ± 2.15 [ABbc]	57.34 ± 3.09 [Ac]
	DFC0.5	61.84 ± 2.25 [Aa]	59.91 ± 4.53 [ABab]	59.87 ± 3.28 [Aab]	59.42 ± 3.41 [Aab]	58.46 ± 3.48 [ABb]	58.06 ± 2.51 [Ab]
	DFC1	62.04 ± 5.08 [Aa]	61.81 ± 3.38 [Aa]	60.73 ± 3.06 [Aab]	60.31 ± 3.32 [Aab]	59.67 ± 1.79 [Aab]	58.89 ± 1.62 [Ab]
	DFC2	62.38 ± 3.18 [Aa]	62.12 ± 2.24 [Aa]	60.52 ± 3.56 [Aab]	59.82 ± 2.80 [Ab]	59.49 ± 2.93 [Ab]	59.04 ± 3.96 [Ab]
	SOC2	61.78 ± 4.73 [Aa]	59.91 ± 3.33 [ABab]	59.71 ± 1.64 [Aab]	58.53 ± 4.10 [ABb]	57.66 ± 3.92 [ABb]	57.51 ± 5.34 [Ab]
a*	NC	3.44 ± 0.63 [a]	3.14 ± 0.55 [ab]	2.93 ± 0.97 [abc]	2.50 ± 0.76 [bc]	2.17 ± 0.50 [cd]	1.37 ± 0.50 [Bd]
	DFC0	3.53 ± 0.50 [a]	3.07 ± 0.33 [ab]	3.01 ± 0.7 [ab]	2.77 ± 0.58 [bc]	2.13 ± 0.57 [cd]	1.62 ± 0.49 [Bd]
	DFC0.5	3.57 ± 0.54 [a]	3.10 ± 0.55 [ab]	2.79 ± 0.45 [ab]	2.78 ± 0.60 [ab]	2.37 ± 0.67 [bc]	1.83 ± 0.66 [Bc]
	DFC1	3.59 ± 0.38	3.10 ± 0.55	2.82 ± 0.65	2.95 ± 0.16	2.67 ± 0.64	2.49 ± 0.48 [A]
	DFC2	3.55 ± 0.58 [a]	3.35 ± 0.69 [b]	2.95 ± 0.39 [bc]	3.01 ± 0.50 [bc]	2.82 ± 0.22 [bc]	2.70 ± 0.52 [Ac]
	SOC2	3.58 ± 0.68	3.19 ± 0.53	2.88 ± 0.56	2.83 ± 0.79	2.71 ± 0.66	2.54 ± 0.56 [A]
b*	NC	11.07 ± 1.18 [b]	11.57 ± 2.01 [b]	11.94 ± 0.79 [b]	12.28 ± 1.36 [b]	12.71 ± 1.24 [ab]	14.17 ± 1.22 [Aa]
	DFC0	11.05 ± 2.04	11.88 ± 2.26	12.07 ± 0.53	12.12 ± 1.31	12.41 ± 1.99	12.97 ± 0.80 [AB]
	DFC0.5	11.18 ± 2.02	11.95 ± 1.51	12.14 ± 1.63	12.70 ± 1.78	12.71 ± 1.99	13.08 ± 1.22 [AB]
	DFC1	10.93 ± 2.06	10.54 ± 1.42	10.97 ± 1.25	11.35 ± 2.30	11.13 ± 2.36	11.68 ± 1.49 [B]
	DFC2	10.81 ± 2.19	10.92 ± 1.12	11.01 ± 2.24	11.09 ± 0.96	11.11 ± 2.37	11.33 ± 0.68 [B]
	SOC2	10.87 ± 1.44	10.70 ± 1.97	10.96 ± 1.98	11.15 ± 1.79	11.55 ± 2.36	11.84 ± 1.94 [B]

[1] NC (non-coated), DFC0 (coated with duck fat, with no chitosan), DFC0.5 (coated with duck fat and 0.5% chitosan), DFC1 (coated with duck fat and 1% chitosan), DFC2 (coated with duck fat and 2% chitosan), SOC2 (coated with soybean oil and 2% chitosan). [A–E] Means values in the same column are significantly different ($p < 0.05$). [a–e] Means values in the same row are significantly different ($p < 0.05$). All values are presented as the mean ± SD of six replicates (n = 6).

3.3. Warner–Bratzler Shear Force of Coated Chicken Meat

The WBSF values are related to the tenderness of meat samples, which is a critical organoleptic property that affects consumer preference [40]. In our study, the WBSF values of coated chicken meat were significantly affected by the presence of edible coatings (Figure 2). WBSF values gradually decreased during storage ($p < 0.05$). DFC2 exhibited the highest WBSF values during storage, whereas WBSF values were lower in the NC group than those in other groups ($p < 0.05$). This result might be attributable to the deterioration of proteins by microorganisms. The WBSF values of samples can decrease due to the degradation of proteins in meat, mainly caused by bacterial or enzymatic processes as storage progresses [41]. Thus, a higher microorganism count in NC may affect the decrease in the WBSF values of meat samples during storage. Collectively, our data suggest that the chitosan–duck fat edible coating for chicken meat can contribute to maintaining meat tenderness by inhibiting microorganisms.

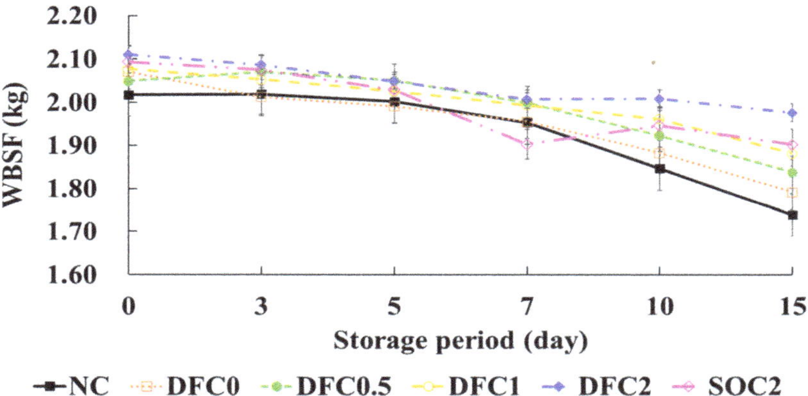

Figure 2. Warner–Bratzler shear force (WBSF) of coated chicken meat during storage at 4 ± 1 °C for 15 days. NC, non-coated; DFC0, coated with duck fat, with no chitosan; DFC0.5, coated with duck fat and 0.5% chitosan; DFC1, coated with duck fat and 1% chitosan; DFC2, coated with duck fat and 2% chitosan; SOC2, coated with soybean oil and 2% chitosan. The error bars indicate SD (*n* = 3).

3.4. TBARS Values and VBN of Coated Chicken Meat

Shelf-life and the quality of meat are highly associated with lipid oxidation and protein deterioration during storage [42]. TBARS values are used for measuring the formation of secondary oxidation products, including malondialdehyde, alkenals, and alkadienals [43]. Variations in the TBARS values of the meat samples are shown in Figure 3A. The difference in TBARS values between DFC treatments and SOC2 can be explained by the fatty acid profiles of duck fat and soybean oil in the coating solution. Duck fat is more stable against lipid oxidation than soybean oil during storage because the main fatty acids in duck fat and soybean oil are oleic acid and linoleic acid, respectively [10]. The oxidative stability of oleic acid in edible oils is almost 10-times greater than that of linoleic acid [16]. Therefore, the higher resistance to oxidation of duck fat could be more suitable as an edible coating material than soybean oil.

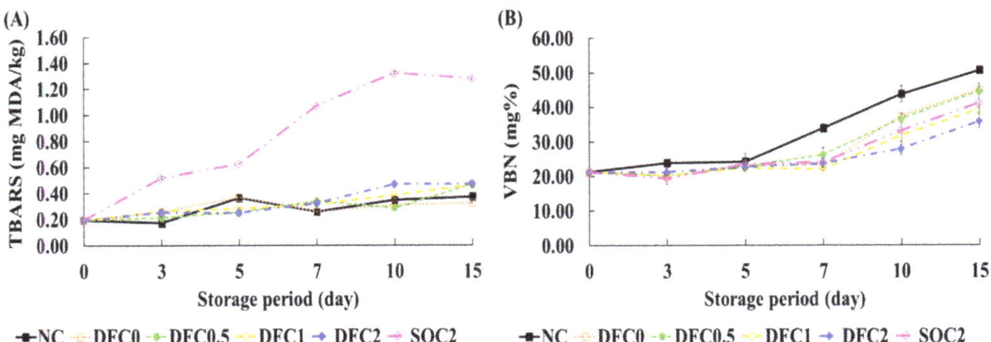

Figure 3. TBARS and VBN values of coated chicken meat during storage at 4 ± 1 °C for up to 15 days. (**A**) Thiobarbituric acid reactive substances (TBARS) and (**B**) volatile basic nitrogen (VBN) values were determined during storage at 4 ± 1 °C for up to 15 days. NC, non-coated; DFC0, coated with duck fat, with no chitosan; DFC0.5, coated with duck fat and 0.5% chitosan; DFC1, coated with duck fat and 1% chitosan; DFC2, coated with duck fat and 2% chitosan; SOC2, coated with soybean oil and 2% chitosan. The error bars indicate SD (*n* = 3).

The VBN value is an important indicator of protein deterioration in meat and meat products [44]. VBN mainly includes ammonia and primary, secondary, and tertiary amines. In general, the VBN value is an indicator of meat spoilage, particularly when it exceeds 25 mg% [34]. The VBN data of the chicken meat samples during storage are presented in Figure 3B. VBN values for all samples increased during storage ($p < 0.05$), while DFC2 had significantly lower VBN values among all groups ($p < 0.05$). The VBN values of most samples were over the standard point (25 mg%) on day 10, while DFC2 showed VBN values over the standard point (25 mg%) on day 15. The increase in VBN is related to the hydrolysis of proteins to amino acids, peptides, biogenic amines, inorganic nitrogen, and the increasing contents of volatile bases due to enzymes and microorganisms during storage [45]. Thus, lower microbial growth might be expected in the DFC groups, particularly in the DFC2 group.

3.5. Growth of Microorganisms on Coated Chicken Meat

The results of TVC, *E. coli*, coliforms, *Listeria* spp., molds, and yeasts are shown in Table 2. Meat decay is generally defined when the TVC exceeds 7 log CFU/g [46]. The TVC of DFC2 was significantly lower than that of other groups during storage ($p < 0.05$), and only DFC2 did not exceed the standard point for meat spoilage until the end of the storage period. The abundance of *E. coli* and coliforms are important hygienic quality indicators for meat and meat products [47]. The microbial counts increased during storage in all groups, and the rate of this increase was significantly lower in DFC2 than that in NC after 15 days of storage. The meat and meat product surfaces are considerably susceptible to mold and yeast growth, which are related to spoilage and have negative effects on organoleptic properties and safety [34]. At the beginning of storage, mold and yeast were not detected in any sample. After 3 days, mold and yeast were detected in all samples, with NC showing the highest counts of mold and yeast at the end of storage ($p < 0.05$). DFC2 showed significantly higher growth inhibitory effects against mold and yeast during storage ($p < 0.05$). *Listeria* spp. are foodborne pathogens in meat and meat products and have increasingly proliferated despite improvements in control measures [48]. *Listeria* spp. were not detected in any samples until day 3, whereas this pathogen was detected on day 5 only in DFC0 and DFC0.5. Meanwhile, *Listeria* spp. were not detected in DFC2 and SOC2 until the end of storage. These data indicate that chitosan inhibits *Listeria* spp. Overall, DFC2 showed the highest antimicrobial activities against TVC, *E. coli*, coliforms, *Listeria* spp., molds, and yeasts during storage. This was due to the strong antimicrobial properties of chitosan. Previous studies have shown that edible coatings with chitosan can inhibit microorganisms in meat and meat products [22,49]. Overall, the 2% chitosan-added duck fat edible coating can improve the shelf-life of chicken meat by inhibiting the growth of microorganisms during refrigerated storage.

Table 2. Microorganisms in coated chicken meat during storage at 4 ± 1 °C for up to 15 days.

Parameter (Log CFU/g)	Treatment [1]	Storage Period (Day)					
		0	3	5	7	10	15
TVC	NC	3.52 ± 0.06 [e]	5.43 ± 0.11 [BCd]	6.78 ± 0.04 [Bc]	8.71 ± 0.02 [Aa]	8.66 ± 0.05 [Aa]	8.15 ± 0.01 [Ab]
	DFC0	3.52 ± 0.06 [e]	5.97 ± 0.02 [Ad]	6.81 ± 0.05 [Bc]	7.62 ± 0.02 [Bb]	8.39 ± 0.22 [Ba]	8.24 ± 0.05 [Aa]
	DFC0.5	3.52 ± 0.06 [f]	5.67 ± 0.02 [ABe]	7.83 ± 0.02 [Ad]	6.66 ± 0.26 [Cc]	7.46 ± 0.09 [Cb]	8.18 ± 0.04 [Aa]
	DFC1	3.52 ± 0.06 [e]	5.32 ± 0.28 [Cd]	5.13 ± 0.02 [Dd]	6.50 ± 0.01 [Cc]	7.66 ± 0.03 [Ca]	6.93 ± 0.04 [Bb]
	DFC2	3.52 ± 0.06 [c]	4.10 ± 0.17 [Db]	4.02 ± 0.03 [Eb]	4.24 ± 0.34 [Eb]	6.32 ± 0.02 [Ea]	6.15 ± 0.21 [Ca]
	SOC2	3.52 ± 0.06 [f]	5.24 ± 0.06 [BCd]	5.48 ± 0.01 [Cc]	4.85 ± 0.01 [De]	7.14 ± 0.04 [Da]	6.78 ± 0.01 [Bb]
E. coli	NC	3.19 ± 0.06 [e]	3.72 ± 0.34 [ABd]	4.18 ± 0.07 [Bc]	5.23 ± 0.12 [Bb]	7.98 ± 0.04 [Aa]	7.85 ± 0.09 [Ba]
	DFC0	3.19 ± 0.06 [f]	3.93 ± 0.04 [Ae]	4.57 ± 0.10 [Ad]	6.24 ± 0.02 [Ac]	7.90 ± 0.02 [Ab]	8.11 ± 0.06 [Aa]
	DFC0.5	3.19 ± 0.06 [e]	3.91 ± 0.19 [Ad]	3.83 ± 0.09 [Dd]	4.71 ± 0.10 [Cc]	6.49 ± 0.02 [Bb]	7.49 ± 0.02 [Ca]
	DFC1	3.19 ± 0.06 [e]	3.54 ± 0.09 [ABCd]	3.92 ± 0.11 [CDc]	4.07 ± 0.16 [Dc]	6.58 ± 0.01 [Ba]	6.32 ± 0.09 [Eb]
	DFC2	3.19 ± 0.06 [d]	3.20 ± 0.01 [Ccd]	3.35 ± 0.01 [Ecd]	3.39 ± 0.12 [Ec]	4.36 ± 0.12 [Db]	5.10 ± 0.04 [Fa]
	SOC2	3.19 ± 0.06 [e]	3.48 ± 0.01 [BCd]	4.07 ± 0.01 [BCc]	3.93 ± 0.21 [Dc]	4.68 ± 0.06 [Cb]	7.30 ± 0.06 [Da]
Coliform	NC	3.10 ± 0.01 [e]	3.72 ± 0.34 [ABd]	4.15 ± 0.02 [Bc]	5.19 ± 0.06 [Bb]	7.95 ± 0.06 [Aa]	7.83 ± 0.08 [Ba]
	DFC0	3.10 ± 0.01 [f]	4.16 ± 0.02 [Ae]	4.71 ± 0.04 [Ad]	6.13 ± 0.07 [Ac]	7.95 ± 0.03 [Ab]	8.08 ± 0.02 [Aa]
	DFC0.5	3.10 ± 0.01 [e]	3.80 ± 0.14 [Ad]	3.86 ± 0.01 [Cd]	4.69 ± 0.03 [Cc]	6.41 ± 0.05 [Bb]	7.42 ± 0.17 [Ca]
	DFC1	3.10 ± 0.01 [f]	3.81 ± 0.05 [Ae]	3.93 ± 0.07 [Cd]	4.18 ± 0.01 [Dc]	6.57 ± 0.01 [Ba]	6.36 ± 0.05 [Db]
	DFC2	3.10 ± 0.01 [c]	3.15 ± 0.21 [Bc]	3.41 ± 0.01 [Dc]	3.24 ± 0.34 [Ec]	4.16 ± 0.17 [Db]	5.14 ± 0.02 [Ea]
	SOC2	3.10 ± 0.01 [d]	3.76 ± 0.40 [Ac]	4.13 ± 0.01 [Bc]	3.78 ± 0.25 [Dc]	4.76 ± 0.01 [Cb]	7.41 ± 0.05 [Ca]
Yeast and molds	NC	N.D.	1.00 ± 0.10 [Cc]	3.77 ± 0.04 [Ab]	4.17 ± 0.01 [Aab]	4.75 ± 0.01 [Aab]	5.47 ± 0.03 [Aa]
	DFC0	N.D.	3.04 ± 0.01 [Ad]	3.88 ± 0.02 [Ac]	3.85 ± 0.01 [Bc]	4.95 ± 0.07 [Ab]	5.32 ± 0.03 [Ba]
	DFC0.5	N.D.	2.15 ± 0.21 [Bc]	3.63 ± 0.19 [Ab]	3.66 ± 0.26 [BCb]	3.69 ± 0.30 [Cb]	4.65 ± 0.03 [Ca]
	DFC1	N.D.	2.24 ± 0.34 [Bc]	3.55 ± 0.03 [Ab]	3.86 ± 0.06 [Bb]	4.24 ± 0.05 [Ba]	4.50 ± 0.03 [Da]
	DFC2	N.D.	1.15 ± 0.36 [Cb]	2.35 ± 0.49 [Bab]	2.39 ± 0.12 [Dab]	3.57 ± 0.05 [Ca]	3.77 ± 0.03 [Fa]
	SOC2	N.D.	2.60 ± 0.01 [Bd]	3.78 ± 0.04 [Ab]	3.41 ± 0.02 [Cc]	3.82 ± 0.02 [Cb]	4.01 ± 0.01 [Ea]
Listeria spp.	NC	N.D.	N.D.	N.D.	2.81 ± 0.47 [Ab]	2.39 ± 0.12 [Ab]	4.12 ± 0.23 [Aa]
	DFC0	N.D.	N.D.	2.48 ± 0.01 [Ab]	2.94 ± 0.34 [Aab]	2.82 ± 0.31 [Ab]	3.35 ± 0.16 [Ba]
	DFC0.5	N.D.	N.D.	1.00 ± 0.10 [Bbc]	3.38 ± 0.33 [Aa]	2.15 ± 0.21 [Aab]	3.35 ± 0.49 [Ba]
	DFC1	N.D.	N.D.	N.D.	N.D.	1.24 ± 0.15 [ABb]	3.93 ± 0.04 [ABa]
	DFC2	N.D.	N.D.	N.D.	N.D.	N.D.	N.D.
	SOC2	N.D.	N.D.	N.D.	N.D.	N.D.	N.D.

[1] NC (non-coated), DFC0 (coated with duck fat, with no chitosan), DFC0.5 (coated with duck fat and 0.5% chitosan), DFC1 (coated with duck fat and 1% chitosan), DFC2 (coated with duck fat and 2% chitosan), SOC2 (coated with soybean oil and 2% chitosan). [A–E] Means values in the same column are significantly different ($p < 0.05$). [a–f] Means values in the same row are significantly different ($p < 0.05$). All values are presented as the mean $\pm$ SD of three replicates ($n = 3$).

4. Conclusions

In this study, the effects of chitosan and duck fat-based emulsion coating on the quality characteristics and microbial stability of chicken breast meat were investigated. The duck fat-based coating solution showed higher apparent viscosity than that of the soybean oil-based coating solution, which resulted in a high coating rate for chicken meat. The physicochemical properties, including pH, color, and WBSF value of the DFC2 group (chicken meat coated with duck fat and 2% chitosan) improved significantly compared to those of other groups ($p < 0.05$). The DFC2 group showed lower lipid oxidation (TBARS value) and protein deterioration (VBN value) during refrigerated storage over 15 days. Furthermore, DFC2 was effective at inhibiting the growth of microorganisms, including TVC, E. coli, coliforms, Listeria spp., molds, and yeasts during storage. Lower lipid oxidation and protein deterioration in DFC2 were owing to the higher apparent viscosity and coating rate in duck fat compared to soybean oil. Here, the higher viscosity and coating rate in DFC2 were probably due to the higher melting point of duck fat. In addition, the higher coating rate of DFC2 made more chitosan concentrations on the coated samples and that resulted in the extending shelf-life of chicken meat. Our data suggest that chitosan/duck fat-based edible coatings can be used to maintain the quality of raw chicken meat during refrigeration. This edible coating solution could be further studied regarding the sensory properties of coated products and its application in a variety of foods, such as meat products, vegetables, and fruits.

Author Contributions: Conceptualization, D.-M.S. and S.-G.H. (Sung-Gu Han); formal analysis, D.-M.S.; investigation, D.-M.S., J.-H.Y., Y.-J.K., D.-H.K., H.-C.K., H.S., S.-G.H. (Seo-Gu Han), J.-H.H. and S.-J.L.; writing—original draft preparation, D.-M.S.; writing—review and editing, Y.-J.K. and S.-G.H. (Sung-Gu Han), supervision, S.-G.H. (Sung-Gu Han). All authors have read and agreed to the published version of the manuscript.

Funding: This paper was supported by Konkuk University Researcher Fund in 2021.

Institutional Review Board Statement: Not applicable.

Informed Consent Statement: Not applicable.

Data Availability Statement: Data is contained within the article.

Conflicts of Interest: The authors declare no conflict of interest.

References

1. Latou, E.; Mexis, S.; Badeka, A.; Kontakos, S.; Kontominas, M. Combined effect of chitosan and modified atmosphere packaging for shelf life extension of chicken breast fillets. *LWT—Food Sci. Technol.* **2014**, *55*, 263–268. [CrossRef]
2. Anang, D.; Rusul, G.; Bakar, J.; Ling, F.H. Effects of lactic acid and lauricidin on the survival of Listeria monocytogenes, Salmonella enteritidis and Escherichia coli O157: H7 in chicken breast stored at 4 C. *Food Control* **2007**, *18*, 961–969. [CrossRef]
3. Langroodi, A.M.; Tajik, H.; Mehdizadeh, T.; Moradi, M.; Kia, E.M.; Mahmoudian, A. Effects of sumac extract dipping and chitosan coating enriched with Zataria multiflora Boiss oil on the shelf-life of meat in modified atmosphere packaging. *LWT—Food Sci. Technol.* **2018**, *98*, 372–380. [CrossRef]
4. Bravin, B.; Peressini, D.; Sensidoni, A. Development and application of polysaccharide–lipid edible coating to extend shelf-life of dry bakery products. *J. Food Eng.* **2006**, *76*, 280–290. [CrossRef]
5. Xiong, Y.; Kamboj, M.; Ajlouni, S.; Fang, Z. Incorporation of salmon bone gelatine with chitosan, gallic acid and clove oil as edible coating for the cold storage of fresh salmon fillet. *Food Control* **2021**, *125*, 107994. [CrossRef]
6. Yousuf, B.; Sun, Y.; Wu, S. Lipid and Lipid-containing Composite Edible Coatings and Films. *Food Rev. Int.* **2021**, 1–24. [CrossRef]
7. Ren, K.; Fei, T.; Metzger, K.; Wang, T. Coating performance and rheological characteristics of novel soybean oil-based wax emulsions. *Ind. Crops Prod.* **2019**, *140*, 111654. [CrossRef]
8. Lehtinen, O.-P.; Nugroho, R.W.N.; Lehtimaa, T.; Vierros, S.; Hiekkataipale, P.; Ruokolainen, J.; Sammalkorpi, M.; Österberg, M. Effect of temperature, water content and free fatty acid on reverse micelle formation of phospholipids in vegetable oil. *Colloids Surf. B Biointerfaces* **2017**, *160*, 355–363. [CrossRef]
9. Vargas, M.; Albors, A.; Chiralt, A. Application of chitosan-sunflower oil edible films to pork meat hamburgers. *Procedia Food Sci.* **2011**, *1*, 39–43. [CrossRef]
10. Shin, D.-M.; Yune, J.H.; Kim, T.-K.; Kim, Y.J.; Kwon, H.C.; Jeong, C.H.; Choi, Y.-S.; Han, S.G. Physicochemical properties and oxidative stability of duck fat-added margarine for reducing the use of fully hydrogenated soybean oil. *Food Chem.* **2021**, *363*, 130260. [CrossRef]
11. Qiao, Y.; Huang, J.; Chen, Y.; Chen, H.; Zhao, L.; Huang, M.; Zhou, G. Meat quality, fatty acid composition and sensory evaluation of Cherry Valley, Spent Layer and Crossbred ducks. *Anim. Sci. J.* **2017**, *88*, 156–165. [CrossRef]
12. Gong, Y.; Weber, P.F.; Richards, M.P. Characterizing quality of rendered duck fat compared to other fats and oils. *J. Food Qual.* **2007**, *30*, 169–186. [CrossRef]
13. Shin, D.-M.; Do Hyun Kim, J.H.Y.; Kwon, H.C.; Kim, H.J.; Seo, H.G.; Han, S.G. Oxidative stability and quality characteristics of duck, chicken, swine and bovine skin fats extracted by pressurized hot water extraction. *Food Sci. Anim. Resour.* **2019**, *39*, 446. [CrossRef]
14. Kang, E.S.; Kim, H.J.; Han, S.G.; Seo, H.G. Duck oil-loaded nanoemulsion inhibits senescence of angiotensin II-treated vascular smooth muscle cells by upregulating SIRT1. *Food Sci. Anim. Resour.* **2020**, *40*, 106. [CrossRef]
15. Ros, E. Health benefits of nut consumption. *Nutrients* **2010**, *2*, 652–682. [CrossRef]
16. Nawade, B.; Mishra, G.P.; Radhakrishnan, T.; Dodia, S.M.; Ahmad, S.; Kumar, A.; Kumar, A.; Kundu, R. High oleic peanut breeding: Achievements, perspectives, and prospects. *Trends Food Sci. Technol.* **2018**, *78*, 107–119. [CrossRef]
17. Taqi, A.; Askar, K.A.; Nagy, K.; Mutihac, L.; Stamatin, L. Effect of different concentrations of olive oil and oleic acid on the mechanical properties of albumen (egg white) edible films. *Afr. J. Biotechnol.* **2011**, *10*, 12963–12972.
18. Ray, C.L.; Gawenis, J.A.; Greenlief, C.M. A New Method for Olive Oil Screening Using Multivariate Analysis of Proton NMR Spectra. *Molecules* **2022**, *27*, 213. [CrossRef]
19. Umaraw, P.; Verma, A.K. Comprehensive review on application of edible film on meat and meat products: An eco-friendly approach. *Crit. Rev. Food Sci. Nutr.* **2017**, *57*, 1270–1279. [CrossRef]

20. Kumarihami, H.P.C.; Kim, Y.-H.; Kwack, Y.-B.; Kim, J.; Kim, J.G. Application of chitosan as edible coating to enhance storability and fruit quality of Kiwifruit: A Review. *Sci. Hortic.* **2022**, *292*, 110647. [CrossRef]
21. Yüksel, Ç.; Atalay, D.; Erge, H.S. The effects of chitosan coating and vacuum packaging on quality of fresh-cut pumpkin slices during storage. *J. Food Processing Preserv.* **2022**, e16365. [CrossRef]
22. Hassanzadeh, P.; Tajik, H.; Rohani, S.M.R.; Moradi, M.; Hashemi, M.; Aliakbarlu, J. Effect of functional chitosan coating and gamma irradiation on the shelf-life of chicken meat during refrigerated storage. *Radiat. Phys. Chem.* **2017**, *141*, 103–109. [CrossRef]
23. Xiong, Y.; Chen, M.; Warner, R.D.; Fang, Z. Incorporating nisin and grape seed extract in chitosan-gelatine edible coating and its effect on cold storage of fresh pork. *Food Control* **2020**, *110*, 107018. [CrossRef]
24. Jasour, M.S.; Ehsani, A.; Mehryar, L.; Naghibi, S.S. Chitosan coating incorporated with the lactoperoxidase system: An active edible coating for fish preservation. *J. Sci. Food Agric.* **2015**, *95*, 1373–1378. [CrossRef] [PubMed]
25. Wardy, W.; Torrico, D.D.; Jirangrat, W.; No, H.K.; Saalia, F.K.; Prinyawiwatkul, W. Chitosan-soybean oil emulsion coating affects physico-functional and sensory quality of eggs during storage. *LWT—Food Sci. Technol.* **2011**, *44*, 2349–2355. [CrossRef]
26. Sathivel, S. Chitosan and protein coatings affect yield, moisture loss, and lipid oxidation of pink salmon (*Oncorhynchus gorbuscha*) fillets during frozen storage. *J. Food Sci.* **2005**, *70*, e455–e459. [CrossRef]
27. Shin, D.-M.; Hwang, K.-E.; Lee, C.-W.; Kim, T.-K.; Park, Y.-S.; Han, S.G. Effect of Swiss chard (*Beta vulgaris* var. *cicla*) as nitrite replacement on color stability and shelf-life of cooked pork patties during refrigerated storage. *Food Sci. Anim. Resour.* **2017**, *37*, 418. [CrossRef] [PubMed]
28. Kim, T.-K.; Hwang, K.-E.; Lee, M.-A.; Paik, H.-D.; Kim, Y.-B.; Choi, Y.-S. Quality characteristics of pork loin cured with green nitrite source and some organic acids. *Meat Sci.* **2019**, *152*, 141–145. [CrossRef]
29. Hwang, J.-K.; Shin, H.-H. Rheological properties of chitosan solutions. *Korea-Aust. Rheol. J.* **2000**, *12*, 175–179.
30. Park, S.-Y.; Kim, H.-Y. Fried pork loin batter quality with the addition of various dietary fibers. *J. Anim. Sci. Technol.* **2021**, *63*, 137. [CrossRef]
31. Abdul-Hamid, A.; Luan, Y.S. Functional properties of dietary fibre prepared from defatted rice bran. *Food Chem.* **2000**, *68*, 15–19. [CrossRef]
32. Liu, L.; Wang, B.; Gao, Y.; Bai, T.-c. Chitosan fibers enhanced gellan gum hydrogels with superior mechanical properties and water-holding capacity. *Carbohydr. Polym.* **2013**, *97*, 152–158. [CrossRef] [PubMed]
33. Tsafrakidou, P.; Sameli, N.; Bosnea, L.; Chorianopoulos, N.; Samelis, J. Assessment of the spoilage microbiota in minced free-range chicken meat during storage at 4 C in retail modified atmosphere packages. *Food Microbiol.* **2021**, *99*, 103822. [CrossRef]
34. Yaghoubi, M.; Ayaseh, A.; Alirezalu, K.; Nemati, Z.; Pateiro, M.; Lorenzo, J.M. Effect of chitosan coating incorporated with Artemisia fragrans essential oil on fresh chicken meat during refrigerated storage. *Polymers* **2021**, *13*, 716. [CrossRef] [PubMed]
35. Liu, F.; Chang, W.; Chen, M.; Xu, F.; Ma, J.; Zhong, F. Tailoring physicochemical properties of chitosan films and their protective effects on meat by varying drying temperature. *Carbohydr. Polym.* **2019**, *212*, 150–159. [CrossRef]
36. Kuswandi, B.; Nurfawaidi, A. On-package dual sensors label based on pH indicators for real-time monitoring of beef freshness. *Food Control* **2017**, *82*, 91–100. [CrossRef]
37. Noori, S.; Zeynali, F.; Almasi, H. Antimicrobial and antioxidant efficiency of nanoemulsion-based edible coating containing ginger (*Zingiber officinale*) essential oil and its effect on safety and quality attributes of chicken breast fillets. *Food Control* **2018**, *84*, 312–320. [CrossRef]
38. Seideman, S.; Cross, H.; Smith, G.; Durland, P. Factors associated with fresh meat color: A review. *J. Food Qual.* **1984**, *6*, 211–237. [CrossRef]
39. Zhang, H.; He, P.; Kang, H.; Li, X. Antioxidant and antimicrobial effects of edible coating based on chitosan and bamboo vinegar in ready to cook pork chops. *LWT—Food Sci. Technol.* **2018**, *93*, 470–476. [CrossRef]
40. Yu, L.; Lee, E.; Jeong, J.; Paik, H.; Choi, J.; Kim, C. Effects of thawing temperature on the physicochemical properties of pre-rigor frozen chicken breast and leg muscles. *Meat Sci.* **2005**, *71*, 375–382. [CrossRef]
41. Sujiwo, J.; Kim, D.; Jang, A. Relation among quality traits of chicken breast meat during cold storage: Correlations between freshness traits and torrymeter values. *Poult. Sci.* **2018**, *97*, 2887–2894. [CrossRef]
42. Domínguez, R.; Pateiro, M.; Gagaoua, M.; Barba, F.J.; Zhang, W.; Lorenzo, J.M. A comprehensive review on lipid oxidation in meat and meat products. *Antioxidants* **2019**, *8*, 429. [CrossRef] [PubMed]
43. Shahidi, F.; Desilva, C.; Amarowicz, R. Antioxidant activity of extracts of defatted seeds of niger (*Guizotia abyssinica*). *J. Am. Oil Chem. Soc.* **2003**, *80*, 443–450. [CrossRef]
44. Rezaei, F.; Shahbazi, Y. Shelf-life extension and quality attributes of sauced silver carp fillet: A comparison among direct addition, edible coating and biodegradable film. *LWT—Food Sci. Technol.* **2018**, *87*, 122–133. [CrossRef]
45. Yu, H.H.; Kim, Y.J.; Park, Y.J.; Shin, D.-M.; Choi, Y.-S.; Lee, N.-K.; Paik, H.-D. Application of mixed natural preservatives to improve the quality of vacuum skin packaged beef during refrigerated storage. *Meat Sci.* **2020**, *169*, 108219. [CrossRef] [PubMed]
46. Peng, Y.; Zhang, J.; Wang, W.; Li, Y.; Wu, J.; Huang, H.; Gao, X.; Jiang, W. Potential prediction of the microbial spoilage of beef using spatially resolved hyperspectral scattering profiles. *J. Food Eng.* **2011**, *102*, 163–169. [CrossRef]
47. Lorenzo, J.M.; Munekata, P.E.; Dominguez, R.; Pateiro, M.; Saraiva, J.A.; Franco, D. Main groups of microorganisms of relevance for food safety and stability: General aspects and overall description. In *Innovative Technologies for Food Preservation*; Elsevier: Amsterdam, The Netherlands, 2018; pp. 53–107.

48. Zwirzitz, B.; Wetzels, S.U.; Dixon, E.D.; Fleischmann, S.; Selberherr, E.; Thalguter, S.; Quijada, N.M.; Dzieciol, M.; Wagner, M.; Stessl, B. Co-occurrence of *Listeria* spp. and spoilage associated microbiota during meat processing due to cross-contamination events. *Front. Microbiol.* **2021**, *12*, 632935. [CrossRef] [PubMed]
49. Kanatt, S.R.; Rao, M.; Chawla, S.; Sharma, A. Effects of chitosan coating on shelf-life of ready-to-cook meat products during chilled storage. *LWT—Food Sci. Technol.* **2013**, *53*, 321–326. [CrossRef]

 *foods*

Article

Biofilm Formation Reduction by Eugenol and Thymol on Biodegradable Food Packaging Material

Pavel Pleva [1] , Lucie Bartošová [1], Daniela Máčalová [1], Ludmila Zálešáková [2], Jana Sedlaříková [3] and Magda Janalíková [1,*]

[1] Department of Environmental Protection Engineering, Faculty of Technology, Tomas Bata University in Zlin, 275 Vavreckova, 76001 Zlin, Czech Republic; ppleva@utb.cz (P.P.); l1_bartosova@utb.cz (L.B.); d_macalova@utb.cz (D.M.)

[2] Department of Food Technology, Faculty of Technology, Tomas Bata University in Zlin, nam. T. G. Masaryka 5555, 76001 Zlin, Czech Republic; lzalesakova@utb.cz

[3] Department of Fat, Surfactant and Cosmetics Technology, Faculty of Technology, Tomas Bata University in Zlin, 275 Vavreckova, 76001 Zlin, Czech Republic; sedlarikova@utb.cz

* Correspondence: mjanalikova@utb.cz; Tel.: +420-57-603-1020

Abstract: Biofilm is a structured community of microorganisms adhering to surfaces of various polymeric materials used in food packaging. Microbes in the biofilm may affect food quality. However, the presence of biofilm can ensure biodegradation of discarded packaging. This work aims to evaluate a biofilm formation on the selected biodegradable polymer films: poly (lactic acid) (PLA), poly (butylene adipate-co-terephthalate) (PBAT), and poly (butylene succinate) (PBS) by selected bacterial strains; collection strains of *Escherichia coli, Staphylococcus aureus*; and *Bacillus pumilus, Bacillus subtilis, Bacillus tequilensis*, and *Stenotrophomonas maltophilia* isolated from dairy products. Three different methods for biofilm evaluation were performed: the Christensen method, 3-(4,5-dimethylthiazol-2-yl)-2,5-diphenyltetrazolium bromide (MTT) assay, and fluorescence microscopy. High biofilm formation was confirmed on the control PBS film, whereas low biofilm formation ability was observed on the PLA polymer sample. Furthermore, the films with incorporated antimicrobial compounds (thymol or eugenol) were also prepared. Antimicrobial activity and also reduction in biofilm formation on enriched polymer films were determined. Therefore, they were all proved to be antimicrobial and effective in reducing biofilm formation. These films can be used to prepare novel active food packaging for the dairy industry to prevent biofilm formation and enhance food quality and safety in the future.

Keywords: antimicrobial activity; biofilm; biodegradable polymers; food packaging

Citation: Pleva, P.; Bartošová, L.; Máčalová, D.; Zálešáková, L.; Sedlaříková, J.; Janalíková, M. Biofilm Formation Reduction by Eugenol and Thymol on Biodegradable Food Packaging Material. *Foods* 2022, 11, 2. https://doi.org/10.3390/foods11010002

Academic Editors: Theodoros Varzakas and Rui M. S. Cruz

Received: 16 November 2021
Accepted: 16 December 2021
Published: 21 December 2021

Publisher's Note: MDPI stays neutral with regard to jurisdictional claims in published maps and institutional affiliations.

1. Introduction

Biofilm is a community of microorganisms attached to a surface and surrounded by an extracellular polymeric matrix. Biofilm provides better living conditions for microorganisms than planktonic form because it maintains the stability of the internal environment, isolates them from the outside, and protects inner cells [1–3]. They are more resistant to harmful effects, such as antimicrobial compounds, UV radiation, bacteriophages, antibiotics, and the human immune system. Biofilm formation is a multi-step process starting with a reversible attachment to a surface aided by intermolecular forces and hydrophobicity [4,5]. In the food industry, microorganisms can adhere to abiotic surfaces, form a biofilm, and reduce the shelf life of food products. Furthermore, it may increase the incidence of foodborne illnesses, which is a significant concern for public health and food quality [6,7].

The critical function of food packaging is to protect products from extrinsic factors, especially gases, temperature, and relative humidity [8]. Therefore, active packaging is a system that actively changes the conditions of packaged food and maintains or extends the product quality [9,10]. Food packaging legislation is defined mainly at a national, European, and worldwide level. In the EU, legislative regulations are divided into food packaging, plastic material and articles intended to direct contact with food, and safe use of active and intelligent packaging [8,11–13]. Regulation (EC) 450/2009 (European Commission 2009) defines specific rules for active and intelligent materials and articles [9,11]. Regulation (EC) 1935/2004 and Regulation (EC) 10/2011 state that materials and articles, including active and intelligent materials and articles, shall be manufactured in compliance with suitable manufacturing practices (Regulation (EC) No. 2023/2006) [8,12]. Due to the deliberate interaction of active packaging, the migration of substances could pose a food safety problem [9]. Regulation (EC) No. 10/2011 specifies rules for the valuation of the release of low molecular weight substances from plastic in the food packaging and a list of substances of specific threshold limits. The European Food Safety Authority (EFSA) recommends these limits [8,12]. In the United States and Canada, the market requires the Food and Drug Administration (FDA) to approve materials and articles for food contact [14].

Despite possible negative impacts, bacterial biofilms may also have beneficial biodegradable effects [14]. High molecular weight polymers are more difficult to degrade because of their poor solubility and hindered penetration into the cell wall where they are enzymatically degraded. Biodegradable polymers are submitted to biodegradation processes in nature easier than synthetic [14,15]. However, a biofilm on biodegradable polymers may influence the change of mechanical properties, including tensile strength. On the other hand, non-biodegradable polymers, such as polyolefins, are not expected to change the mechanical characteristics [14,16].

With a growing negative view of environmental pollution, biodegradable polymers attract increasing attention, mainly in the fields of packaging materials, agriculture, and medicine [16,17]. Many types of biodegradable polymers can be used in food packaging. They may include conventional polylactide (PLA), polybutylene succinate (PBS), polyamides (PA), poly (butylene adipate-co-terephthalate) (PBAT), polypropylene (PP), and many others [18,19]. They are positively evaluated in the food industry, especially for their diverse mechanical properties, tear resistance, and health safety [20,21]. Chemical and biological additives, such as antioxidants, lubricants, stabilizers, pigments, are often applied to polymeric materials to enhance their properties [22]. Eugenol and thymol, natural phenolic compounds occurring in plant essential oils, could be used for their known antimicrobial properties [23,24]. These added compounds may be rapidly degraded due to their instability and high volatility [25].

Microbial biofilm on packaging material may threaten food safety and quality; what is more, it is also required for their biodegradation after disposal. This study aims to investigate biofilm formation by food isolates on biodegradable polymers used as food packaging and enhance the antibacterial properties of these polymers by enriching them with thymol and eugenol.

2. Materials and Methods

2.1. Materials and Chemicals

Biodegradable polymers: poly(lactic) acid—PLA (NatureWorks, Minnetonka, MN, USA), poly(butylene adipate-co-terephthalate)—PBAT Ecoflex® (BASF, Ludwigshafen, Germany), and poly(butylene succinate)—PBS G4560 (IRe Chemical Ltd., Seoul, Korea) were tested in this study.

Bacterial strains of *Bacillus tequilensis* R23, *Bacillus subtilis* R25, *Bacillus pumilus* R34, *Stenotrophomonas maltophilia* GK CIP 1/1 were obtained from the Dairy Research Institute in Prague (Czech Republic). These strains were isolated from dairy products. Bacterial strains *Escherichia coli* ATCC 25922 and *Staphylococcus aureus* ATCC 25923 were provided by the Czech Collection of Microorganisms (CCM, Brno, Czech Republic). *Escherichia coli* ATCC 25922 and *Staphylococcus aureus* ATCC 25923 were selected for testing as typical model organisms representing both Gram-negative and Gram-positive bacteria. Isolates from dairy products were chosen as representatives of natural contaminants due to the possibility of future application of tested materials as food packaging. All strains were cultivated in BHI medium (brain heart infusion broth, HiMedia, Mumbai, India). In addition, the culture medium was enriched with 5% sucrose (HiMedia, Mumbai, India) to test biofilm formation. *E. coli* and *S. aureus* strains were cultivated at 37 °C/24 h, other strains were grown at 30 °C/24 h. Mueller Hinton agar (HiMedia Laboratories Pvt. Ltd., Mumbai, India) was used to determine antibacterial activity.

Eugenol and thymol were provided by Sigma-Aldrich (St. Louis, MO, USA). Other used chemicals were obtained from specified companies: chloroform (Penta chemicals, Chrudim, Czech Republic); 96% ethanol (Lach-Ner, Neratovice, Czech Republic); crystal violet (Penta, Prague, Czech Republic); 3-(4,5-dimethylthiazol-2-yl)-2,5-diphenyltetrazolium bromide (MTT) powder (Serva, Heidelberg, Germany); DMSO (Sigma-Aldrich, St. Louis, MO, USA).

2.2. Film Preparation

Tested biodegradable polymers (PLA, PBAT, PBS) were prepared in the form of films. The mixture of polymer and chloroform in the concentration of 14% w/v was homogenized under the continuous stirring (500 rpm, 48 h) at 25 $\pm$ 1 °C (except PBS/50 $\pm$ 1 °C). The active compounds (eugenol—E, thymol—T) were added into the polymer solution in the final concentration of 3% w/v and stirred (500 rpm, 1 h) at 25 $\pm$ 1 °C. The samples of PLA/E; PLA/T; PBAT/E; PBAT/T; PBS/E; PBS/T were prepared. The solution without active compounds was used as the control. The prepared mixture (10 mL) was poured into Petri dishes (6 cm in diameter) and left to dry in an air-circulated oven (30 °C, 24 h) [26]. A higher concentration of active compounds was chosen to ensure a more extended provision of antimicrobial properties. Tested active compounds in the concentration of 3% w/v should provide antimicrobial protection for several months for water-based foods, using the result from Narayanan et al. [27].

2.3. Antibacterial Activity

Antibacterial activities of PLA, PBS, and PBAT control films and films with eugenol and thymol were tested by the standard agar diffusion test with the following modifications [28]. The films were cut into disks (5 mm in diameter). Before testing, the samples were surface sterilized by UV radiation for 20 min without any sample damage [29]. They were placed on Mueller Hinton agar plates inoculated with 1 mL of 0.5 McF turbid bacterial suspension (*Bacillus tequilensis* R23, *Bacillus subtilis* R25, *Bacillus pumilus* R34, *Stenotrophomonas maltophilia* GK CIP 1/1, *Escherichia coli* ATCC 25922 and *Staphylococcus aureus* ATCC 25923) in the sterile saline solution. *E. coli* and *S. aureus* strains were cultivated at 37 °C/24 h, other strains were grown at 30 °C/24 h. The inhibition zones were evaluated. All experiments were repeated in triplicate.

2.4. Biofilm Formation

2.4.1. MTT Assay

The films were cut into squares of 25 mm^2, placed in a sterile tube with 4.5 mL BHI broth, and inoculated with 50 µL of a 24 h bacterial culture suspension with bacterial turbidity of 1 McF. The cultivation of *E. coli* and *S. aureus* was performed at 37 °C/48 h. For the other strains, it was at 30 °C/48 h with constant stirring on a shaker. The solution was aspirated, and then the samples were rinsed with sterile distilled water to remove the planktonic cells. MTT powder was dissolved in ultrapure water at a 5 mg/mL concentration. The rinsed material was placed into individual wells of a microplate together with 180 µL of BHI broth, and MTT solution was added to the final concentration of 0.5 mg/mL. After the treatment with MTT for 4 h at 30 °C and shaking, the solution was aspirated and replaced with 200 µL of dimethyl sulfoxide (DMSO), which dissolved the formed formazan. Subsequently, 100 µL of formazan solution was placed into each well of a microplate, the absorbance (optical density) at 690 nm was measured, and the background was read at 570 nm using a Tecan Infinite® 200 PRO (Tecan, Mannedorf, Switzerland). The average and standard deviation of the optical density (OD) of negative controls (OD_{NC}) were calculated from the measured values. According to Stepanović et al., the cut-off value for positivity (OD_p) was calculated as the sum of three times the standard deviation and OD_{NC} ($OD_p = OD_{NC} + 3 \times$ standard deviation OD_{NC}) [30]. The resulting biofilm formation was evaluated into 3 categories by limit values:

- Non-biofilm formation (−): $OD \leq OD_P$;
- Weak biofilm formation (+): $OD_P < OD \leq 2OD_P$;
- Strong biofilm formation (++): $2OD_P < OD$.

2.4.2. Christensen Method

The films (25 mm^2) were placed in a sterile tube with 4.5 mL BHI and inoculated with 50 µL of 1 McF turbid bacterial suspension. After the cultivation (see Section 2.4.1), the material was rinsed thoroughly with sterile distilled water to remove the adhering planktonic cells. Subsequently, 200 µL of 96% ethanol was added for 20 min. The ethanol was then washed and replaced with 200 µL of crystal violet, leaving it to act for 20 min. The stained sample was placed in a test tube with 200 µL of 96% ethanol, which dissolved the bound dye. Next, 100 µL of colored ethanol solution was transferred to each well of a microplate, and then the absorbance at 600 nm was measured using a Tecan Infinite$^{\circledR}$ 200 PRO (Tecan, Männedorf, Switzerland). The limit values for the evaluation of a biofilm formation were determined in the same way as for the MTT assay (see Section 2.4.1).

2.4.3. Fluorescence Microscopy

The samples (25 mm^2) were washed with sterile saline solution after the cultivation (see Section 2.4.1) and placed on a glass slide. They were dyed by fluorescence dye (SYTO$^{\circledR}$9 and propidium iodide) for 10 s and then covered with a square coverslip. Fluorescence microscopy was performed using a fluorescence microscope Olympus BX53 (Olympus, Tokyo, Japan) equipped with Microscope Digital Camera DP73 (Olympus, Tokyo, Japan) and the cell Sens Standard 1.18 (Olympus, Tokyo, Japan) software. The analysis was performed on a minimum of 20 positions in three replicates. LIVE/DEAD™ BacLight™ Bacterial Viability Kit (Thermo Fisher, Waltham, MA, USA), based on the protocol [31], was executed using slight modifications. SYTO$^{\circledR}$9 dyed plasma membranes of all bacteria, while propidium iodide can color DNA of only dead cells. The excitation/emission maxima for these dyes are 480/500 nm for SYTO$^{\circledR}$9 stain and 490/635 nm for propidium iodide. Thus, bacteria with intact cell membranes stain fluorescent green, whereas bacteria with damaged membranes (dead) stain fluorescent red.

2.5. Material Properties

2.5.1. FTIR-ATR Analysis

The chemical composition of prepared films was characterized by Fourier transform infrared spectroscopy (FTIR) on Nicolet 6700 spectrometer (Thermo Fisher Scientific, Waltham, MA, USA) set to ATR mode and fitted with a diamond crystal equipped with OMNIC Paradigm software. Measurement conditions comprised 64 scans at the resolution of 2 cm^{-1} and the range of 4000 to 400 cm^{-1}.

2.5.2. Contact Angle Measurement

The wettability of polymer films was measured by the sessile drop method at ambient temperature on a Theta optical tensiometer (Biolin Scientific, Göteborg, Sweden) equipped with OneAttension software. Distilled water of the volume equaling 3 µL was applied as the reference liquid.

2.6. Statistical Analysis

The statistical evaluation of the contact angles employed one-way analysis of variance (ANOVA) using Statistica software (version 10, StatSoft, Inc., Tulsa, OK, USA), at the significance level of $p < 0.05$.

3. Results

3.1. Antibacterial Activity

A method based on the diffusion of active compounds, the disk diffusion method, was used to determine the antimicrobial activity. The inhibition zones were measured, and the values, including the disk diameter (5 mm), are shown in Table 1.

Table 1. Antimicrobial activity determined by disk diffusion method (sample 5 mm in diameter).

Samples	*B. tequilensis* (mm)	*B. subtilis* (mm)	*B. pumilus* (mm)	*S. maltophilia* (mm)	*E. coli* (mm)	*S. aureus* (mm)
PLA	*	*	*	*	*	*
PLA/T	*	*	7.8 ± 1.2	12.5 ± 0.3	*	8.0 ± 0.4
PLA/E	9.5 ± 0.5	9.8 ± 0.3	7.3 ± 0.5	13.3 ± 0.3	*	9.0 ± 0.4
PBS	*	*	*	*	*	*
PBS/T	10.0 ± 0.4	9.5 ± 0.3	7.0 ± 0.4	*	10.5 ± 1.2	15.8 ± 0.5
PBS/E	11.8 ± 0.5	13.0 ± 0.7	12.3 ± 0.9	13.0 ± 0.4	15.8 ± 1.1	17.8 ± 0.5
PBAT	*	*	*	*	*	*
PBAT/T	*	7.3 ± 0.3	7.3 ± 0.3	6.3 ± 0.3	9.3 ± 0.3	10.3 ± 0.3
PBAT/E	7.8 ± 0.5	7.5 ± 0.3	9.5 ± 0.3	10.8 ± 0.3	6.3 ± 0.3	8.8 ± 0.5

PLA: poly(lactic) acid, PBAT: poly(butylene adipate-co-terephthalate), PBS: poly(butylene succinate), E: 3% *w/v* eugenol, T: 3% *w/v* thymol. * —no inhibition zone.

The samples without incorporated active compounds did not inhibit the tested bacteria. In contrast, films enriched with antimicrobial compounds, eugenol, and thymol, have formed the inhibition zones confirming the antimicrobial activity (Table 1). These results corresponded with the published results of eugenol and thymol inhibition effect against bacteria [23,24]. The best inhibitory effect of synthetic polymers PLA, PBS, and PBAT with added phenolic compounds were performed by films enriched with 3% *w/v* eugenol. Except for Gram-negative bacteria *E. coli*, all tested isolates from dairy products were inhibited by PLA/E. The samples of PBS/E and PBAT/E performed a complete inhibitory effect against all tested bacteria. The sample of PBS/E showed the largest inhibition zone against *S. aureus*, with the zone reaching the diameter of 17.8 ± 0.5 mm. The polymer films with 3% *w/v* thymol presented generally lower antibacterial activities than eugenol films. PLA/T produced the smallest inhibition zones against the tested bacteria with the observed activity only against *B. pumilus*, *S. maltophilia*, and *S. aureus*. The internal structure of the polymer may have an important impact on the release and diffusion of the active compound to the external environment. It may be concluded that incorporating eugenol or thymol into biodegradable polymer films enhances their antimicrobial properties.

3.2. Biofilm Formation

Biodegradable polymers are used as an alternative to disposable packaging in the food industry [32]. Bacterial biofilm can grow on these polymers. What is more, it is necessary for their biodegradation [33]. Therefore, it is important to identify a suitable active compound with the optimal concentration not only for extending the shelf life of food but also for prolonged biodegradation. This study examined three methods (MTT assay, Christensen method, and fluorescence microscopy) to monitor biofilm formation on packaging materials.

MTT assay and Christensen method are spectrophotometric methods based on the biofilm staining. In the Christensen method, biofilm is stained with crystal violet. Subsequently, the color is leached with ethanol, corresponding to the strength of a biofilm formation [34]. In contrast, in MTT assay, only the number of living cells in a biofilm is monitored based on their metabolic activity. In MTT assay, living cells reduce yellow MTT to violet formazan. Subsequently, formazan is dissolved in dimethyl sulfoxide (DMSO), and the color of the extract is directly proportional to the number of viable cells [35]. Fluorescence microscopy allows the observation of biofilm viability. Differentiation of living cells from dead cells can be accomplished by staining with propidium iodide, which penetrates the disrupted cytoplasmic membrane of dead cells [36]. Propidium iodide is a red fluorescent nucleic acid that excites at 490 nm and emits 635 nm. SYTO® 9 green-fluorescent nucleic acid, which excites at 480 nm and is emitted at 500 nm, could be used to stain all cells. Using propidium iodide with SYTO® 9 reduces the fluorescence of the SYTO® 9 dye, thereby distinguishing red fluorescent dead cells from green living cells [31].

For both, Christensen method and MTT assay, a biofilm formation was examined by determining the limit values for prepared materials. The limit values were determined from the reference value, i.e., without inoculation with a microorganism and three standard deviations. These values are unique for every single prepared polymer film. The limit value was selected based on a normal (Gaussian) distribution, i.e., the average together with three standard deviations cover almost all probable values (99.7%) [37]. A p-value of <0.003 was considered to be statistically significant. The resulting biofilm formation was classified into three categories (non-biofilm formation, weak biofilm formation, and strong biofilm formation). The non-forming biofilm showed an average absorbance value lower than the limit values. Weak biofilm formation achieved values twice the limit value, and above this value, the strains were considered to form a strong biofilm. In the Christensen method, PLA and PBS were stained too much. Strong staining of the surfaces of the tested samples is inappropriate due to the requirement for the subsequent dilution for the samples with a stronger biofilm. It complicates the method and prolongs the time required for the test. Therefore, the Christensen method may be inappropriate for many polymers.

Three methods for the detection of biofilm formation on biodegradable polymer films by *Bacillus tequilensis*, *Bacillus subtilis*, *Bacillus pumilus*, *Stenotrophomonas maltophilia*, *Escherichia coli*, and *Staphylococcus aureus* were compared in order to select the most suitable one (Table 2 for pure materials, Supplementary Materials Table S1 for all tested materials).

Table 2. Comparison of methods for evaluating biofilm formation for pure materials.

Materials	Methods	*B. tequilensis*	*B. subtilis*	*B. pumilus*	*S. maltophilia*	*E. coli*	*S. aureus*
PLA	MTT assay	−	−	−	−	−	−
	Christensen method	−	−	−	−	−	−
	Fluorescence microscopy (LIVE)	+++	+++	+++	+	+++	+++
	Fluorescence microscopy (DEAD)	+	−	−	+	++	++
PBS	MTT assay	+	+	+	+	+	+
	Christensen method	−	−	+	+	−	−
	Fluorescence microscopy (LIVE)	−	++	−	++	−	+
	Fluorescence microscopy (DEAD)	++	+	++	+	+++	+
PBAT	MTT assay	−	−	−	−	−	−
	Christensen method	+	+	+	+	+	+
	Fluorescence microscopy (LIVE)	+	+	+	+	+++	+
	Fluorescence microscopy (DEAD)	−	−	−	−	+++	−

PLA: poly(lactic) acid, PBAT: poly(butylene adipate-co-terephthalate), PBS: poly(butylene succinate). MTT assay and Christensen method: −: non-biofilm formation, +: with weak biofilm formation, ++: with strong biofilm formation ($p < 0.003$). Fluorescence microscopy: −: without microorganisms, +: 1–10 microorganisms, ++: 10–50 microorganisms, +++: >50 microorganisms.

As all tested strains are able to form a biofilm, its production was expected on all tested biodegradable polymers. Neither the MTT method nor the Christensen test performed relevant results on PLA. Limoli et al. described that the absence of a detectable biofilm could be caused by the smooth surface of the sample [38]. The results indicate that the MTT method was more suitable than Christensen for PBS, test in a biofilm detection in vitro (Table 2). Nevertheless, the microplate adherence test described by Christensen provided better results on PBAT.

The presence of adherent cells on PLA was confirmed by fluorescence microscopy. MTT method proved the production of a weak biofilm on PBS within all tested strains. On the other hand, only two isolates (*Bacillus pumilus*, *Stenotrophomonas maltophilia*) demonstrated a biofilm using the Christensen method on PBS. On PBAT, a weak biofilm was detected by the Christensen method and fluorescence microscopy. However, no biofilm was detected by MTT assay. These results demonstrate a weak biofilm formation with low metabolic activity on PBAT [35].

When comparing the tested methods, Christensen's method for detecting a biofilm formation on polymers appeared to be unsuitable due to the strong dyeability of polymers themselves. Another disadvantage is that MTT assay and Christensen method are time-consuming. What might be considered beneficial of these methods is that the samples are not rinsed during this test. A biofilm may be destroyed when rinsed, causing a false negative result [5].

Based on the obtained results, it could be concluded that a biofilm on the polymer films was precisely detected using fluorescence microscopy. LIVE/DEAD bacterial viability assays were performed by fluorescence microscopy, as can be seen in Figure 1. Live bacterial cells are displayed green, and dead bacterial cells are red. It seems that an insignificant difference was found between all neat polymers. However, the spherulite-forming PBS does not provide a suitable surface for bacterial adhesion. Thus, the bacteria accumulate on the amorphous portion of the semicrystalline polymer [39]. Fluorescence microscopy of the modified films enriched with active phenolic agents proved no bacterial cells on the surface. This experiment confirmed that biodegradable polymer films with eugenol or thymol prevent a biofilm formation by bacterial isolates from the dairy products. These results correspond with proven antibacterial properties (see Section 3.1).

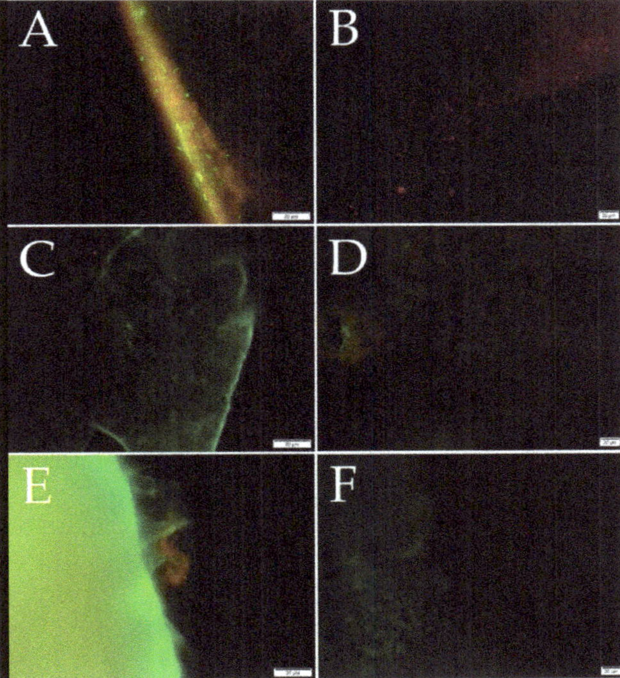

Figure 1. Fluorescence microscopy of LIVE/DEAD bacterial viability assay with *Stenotrophomonas maltophilia*. (**A**): PLA, (**B**): PLA/T, (**C**): PBS, (**D**): PBS/T, (**E**): PBAT, (**F**): PBAT/T. PLA: poly(lactic) acid, PBAT: poly(butylene adipate-co-terephthalate), PBS: poly(butylene succinate), T: 3% *w/v* thymol.

3.3. Material Properties

The prepared polymer films were homogeneous and compact with a smooth surface. PLA films were clear and elastic with a slight shade of yellow. Modified PLA films were visually darker yellow due to the color of eugenol and thymol. PBAT films exhibited higher elasticity in comparison to PLA samples. Pure PBAT films were white, while eugenol and thymol also caused the change in the film to light yellow. On the other hand, all prepared PBS films were fragile with a similar yellowish color tint as PBAT films.

3.3.1. Contact Angle

The wettability of polymer surfaces evidences their hydrophilicity or hydrophobicity, significantly affecting their potential applications. The wetting properties of the samples without and with active agents were analyzed by the sessile drop method. As can be seen from the values of contact angles (Table 3), prepared samples exhibited contact angles up to 90°. The base polymer films showed values ranging from 55 to 75°. More hydrophobic surfaces (103 to 120°) were observed in PBS membrane or PBS-based foam [40].

Table 3. Contact angles values for tested polymer films.

Active Compounds	PLA (°)	PBS (°)	PBAT (°)
*	75 ± 4 [aA]	74 ± 2 [aA]	56 ± 4 [aB]
3% w/v thymol	67 ± 3 [aAB]	75 ± 4 [aA]	60 ± 4 [aB]
3% w/v eugenol	66 ± 2 [aA]	74 ± 2 [aB]	63 ± 3 [aA]

PLA: poly(lactic) acid, PBAT: poly(butylene adipate-co-terephthalate), PBS: poly(butylene succinate), *—without active compounds. Different lower-case/upper-case letters in the same column/line indicate significant differences, respectively ($p < 0.05$).

The modification of PLA film with eugenol and thymol resulted in a slight increase in wettability. A rather opposite trend, an increase in hydrophobicity, was shown in PBS and PBAT samples even though the differences were not statistically significant at the p level of 0.05. In spite of this fact, ATR-FTIR analysis (Figure 2) proved the presence of phenolic groups, and the results of microbiological testing clearly revealed the antibacterial properties and biofilm reduction in modified samples (Sections 3.1 and 3.2). No significant changes ($p < 0.05$) in wettability were observed when thymol and eugenol active agents were compared, which is probably due to the similar molecular structures of these phenolic major components of thyme and clove essential oils, respectively. A lower contact angle value in pure PBAT film could consist of its molecular structure containing more oxygen substituents with a hydrophilic character. In contrast, PLA is composed of polar and non-polar substituents in its molecular structure resulting in higher wetting angle values than those of PBAT. The incorporation of non-polar substances (thymol, eugenol) increases the overall hydrophobicity of the PBAT polymer [41]. Similarly, in the study of Moustafa et al. [42], who investigated the PBAT/coffee grounds composites for food packaging applications, a higher contact angle with the addition of torrefied coffee grounds was observed. It is worth stating that the values could be strongly affected by several factors, such as the surface roughness of the samples and the location of a droplet's placement.

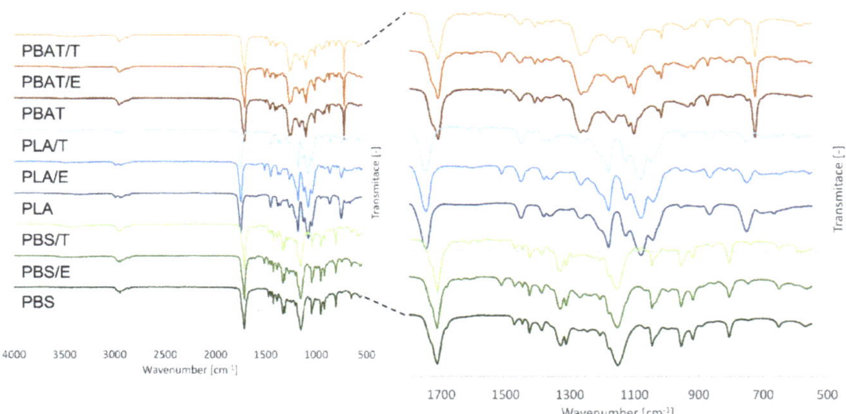

Figure 2. The FTIR spectra of pure and modified polymers; PLA: poly(lactic) acid, PBAT: poly(butylene adipate-co-terephthalate), PBS: poly(butylene succinate), E: 3% w/v eugenol, T: 3% w/v thymol.

3.3.2. FTIR-ATR

Infrared spectroscopy was used to characterize the main functional groups of the base polymers and reveal potential interactions between them and added active phenolic compounds. The FTIR spectra of pure and modified polymers are shown in Figure 2. The base polymers revealed the peaks corresponding to their main characteristic bonds [43,44]. Regarding PBS polymer, a very weak absorption band in the range of 3430–3500 cm^{-1} is assigned to polymer chain terminal –OH groups; -CH-groups were identified in the range of 3000–2800. Several overlapping peaks were observed in the –C=O region, relating to ester carbonyl groups of semicrystalline polyesters. In PBAT polymer spectra, the main absorption peaks at 2958 and 2869 cm^{-1} occurred correspondingly to -CH$_2$ groups [43]. The peaks of PLA in the range of 3000 to 2900 cm^{-1} were assigned to –C-H vibrations of the CH$_3$ side chains groups. PLA spectrum showed characteristic peaks at 1747, 2945, 2995 cm^{-1} assigned to C=O, -CH$_3$ asymmetric, and -CH$_3$ symmetric, respectively [44,45].

When active compounds were added, new peaks in the range from 1600 to 500 cm^{-1} were observed, indicating their interactions with polymer.

All tested polymer films showed the peaks associated with thymol presence at 1618, 1585, and 1519 cm^{-1} referring to characteristics C=C peaks corresponding to thymol aromatic ring [46,47]. The peaks about 1638, 1610, and 1514 cm^{-1} (in PLA/E film) are in suitable agreement with the characteristic structural spectrum of eugenol [48]. In a detailed analysis of IR spectra (Supplementary Materials Figure S1), a change in the form of more intensive peaks in the range from 3500 to 3400 cm^{-1} can be observed in modified polymer films. These could be assigned to phenolic groups of thymol and eugenol active agents.

4. Discussion

In this experiment, polymer films were prepared from synthetic polymers: PLA, PBAT, PBS. A biofilm formation was investigated on all films by various detection techniques, such as chemical (Christensen method), biological (MTT assay), and microscopic (fluorescence microscopy) [5,49]. The staining techniques belong to the least sensitive reproducible methods for determining total biofilm biomass. A safranin dye can be used for biofilm detection, as described by Ojima et al. [50] and Nguyen et al. [51]. This staining technique, similar to the Christensen method, was applied to the strain of *E. coli*. Marcos-Zambrano et al. [52] used crystal violet as in this study. Considering different biomass staining methods, it could be concluded that these staining methods are often inappropriate due to their low sensitivity [49,53]. The basic technique for visual detection of biofilm formation is light microscopy. It is a simple, inexpensive, and easy method; however, its resolution is low in comparison with other microscopic techniques [5]. Nevertheless, according to the previous studies by Azaredo et al. [5] and Hassan et al. [54], biofilm-based staining methods are one of the cheapest and most effective methods. Neither the Christensen method nor MTT assay confirmed the growth of a biofilm on PLA. Morohoshi et al. did not confirm the presence of a biofilm on PLA after two weeks of the cultivation in the seawater by DNA analysis of a biofilm [55]. On the other hand, fluorescence microscopy proved the highest sensitivity and biofilm formation. The presence of a biofilm was also confirmed on PBS and PBAT [55], as monitored in this study.

Polymer films from this study could be applied in food packaging. Substances with a potential antimicrobial activity may be added to ensure greater food safety and quality. Many compounds are used to increase antimicrobial protection in the food industry, such as essential oils or fatty acids [56,57]. Two compounds of essential oils, eugenol, and thymol were tested and proved favorable effects for preventing biofilm formation [58]. Several studies have confirmed the broad-spectrum antimicrobial activity of thymol and eugenol against bacteria and fungi [59–61]. Thymol incorporated in poly(lactide-co-glycolide) nanofibers proved a suitable preservative effect in wrapping strawberries in the study of Zhang et al. Therefore, thymol is beneficial to prolong the shelf life and keep the food fresh without affecting the flavor [62]. Hamzah et al. investigated various concentrations of these phenolic compounds against *Escherichia coli*, *Staphylococcus aureus*, *Pseudomonas aeruginosa* biofilm. The concentration of 0.96% w/v proved a high inhibition against the selected bacteria, with thymol showing the highest ability to cease a biofilm formation [63,64]. The antibacterial activity of phenolic compounds is attributed to phenolic hydroxyl in their structure [65]. Eugenol with the concentration of 0.625 µg/mL does not affect the bacterial viability; nonetheless, it downregulates the expression of virulence genes involved in the adhesion and biofilm formation (*brp*A, *com*DE, *ftf*, *gbp*B, *gtf*B, *gtf*C, *rel*A, *smu*630, *spa*P, and *vic*R) [66].

Therefore, eugenol and thymol in the concentration of 3% w/v were selected for incorporation into biodegradable polymeric films (PLA, PBAT, PBS) to prevent bacterial biofilm formation and possible occurrence of pathogenic microorganisms. These phenolic monoterpenes play an important role in the enhancement of the antimicrobial properties of polymers. They are released through the polymer matrix over time, continuously available, and diffused through the bacterial cell membrane, thereby providing the antimicrobial effect [67]. In this study, the disk diffusion method proved the antimicrobial properties of eugenol and thymol. The presence of eugenol or thymol in the tested materials was confirmed by FTIR analysis, and hydrophobic characteristics of their surface were described. The films containing eugenol showed higher antimicrobial activity than the films with thymol, similarly as published by De Morais et al. [64].

The improvement of the antimicrobial properties of PLA was achieved by the addition of thymol in the concentration of 8% w/v in the study by Ramos et al. [68]. Moderate antimicrobial activity for *Bacillus cereus* and *Staphylococcus aureus* was also detected for eugenol-grafted PLA. However, as in this work, the antimicrobial activity for *Escherichia coli* was not confirmed [69]. The advancement of the antimicrobial properties of biodegradable eugenol-grafted polymers was demonstrated in other studies. Grafted eugenol and carvacrol in chitosan nanoparticles improved the antimicrobial properties against *Escherichia coli* and *Staphylococcus aureus*, conferred antioxidant properties to the nanoparticles, and reduced cytotoxicity over pure essential oils [70]. The minimum inhibitory concentration of incorporated eugenol required for the film to exhibit antimicrobial activity was ≥ 40 µg/g for bacteria and ≥ 80 µg/g polyhydroxybutyrate (PHB) for fungi. The increase in eugenol concentration, i.e., ≥ 200 µg/g PHB, resulted in the complete growth inhibition of both bacteria and fungi in Narayanan et al. However, the antimicrobial activity of eugenol in the film decreased in the following order: water >3.0% acetic acid > n-hexane > 50% ethanol due to the specific migration [27]. Therefore, the concentration of 3% w/v of active substances was selected to ensure more significant antimicrobial properties of the incorporated films.

The addition of active ingredients to the polymer films has demonstrated biofilm inhibition and antimicrobial properties. Therefore, this could cause a prolongation of the biodegradation time as the enzymatic activity of the microorganisms initiates this. Pure polymers have been determined to biodegrade under various abiotic conditions. Polymer PLA is biodegradable under industrial composting conditions (58 °C, high humidity, aerobic). Polymers PBAT and PBS are biodegradable not only in composting but also in the soil (25 °C, humidity, aerobic) [71–73].

5. Conclusions

This study investigated biofilm formation on the polymer films with a perspective application in the food industry as the packaging material. PLA, PBS, and PBAT polymer films were tested for biofilm formation using various detection techniques (Christensen method, MTT assay, and fluorescence microscopy). The individual methods were compared, and their effectiveness was determined. Fluorescence microscopy proved to be the most sensitive for biofilm detection. The incorporation of the active compounds (thymol and eugenol) into the polymers mentioned above was also performed to prevent a biofilm formation by microorganisms that may affect the food quality. The antibacterial properties have been achieved by this modification, and biofilm formation was not detected on polymer films with incorporated antimicrobial compounds. These active packaging may enhance food safety and protect the consumer's health. Consequently, a biofilm could form on the surface after releasing antimicrobial volatile compounds, which contribute to its biodegradation. Thus, future experiments will assess the release of antimicrobial compounds from the prepared polymers and the biodegradability test.

Supplementary Materials: The following are available online at https://www.mdpi.com/article/10.3390/foods11010002/s1, Figure S1: The FTIR spectra of pure and modified polymers in the range 4000 to 3000 cm^{-1}; PLA: poly(lactic) acid, PBAT: poly(butylene adipate-co-terephthalate), PBS: poly(butylene succinate), E: 3% w/v eugenol, T: 3% w/v thymol., Table S1: Comparison of methods for evaluating biofilm formation.

Author Contributions: Conceptualization, P.P. and M.J.; methodology D.M.; software P.P., validation M.J.; formal analysis, J.S.; investigation, L.B. and D.M.; resources, D.M. and L.Z.; data curation P.P. and J.S.; writing—original draft preparation D.M., L.B. and L.Z.; writing—review and editing, P.P., J.S. and M.J.; visualization, L.B.; supervision, P.P.; project administration, P.P. and M.J. All authors have read and agreed to the published version of the manuscript.

Funding: The authors acknowledge the support given by the Internal Grant Agency of Tomas Bata University in Zlin (project no. IGA/FT/2021/009).

Data Availability Statement: Not applicable.

Conflicts of Interest: The authors declare no conflict of interest.

References

1. Donlan, R.M.; Costerton, J.W. Biofilms: Survival Mechanisms of Clinically Relevant Microorganisms. *Clin. Microbiol. Rev.* **2002**, *15*, 167–193. [CrossRef] [PubMed]
2. Marshall, K.C. (Ed.) *Microbial Adhesion and Aggregation*, 1st ed.; Springer: Berlin, Heidelberg, 1984; ISBN 9783642701399.
3. Li, Y.-H.; Tian, X. Quorum Sensing and Bacterial Social Interactions in Biofilms. *Sensors* **2012**, *12*, 2519–2538. [CrossRef] [PubMed]
4. Cerca, N.; Pier, G.B.; Vilanova, M.; Oliveira, R.; Azeredo, J. Quantitative analysis of adhesion and biofilm formation on hydrophilic and hydrophobic surfaces of clinical isolates of Staphylococcus epidermidis. *Res. Microbiol.* **2005**, *156*, 506–514. [CrossRef]
5. Azeredo, J.; Azevedo, N.F.; Briandet, R.; Cerca, N.; Coenye, T.; Costa, A.R.; Desvaux, M.; Di Bonaventura, G.; Hébraud, M.; Jaglic, Z.; et al. Critical review on biofilm methods. *Crit. Rev. Microbiol.* **2017**, *43*, 313–351. [CrossRef]
6. Shi, X.; Zhu, X. Biofilm formation and food safety in food industries. *Trends Food Sci. Technol.* **2009**, *20*, 407–413. [CrossRef]
7. Bridier, A.; Sanchez-Vizuete, P.; Guilbaud, M.; Piard, J.-C.; Naïtali, M.; Briandet, R. Biofilm-associated persistence of food-borne pathogens. *Food Microbiol.* **2015**, *45*, 167–178. [CrossRef]
8. Wyrwa, J.; Barska, A. Innovations in the food packaging market: Active packaging. *Eur. Food Res. Technol.* **2017**, *243*, 1681–1692. [CrossRef]

9. De Abreu, D.A.P.; Cruz, J.M.; Losada, P.P. Active and Intelligent Packaging for the Food Industry. *Food Rev. Int.* **2012**, *28*, 146–187. [CrossRef]

10. Yildirim, S.; Röcker, B.; Pettersen, M.K.; Nilsen-Nygaard, J.; Ayhan, Z.; Rutkaite, R.; Radusin, T.; Suminska, P.; Marcos, B.; Coma, V. Active packaging applications for food. *Compr. Rev. Food Sci. Food Saf.* **2018**, *17*, 165–199. [CrossRef] [PubMed]

11. EC. *Regulation (EC) No. 1935/2004 of the European Parliament and of the Council of 27 October 2004, on Materials and Articles Intended to Come into Contact with Food and Repealing Directives 80/509/EEC and 89/109/EEC*; European Commission: Brussels, Belgium, 2004; Volume 338, pp. 4–17.

12. EC. *Commission Regulation (EC) No. 450/2009 on Active and Intelligent Materials and Articles Intended to Come into Contact with Food*; European Commission: Brussels, Belgium, 2009; Volume 135, pp. 3–11.

13. Beneventi, E.; Tietz, T.; Merkel, S. Risk Assessment of Food Contact Materials. *EFSA J.* **2020**, *18*, e181109. [CrossRef]

14. Irankhah, S.; Ali, A.A.; Mallavarapu, M.; Soudi, M.R.; Subashchandrabose, S.R.; Gharavi, S.; Ayati, B. Ecological role of Acinetobacter calcoaceticus GSN3 in natural biofilm formation and its advantages in bioremediation. *Biofouling* **2019**, *35*, 377–391. [CrossRef] [PubMed]

15. Gu, J.-D. Microbiological deterioration and degradation of synthetic polymeric materials: Recent research advances. *Int. Biodeterior. Biodegrad.* **2003**, *52*, 69–91. [CrossRef]

16. Barron, A.; Sparks, T.D. Commercial Marine-Degradable Polymers for Flexible Packaging. *iScience* **2020**, *23*, 101353. [CrossRef]

17. Vroman, I.; Tighzert, L. Biodegradable Polymers. *Materials* **2009**, *2*, 307–344. [CrossRef]

18. Chandra, R.; Rustgi, R. Biodegradable polymers. *Prog. Polym. Sci.* **1998**, *23*, 1273–1335. [CrossRef]

19. Sharma, S.; Jaiswal, A.K.; Duffy, B.; Jaiswal, S. Food Contact Surfaces: Challenges, Legislation and Solutions. *Food Rev. Int.* **2021**, 1–24. [CrossRef]

20. Sin, L.T.; Rahmat, A.R.; Rahman, W.A.W.A. Overview of Poly(lactic Acid). In *Handbook of Biopolymers and Biodegradable Plastics*; Elsevier: Amsterdam, The Netherlands, 2013; pp. 11–54, ISBN 9781455728343. [CrossRef]

21. Muthuraj, R.; Misra, M.; Mohanty, A.K. Hydrolytic degradation of biodegradable polyesters under simulated environmental conditions. *J. Appl. Polym. Sci.* **2015**, *132*. [CrossRef]

22. Brocca, D.; Arvin, E.; Mosbæk, H. Identification of organic compounds migrating from polyethylene pipelines into drinking water. *Water Res.* **2002**, *36*, 3675–3680. [CrossRef]

23. Zhang, H.; Dudley, E.G.; Davidson, P.M.; Harte, F. Critical Concentration of Lecithin Enhances the Antimicrobial Activity of Eugenol against Escherichia coli. *Appl. Environ. Microbiol.* **2017**, *83*, e03467-16. [CrossRef]

24. Hu, Y.; Du, Y.; Wang, X.; Feng, T. Self-aggregation of water-soluble chitosan and solubilization of thymol as an antimicrobial agent. *J. Biomed. Mater. Res. Part A* **2008**, *90A*, 874–881. [CrossRef] [PubMed]

25. Khayyat, S.A.; Roselin, L.S. Recent progress in photochemical reaction on main components of some essential oils. *J. Saudi Chem. Soc.* **2018**, *22*, 855–875. [CrossRef]

26. Khatsee, S.; Daranarong, D.; Punyodom, W.; Worajittiphon, P. Electrospinning polymer blend of PLA and PBAT: Electrospinnability-solubility map and effect of polymer solution parameters toward application as antibiotic-carrier mats. *J. Appl. Polym. Sci.* **2018**, *135*, 46486. [CrossRef]

27. Narayanan, A.; Neera; Mallesha; Ramana, K.V. Synergized Antimicrobial Activity of Eugenol Incorporated Polyhydroxybutyrate Films Against Food Spoilage Microorganisms in Conjunction with Pediocin. *Appl. Biochem. Biotechnol.* **2013**, *170*, 1379–1388. [CrossRef]

28. CLSI. *Performance Standards for Antimicrobial Disk Susceptibility Tests*, 13th ed.; CLSI Standard M02; Clinical and Laboratory Standards Institute: Wayne, PA, USA, 2018.

29. Zanoaga, M.; Tanasa, F. Photochemical Behavior of Synthetic Polymeric Multicomponent Materials Composites and Nanocomposites. In *Photochemical Behavior of Multicomponent Polymeric-Based Materials; Advanced Structured Materials*; Springer International Publishing: Cham, Switzerland, 2016; pp. 109–164, ISBN 978-3-319-25194-3.

30. Stepanović, S.; Vuković, D.; Dakić, I.; Savić, B.; Švabić-Vlahović, M. A modified microtiter-plate test for quantification of staphylococcal biofilm formation. *J. Microbiol. Methods* **2000**, *40*, 175–179. [CrossRef]

31. Molecular Probes. *LIVE/DEAD® BacLight Bacterial Viability Kits*; Invitrogen: Waltham, MA, USA, 2004; pp. 1–8.

32. Ncube, L.K.; Ude, A.U.; Ogunmuyiwa, E.N.; Zulkifli, R.; Beas, I.N. Environmental Impact of Food Packaging Materials: A Review of Contemporary Development from Conventional Plastics to Polylactic Acid Based Materials. *Materials* **2020**, *13*, 4994. [CrossRef] [PubMed]

33. Galiè, S.; García-Gutiérrez, C.; Miguélez, E.M.; Villar, C.J.; Lombó, F. Biofilms in the Food Industry: Health Aspects and Control Methods. *Front. Microbiol.* **2018**, *9*, 898. [CrossRef] [PubMed]

34. Christensen, G.D.; Simpson, W.A.; Younger, J.J.; Baddour, L.M.; Barrett, F.F.; Melton, D.M.; Beachey, E.H. Adherence of coagulase-negative staphylococci to plastic tissue culture plates: A quantitative model for the adherence of staphylococci to medical devices. *J. Clin. Microbiol.* **1985**, *22*, 996–1006. [CrossRef] [PubMed]

35. Riss, T.L. Cell Viability Assays. In *Assay Guidance Manual*; Moravec, R.A., Niles, A.L., Duellman, S., Benink, H.A., Worzella, T.J., Minor, L., Eds.; Eli Lilly & Company and the National Center for Advancing Translational Sciences: Bethesda, MD, USA, 2004; pp. 357–388.

36. Williams, S.C.; Hong, Y.; Danavall, D.C.A.; Howard-Jones, M.H.; Gibson, D.; Frischer, M.; Verity, P.G. Distinguishing between living and nonliving bacteria: Evaluation of the vital stain propidium iodide and its combined use with molecular probes in aquatic samples. *J. Microbiol. Methods* **1998**, *32*, 225–236. [CrossRef]
37. Williams, G.W.; Schork, M.A.; Brunden, M.N. Basic Statistics for Quality Control in the Clinical Laboratory. *CRC Crit. Rev. Clin. Lab. Sci.* **1982**, *17*, 171–199. [CrossRef]
38. Limoli, D.H.; Jones, C.J.; Wozniak, D.J. Bacterial Extracellular Polysaccharides in Biofilm Formation and Function. *Microbiol. Spectr.* **2015**, *3*, 1–19. [CrossRef]
39. Vacheethasanee, K.; Marchant, R.E. Surfactant polymers designed to suppress bacterial (Staphylococcus epidermidis) adhesion on biomaterials. *J. Biomed. Mater. Res.* **2000**, *50*, 302–312. [CrossRef]
40. Chen, S.; Peng, X.; Geng, L.; Wang, H.; Lin, J.; Chen, B.; Huang, A. The effect of polytetrafluoroethylene particle size on the properties of biodegradable poly(butylene succinate)-based composites. *Sci. Rep.* **2021**, *11*, 6802. [CrossRef] [PubMed]
41. Bhole, Y.S.; Karadkar, P.B.; Kharul, U.K. Effect of substituent polarity, bulk, and substitution site toward enhancing gas permeation in dibromohexafluorobisphenol-a based polyarylates. *J. Polym. Sci. Part B Polym. Phys.* **2007**, *45*, 3156–3168. [CrossRef]
42. Moustafa, H.; Guizani, C.; Dupont, C.; Martin, V.; Jeguirim, M.; Dufresne, A. Utilization of Torrefied Coffee Grounds as Reinforcing Agent To Produce High-Quality Biodegradable PBAT Composites for Food Packaging Applications. *ACS Sustain. Chem. Eng.* **2016**, *5*, 1906–1916. [CrossRef]
43. De Matos Costa, A.R.; Crocitti, A.; de Carvalho, L.H.; Carroccio, S.C.; Cerruti, P.; Santagata, G. Properties of Biodegradable Films Based on Poly(butylene Succinate) (PBS) and Poly(butylene Adipate-*co*-Terephthalate) (PBAT) Blends. *Polymers* **2020**, *12*, 2317. [CrossRef]
44. Jasim, S.M.; Ali, N.A. Properties characterization of plasticized polylactic acid /Biochar (bio carbon) nano-composites for antistatic packaging. *Iraqi J. Phys.* **2019**, *17*, 13–26. [CrossRef]
45. Chieng, B.W.; Ibrahim, N.A.B.; Yunus, W.M.Z.W.; Hussein, M.Z. Poly(lactic acid)/Poly(ethylene glycol) Polymer Nanocomposites: Effects of Graphene Nanoplatelets. *Polymers* **2013**, *6*, 93–104. [CrossRef]
46. Kumari, S.; Kumaraswamy, R.V.; Choudhary, R.C.; Sharma, S.S.; Pal, A.; Raliya, R.; Biswas, P.; Saharan, V. Thymol nanoemulsion exhibits potential antibacterial activity against bacterial pustule disease and growth promotory effect on soybean. *Sci. Rep.* **2018**, *8*, 6650. [CrossRef]
47. Zamani, Z.; Alipour, D.; Moghimi, H.R.; Mortazavi, S.A.R.; Saffary, M. Development and Evaluation of Thymol Microparticles Using Cellulose Derivatives as Controlled Release Dosage Form. *Iran. J. Pharm. Res.* **2015**, *14*, 1031–1040. [CrossRef] [PubMed]
48. Pramod, K.; Suneesh, C.V.; Shanavas, S.; Ansari, S.H.; Ali, J. Unveiling the compatibility of eugenol with formulation excipients by systematic drug-excipient compatibility studies. *J. Anal. Sci. Technol.* **2015**, *6*, 34. [CrossRef]
49. Achinas, S.; Yska, S.K.; Charalampogiannis, N.; Krooneman, J.; Euverink, G.J.W. A Technological Understanding of Biofilm Detection Techniques: A Review. *Materials* **2020**, *13*, 3147. [CrossRef]
50. Ojima, Y.; Nunogami, S.; Taya, M. Antibiofilm effect of warfarin on biofilm formation of Escherichia coli promoted by antimicrobial treatment. *J. Glob. Antimicrob. Resist.* **2016**, *7*, 102–105. [CrossRef] [PubMed]
51. Nguyen, T.; Roddick, F.A.; Fan, L. Biofouling of Water Treatment Membranes: A Review of the Underlying Causes, Monitoring Techniques and Control Measures. *Membranes* **2012**, *2*, 804–840. [CrossRef]
52. Marcos-Zambrano, L.J.; Escribano, P.; Bouza, E.; Guinea, J. Production of biofilm by *Candida* and non-*Candida* spp. isolates causing fungemia: Comparison of biomass production and metabolic activity and development of cut-off points. *Int. J. Med Microbiol.* **2014**, *304*, 1192–1198. [CrossRef]
53. Rajamani, S.; Sandy, R.; Kota, K.; Lundh, L.; Gomba, G.; Recabo, K.; Duplantier, A.; Panchal, R.G. Robust biofilm assay for quantification and high throughput screening applications. *J. Microbiol. Methods* **2019**, *159*, 179–185. [CrossRef]
54. Hassan, A.; Usman, J.; Kaleem, F.; Omair, M.; Khalid, A.; Iqbal, M. Evaluation of different detection methods of biofilm formation in the clinical isolates. *Braz. J. Infect. Dis.* **2011**, *15*, 305–311. [CrossRef]
55. Morohoshi, T.; Oi, T.; Aiso, H.; Suzuki, T.; Okura, T.; Sato, S. Biofilm Formation and Degradation of Commercially Available Biodegradable Plastic Films by Bacterial Consortiums in Freshwater Environments. *Microbes Environ.* **2018**, *33*, 332–335. [CrossRef]
56. Gzyra-Jagieła, K.; Sulak, K.; Draczyński, Z.; Podzimek, S.; Gałecki, S.; Jagodzińska, S.; Borkowski, D. Modification of Poly(lactic acid) by the Plasticization for Application in the Packaging Industry. *Polymers* **2021**, *13*, 3651. [CrossRef] [PubMed]
57. Al-Ghani, M.M.A.; Azzam, R.A.; Madkour, T.M. Design and Development of Enhanced Antimicrobial Breathable Biodegradable Polymeric Films for Food Packaging Applications. *Polymers* **2021**, *13*, 3527. [CrossRef] [PubMed]
58. Hamzah, H.; Pratiwi, S.U.T.; Hertiani, T. Efficacy of Thymol and Eugenol Against Polymicrobial Biofilm. *Indones. J. Pharm.* **2018**, *29*, 214. [CrossRef]
59. Marchese, A.; Barbieri, R.; Coppo, E.; Orhan, I.E.; Daglia, M.; Nabavi, S.M.; Izadi, M.; Abdollahi, M.; Ajami, M. Antimicrobial activity of eugenol and essential oils containing eugenol: A mechanistic viewpoint. *Crit. Rev. Microbiol.* **2017**, *43*, 668–689. [CrossRef] [PubMed]
60. Marchese, A.; Orhan, I.E.; Daglia, M.; Barbieri, R.; Di Lorenzo, A.; Nabavi, S.F.; Gortzi, O.; Izadi, M. Antibacterial and antifungal activities of thymol: A brief review of the literature. *Food Chem.* **2016**, *210*, 402–414. [CrossRef] [PubMed]
61. Memar, M.Y.; Raei, P.; Alizadeh, N.; Aghdam, M.A.; Kafil, H.S. Carvacrol and thymol: Strong antimicrobial agents against resistant isolates. *Rev. Med Microbiol.* **2017**, *28*, 63–68. [CrossRef]

62. Zhang, Y.; Zhang, Y.; Zhu, Z.; Jiao, X.; Shang, Y.; Wen, Y. Encapsulation of Thymol in Biodegradable Nanofiber via Coaxial Eletrospinning and Applications in Fruit Preservation. *J. Agric. Food Chem.* **2019**, *67*, 1736–1741. [CrossRef] [PubMed]
63. Hertiani, T.; Utami, D.T.; Pratiwi, S.U.T.; Haniastuti, T. Eugenol and thymol as potential inhibitors for polymicrobial oral biofilms: An in vitro study. *J. Int. Oral Heal.* **2021**, *13*, 45. [CrossRef]
64. de Morais, S.M.; Vila-Nova, N.S.; Bevilaqua, C.; Rondon, F.C.; Lobo, C.H.; de Alencar Araripe Noronha Moura, A.; Sales, A.D.; Rodrigues, A.P.R.; de Figuereido, J.R.; Campello, C.C.; et al. Thymol and eugenol derivatives as potential antileishmanial agents. *Bioorganic Med. Chem.* **2014**, *22*, 6250–6255. [CrossRef] [PubMed]
65. Veldhuizen, E.J.A.; Tjeerdsma-van Bokhoven, J.L.M.; Zweijtzer, C.; Burt, S.A.; Haagsman, H.P. Structural Requirements for the Antimicrobial Activity of Carvacrol. *J. Agric. Food Chem.* **2006**, *54*, 1874–1879. [CrossRef]
66. Adil, M.; Singh, K.; Verma, P.K.; Khan, A.U. Eugenol-induced suppression of biofilm-forming genes in Streptococcus mutans: An approach to inhibit biofilms. *J. Glob. Antimicrob. Resist.* **2014**, *2*, 286–292. [CrossRef]
67. Liu, D.; Li, H.; Jiang, L.; Chuan, Y.; Yuan, M.; Chen, H. Characterization of Active Packaging Films Made from Poly(Lactic Acid)/Poly(Trimethylene Carbonate) Incorporated with Oregano Essential Oil. *Molecules* **2016**, *21*, 695. [CrossRef]
68. Ramos, M.; Fortunati, E.; Beltrán, A.; Peltzer, M.; Cristofaro, F.; Visai, L.; Valente, A.J.M.; Jiménez, A.; Kenny, J.M.; Garrigós, M.C. Controlled Release, Disintegration, Antioxidant, and Antimicrobial Properties of Poly (Lactic Acid)/Thymol/Nanoclay Composites. *Polymers* **2020**, *12*, 1878. [CrossRef]
69. Gazzotti, S.; Todisco, S.A.; Picozzi, C.; Ortenzi, M.A.; Farina, H.; Lesma, G.; Silvani, A. Eugenol-grafted aliphatic polyesters: Towards inherently antimicrobial PLA-based materials exploiting OCAs chemistry. *Eur. Polym. J.* **2019**, *114*, 369–379. [CrossRef]
70. Chen, F.; Shi, Z.; Neoh, K.G.; Kang, E.T. Antioxidant and antibacterial activities of eugenol and carvacrol-grafted chitosan nanoparticles. *Biotechnol. Bioeng.* **2009**, *104*, 30–39. [CrossRef] [PubMed]
71. Husárová, L.; Pekařová, S.; Stloukal, P.; Kucharzcyk, P.; Verney, V.; Commereuc, S.; Ramone, A.; Koutny, M. Identification of important abiotic and biotic factors in the biodegradation of poly(l-lactic acid). *Int. J. Biol. Macromol.* **2014**, *71*, 155–162. [CrossRef] [PubMed]
72. Šerá, J.; Stloukal, P.; Jančová, P.; Verney, V.; Pekařová, S.; Koutný, M. Accelerated Biodegradation of Agriculture Film Based on Aromatic–Aliphatic Copolyester in Soil under Mesophilic Conditions. *J. Agric. Food Chem.* **2016**, *64*, 5653–5661. [CrossRef] [PubMed]
73. Šerá, J.; Serbruyns, L.; De Wilde, B.; Koutný, M. Accelerated biodegradation testing of slowly degradable polyesters in soil. *Polym. Degrad. Stab.* **2019**, *171*, 109031. [CrossRef]

*foods*

Article

Developing Edible Starch Film Used for Packaging Seasonings in Instant Noodles

Hui Chen [1], Mahafooj Alee [1], Ying Chen [1,2], Yinglin Zhou [1], Mao Yang [1], Amjad Ali [1,3], Hongsheng Liu [1,4], Ling Chen [1] and Long Yu [1,4,*]

[1] Collage of Food Science and Engineering, South China University of Technology, Guangzhou 510640, China; 201920125371@mail.scut.edu.cn (H.C.); 201912800038@mail.scut.edu.cn (M.A.); 201620119572@mail.scut.edu.cn (Y.C.); 202020126236@mail.scut.edu.cn (Y.Z.); 202021027102@mail.scut.edu.cn (M.Y.); dr.amjadali@kiu.edu.pk (A.A.); hs.liu@mail.scut.edu.cn (H.L.); felchen@scut.edu.cn (L.C.)
[2] Department of Food Science and Technology, National University of Singapore, Science Drive 2, Singapore 117542, Singapore
[3] Department of Agriculture and Food Technology, Karakorum International University, Gilgit 15100, Pakistan
[4] China-Singapore International Joint Research Institute, Knowledge City, Guangzhou 510663, China
* Correspondence: felyu@scut.edu.cn; Tel.: +86-21-87111971

Abstract: Edible starch-based film was developed for packaging seasoning applied in instant noodles. The edible film can quickly dissolve into hot water so that the seasoning bag can mix in the soup of instant noodles during preparation. To meet the specific requirements of the packaging, such as reasonable high tensile properties, ductility under arid conditions, and low gas permeability, hydroxypropyl cornstarch with various edible additives from food-grade ingredients were applied to enhance the functionality of starch film. In this work, xylose was used as a plasticizer, cellulose crystals were used as a reinforcing agent, and laver was used to decrease gas permeability. The microstructures, interface, and compatibility of various components and film performance were investigated using an optical microscope under polarized light, scanning electron microscope, gas permeability, and tensile testing. The relationship was established between processing methodologies, microstructures, and performances. The results showed that the developed starch-based film have a modulus of 960 MPa, tensile strength of 36 Mpa with 14% elongation, and water vapor permeability less than 5.8 g/m^2.h under 20% RH condition at room temperature (25 °C), which meets the general requirements of the flavor bag packaging used in instant noodles.

Keywords: starch film; packaging; edible; reinforcement; interface; instant noodle

Citation: Chen, H.; Alee, M.; Chen, Y.; Zhou, Y.; Yang, M.; Ali, A.; Liu, H.; Chen, L.; Yu, L. Developing Edible Starch Film Used for Packaging Seasonings in Instant Noodles. *Foods* **2021**, *10*, 3105. https://doi.org/10.3390/foods10123105

Academic Editors: Theodoros Varzakas and Rui M. S. Cruz

Received: 10 November 2021
Accepted: 8 December 2021
Published: 14 December 2021

Publisher's Note: MDPI stays neutral with regard to jurisdictional claims in published maps and institutional affiliations.

1. Introduction

Developing eco-friendly packaging materials is no longer an option, but it has become an urgent necessity since many countries have restricted non-biodegradable materials, particularly for disposable packaging materials. The simplest way to treat the food packaging is to eat them with the foods packaged in. Edible packaging has attracted increasing attention [1–4], and is mainly based on polysaccharide and protein materials. Starch is the most promising material because of its natural edibility, overwhelming abundance, and annual renewability. Edible starch-based films have been developed and widely used in food and medicine packaging [5–8], such as applications in candy wrappers and medicine capsules [9–12].

The bag used for packaging seasonings has been widely used in instant noodles, easy soup, and various pre-prepared ingredients for cooking. It was reported that, on average, everyone ate about 13.6 bags of instant noodles, and the market was more than one myriad bag in 2018 in the world [13]. Traditionally, the films used for flavor bags are bi-orientation polypropylene (BOPP) or polyamide (BOPA) with reasonably excellent mechanical performance and gas barrier properties. Ideally, the packaging film used for the flavor bag would be edible. It could easily dissolve into the hot water during the noodle preparation so that all the treatments before (carefully tearing and pressuring the bag) and after (waste) eating can be omitted, which avoids any environmental issues. An edible starch-based film should be an ideal candidate.

Food and non-food packaging have benefited from starch films' development [5,10–12,14,15]. Improvement of the mechanical properties of starch-based materials is an ongoing challenge due to its poor mechanical performance, particularly tensile strength [16–19]. Several improvements to the mechanical properties of starch-based materials is an ongoing challenge due to its poor mechanical performance, and tensile strength compositing and blending strategies have been invented to increase these mechanical qualities, such as reinforcing with mineral and natural fillers, or mixing with various decomposable polyesters [20–29]. However, any additive in edible packaging films is sensitive because of a safety issue or a hazardous risk. All the additives, including plasticizers and reinforcing agents, must be food-grade ingredients.

The aim of this work is to develop edible starch-based film used for packaging seasoning in instant noodles through considering and novelly combining multiple factors, including plasticing, reinforcing, and barriering. The film used for packaging seasonings would keep with the noodles constantly under low humidity conditions (20–30% RH) in a plastic bag or container to maintain the shelf-life of the noodles. The starch-based film needs to be plasticized and reinforced to meet these requirements. Another challenge is to boost the gas barrier function of the film to maintain the flavors. Because it possesses appropriate mechanical and processing capabilities, this investigation used a commercially available food-grade modified (hydroxypropyl) cornstarch (HPCS) as a matrix. It has previously been used to make pharmaceutical capsules and edible films for food packaging [5,11,12]. In order to further improve the performance of the film, multiple additives, including plasticizers, reinforcing agents, and barrier fillers were used, where xylose was used as the predominant plasticizer, cellulose crystals were used as a reinforcing agent, and laver was used to decrease gas permeability. Laver is a very popular seaweed food in Asian countries, such as in Japan, China, Korea, etc. Its popular name in China is Zicai (purple vegetable). Since it has the natural structure of a thin film, the laver can be used as a gas barrier efficiently [5]. The effect of these added ingredients on the starch film functionality was analyzed by an optical microscope under polarized light, electronic scanning microscope, gas permeability, and tensile testing. The multiple relationships among the starch and different additives were studied and used to guide the film development.

2. Materials and Methods

2.1. Materials

The ingredients utilized in the current study are all readily accessible in the marketplace. Hengrui Starch Company, Luohe, China, provided commercially accessible hydroxypropyl cornstarch (HPCS) (DS 0.4%, moisture content 13 wt%, 23% amylose content). It has good mechanical characteristics [5,11,12]. Food-grade cellulose crystals were procured from Qianrun Bioengineer (Wuhan, China), with a particle size of about 5 mm. Figure 1 shows images of the cellulose crystals under SEM and optical microscope polarized lights. Laver was acquired at a nearby market (Guangzhou, China). The laver, with about 6 mm thickness, was firstly crushed into around 0.5 mm diameter fine particles to distribute it homogeneously in the starch suspension. The laver has a protein content of around 31.3% d.w. Figure 2 depicts images of the laver taken with normal and polarized lights under an optical microscope. Tianjin Kemeou Chemical Reagent Company (Nanjing, China) provided xylose (99.8% pure). Sinopharm Chemical Reagent Company Limited (Shanghai, China) provided 99.5% pure glycerol.

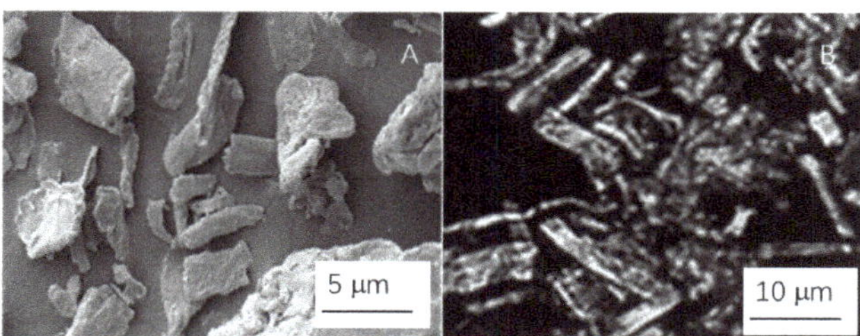

Figure 1. Cellulose crystals under SEM (**A**) and optical microscope polarized lights (**B**).

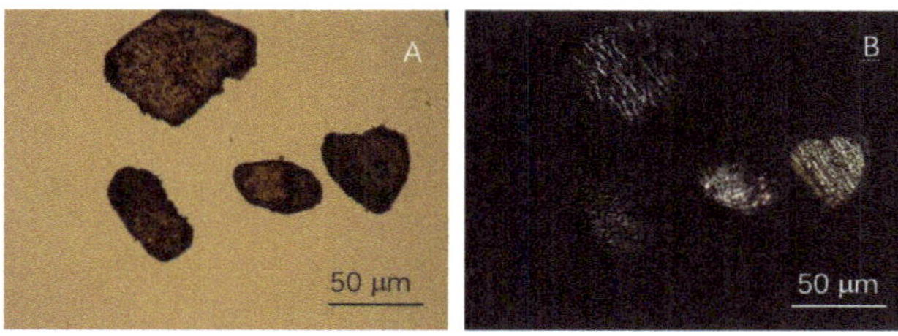

Figure 2. Images of the laver taken with natural (**A**) and polarized (**B**) lights under an optical microscope.

2.2. Sample Preparation

The final and optimized starch film contains plasticizer (water, glycerol, and xylose), reinforced cellulose crystals, and laver as the barrier filler. The effect of each additive on the starch-based film was initially studied separately, then mixed according to optimized formulations. The baseline was established as the starch-based film plasticized with water and glycerol. The specimen codes and compositions are included in Table 1. Film thickness and moisture content were measured after the films were kept under 20% RH, and 25 °C temperature.

Table 1. Sample code, formulations, as well as film thickness and moisture content.

Sample Code	Cellulose Cryst (*w/w*)	Laver (*w/w*)	Xylose (*w/w*)	Thickness (mm)	Moisture Content (%) [2]
Starch Film [1]	-	-	-	146 ± 4	7.01
C-1	2	-	-	160 ± 8	8.62
C-2	5	-	-	171 ± 9	8.53
C-3	**7**	-	-	**173 ± 8**	**8.73**
C-4	10	-	-	181 ± 9	8.87
L-1	-	10	-	167 ± 7	9.37
L-2	-	**20**	-	**171 ± 6**	**9.77**
L-3	-	30	-	177 ± 8	9.91
X-1	-	-	5	152 ± 8	9.67
X-2	-	-	10	168 ± 6	10.17
X-3	-	-	**15**	**167 ± 7**	**10.41**
X-4	-	-	20	165 ± 6	10.85
CX film	**7**		**15**	**171 ± 7**	**9.71**
CLX film	**7**	**20**	**15**	**178 ± 9**	**9.55**

Notice: [1] All of the mentioned additives are made up of 0.5% glycerol (0.5 g), 10% dried starch (10 g), and 90% water (90 g). The ratio to starch (*w/w*). [2] The films were stored at a temperature of 25 °C, and a humidity of 20%. Sample with C- means containing cellulose, L- means containing laver, X- means containing xylose. Bold means these formulations were used as baseline for CX and CLX.

The starch solutions were made in a beaker using the solution-casting methodology (10% *w/w*), in which 10 g (10% *w/w*) of starch on a dry basis was dissolved in 90 g water (90% *w/w*). Glycerol was added at a constant weight of 0.5 g (5%) (*w/w*) in proportion to the 10 g of starch on a dry basis. A referee starch film was made by pre-mixing the solution, then heating it 25 °C to 99 °C, where that was kept for 1 h while continually stirring it with a magnet. The gelatinized starch suspension was rapidly stirred for 45 min before being put onto a polystyrene plate (diameter 10 cm). The film was then dried in an oven for 10–12 h at around 35 degrees Celsius to get a consistent weight. The film containing xylose was prepared by adding concentrations of 5%, 10%, 15%, and 20% (0.5, 1.0, 1.5, and 2.0 g, respectively) into the hot gelatinized suspension of the starch prepared reference sample. The cellulose crystal films were prepared by adding concentrations of 2%, 5%, 7%, and 10% (0.2, 0.5, 0.7, and 1.0 g, respectively) into the hot gelatinized suspension of the starch prepared, the same as the reference sample. The films containing laver were prepared by adding concentrations of 10%, 20%, and 30% (1.0, 2.0, and 3.0 g, respectively) into the hot gelatinized suspension of the starch prepared, the same as the reference sample. Cast films had a final thickness of around 150 μm, and this was measured by estimating how much starch suspension was placed on the plate. A micrometer (μm) was used to determine the actual thickness of the dried film.

2.3. Tensile Characteristics

The tensile properties were tested using the standards of ASTM D882-12. The prepared starch films were cut into uniformly-sized tensile bar-shaped specimens. The tensile testing apparatus (Instron 5565) was used at 25 °C with the cross-head at 5 mm/min^{-1} stretching speed. All specimens were conditioned for 48 h before testing in a temperature-humidity box (Lab Companion, Qingdao, China) with a humidity of 20% RH. The data are based on the mean value of the seven samples.

2.4. Moisture Effect and Permeability

The moisture effect on the film was tested using contact angle changes with water droplets. The sessile drop methodology was used to analyze water's liquid/solid/contact angles on sample surfaces using a contact angle goniometer: model OCA 20 from Dataphysics (Dataphysics instrument Co., Filderstadt, Germany). Based on the results of three separate drops, average contact angles were calculated.

As per the ASTM E96/E96M-14 standardized process [30], the water vapor transfer rate (WVTR) was measured in triplicates using a thermos hygrometer and deionized water. Distinct glass cups with a depth of 2.5 cm and a diameter of 4 cm were used to measure the film's WVTR. A scissor cut the films into a circular form with a length marginally greater than the cup diameter. Each cup was covered with film samples after being filled with anhydrous $CaCl_2$. Every cup was placed in a desiccator with a small beaker containing saturated NaCl solution at the surface. To ensure that the saturated solution remained saturated at all times, a small quantity of solid NaCl was left at the bottom. At room temperature, the saturated NaCl solution in the desiccators maintained a constant RH of 75%. Increment in the weight of the cup was used to determine the water vapor transport. To calculate the water vapor transmission rate (WVTR), the slope (g/h) was divided by the transfer area (m^2).

2.5. Characterization of Microstructure and Morphology

2.5.1. Scanning Electron Microscope (SEM)

To examine the film's surface and the contact between the laver and starch matrix, a scanning electron microscope (SEM, Thermo Fisher Scientific Inc., (NYSE: TMO), Waltham, MA, USA) was utilized. All specimens were gold-coated under vacuum using an Eiko Sputter Coater, and mounted on metal stubs that had been previously coated with double-sided glue. In this experiment, a low voltage of 3 kV was used to minimize the risk of damaging the surface.

2.5.2. Optical Microscope (OM)

The morphology of scattered lavers in the matrix of starch was observed through an optical microscope (Axioskop 40 pol/40 A Pol, ZEISS) linked with a 35 mm SLA camera. XTZ

Optical Instrument Factory's stage micrometer Type C1 slide (1 div = 0.01 mm) was used to calibrate the microscope. Each sample was examined using a microscope at a magnification of 500×, where the photographs were 1.3 times larger than their original size. The crystalline structures of cellulose contained in the laver were studied using polarized light.

2.6. Statistical Analysis

All quantitative date were statistically analyzed and presented as means ± standard deviation (SD). The data were analyzed by one way analysis of variance (ANOVA) using SPSS 17.0 package (IBM, Armonk, NY, USA), and when $p < 0.05$, samples were considered to have significant differences according to Duncan's multiple range tests.

3. Results

3.1. Effect of Various Additives on Performance of Starch-Based Film

The instant noodles were manufactured in less than 50 RH% conditions, and the water content contained in the noodles itself was less than 5%. The relative humidity in the plastic bag or container is about 20–30% RH. Thus, all the performances of the film were evaluated under 20% RH to meet the conditions. Table 2 lists the effect of various additives on starch-based film, including tensile properties, WVTR, and contact angle. The effect of each additive was initially investigated individually, and then optimized based on the results.

Table 2. Effect of various ingredients on the starch film mechanical performances, WVTR, and contact angle.

Sample Code	Modulus (MPa)	Tensile Str (MPa)	Elongation (%)	WVTR (g/(m^2 h))	Contact Angle (θ)
Starch Film	1352 ± 119 [a]	45.3 ± 2.9 [a]	3.9 ± 0.7 [b]	16.2 ± 3.4 [a]	82.0 ± 2.9 [c]
C-1	1387 ± 102 [b]	45.5 ± 4.8 [b]	3.7 ± 0.7 [a]	16.0 ± 2.8 [ab]	83.6 ± 2.7 [b]
C-2	1429 ± 118 [ab]	47.2 ± 4.7 [ab]	3.4 ± 1.1 [ab]	15.6 ± 1.9 [ab]	85.5 ± 3.3 [ab]
C-3	1492 ± 72 [a]	48.2 ± 5.2 [ab]	3.3 ± 0.8 [ab]	15.3 ± 3.5 [b]	85.7 ± 3.6 [ab]
C-4	1533 ± 126 [a]	51.4 ± 5.6 [a]	2.6 ± 0.6 [b]	16.1 ± 2.7 [a]	87.9 ± 4.1 [a]
L-1	1382 ± 122 [b]	46.8 ± 3.9 [b]	3.4 ± 2.1 [a]	6.7 ± 0.5 [a]	91.4 ± 3.6 [a]
L-2	1410 ± 116 [ab]	47.7 ± 4.8 [ab]	3.1 ± 1.6 [ab]	5.7 ± 0.4 [b]	105.8 ± 4.2 [b]
L-3	1432 ± 104 [a]	51.1 ± 5.9 [a]	2.7 ± 0.2 [b]	5.3 ± 0.6 [b]	119.9 ± 3.8 [c]
X-1	1242 ± 79 [a]	42.6 ± 3.2 [a]	8.6 ± 2.2 [d]	13.2 ± 0.8 [b]	87.7 ± 3.9 [a]
X-2	967 ± 74 [b]	36.1 ± 3.1 [b]	11.3 ± 1.5 [c]	14.8 ± 1.3 [ab]	94.2 ± 3.4 [a]
X-3	818 ± 81 [c]	27.0 ± 2.3 [c]	19.9 ± 2.1 [b]	15.7 ± 1.1 [a]	102.4 ± 2.8 [b]
X-4	620 ± 66 [d]	21.3 ± 1.7 [d]	28.4 ± 2.3 [a]	15.5 ± 1.6 [a]	103.9 ± 2.7 [c]
CX film	925 ± 73 [b]	37 ± 3.6 [b]	16.1 ± 2.6 [a]	15.6 ± 0.5 [a]	91.8 ± 2.7 [b]
CLX film	965 ± 81 [b]	36 ± 2.3 [b]	14.2 ± 2.1 [a]	5.8 ± 0.4 [b]	97.9 ± 2.9 [a]

The films were kept at room temperature under 20% RH. The data were analyzed by one way analysis of variance (ANOVA) and marked as [abcd].

3.1.1. Effect of Cellulose Crystals

The inclusion of the cellulose crystals considerably enhanced the starch film's strength, as depicted in Table 2. After the crystals were added, both modulus and tensile strength rose. Because hard crystal particles worked as reinforcing agents, this was predicted. The reinforcing process of stiff particles in a matrix is widely understood, and crystals behave similarly. The modulus and tensile strength of the crystals steadily rose as the crystal content increased. On the other hand, the results demonstrate that adding crystals to the films has a minor effect on the elongation at the break. It has to be pointed out that both starch film and the film reinforced by the cellulose crystals are generally very brittle since the elongations are very low, with a value of only about 3–4%. Because of this brittle behavior, the standard dividends (±) of the elongation is higher (>20%).

The starch film WVTP was not affected by the cellulose crystals. It was not expected that a rigid particle could reduce the gas permeability if it has much low gas permeability, since the gap or weak interface between the particle and matrix significantly increases gas permeability. The results indicate that the interface between the starch and cellulose particle is excellent. CA of the starch film improved marginally after adding the cellulose crystals, indicating the moisture sensitivity of the film decreased.

3.1.2. Effect of Laver

The modulus elevated when adding more laver, and steadily increased when enhancing laver concentration, indicating that laver improves the films' stiffness as predicted. The tensile strength rose slightly as well, especially when the laver loading was more than 20%. One theory is that the stiff fibers strengthen the coatings in the laver cell walls. As with how most other fillers behaved, elongations of starch-based films reduced marginally with laver content, and steadily declined when increasing laver content. Similar to the reinforcement by the cellulose crystals, the elongation standard dividends ($\pm$) of the starch-based film comprising laver are also reasonably high.

CA increased as laver content increased, implying that the starch-based film's moisture sensitivity diminished. The findings explain why the laver is moisture resistant. The laver's moisture resistance is due to its semi-crystalline cellulose and protein [5].

After adding laver to starch-based films, the WVTP declined significantly, and the WVTP decreased even more when the laver content was raised. Creating a channel with some twists in the film matrix that help water vapors go through reduces WVTP when laver is induced [5]. Furthermore, creating the inter-link between protein and starch [30] may lower gas permeability because the tight interlinking between protein and starch has a lower free volume, and allows for less diffusion of moisture. Edible films produced by a combination of whey protein and starch have shown similar results [31,32].

3.1.3. Effect of Xylose

The mechanical performance of starch-based materials relies on plasticizer content (water or others). To achieve reasonable flexibility (toughness) under arid conditions (20% RH), different plasticizers were used in this work. The most common plasticizer system used for starch-comprised products is water with glycerol, since glycerol has assertive moisture absorption behavior and is used to maintain the water in the materials. However, the water is still not stable under low RH conditions. A constant concentration of water/glycerol 10/3 was used in this work, and additional plasticizer xylose from food ingredients was added. Xylose is a monosaccharide with a linear structure which has shown remarkably high efficiency in destroying the order structures, and enhancing the movement of polymer chains. As a result, plasticization efficiency improves. [31] Table 2 displays that xylose considerably enhanced the toughness of the starch film, with elongation increasing by up to eight times. With more xylose, the elongation rate increased. However, when the xylose level increased, the modulus and tensile strength of the material declined.

WVTR slightly decreased after the additional xylose. CA was slightly enhanced by xylose, indicating that the hydrophobicity of the starch film marginally increased. The hydroxyl groups in linear xylose are more flexible and face out, which makes sense [33].

3.2. Morphologies and Interface

3.2.1. Surface Morphologies

The linkage between cellulose crystals or laver and the starch matrix was investigated using SEM (see Figure 3). It is visible that the pure starch film has a relatively smooth surface (Figure 3A). The cellulose crystals and laver can be seen on the film's surface after the film's dried and shrinking matrix. The interface between the starch matrix and cellulose crystalline (B) or laver (C) was homogeneous, and there was no space between the particles and the matrix, which implies that polysaccharide crystals are fully compatible inside the starch matrix. The mechanism may be applied to describe why the films have better mechanical characteristics.

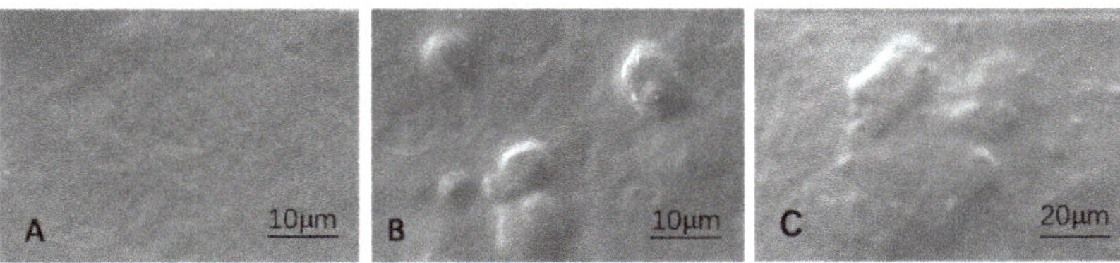

Figure 3. The surface images using SEM. (**A**) The pure starch film, (**B**) film containing cellulose crystals, (**C**) film with the laver.

3.2.2. Interface

The cross-section images of the films using SEM were used to investigate the linkage between the fillers and the starch matrix (see Figure 4). The pure starch film's cross-section picture has a smooth surface (a). Even after being sliced, the crystals and laver may still be detected in the starch matrix (b). At higher temperatures, the laver's cell structure has been disrupted and expanded. These findings were in accordance with the surface morphology investigation (Figure 4), and may be utilized to elaborate mechanical characteristics and moisture barriers. The thin laver also formed a twisted gas channel in the film matrix, which reduced gas permeability.

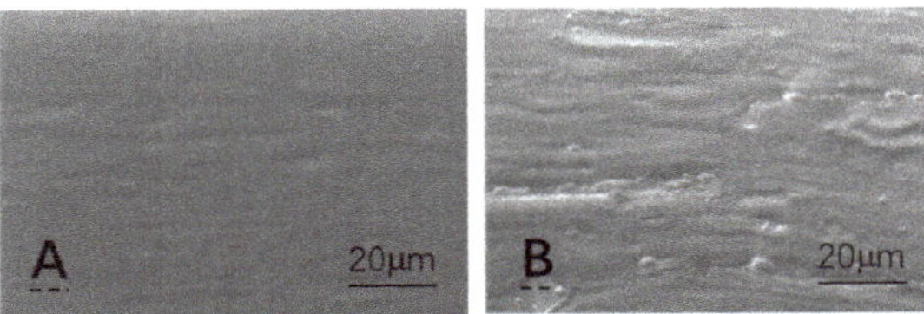

Figure 4. Image of the pure starch film in cross-section using SEM (**A**). The film containing cellulose crystals and laver (**B**) (sample CLX).

3.3. Developing Flavor Bag Packaging

The above results helped to develop coating, and extrude the starch-based films and various flavor bag packagings. Figure 5 depicts the images of extruded starch-comprised film and flavor bag packaging without laver (A,B), since some industries require high gas barriers and others do not. The films were developed based on the CX and CLX films presented in Table 1, respectively. All the starch-based flavor bags dissolve in hot water (85 °C) within a few seconds.

Figure 5. Photos of starch-based film and flavor bag packaging without (**A,B**) laver.

4. Discussion

It is well established that the functionality of starch-based products strongly depends on plasticizers and environmental conditions. One of the key challenges of applying starch-comprised film in flavor bag packaging is keeping the film with a good strength and toughness. Improving gas permeability is an additional advantage. In this work, xylose was used as a plasticizer, cellulose crystals were used as a reinforcing agent, and laver was used to decrease gas permeability. Tensile strength improved, according to the findings, after the addition of both cellulose crystals and laver. The laver also decreased WVTR significantly. But the flexibility or toughness of the starch-based film became worse. The films are generally brittle since they have very low elongation at the broken point with or without cellulose crystals and laver under low RH conditions. Additional xylose significantly improved the toughness, and the elongation increased up to 7–8 times. The mechanical properties could be balanced through designing and optimizing various additives. Moisture sensitivity also decreased after adding cellulose crystals and laver, which is crucial for food packaging.

The morphologies of the starch-based film have been investigated using SEM to explain the results. Both surface morphologies and cross-section images showed that starch matrix was compatible with cellulose crystals and laver. That was an excellent indication of a linkage between the strengthening components and the matrix. This was expected, as both starch and cellulose contained in crystals and laver have the same chemical unit, glucose. Actually, that is why they were selected as suitable materials for this work.

5. Conclusions

The edible starch-based film has been successfully developed for packaging seasonings applied in instant noodles. To meet the specific requirements of the packaging, such as reasonable high tensile properties, ductility under arid conditions, and low gas permeability, several edible additives from food-grade ingredients were applied to enhance the functionality of the starch film. Tensile strength increased after additional cellulose crystals and laver, whereas toughness significantly improved with additional xylose. The laver also decreased WVTR significantly. Based on the particular function of each additive, the balanced and optimized starch-based film for flavor bag packaging has been successfully developed.

Author Contributions: Conceptualization, L.Y.; methodology and investigation, Y.C., A.A., M.A., Y.Z., M.Y., H.C.; formal analysis, H.C.; resources, L.Y.; data curation, H.C.; writing—original draft preparation, L.Y., H.C.; writing—review and editing, H.C., M.A.; visualization, L.C.; supervision, L.Y., H.L.; project administration, H.L.; funding acquisition, L.Y. All authors have read and agreed to the published version of the manuscript.

Funding: This study received support from the Natural Science Foundation of China (22178124), the 111 projects (B17018), and the Guangzhou Science and Technology Plan Project (202102080150). South China University's National Undergraduate Training Program for Innovation and Entrepreneurship. The State Key Laboratory of Pulp and Paper Engineering also contributed to the project (202107).

Institutional Review Board Statement: Not applicable.

Informed Consent Statement: Not applicable.

Acknowledgments: Y. Chen wants to extend her gratitude to the China Scholarship Council for supporting her studies in Singapore. M. Alee wants to show his gratitude to the China Scholarship Council and the South China University of Technology for their assistance in providing him with opportunities and assistance while studying in China.

Conflicts of Interest: There are no conflicts of interest disclosed by the authors.

References

1. Patel, P. Edible packaging. *ACS Cent. Sci.* **2019**, *5*, 1907–1910. [CrossRef] [PubMed]
2. Zhao, Y.; Li, B.; Li, C.; Xu, Y.; Luo, Y.; Liang, D.; Huang, C. Comprehensive Review of Polysaccharide-Based Materials in Edible Packaging: A Sustainable Approach. *Foods* **2021**, *10*, 1845. [CrossRef]
3. Brody, A.L. Packaging, food. *Kirk-Othmer Encycl. Chem. Technol.* **2005**, *2*, 65–66. [CrossRef]
4. Debeufort, F.; Quezada-Gallo, J.; Voilley, A. Edible films, and coatings: Tomorrow's packaging: A review. *Crit Rev. Food Sci. Nutr.* **1998**, *38*, 299–313. [CrossRef] [PubMed]

5. Chen, Y.; Yu, L.; Ge, X.; Liu, H.; Ali, A.; Wang, Y.; Chen, L. Preparation and characterization of edible starch film reinforced by laver. *Int. J. Biol. Macromol.* **2019**, *129*, 944–951. [CrossRef] [PubMed]
6. Zhu, J.; Chen, H.; Lu, K.; Liu, H.-S.; Yu, L. Recent progress on starch-based biodegradable materials. *Acta Polym. Sin.* **2020**, *51*, 983–995. [CrossRef]
7. Ali, A.; Xie, F.; Yu, L.; Liu, H.; Meng, L.; Khalid, S.; Chen, L. Preparation and characterization of starch-based composite films reinforced by polysaccharide-based crystals. *Compos. Part B* **2018**, *133*, 122–128. [CrossRef]
8. Ali, A.; Yu, L.; Liu, H.; Khalid, S.; Meng, L.; Chen, L. Preparation and characterization of starch-based composite films reinforced by corn and wheat hulls. *J. Appl. Polym. Sci.* **2017**, *134*, 45159. [CrossRef]
9. Deore, U.V.; Mahajan, H.S. Isolation and characterization of natural polysaccharide from Cassia Obtustifolia seed mucilage as film-forming material for drug delivery. *Int. J. Biol. Macromol.* **2018**, *115*, 1071–1078. [CrossRef]
10. Devi, P.; Bajala, V.; Garg, V.; Mor, S.; Ravindra, K. Heavy metal content in various types of candies and their daily dietary intake by children. *Environ. Monit. Assess.* **2016**, *188*, 86. [CrossRef]
11. Zhang, L.; Wang, Y.; Liu, H.; Yu, L.; Liu, X.; Chen, L.; Zhang, N. Developing hydroxypropyl methylcellulose/hydroxypropyl starch blends for use as capsule materials. *Carbohydr. Polym.* **2013**, *98*, 73–79. [CrossRef]
12. Zhang, L.; Wang, Y.-F.; Liu, H.-S.; Zhang, N.-Z.; Liu, X.-X.; Chen, L.; Yu, L. Development of capsules from natural plant polymers. *Acta Polym. Sinica* **2013**, 1–10. [CrossRef]
13. China Commercial Information Web. Available online: www.askci.com (accessed on 20 July 2021).
14. Yusof, N.L.; Mutalib, N.-A.A.; Nazatul, U.; Nadrah, A.; Aziman, N.; Fouad, H.; Jawaid, M.; Ali, A.; Kian, L.K.; Sain, M. Efficacy of Biopolymer/Starch-Based Antimicrobial Packaging for Chicken Breast Fillets. *Foods* **2021**, *10*, 2379. [CrossRef] [PubMed]
15. Bangar, S.P.; Purewal, S.S.; Trif, M.; Maqsood, S.; Kumar, M.; Manjunatha, V.; Rusu, A.V. Functionality and applicability of starch-based films: An eco-friendly approach. *Foods* **2021**, *10*, 2181. [CrossRef]
16. Yu, L.; Dean, K.; Li, L. Polymer blends and composites from renewable resources. *Prog. Polym. Sci.* **2006**, *31*, 576–602. [CrossRef]
17. Nasseri, R.; Mohammadi, N. Starch-based nanocomposites: A comparative performance study of cellulose whiskers and starch nanoparticles. *Carbohy Polym.* **2014**, *106*, 432–439. [CrossRef] [PubMed]
18. Azizi Samir, M.A.S.; Alloin, F.; Dufresne, A. Review of recent research into cellulosic whiskers, their properties and their application in nanocomposite field. *Biomacromolecules* **2005**, *6*, 612–626. [CrossRef]
19. Bras, J.; Dufresne, B. Starch nanoparticles: A review. *Biomacromolecules* **2010**, *11*, 1139–1153. [CrossRef]
20. Gao, C.; Yu, L.; Liu, H.; Chen, L. Development of self-reinforced polymer composites. *Prog. Polym. Sci.* **2012**, *3*, 767–780. [CrossRef]
21. Bodirlau, R.; Teaca, C.-A.; Spiridon, I. Influence of natural fillers on the properties of starch-based biocomposite films. *Compos. Part B Eng.* **2013**, *44*, 575–583. [CrossRef]
22. Wan, Y.; Luo, H.; He, F.; Liang, H.; Huang, Y.; Li, X. Mechanical, moisture absorption, and biodegradation behaviours of bacterial cellulose fibre-reinforced starch biocomposites. *Compo. Sci. Technol.* **2009**, *69*, 1212–1219. [CrossRef]
23. Marvizadeh, M.M.; Oladzadabbasabadi, N.; Nafchi, A.M.; Jokar, M. Preparation, and characterization of bionanocomposite film based on tapioca starch/bovine gelatin/nanorod zinc oxide. *Int. J. Biol. Macromol.* **2017**, *99*, 1–7. [CrossRef] [PubMed]
24. Wilhelm, H.-M.; Sierakowski, M.-R.; Souza, G.; Wypych, F. Starch films reinforced with mineral clay. *Carbohydr. Polym.* **2003**, *52*, 101–110. [CrossRef]
25. Ochoa, T.A.; Almendárez, B.E.G.; Reyes, A.A.; Pastrana, D.M.R.; López, G.F.G.; Belloso, O.M.; Regalado-González, C. Design and characterization of corn starch edible films including beeswax and natural antimicrobials. *Food Bioproc. Technol.* **2016**, *10*, 103–114. [CrossRef]
26. Aldana, D.S.; Andrade-Ochoa, S.; Aguilar, C.N.; Contreras-Esquivel, J.C.; Nevárez-Moorillón, G.V. Antibacterial activity of pectic-based edible films incorporated with Mexican lime essential oil. *Food Control* **2015**, *50*, 907–912. [CrossRef]
27. Chaichi, M.; Hashemi, M.; Badii, F.; Mohammadi, A. Preparation and characterization of a novel bionanocomposite edible film based on pectin and crystalline nanocellulose. *Carbohydr. Polym.* **2017**, *157*, 167–175. [CrossRef]
28. Ali, A.; Chen, Y.; Liu, H.; Yu, L.; Baloch, Z.; Khalid, S.; Zhu, J.; Chen, L. Starch-based antimicrobial films functionalized by pomegranate peel. *Int. J. Biol. Macromol.* **2019**, *129*, 1120–1129. [CrossRef] [PubMed]
29. Duan, Q.; Jiang, T.; Xue, C.; Liu, H.; Liu, F.; Alee, M.; Ali, A.; Chen, L.; Yu, L. Preparation and characterization of starch/enteromorpha/nano-clay hybrid composites. *Int. J. Biol. Macromol.* **2020**, *150*, 16–22. [CrossRef]
30. Sukhija, S.; Singh, S.; Riar, C.S. Analyzing the effect of whey protein concentrate and psyllium husk on various characteristics of biodegradable film from lotus (Nelumbo nucifera) rhizome starch. *Food Hydrocolloid.* **2016**, *60*, 128–137. [CrossRef]
31. Munoz, L.; Aguilera, J.; Rodriguez-Turienzo, L.; Cobos, A.; Diaz, O. Characterization and microstructure of films made from mucilage of salvia hispanica and whey protein concentrate. *J. Food Eng.* **2012**, *111*, 511–518. [CrossRef]
32. Sun, Q.; Sun, C.; Xiong, L. Mechanical, barrier and morphological properties of pea starch and peanut protein isolate blend films. *Carbohydr. Polym.* **2013**, *98*, 630–637. [CrossRef] [PubMed]
33. Alee, M.; Duan, Q.; Chen, Y.; Liu, H.; Ali, A.; Zhu, J.; Jiang, T.; Rahaman, A.; Chen, L.; Yu, L. Engineering, Plasticization Efficiency and Characteristics of Monosaccharides, Disaccharides, and Low-Molecular-Weight Polysaccharides for Starch-Based Materials. *ACS Sustain. Chem. Eng.* **2021**, *9*, 11960–11969. [CrossRef]

Review

Recyclability and Redesign Challenges in Multilayer Flexible Food Packaging—A Review

Anna-Sophia Bauer [1], Manfred Tacker [1,2], Ilke Uysal-Unalan [3,4], Rui M. S. Cruz [5,6], Theo Varzakas [7] and Victoria Krauter [1,*]

1 Packaging and Resource Management, Department Applied Life Sciences, FH Campus Wien, University of Applied Sciences, Helmut-Qualtinger-Gasse 2/2/3, 1030 Vienna, Austria; anna-sophia.bauer@fh-campuswien.ac.at (A.-S.B.); manfred.tacker@fh-campuswien.ac.at (M.T.)
2 Circular Analytics TK GmbH, Otto-Bauer-Gasse 3/13, 1060 Vienna, Austria
3 Department of Food Science, Aarhus University, Agro Food Park 48, 8200 Aarhus, Denmark; iuu@food.au.dk
4 CiFOOD—Center for Innovative Food Research, Aarhus University, Agro Food Park 48, 8200 Aarhus, Denmark
5 Department of Food Engineering, Institute of Engineering, Campus da Penha, Universidade do Algarve, 8005-139 Faro, Portugal; rcruz@ualg.pt
6 MED—Mediterranean Institute for Agriculture, Environment and Development, Faculty of Sciences and Technology, Campus de Gambelas, Universidade do Algarve, 8005-139 Faro, Portugal
7 Department of Food Science and Technology, University of Peloponnese, 24100 Kalamata, Greece; t.varzakas@uop.gr
* Correspondence: victoria.krauter@fh-campuswien.ac.at; Tel.: +43-1-606-68-77-3592

Citation: Bauer, A.-S.; Tacker, M.; Uysal-Unalan, I.; Cruz, R.M.S.; Varzakas, T.; Krauter, V. Recyclability and Redesign Challenges in Multilayer Flexible Food Packaging—A Review. *Foods* **2021**, *10*, 2702. https://doi.org/10.3390/foods10112702

Academic Editor: Qin Wang

Received: 29 September 2021
Accepted: 30 October 2021
Published: 5 November 2021

Publisher's Note: MDPI stays neutral with regard to jurisdictional claims in published maps and institutional affiliations.

Abstract: Multilayer flexible food packaging is under pressure to redesign for recyclability. Most multilayer films are not sorted and recycled with the currently available infrastructure, which is based on mechanical recycling in most countries. Up to now, multilayer flexible food packaging was highly customizable. Diverse polymers and non-polymeric layers allowed a long product shelf-life and an optimized material efficiency. The need for more recyclable solutions asks for a reduction in the choice of material. Prospectively, there is a strong tendency that multilayer flexible barrier packaging should be based on polyolefins and a few recyclable barrier layers, such as aluminium oxide (AlOx) and silicon oxide (SiOx). The use of ethylene vinyl alcohol (EVOH) and metallization could be more restricted in the future, as popular Design for Recycling Guidelines have recently reduced the maximum tolerable content of barrier materials in polyolefin packaging. The substitution of non-recyclable flexible barrier packaging is challenging because only a limited number of barriers are available. In the worst case, the restriction on material choice could result in a higher environmental burden through a shortened food shelf-life and increased packaging weights.

Keywords: multilayer packaging; flexible packaging; polyolefin; recyclability; redesign; mono-material; shelf-life of foods

1. Introduction

Packaging is essential for maintaining the quality, safety, and security of many food products [1,2]. Robertson [1,3] described its basic functions as protection, containment, convenience, and communication. In addition to these functions, packaging should be recyclable but often faces end-of-life challenges. Recycling rates, particularly for plastic packaging, are low (42% on average throughout the European Union in 2018) [4]. Politics at the European level demand a stepwise increase in recycling rates for packaging [5]. This induces pressure on certain packaging solutions. Trend analysis shows that non-recyclable plastic packaging will no longer be tolerated by brand owners and retail chains [6]. Until 2030, all plastic packaging must be reusable or recyclable [5]. To reach this goal in the EU, most countries need investments to upgrade the collection, sorting, and recycling infrastructure, and principles of design for recycling must be comprehensively applied [7–9].

Guidelines from industry and academia support this transformation. They give guidance on material choice and design for packaging, packaging aid, and decoration, mostly in relation to established collection, sorting, and recycling infrastructure of specific regions or countries [10–12].

Multilayer food packaging is especially under pressure since it combines various materials such as polymers, paper, aluminium, and organic or inorganic coatings [13–15]. Considering environmental effects measured by Life Cycle Assessments, these packaging solutions are highly efficient [16,17]. The main problem, however, is that they are hardly recycled in the existing waste management infrastructure, as Europe widely relies on traditional approaches of mechanical recycling in regranulation processes, which generally means combined processing of materials [4,13,18,19]. The thermal incompatibility of the diverse combined materials is one major obstacle in reprocessing [20]. New technologies such as chemical recycling show promising results, but they need further and deep investigation and up-scaling [21,22]. Currently, a great deal of effort is put on the redesign of multilayer flexible packaging to improve the recyclability in the existing collection, sorting, and recycling infrastructure [21]. Recyclable film solutions based on polyolefins (polyethylene (PE) and polypropylene (PP)) have already been achieved, as packaging waste material streams exist for these films, at least for mixed polyolefin streams [23–25]. As polyolefins already dominate flexible food packaging, the restriction of the use of certain polymers such as polyethylene terephthalate (PET) or polyamide (PA), which are not compatible with polyolefin recycling, is tangible [11,26–29].

A challenge is posed by the fact that enhancing the recyclability of multilayer films often goes hand in hand with a reduction of packaging efficiency. Current solutions on the market have been optimized over the last decades for resource efficiency and product protection. Reducing the complexity of these films would likely lead to thicker films and therefore heavier packaging solutions would be required [30,31]. This goes against the goals of a circular economy to reduce resource consumption and environmental impacts [7].

Multilayer flexibles by weight account for 10% of all packaging solutions [21]. The relative amount may not seem huge, but at least 40% of food products are packed in flexible solutions [32]. This induces the need to review redesign suggestions. Their comparison should allow the implementation of redesign approaches throughout and be supported by the European packaging branch.

A brief overview of the characteristics of multilayer flexibles, their contribution to sustainability, and their incompatibility in widely applied recycling technology make it possible to discuss the future design of this type of packaging. Research is necessary to bring recyclability and overall sustainability together in barrier packaging. Material combinations and recycling options with a clear benefit for the environment have to be developed.

The main objective of this review is to gather information on the benefits of multilayer flexible food packaging and show the negative recyclability trade-offs, especially for food technologists. The whole food-producing industry is under pressure to apply recyclable, at best circular packaging solutions throughout. To get there, we have to raise consciousness about what is considered as recyclable, and which negative effects might come along with redesign if we strive for circularity to enhance the packaging sustainability of specific products. This work mainly focuses on literature back to 2009, as the very first collection of hurdles (Figure 1) started in 2019, collecting evidence on a topic that gained momentum in the last decade.

2. Multilayer Flexible Food Packaging

Multilayer food packaging is a tailored packaging application. Beneficial properties of diverse materials are combined into one packaging solution. Flexible packaging like pouches, bags, lidding as well as rigid packaging like trays, cups, and bottles consist of variable material, sometimes combined in layers. Through the approach to combine materials, these products offer technical and systemic strengths but also weaknesses along the life cycle stages, from production to use phase and end-of-life scenarios [13,19,33,34].

Figure 1 shows a collection of hurdles in relation to circular packaging, with a focus on multilayer flexible packaging, but not solely limited to it, encompassing literature research via Science Direct, Google Scholar, and Scopus, following the keywords "circular multilayer packaging", "recycling flexible packaging", "circular economy multilayer", "multilayer recycling", "polymer film food", as well as secondary sources therein. Most mentioned hurdles, for example, the coordination along the supply chain, costs, and profitability, or the separation of materials, were collected and assigned to life cycle stages.

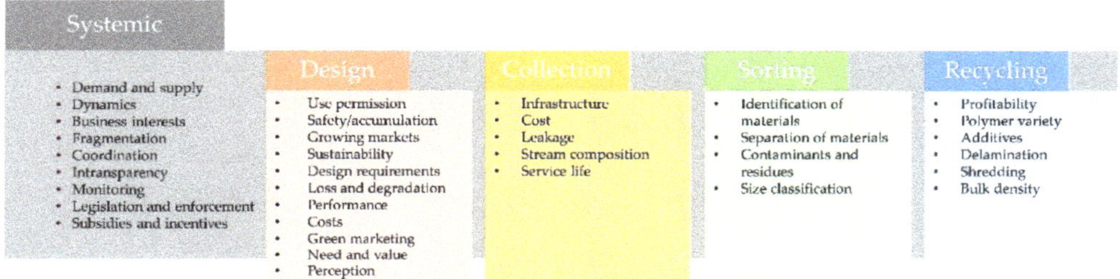

Figure 1. Hurdles to circularity of packaging focused on, but not limited to multilayer flexible packaging [7–9,13,14,19,21, 26,30,33,35–76].

The improvement of barrier properties of packaging to control the food quality and safety is one main intention of combining materials. The permeability against relevant gases (oxygen (O_2), carbon dioxide (CO_2), nitrogen (N_2), water vapor (WV)), the transmittance of light, the barrier against grease or oil, as well as odors/aromas are important elements controlled by packaging. Depending on mainly fat, carbohydrates, and protein contents of food commodities, diverse permeability, and light transmission is acceptable to reduce negative changes in food quality or safety [77].

2.1. Production, Characteristics, and Application

The characteristics of multilayer flexible packaging are related to the molecular properties of the used materials. The polymer type, its crystallinity, branching, tacticity, and polarity influence the gas permeability and light transmission of the film. In order to reach the required packaging specifications, a combination of polymers or the introduction of other non-polymeric layers like paper or aluminium is frequently applied. This could be taken as a point for differentiation. Some multilayer flexible packaging solutions solely include polymeric layers. In other cases, also stiffer material like paper is included [77,78]. There is hardly any limit to imaginable combinations. Even 24 layers of material combined into one film are marketed, found in a cheese packaging solution through polarization microscopy [79].

The production of multilayer packaging film mainly relies on extrusion or lamination processes. While extrusion (coextrusion) is reported to dominate the production of multilayers for inter alia practicability and economic reasons, lamination is necessary to combine material that cannot be coextruded (for example the combination of polymers with non-polymers) [80]. Next to these two basic production methods, coatings allow the integration of even more beneficial properties to one packaging film, e.g., more functional layers. Whereas Selke and Hernandez [34] discussed metallization through vacuum deposition or SiOx (silicon oxide) as examples for coating, Farris [80] refers to the application of melts and liquids. Recent developments in this area include the development of nanocoatings applied at levels below a critical concentration. Coatings can form a thin layer of material that can be deposited directly on a surface, applied in liquid form (film-forming solution/dispersion), by immersion, homogeneous spreading, or spraying [81,82]. In either way, customization to enhance barrier properties is possible. As an example, polyolefins as non-polar polymers, show low water vapor but high oxygen permeability. To reduce the oxygen permeability, barrier layers are introduced [77].

Based on the quantity, polyolefins, PET, PS, and polyvinylidene chloride (PVC) are in general the leading polymeric materials in packaging applications in Europe [4]. Häsänen [79] stated in a case study, that in samples of purely polymeric multilayer flexible packaging, polyolefins, polyamides, and PET, followed by ethylene vinyl acetate (EVA) and ethylene vinyl alcohol (EVOH) dominate this type of packaging.

Polyethylene in general is used as a moisture barrier and for its toughness. low-density polyethylene (LDPE) and linear low-density polyethylene (LLDPE) are used as sealants, bonding layers, tie resins, adhesives, or structural layers. LDPE and LLDPE are found in several applications of flexible food packaging. Exemplary for bakery products, mono- or multilayer solutions including LDPE or LLDPE are widely marketed. With increasing barriers needs against moisture, high-density polyethylene (HDPE) is of higher interest in flexible packaging. The crystallinity of HDPE induces strength and stiffness, which allows its use as a structural layer. One prominent example of the use of HDPE in multilayer flexibles is cereal packaging, possibly in a combination with EVOH for the enhancement of the oxygen barrier [15,34,78,83,84].

Polypropylene in packaging is referred to as a moisture barrier, connected to benefits through its crystallinity. Reflected in the high melting point, it offers strength and is stable against exposures to higher temperatures. It shows clarity and stiffness and is also used as a sealant. Specifically in multilayer flexible packaging, it is frequently combined with PE. Metallization of PP is common for dry food products, requiring high oxygen barriers [13,15,34,78].

Polyamide is used for its mechanical properties, as an oxygen as well as oil, grease, and aroma barrier. Beneficial optical and thermal properties also lead to its use in multilayer food packaging. PA is also used in vacuum packaging or applications with modified atmospheres, for example in the food group of meat. Addressing possible polymer combinations with PA for meat, PE is common [15,34,77,83].

Oil and grease resistance is also known to be a beneficial property of PET, not only for PA. Its printability, thermal, mechanical as well as optical properties are the reasons for its use in multilayer packaging solutions, similarly for example in meat packaging [15,34,77,84].

In addition to the commonly used polymers in packaging (PE, PP, PET), EVOH and EVA can broaden the attributes of the bulk plastics [4]. EVOH finds use predominantly as a barrier material against oxygen, oil, and grease. In multilayer flexible packaging, it is widely used for food products, which in contact with oxygen, would face quality degradation. This includes a variety of possible applications, for example, snacks products. Contrary to metallization, it offers transparency [15,34,77,78]. EVA is used as a sealant and adhesive in multilayer food packaging. Furthermore, also its optical properties are said to promote its use. Applications in multilayer flexibles include inter alia combinations with polyolefins, for example in fresh convenience products like pre-cut salads [15,34,78]. Another adhesive used in packaging is Polyurethane (PU) [78].

Polyvinylidene dichloride (PVDC) is used for its barrier properties against oxygen and moisture, its optical properties, as well as a layer resistant to oil and grease. Its stiffness or softness is highly customizable. Food products in multilayers with PVDC are for example snacks. Its use in shrink films or stretch wraps, mono- or multilayer variations, can also be found [15,34,77,78].

Next to the above-specified polymers, aluminium is used to protect the food from moisture, oxygen, and light. Optical properties too account for its use in multilayer flexible packaging. One multilayer example with aluminium foil is packaging of food with sterilization steps in production, for example, ready-to-eat meals [15,34,77].

Coatings such as aluminium oxide (AlOx) and silicon oxide both facilitate highly enhanced barrier properties against oxygen and moisture while offering transparency at thicknesses in the nanometer range, compared to several micrometers for polymer-based barrier layers. In multilayer flexibles, one can find for example combinations with PET. However, these coatings are discussed as being prone to cracks affecting the barrier properties, inter alia when used on flexible substrate materials. In general, the coatings can be used between layers to enhance the durability [13,51,77,78,80].

Furthermore, paper is a material commonly used in multilayer flexible food packaging, including non-polymeric layers. Depending on the paper, it can increase the rigidity/stiffness of multilayer packaging. Marketed solutions include combinations with PE, also EVOH or foil. Paper is beneficial in the context of printability and shows different possible haptics and optics compared to polymer packaging. It is also used as a light barrier [34,51,78,80].

Within this multitude of possible materials to combine, Kaiser et al. [13] gave an overview of widely used polymers in multilayer flexible packaging and their associated application purpose. A modified version is shown in Table 1. It points up that single multilayer structures are hard to exemplify as "typical".

Table 1. Properties and materials in multilayer flexible food packaging. Modified after the work in [13] based on the works in [51,85,86].

Mechanical Stability	Oxygen Barrier	Moisture Barrier	Light Barrier	Tie Layer	Sealant
PO	EVOH	PO	Aluminium	PU	PO
PET	PVDC	EVA	Paper	PO	EVA
PS	PA	PVDC			PA
Paper	PET	Aluminium			PET
	SiOx				
	AlOx				
	PVOH				
	Aluminium				

Abbreviations: PO (polyolefins: polyethylene, polypropylene), PET (polyethylene terephthalate), PS (polystyrene), EVOH (ethylene vinyl alcohol), PVDC (polyvinylidene dichloride), PA (polyamide), SiOx (silicon oxide), AlOx (aluminium oxide), PVOH (poly vinyl alcohol), EVA (ethylene vinyl acetate), PU (polyurethane).

Other than the properties overview from Kaiser et al. [13], Morris [15] describes multilayer film structures depending on food products. Two multilayer flexible packaging solutions for meat products are illustrated in Figure 2, irrespective of layer thicknesses. To avoid oxygen ingress as one major quality determinant in processed meat products, PVDC and EVOH offer enhanced barrier properties in these examples. The use of PA for meat products is mainly referred as beneficial, on one hand for printability and on the other hand, thermal stability.

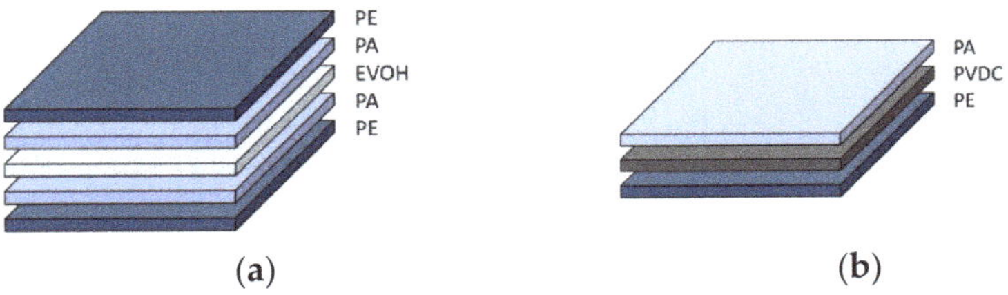

(a)   (b)

Figure 2. Two exemplary multilayer solutions for meat packaging: (**a**) 5-layered packaging solution and (**b**) 3-layered packaging solution. Abbreviations: PE (polyethylene), PA (polyamide), EVOH (ethylene vinyl alcohol), PVDC (polyvinylidene dichloride), PE (polyethylene). Figure adapted from the works in [15,87].

It is not only meat products that require enhanced barrier properties and profit from the use of combined materials; most products that are sensitive to water loss or uptake, oxygen ingress, light, and possible loss of aroma, require barriers to maintain quality over long periods of shelf-life. The shelf-lives of, for example, specific dairy products, sweets and confectionary, cereals, or processed fruits and vegetables are related to barrier properties of the applied packaging solutions. Various degradation mechanisms (e.g., biological) can be slowed down by proper packaging, as the products are for example subjected to ripening, wilting and oxidation processes, just to name a few. Next to that, the microbiological safety stands in relation to gas/vapor permeability. Packaging can be one key within a hurdle concept, to keep food products at high quality [3,77]. However, the matching of product needs together with the possible barrier ranges of packaging material in Figure 3 shows, that often, one material alone, cannot serve the barrier requirements (water vapor transmission rate (WVTR)/oxygen transmission rate (OTR)) of specific products [88].

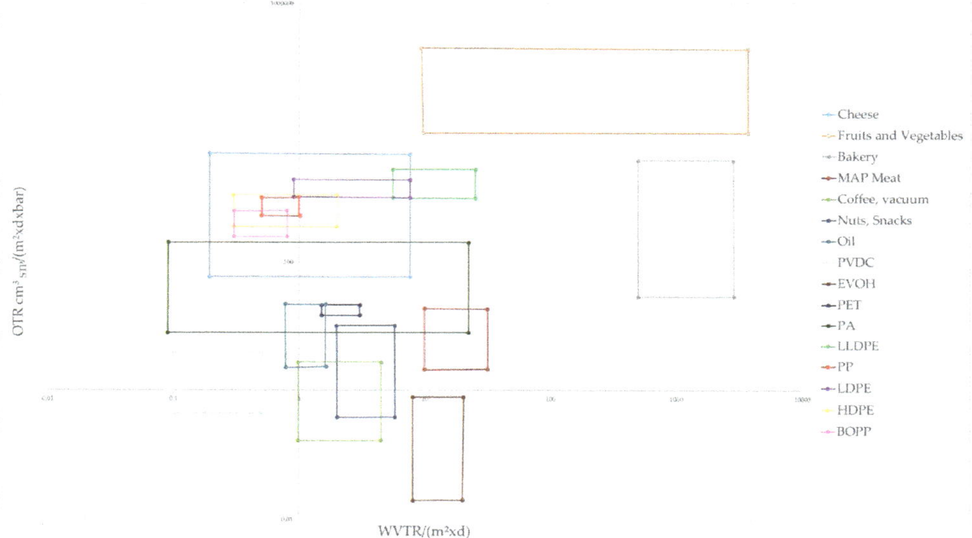

Figure 3. Water vapor and oxygen transmission rates versus the barrier requirements of food products and barrier ranges of polymers for packaging. Modified after the work in [88].

Robertson [1,3] and Morris [77] thoroughly describe the needs of specific food groups through associated quality determining intrinsic and extrinsic factors that can be influenced by packaging applications. They show the tolerable levels of permeation and describe the required storage conditions for a broad range of fresh as well as processed food products.

2.2. Efficiency and Sustainability—Trade-Offs Regarding Recycling

The protection of food through combinations of materials with desired characteristics is highly effective. Thin layers of materials in multilayer flexibles suffice to make use of beneficial properties. This allows the development of lightweight, efficient packaging solutions, which is related to questions of overall sustainability of packaging solutions. Mono-material flexible packaging can also be such lightweight solutions, however, having often inferior barrier properties. This goes hand in hand with packaging efficiency and effectiveness. The complexity of the material can be reduced. Still, thickness and therefore weight increases have also negative environmental consequences. The main question is, what the environmentally favorable solution is when overall, a resource reduction is the goal. In the discussion about multilayer versus mono-material, the focus lays on recyclability trade-offs. One less complex solution might show better recyclability; however, it is probably linked to higher material inputs what is neither environmental favorable [30,51].

In Wellenreuther [16], the comparison of environmental effects (energy demand, raw material demand, and waste) connected to multilayer flexible pouches versus rigid solutions shows beneficial properties in the context of efficiency on the side of the multilayer solution. Branch reports and communication charts highlight optimized product-to-packaging ratios stated as, at a maximum, 10 times lower compared to rigid packaging solutions. In the longer term, this is related to benefits in transportation (weight, space) and in general, a reduction of used associated resources [32,89]. The Flexible Packaging Association [89] summarizes these factors as "beneficial life cycle metrics" of flexible packaging, referring to a reduction of water use, fuel use, and a reduced carbon footprint of products. According to Flexible Packaging Europe [32], flexible packaging makes on average less than 10% of a packaged food products CO_2 footprint. Despite these benefits that are often highlighted through industry near associations, which is part of critical voices arguments, it is clear that the optimal point of packaging between the protection of used resources in food and the resources used for the packaging material itself is where the least possible environmental impact occurs. Food products are expected to keep their quality until consumption and therefore prevent food losses and waste. Parallel, quantitative packaging material input is kept at a low level [2,90]. Both aspects are of high importance in reaching sustainable production and consumption. This is shown by projects and publications analyzing and evaluating the effects of zero packaging as well as the environmental burden of unconsumed food residues. These scenarios clearly show that the protection of the filling good is key to sustainable consumption and still, the input of packaging material should be kept at a minimum. This is an important argument/feedback loop to make use of the highly improved, customized, and often combined material flexible packaging solutions [91]. Efficiency and low carbon footprints are the major benefits of multilayer packaging in comparison to other packaging solutions [14,30,51].

However, the weak spot of multilayer packaging is, that it is difficult to recycle, and its recycling rate is very low [13,18,19]. Ellen MacArthur [21] estimated in 2017 that 26 weight percent of flexible packaging is multi-material, representing 10% of global plastic packaging. Worst case, these 10% are lost for the aspired circular economy, as with the current infrastructure, the properties of the materials cannot reach the ones of virgin material again.

Currently applied mechanical recycling technology consists of shredding, sorted, and washed plastic input material and its re-granulation [4]. The incompatibility/immiscibility of diverse plastic materials in the melting process limits this approach to pure waste streams/fractions, so that many material combinations present in multilayer materials cannot be processed, due to different melting points and thermal stability [43,92]. These material combinations in flexible packaging are therefore considered as non-recyclable with the current sorting and mechanical recycling infrastructure [12]. The incompatibility of polymers in thermal processes is not a new discovery as already described by Nickel [20] more than 25 years ago (Table 2). Describing the incompatibility, differences can be found in the literature [20,93]. However, redesigning to fit the existing infrastructure is currently an absolute priority [7–12].

Although compatibilizing agents can partly solve this problem, they are only used for some applications, inter alia due to high costs [19,48,94]. Uehara et al. [94] described for example the use of maleic anhydride and glycidyl methacrylate. To enable the blending of polymers by compatibilizers, the unknown composition of the material stream is another obstacle to be faced [95]. Pinzón and Saron [96] showed for example the blending of post-industrial LDPE multilayers with up to 20% PA through compatibilization. Furthermore, the potential of blending PET/PE multilayers with compatibilizers was already assessed and described as useful, considering the recyclability of incompatible polymers [94]. More recently, Jönkkäri et al. [92] tested the compatibilization of input material from post-consumer multilayers with virgin LDPE, excluding packaging with non-polymeric layers (paper, cardboard, aluminium). Secondary material thereof is described to be suitable for applications not requiring specific optical properties or high thermal stability. Without the use of compatibilizers, the extrusion of different types of polymers shows mainly incompatibility towards homogeneous blends and the deterioration of visual and mechanical properties of the secondary material [97].

Table 2. Compatibility of polymers in recycling. Modified after works in [20]. * indicates differences in the comparison to the works in [93,98].

	PE	PP	PVC	PS	PA	PET
PE	+	~ (+ *)	−	−	~ (− *)	−
PP	~	+	−	−	~ (− *)	−
PVC	−	−	+	~ (- *)	−	−
PS	−	−	−	+	~ (− *)	~ (− *)
PA	−	−	−	~	+	~
PET	−	−	−	− (~ *)	~	+

+ (compatible), ~ (partly compatible), − (incompatible). Abbreviations: PE (polyethylene), PP (polypropylene), PVC (polyvinylidene chloride), PS (polystyrene), PA (polyamide), PET (polyethylene terephthalate).

Furthermore, the available waste management infrastructure in collection and sorting is country-specific and influences the recyclability of food packaging [10]. Flexible packaging itself is a heterogeneouswaste fraction, which is, although dominated by polyolefins, frequently accompanied by other polymers and non-polymeric material [62,78]. One other reason for the heterogeneity is due to the collection in mixed fractions or "undifferentiated garbage" [14]. Regarding specific material fractions of collected flexible packaging, PE dominates, whereas flexible PP and PET, according to the flow charts in van Eygen [23], are not separately considered in the film category for the widely available mechanical recycling processes. Marrone and Tamarindo [14] supports this perspective: not only multilayer flexibles but also mono-material films are not collected consistently.

Referring further to the lightweight character of flexible packaging, proportionately large amounts of impurities from food residues accompany collected post-consumer flexibles. This leads to possibly high ratios of impurities per packaging weight [43,60,69]. Irrespective of the already high level of contamination through the diverse materials used, major cleaning efforts might be necessary prior to extrusion processes [50,59,64].

Moreover, the typical sorting procedures are not widely optimized for a high-quality sorting of flexible films, although NIR detector (near-infrared) technology could detect material layers [30,97,99]. That flexible packaging is collected separately, then sorted and recycled, therefore depends on economic considerations, related to the mentioned hurdles [19,51]. New approaches to optimize the sorting for this fraction are sought, as this process is a vital pre-request to enhance recyclability and circularity [21].

3. Discussion

As the situation described above shows, multiple criteria are leading to a strong tendency in the European Union, to substitute non-recyclable multilayer barrier films with recyclable solutions based on polyolefins. Taken together, three main factors are found to build the core of the redesign suggestions. The first, for sure, is the mentioned ban of all non-recyclable plastic packaging from the European Market from 2030 on and the even stricter commitments from parts of the food and packaging supply chain [5,100]. The second determining factor is the currently available waste management infrastructure in collection, sorting, and recycling. As many material combinations are incompatible, this prevents the recycling of polymer combinations such as PET or PA with polyolefins, as the layers, in general, are not separated before the melting process [4,13,18–20]. This brings economic factors into play. The waste stream of post-consumer flexibles is dominated by polyolefins, with PE and PP constituting more than 60% of the weight of flexible packaging [26]. The level of other polymer types is small and therefore the establishment of separate recycling streams for PET or PA-based films is not profitable [4,48,49,66]. Decontamination steps to clean plastic waste from residuals such as food, and the small size of many flexible packaging, makes sorting even more demanding [21,84]. In addition, as incineration is widely accessible, recycling of this fraction is often not profitable [5,7,60,64]. It seems beneficial, that the substitution of polymers incompatible with the mechanical reprocessing of polyolefins could lead to higher market shares of polyolefins, which might increase the efficiency and the economics of the recycling process of flexibles, as the variability of material might find reduction [45,46,70].

3.1. Redesign and Trade-Offs to Fit the Actual Recycling Technology

Suggestions to reduce the material variability to mainly polyolefin material, tolerating EVOH, metallized aluminium layers as well as coatings to a certain extent, have been published widely [10–12,27–29,101–103]. That polyolefins show the best compatibilities with other polyolefins entails the theoretical basis for published redesign options. Moreover, it is possible to blend different grades of post-consumer polyolefins in certain percentages, however, it results in lower quality recyclates. The content of polyolefins should at least reach 90% to be considered as mono-material, which is seen as beneficial composition for recycling [10]. Combinations of polyolefins with other polymer types such as PET or PA are not considered as recyclable in traditionally applied mechanical recycling processes [11,27–29]. Looking at the available infrastructure in Europe, the incompatibility of most polymers in traditional approaches of mechanical recycling and the complex sorting of multilayer flexible packaging, the step to return to already recyclable solutions seems obvious.

Economies of scale for potential valorization are in favor of polyolefins as they dominate packaging applications [4,101].

Considering the need for enhanced barrier properties in the substitution of multilayers, the consensus on redesign suggestions includes the following material combinations:

- mono-polyolefins with EVOH,
- mono-polyolefins with SiOx or AlOx,
- mono-polyolefins metallized [10,11,27–29,103].

The details on how combinations should look vary slightly between the guidelines. Some suggestions are more restrictive than others. The optimal flexible packaging from the recycling point of view is unpigmented/transparent mono-polyolefin material. The use of EVOH and SiOx and AlOx layers does not significantly reduce the quality of secondary materials if these contaminations do not surpass certain critical thresholds. Aluminium laminated and metallized does lead to greying of the recyclate and is therefore not considered as an optimal barrier material to choose. Nevertheless, metallization is mostly tolerated to a certain extent. Possible negative interactions with sorting infrastructure are addressed and discussed in guidelines what leads to stated limitations or investigation needs, for example in cases of surface metallization. The combination of polyolefins with PET, PS, Polylactic acid (PLA), paper, PVC, PVDC, and PA is not recommended. However, you can find statements that PA layers and PVDC coatings as barrier material are under investigation [10,11,27–29]. Table 3 shows slight differences in recommendations for barrier layers for polyolefin films between two popular guidelines. Where one excludes most combinations of polyolefins with common barrier materials, the other allows more options according to weight percent in a certain packaging solution. One interesting point is that EVOH content is not fully harmonized. The information on EVOH levels tolerated in PP film is in one guideline stated as 5% (Ceflex), whereas Recyclass lists it in "conditional–limited compatibility" in tables of 2020 as "may be suitable". As the recommendations for EVOH changed quite drastic from accepted 10% to 5% to 1% and in the meantime even "no–low compatibility" (at least for rigid PP packaging—until 2021— back to 6% for specific cases, questions on further developments arise in the case of PP film [12,104–107].

Against this background, Table 4 shows the remaining materials to be used in future multilayer packaging design and highlights design restrictions with regard to mechanical and barrier properties. Future solutions for multilayers are technically still not only one material. However, combinations can be categorized as mono-material, if the amounts of barrier materials stay under tolerated levels. It is evident that, in comparison to Table 1, only a few materials remain for recyclable design.

Table 3. Tolerated materials in multilayer barrier flexible packaging modified after the works in [10,11,27–29].

	EVOH	Metallization	SiOx	AlOx	Acrylic Coatings	PVOH	PVDC	References
PP-film	Conditional–limited	Conditional–limited	Fully compatible	Fully compatible	"any other barrier" no–low compatibility		No–low compatibility	[28,29]
	<5%	Compatible with PE or PP mechanical recycling	<5%	<5%	<5%	<5%	Further investigation	[10]
PE-film	<5%	Conditional–limited	Fully compatible	Fully compatible	"any other barrier" no–low compatibility		No–low compatibility	[11,27]
	<5%	Compatible with PE or PP mechanical recycling	<5%	<5%	<5%	<5%	Further investigation	[10]

Abbreviations: PP (polypropylene), PE (polyethylene), EVOH (ethylene vinyl alcohol), SiOx (silicon oxide), AlOx (aluminium oxide), PVOH (poly vinyl alcohol), PVDC (polyvinylidene dichloride).

Table 4. Materials suggested for recyclable multilayer flexible food packaging. Modified after the works in [10,11,13,27–29,51,85,86].

Mechanical Stability	Oxygen Barrier	Moisture Barrier	Light Barrier	Tie Layer	Sealant
PO	EVOH	PO	Aluminium (metallised)	~~PU~~	PO
~~PET~~	~~PVDC~~	~~EVA~~	~~Paper~~	PO	~~EVA~~
~~PS~~	PA	~~PVDC~~			~~PA~~
~~Paper~~	~~PET~~	Aluminium (metallised)			~~PET~~
	SiOx				
	AlOx				
	~~PVOH~~				
	Aluminium (metallised)				

Strikethrough indicates design restrictions. Abbreviations: PO (polyolefins: polyethylene, polypropylene), PET (polyethylene terephthalate), PS (polystyrene), EVOH (ethylene vinyl alcohol), PVDC (polyvinylidene dichloride), PA (polyamide), SiOx (silicon oxide), AlOx (aluminium oxide), PVOH (poly vinyl alcohol), EVA (ethylene vinyl acetate), PU (polyurethane).

The trends for design for recycling also induce trade-offs concerning the substitution of specific material properties, the barrier requirements, the related shelf-life as the further connected products sustainability [21,30,73]. Due to the pressure to reduce EVOH-content and metallization to avoid quality impairment in secondary material properties, the development of novel recyclable barriers, mainly against oxygen, is needed. It must be assured, however, that the redesigned flexible packaging protects the food correctly and that reduced shelf-life does not result from inferior oxygen or water vapor barriers. Many confectionary products for example hardly tolerate the ingress of water vapor or oxygen resulting in rancidity and loss of crispness [21,77].

Thus, a strong research need is present to develop recyclable barriers substituting EVOH and other barrier polymers such as PA and PVDC. A clear tendency is visible that the percentage of allowed EVOH in recyclable packaging solutions is one focus of discussion, as could be seen in the case of rigid PP packaging in 2020 and 2021 [105,107]. The range of currently available barrier options is small with SiOx and AlOx, and most SiOx- and AlOx coatings are currently neither generally suitable for sterilizable packaging nor deep drawing applications, which is of importance in the sector of, for example, convenience foods [77].

The focus on mostly mono-polyolefins with certain tolerated barrier layers for enhanced recyclability of multilayer flexibles should not lead to higher resource consumption, as this would increase the environmental burden. This is particularly important in the specific case of flexible packaging where in recent decades, lightweight solutions have been developed and optimized [30,31].

The elimination of PA, PET, and other polymers in this context also induces the need for further developments of satisfactory substitutions for puncture-resistant materials. Another point to consider is to optimize the sealability of PP-films. The combination of PET on the external side and polyolefins as a sealing layer on the internal side has been used very often. PET (or PA) shows higher melting points than polyolefins, which in general allows good sealing properties [13,15,83].

3.2. Harmonization of Recyclability Guidelines in Europe

Multilayer flexibles are considered as a sustainable packaging solution due to low resource consumption and low carbon footprint but are being difficult to recycle with the collection and recycling infrastructure currently in place. Thus, there is this clear and urgent need for a redesign that balances recyclability and sustainability [16,17,21,84]. The switch from non-recyclable multilayer flexible to easily recyclable, predominantly mono-material packaging solutions, within the intention to increase recycling rates, however, leaves questions for discussion: If all rigid packaging (excluding beverage packaging) was 100% recyclable but substituted by non-recyclable flexible packaging, the global warming potential would decrease [17]. Questions arise referring to the intended goals of packaging redesign, underlying the increase of recycling rates.

Although replacing one material with another is already not a simple task, employing the best material for each food system is also still necessary. This is a true challenge that only when addressed, will result in its implementation. However, there are already commercial applications of flexible packaging available, that seem to close the gap between recyclability and enhanced barrier needs through, for example, improved orientation processes of mono-polyolefin films, which can be found in web search.

Still, currently, recycling is not the best solution for all types of packaging, if enhanced sustainability is the target of increased recycling rates [25].

To compare future packaging options, a holistic sustainability assessment is necessary. The harmonization of guidelines must build the basis for global standards. It should proactively include changes in shelf-life due to changes in barrier properties and therefore food waste as well as aspects of littering. Holistic and harmonized approaches are vital for the sustainability assessment and the perspective of a common market. The understanding of recyclability must be the same, at least in all European countries. The implementation of a redesign for recyclability needs the support of the packaging industry. This includes the improvement of collection, sorting, and recycling infrastructure to allow a high-quality secondary material production [21,22,108,109]. The discussion currently shows a highly Eurocentric perspective, however, other global regions like the US and Australia are already following [110].

3.3. Novel Recycling Technologies and Secondary Material for Food Contact

Next to the option to fit packaging design into existing infrastructure, other recycling technologies or collection logistics can cope with multilayer films/material combinations. The developments in chemical recycling could lead more quickly to improved secondary materials. Delamination technologies of the single materials from multilayers as a pre-treatment is promising, as it could allow the further use of traditional mechanical recycling. Developments include inter alia

- chemically separating the layers of multi-material,
- recovering the aluminium content of multilayer food packaging by microwave-induced pyrolysis, and
- separate collection of specific multilayers for regranulation with compatibilizers [13, 42,75,92,95,111].

Nevertheless, these exemplary solutions are either in development or not yet widely introduced, and thus, the focus on the available instead of new recycling technology, still asks for the development of mono-material solutions [13].

Even if the redesign and recycling of flexible packaging becomes successful to a high degree, closing the material cycle faces another obstacle. Apart from a very few exceptions such as HDPE from milk bottles, secondary post-consumer polyolefins are currently not permitted for use in food contact materials [112]. Due to a more complex decontamination in comparison to PET, as well as degradation in reprocessing, polyolefins lag behind as available secondary material. Cecon et al. [113] resumed the hurdles, but also new approaches in recycling technologies that could enable the use of polyolefins as secondary material in food contact in the future.

Still, in the current infrastructure, this above all is one knock-out criterion inhibiting the attempts to achieve truly circular flexibles for food packaging at present.

4. Conclusions

Multilayer flexible packaging is efficient. It combines the properties of polymers and non-polymeric materials to thin, lightweight packaging solutions for foods with and without barrier needs. The main problem is that it is rarely recycled in the existing waste management infrastructure. This is caused by multiple circumstances. The variability of used materials, the collection infrastructure, the complex sorting, and high levels of food residues outline the situation. Furthermore, the focus on mechanical recycling through combined processing complicates the situation. New solutions in recycling technology exist but are not yet available on a larger scale. This leads to a concentration on mono-material solutions to fit into the existing recycling infrastructure and diminishes the material choice to overcome thermal incompatibilities. The maximum tolerated levels of barrier materials are widely discussed and are in the process of being reduced. The substitution of a specific material is challenging, as only a limited number of barriers are available. In relation to the main purpose of packaging, the products' protection, this could result in negative side effects. A reduction of food shelf-life, higher packaging weights, and derived increased environmental burden are imaginable consequences that need to be considered when taking steps towards the goal of packaging redesign for holistic sustainability.

Author Contributions: Conceptualization and methodology, A.-S.B., M.T. and V.K.; writing—original draft preparation, A.-S.B.; writing—review and editing, A.-S.B., M.T., V.K., I.U.-U., R.M.S.C. and T.V.; supervision, M.T.; project administration, M.T.; funding acquisition, M.T. All authors have read and agreed to the published version of the manuscript.

Funding: This research was supported by Vienna Business Agency, grant number 2658497. The funding source was not involved with the collection, analysis, and interpretation of the information. Further, this publication was supported by the network project COST Action 19124, RETHINKING PACKAGING FOR CIRCULAR AND SUSTAINABLE FOOD SUPPLY CHAINS OF THE FUTURE.

Acknowledgments: Mary Grace Wallis provided comments on the manuscript.

Conflicts of Interest: The authors declare no conflict of interest.

References

1. Robertson, G.L. Food packaging and shelf life. In *Food Packaging and Shelf Life: A Practical Guide*, 1st ed.; Robertson, G.L., Ed.; CRC Press/Taylor & Francis Group: Boca Raton, FL, USA, 2009. [CrossRef]
2. Food and Agriculture Organization of the United Nations. The State of Food and Agriculture. 2019. Available online: http://www.fao.org/publications/sofa/2019/en/ (accessed on 16 February 2021).
3. Robertson, G.L. *Food Packaging: Principles and Practice*, 3rd ed.; CRC Press/Taylor & Francis Group: Boca Raton, FL, USA, 2012.
4. PlasticsEurope. Plasctics—The Facts 2020. Brussels, Belgium. 2020. Available online: https://www.plasticseurope.org/en/resources/publications/4312-plastics-facts-2020 (accessed on 16 February 2021).
5. European Commission. *A European Strategy for Plastics in a Circular Economy*; European Commission: Brussels, Belgium, 2018. Available online: https://eur-lex.europa.eu/legal-content/EN/TXT/?qid=1516265440535&uri=COM:2018:28:FIN (accessed on 16 February 2021).
6. Smithers. Brand Owners and Converters Drive Packaging Recycling Growth. Available online: https://www.smithers.com/resources/2019/mar/brand-owners-drive-packaging-recycling-growth (accessed on 4 January 2021).
7. European Commission. *Towards a Circular Economy: A Zero Waste Programme for Europe*; European Commission: Brussels, Belgium, 2014. Available online: https://eur-lex.europa.eu/legal-content/EN/TXT/?uri=celex%3A52014DC0398 (accessed on 16 February 2021).
8. Bicket, M.; Guilcher, S.; Hestin, M.; Hudson, C.; Razzini, P.; Tan, A.; ten Brink, P.; van Dijl, E.; Vanner, R.; Watkins, E. *Scoping Study to Identify Potential Circular Economy Actions, Priority Sectors, Material Flows and Value Chains*; European Union: Luxembourg, 2014. Available online: https://op.europa.eu/de/publication-detail/-/publication/0619e465-581c-41dc-9807-2bb394f6bd07 (accessed on 16 February 2021). [CrossRef]
9. European Commission. *Closing the Loop—An EU Action Plan for the Circular Economy*; European Commission: Brussels, Belgium, 2015. Available online: https://eur-lex.europa.eu/legal-content/EN/TXT/?uri=CELEX%3A52015DC0614 (accessed on 16 February 2021).
10. Ceflex. Designing for a Circular Economy: Recyclability of Polylefin-Based Flexible Packaging. 2020. Available online: https://guidelines.ceflex.eu/ (accessed on 16 February 2021).
11. RecyClass. PE Natural Flexible Film Guideline. 2020. Available online: https://recyclass.eu/wp-content/uploads/2020/07/PE-natural-films_guideline-1.pdf (accessed on 6 October 2020).
12. FH Campus Wien. *Circular Packaging Design Guideline: Empfehlungen für Recyclinggerechte Verpackungen*; FH Campus Wien: Vienna, Austria, 2020. Available online: https://pub.fh-campuswien.ac.at/urn:nbn:at:at-fhcw:3-757 (accessed on 16 February 2021). [CrossRef]

13. Kaiser, K.; Schmid, M.; Schlummer, M. Recycling of Polymer-Based Multilayer Packaging: A Review. *Recycling* **2018**, *3*, 1. [CrossRef]
14. Marrone, M.; Tamarindo, S. Paving the sustainability journey: Flexible packaging between circular economy and resource efficiency. *J. Appl. Packag. Res.* **2018**, *10*, 53–60.
15. Morris, B. Appendix B: Examples of flexible packaging film structures. In *The Science and Technology of Flexible Packaging*; Elsevier: Oxford, UK; Cambridge, MA, USA, 2017; pp. 697–709. [CrossRef]
16. Wellenreuther, F. *Resource Efficient Packaging*; IFEU (Institut für Energie- und Umweltforschung Heidelberg): Heidelberg, Germany, 2016. Available online: https://www.flexpack-europe.org/files/FPE/sustainability/IFEU_Resource%20Efficient%20Packaging_summary_2016.pdf (accessed on 16 February 2021).
17. Wellenreuther, F. *Potential Packaging Waste Prevention by the Usage of Flexible Packaging and Its Consequences for the Environment*; IFEU (Institut für Energie- und Umweltforschung Heidelberg): Heidelberg, Germany, 2019. Available online: https://www.flexpack-europe.org/files/FPE/sustainability/2020/FPE-ifeu_Study_Update_2019_Executive_Summary.pdf (accessed on 16 February 2021).
18. Gandenberger, C.; Orzanna, R.; Klingenfuß, S.; Sartorius, C. The Impact of Policy Interactions on the Recycling of Plastic Packaging Waste in Germany. Working Paper Sustainability and Innovation. 2014. Available online: https://www.econstor.eu/handle/10419/100033 (accessed on 16 February 2021).
19. Hopewell, J.; Dvorak, R.; Kosior, E. Plastics recycling: Challenges and opportunities. *Philos. Trans. R. Soc. B* **2009**, *364*, 2115–2126. [CrossRef] [PubMed]
20. Nickel, W. Materialeinsatz. In *Recycling-Handbuch: Strategien—Technologien—Produkte*, 1st ed.; Nickel, W., Ed.; VDI Verlag: Düsseldorf, Germany, 1996; p. 83. [CrossRef]
21. Ellen MacArthur Foundation. The New Plastics Economy: Catalysing Action. 2017. Available online: https://www.ellenmacarthurfoundation.org/publications/new-plastics-economy-catalysing-action (accessed on 16 February 2021).
22. European Commission. *A New Circular Economy Action Plan: For a Cleaner and More Competitive Europe*; European Commission: Brussels, Belgum, 2020. Available online: https://eur-lex.europa.eu/legal-content/EN/TXT/?qid=1583933814386&uri=COM:2020:98:FIN (accessed on 16 February 2021).
23. Van Eygen, E.; Laner, D.; Fellner, J. Circular economy of plastic packaging: Current practice and perspectives in Austria. *Waste Manag.* **2018**, *72*, 55–64. [CrossRef] [PubMed]
24. OÖ Landesabfallverband Umweltprofis. Der Gelbe Sack—Was Darf Hinein und Was Darf Nicht Hinein? 2018. Available online: https://www.umweltprofis.at/eferding/aktuelles/nachrichten_detail/n/detail/News/der_gelbe_sack_was_darf_hinein_und_was_darf_nicht_hinein.html (accessed on 16 February 2021).
25. Institute Cyclos-HTP. *Verification and Examination of Recyclability: Requirements and Assessment Catalogue of the Institute Cyclos-HTP for EU-Wide Certification (CHI-Standard)*; Institute Cyclos-HTP: Aachen, Germany, 2019. Available online: https://www.cyclos-htp.de/publications/r-a-catalogue/ (accessed on 16 February 2021).
26. Nonclercq, A. *Mapping Flexible Packaging in a Circular Economy [F.I.A.C.E.]*; Delft University of Technology: Delft, Netherland, 2018. Available online: https://www.google.com.hk/url?sa=t&rct=j&q=&esrc=s&source=web&cd=&ved=2ahUKEwiq5tKBloD0AhVHsaQKHZIQDscQFnoECAMQAQ&url=https%3A%2F%2Fceflex.eu%2Fpublic_downloads%2FFIACE-Final-report-version-24-4-2017-non-confidential-version-Final.pdf&usg=AOvVaw1tmuElrEtQUh4PXQMnDH9s (accessed on 16 February 2021).
27. RecyClass. PE Colored Flexible Films Guideline. 2020. Available online: https://recyclass.eu/wp-content/uploads/2020/07/PE-coloured-films_guideline.pdf (accessed on 6 October 2020).
28. RecyClass. PP Colored Films Guideline. 2020. Available online: https://recyclass.eu/wp-content/uploads/2020/07/PP-colored-films_guideline-2.pdf (accessed on 6 October 2020).
29. RecyClass. PP Natural Films Guideline. 2020. Available online: https://recyclass.eu/wp-content/uploads/2020/07/PP-natural-films_guideline-3.pdf (accessed on 6 October 2020).
30. Barlow, C.; Morgan, D. Polymer film packaging for food: An environmental assessment. *Resour. Conserv. Recycl.* **2013**, *78*, 74–80. [CrossRef]
31. van Sluisveld, M.; Worrell, E. The paradox of packaging optimization—A characterization of packaging source reduction in the Netherlands. *Resour. Conserv. Recycl.* **2013**, *73*, 133–142. [CrossRef]
32. Flexible Packaging Europe. *Fact Sheet: Flexible Packaging Supports Sustainable Consumption and Production*; Flexible Packaging Europe: Düsseldorf, Germany, 2018. Available online: https://www.flexpack-europe.org/en/toolkit-downloads.html (accessed on 17 February 2021).
33. Morris, B. End-use factors influencing the design of flexible packaging. In *The Science and Technology of Flexible Packaging*; Elsevier: Oxford, UK; Cambridge, MA, USA, 2017; pp. 617–654. [CrossRef]
34. Selke, S.; Hernandez, R. Packaging: Polymers in flexible packaging. In *Reference Module in Materials Science and Materials Engineering*; Elsevier: Oxford, UK; Cambridge, MA, USA, 2016.
35. Milios, L.; Holm Christensen, L.; McKinnon, D.; Christensen, C.; Rasch, M.K.; Hallstrøm Eriksen, M. Plastic recycling in the Nordics: A value chain market analysis. *Waste Manag.* **2018**, *76*, 180–189. [CrossRef]
36. World Economic Forum. Towards the Circular Economy: Accelerating the Scale-Up Across Global Supply-Chains. 2014. Available online: http://www3.weforum.org/docs/WEF_ENV_TowardsCircularEconomy_Report_2014.pdf (accessed on 27 September 2021).

37. Ellen MacArthur Foundation. *Towards the Circular Economy Vol. 2: Opportunities for the Consumer Goods Sector*; Ellen MacArthur Foundation: Isle of Wight, UK, 2013. Available online: https://www.ellenmacarthurfoundation.org/publications/towards-the-circular-economy-vol-2-opportunities-for-the-consumer-goods-sector (accessed on 17 February 2021).
38. Geueke, B.; Groh, K.; Muncke, J. Food packaging in the circular economy: Overview of chemical safety aspects for commonly used materials. *J. Clean. Prod.* **2018**, *193*, 491–505. [CrossRef]
39. Geissdoerfer, M.; Savaget, P.; Bocken, N.; Hultink, E.J. The Circular Economy—A new sustainability paradigm? *J. Clean. Prod.* **2017**, *143*, 757–768. [CrossRef]
40. Fellner, J.; Lederer, J.; Scharff, C.; Laner, D. Present Potentials and Limitations of a Circular Economy with Respect to Primary Raw Material Demand. *J. Ind. Ecol.* **2017**, *21*, 494–496. [CrossRef]
41. Singh, J.; Ordoñez, I. Resource recovery from post-consumer waste: Important lessons for the upcoming circular economy. *J. Clean. Prod.* **2016**, *134A*, 342–353. [CrossRef]
42. Ragaert, K.; Delva, L.; van Geem, K. Mechanical and chemical recycling of solid plastic waste. *Waste Manag.* **2017**, *69*, 24–58. [CrossRef] [PubMed]
43. Haig, S.; Morrish, L.; Mortin, R.; Wilkinson, S. *Final Report: Film Reprocessing Technologies and Collection Schemes*; The Waste and Resources Action Programme: Banbury, UK, 2012. Available online: https://www.wrap.org.uk/sites/files/wrap/Film%20reprocessing%20technologies%20and%20collection%20schemes.pdf (accessed on 4 October 2020).
44. Kampmann Eriksen, M.; Damgaard, A.; Boldrin, A.; Astrup Fruergaard, T. Quality Assessment and Circularity Potential of Recovery Systems for Household Plastic Waste. *J. Ind. Ecol.* **2019**, *23*, 156–168. [CrossRef]
45. Dainelli, D. 12-Recycling of food packaging materials: An overview. In *Environmentally Compatible Food Packaging*; Chiellini, E., Ed.; (Woodhead Publishing Series in Food Science, Technology and Nutrition); Elsevier: Amsterdam, The Netherlands, 2008; pp. 294–325. [CrossRef]
46. Bartl, A. Moving from recycling to waste prevention: A review of barriers and enables. *Waste Manag. Res.* **2014**, *32*, 3–18. [CrossRef] [PubMed]
47. Carey, J. On the brink of a recycling revolution?: We're awash in plastics, many of which are hard to recycle. Could innovations, girded by the right incentives, finally whittle down the piles of plastic waste? *Proc. Natl. Acad. Sci. USA* **2017**, *114*, 612–616. [CrossRef] [PubMed]
48. Horodytska, O.; Valdés, F.; Fullana, A. Plastic flexible films waste management–A state of art review. *Waste Manag.* **2018**, *77*, 413–425. [CrossRef]
49. Iacovidou, E.; Gerassimidou, S. Sustainable packaging and the circular economy: An EU perspective. In *Reference Module in Food Science*; Elsevier: Amsterdam, The Netherlands, 2018. [CrossRef]
50. House of Commons Environmental Audit Committee. *Growing a Circular Economy: Ending the Throwaway Society*; House of Commons: London, UK, 2014. Available online: https://publications.parliament.uk/pa/cm201415/cmselect/cmenvaud/214/214.pdf (accessed on 27 September 2021).
51. Dixon, J. *Packaging Materials 9: Multilayer Packaging for Food and Beverages*; ILSI Europe Report Series; ILSI Europe Packaging Materials: Washington, DC, USA, 2011. Available online: https://ilsi.eu/publication/packaging-materials-9-multilayer-packaging-for-food-and-beverages/ (accessed on 17 February 2011).
52. Ellen MacArthur Foundation. *Towards the Circular Economy Vol.1: Economic and Business Rationale for an Accelerated Transition*; Ellen MacArthur Foundation: Isle of Wight, UK, 2013. Available online: https://ellenmacarthurfoundation.org/towards-the-circular-economy-vol-1-an-economic-and-business-rationale-for-an (accessed on 27 September 2021).
53. Selke, S. Recycling: Polymers. In *Reference Module in Materials Science and Materials Engineering*; Elsevier: Amsterdam, The Netherlands, 2016. [CrossRef]
54. PlasticsEurope. *Plasctics—The Facts 2017*; PlasticsEurope: Brussels, Belgium, 2020. Available online: https://www.plasticseurope.org/application/files/5715/1717/4180/Plastics_the_facts_2017_FINAL_for_website_one_page.pdf (accessed on 27 September 2021).
55. European Commission. *REGULATION (EC) No 282/2008 of 27 March 2008 on Recycled Plastic Materials and Articles Intended to Come into Contact with Foods and Amending Regulation (EC). No 2023/2006*; European Commission: Brussels, Belgium, 2008.
56. Faraca, G.; Astrup, T. Plastic waste from recycling centres: Characterisation and evaluation of plastic recyclability. *Waste Manag.* **2019**, *95*, 388–398. [CrossRef]
57. Hultman, J.; Corvellec, H. The European Waste Hierarchy: From the sociomateriality of waste to a politics of consumption. *Environ. Plan. A* **2012**, *44*, 2413–2427. [CrossRef]
58. European Commission. *Green Paper: On a European Strategy on Plastic Waste in the Environment*; European Commission: Brussels, Belgium, 2013. Available online: https://eur-lex.europa.eu/legal-content/EN/TXT/?uri=celex%3A52013DC0123 (accessed on 27 September 2021).
59. Iacovidou, E.; Millward-Hopkins, J.; Busch, J.; Purnell, P.; Velis, C.; Hahladakis, J.; Zwirner, O.; Brown, A. A pathway to circular economy: Developing a conceptual framework for complex value assessment of resources recovered from waste. *J. Clean. Prod.* **2017**, *168*, 1279–1288. [CrossRef]
60. Brems, A.; Baeyens, J.; Dewil, R. Recycling and recovery of post-consumer plastic solid waste in a European context. *Therm. Sci.* **2012**, *16*, 669–685. [CrossRef]

61. Dahlbo, H.; Poliakova, V.; Mylläri, V.; Sahimaa, O.; Anderson, R. Recycling potential of post-consumer plastic packaging waste in Finland. *Waste Manag.* **2018**, *71*, 52–61. [CrossRef]
62. Allwood, J. Squaring the circular economy: The role of recycling within a hierarchy of material management strategies. In *Handbook of Recycling: State-of-the-Art for Practitioners, Analysts, and Scientists*; Worrell, E., Reuter, M., Eds.; Elsevier: Amsterdam, The Netherlands, 2014; pp. 445–477. [CrossRef]
63. Al-Salem, S.; Lettieri, P.; Baeyens, J.; Al-Salem, S.M.; Lettieri, P.; Baeyens, J. Recycling and recovery routes of plastic solid waste (PSW): A review. *Waste Manag.* **2009**, *29*, 2625–2643. [CrossRef]
64. Cooper, T. Developments in plastic materials and recycling systems for packaging food, beverages and other fast-moving consumer goods. In *Trends in Packaging of Food, Beverages and Other Fast-Moving Consumer Goods (FMCG)*, 1st ed.; Farmer, N., Ed.; Woodhead Publishing Limited: Southston, UK, 2013; pp. 58–107.
65. Luijsterburg, B.; Goossens, H. Assessment of plastic packaging waste: Material origin, methods, properties. *Resour. Conserv. Recycl.* **2014**, *85*, 88–97. [CrossRef]
66. Tartakowski, Z. Recycling of packaging multilayer films: New materials for technical products. *Resour. Conserv. Recycl.* **2010**, *55*, 167–170. [CrossRef]
67. Veelaert, L.; Du Els, B.; Hubo, S.; van Kets, K.; Ragaert, K. Design from recycling. In Proceedings of the International Conference on Experiential Knowledge and Emerging Materials, Delft, The Netherlands, 19–20 June 2017. Available online: https://www.eksig.org/PDF/EKSIG2017Proceedings.pdf (accessed on 27 September 2021).
68. Favaro, S.L.; Pereira, A.G.B.; Fernandes Rodrigues, J.; Baron, O.; da Silva, C.T.P.; Moisés, M.P.; Radovanovic, E. Outstanding Impact Resistance of Post-Consumer HDPE/Multilayer Packaging Composites. *Mater. Sci. Appl.* **2017**, *8*, 15–25. [CrossRef]
69. Drzyzga, O.; Prieto, A. Plastic waste management, a matter for the 'community'. *Microb. Biotechnol.* **2019**, *12*, 66–68. [CrossRef] [PubMed]
70. Grosso, M.; Niero, M.; Rigamonti, L. Circular economy, permanent materials and limitations to recycling: Where do we stand and what is the way forward? *Waste Manag. Res.* **2017**, *35*, 793–794. [CrossRef]
71. Haas, W.; Krausmann, F.; Wiedenhofer, D.; Heinz, M. How Circular is the Global Economy?: An Assessment of Material Flows, Waste Production, and Recycling in the European Union and the World in 2005. *J. Ind. Ecol.* **2015**, *19*, 765–777. [CrossRef]
72. European Parliament and Council. Directive 2008/98/EC of 19 November 2008 on Waste and Repealing Certain Directives (Text with EEA Relevance). Available online: https://eur-lex.europa.eu/legal-content/EN/TXT/?uri=celex%3A32008L0098 (accessed on 27 September 2021).
73. Briassoulis, D.; Tserotas, P.; Hiskakis, M. Mechanical and degradation behaviour of multilayer barrier films. *Polym. Degrad. Stab.* **2017**, *143*, 214–230. [CrossRef]
74. Velis, C.; Brunner, P. Recycling and resource efficiency: It is time for a change from quantity to quality. *Waste Manag. Res.* **2013**, *31*, 539–540. [CrossRef]
75. Garcia, J.; Robertson, M. The future of plastics recycling: Chemical advances are increasing the proportion of polymer waste that can be recycled. *Science* **2017**, *358*, 870–872. [CrossRef]
76. Soto, J.M.; Blázquez, G.; Calero, M.; Quesada, L.; Godoy, V.; Martín-Lara, M.Á. A real case study of mechanical recycling as an alternative for managing of polyethylene plastic film presented in mixed municipal solid waste. *J. Clean. Prod.* **2018**, *203*, 777–787. [CrossRef]
77. Morris, B. Barrier. In *The Science and Technology of Flexible Packaging*; Elsevier: Oxford, UK; Cambridge, MA, USA, 2017; pp. 259–308. [CrossRef]
78. Morris, B. Commonly used resins and substrates in flexible packaging. In *The Science and Technology of Flexible Packaging*; Elsevier: Oxford, UK; Cambridge, MA, USA, 2017; pp. 69–119. [CrossRef]
79. Häsänen, E. Composition Analysis and Compatibilization of Post-Consumer Recycled Multilayer Plastic Films. Master's Thesis, Tampere University of Technology, Tampere, Finland, 2016.
80. Farris, S. Main manufacturing processes for food packaging materials. In *Reference Module in Food Science*; Elsevier Inc.: Amsterdam, The Netherlands, 2016. [CrossRef]
81. Li, F.; Biagioni, P.; Finazzi, M.; Tavazzi, S.; Piergiovanni, L. Tunable green oxygen barrier through layer-by-layer self-assembly of chitosan and cellulose nanocrystals. *Carbohydr. Polym.* **2013**, *92*, 2128–2134. [CrossRef] [PubMed]
82. Farris, S.; Uysal Unalan, I.; Introzzi, L.; Fuentes-Alventosa, J.M.; Cozzolino, C.A. Pullulan-based films and coatings for food packaging: Present applications, emerging opportunities, and future challenges. *J. Appl. Polym. Sci.* **2014**, *131*, 40539. [CrossRef]
83. Siracusa, V.; Ingrao, C.; Lo Giudice, A.; Mbohwa, C.; Dalla Rosa, M. Environmental assessment of a multilayer polymer bag for food packaging and preservation: An LCA approach. *Food Res. Int.* **2014**, *62*, 151–161. [CrossRef]
84. Reclay StewardEdge. Analysis of Flexible Film Plastics Packaging Diversion Systems. 2013. Available online: https://thecif.ca/projects/documents/714-Flexible_Film_Report.pdf (accessed on 16 February 2021).
85. Morris, B. *The Science and Technology of Flexible Packaging*; Elsevier: Oxford, UK; Cambridge, MA, USA, 2017. [CrossRef]
86. Ebnesajjad, S. *Plastic Films in Food Packaging: Materials, Technology, and Applications*; William Andrew: Oxford, UK; Waltham, MA, USA, 2013.
87. Butler, T.I.; Morris, B. PE-based multilayer film structures. In *Multilayer Flexible Packaging*, 2nd ed.; Wagner, J., Ed.; William Andrew: Oxford, UK, 2016; pp. 281–310.

88. Detzel, A.; Bodrogi, F.; Kauertz, B.; Bick, C.; Welle, F.; Schmid, M.; Schmitz, K.; Müller, K.; Käb, H. *Biobasierte Kunststoffe als Verpackung von Lebensmitteln*; Bundesministerium für Ernährung und Landwirtschaft: Endbericht, Heidelberg, Germany, 2018. Available online: https://www.ifeu.de/fileadmin/uploads/Endbericht-Bio-LVp_20180612.pdf (accessed on 27 September 2021).
89. Flexible Packaging Association. *A Holistic View of the Role of Flexible Packaging in a Sustainable World: A Flexible Packaging Association Report*; Flexible Packaging Association: Annapolis, MD, USA, 2018. Available online: https://www.flexpack.org/resources/sustainability-resources#a-holistic-view-of-the-role-of-flexible-packaging-in-a-sustainable-world (accessed on 17 February 2021).
90. Clark, D. Food packaging and sustainability: A manufacturer's view. In *Reference Module in Food Science*; Elsevier Inc.: Amsterdam, The Netherlands, 2018. [CrossRef]
91. Ecoplus, BOKU, Denkstatt, OFI. *Lebensmittel-Verpackungen-Nachhaltigkeit: Ein Leitfaden für Verpackungshersteller, Handel, Politk & NGOs*; Enstanden aus den Ergebnissen des Projektes "STOP waste–SAVE Food": Wien, Austria, 2020. Available online: https://www.ecoplus.at/media/20682/leitfaden_stopwaste_de.pdf (accessed on 27 September 2021).
92. Jönkkäri, I.; Poliakova, V.; Mylläri, V.; Anderson, R.; Andersson, M.; Vuorinen, J. Compounding and characterization of recycled multilayer plastic films. *J. Appl. Polym. Sci.* **2020**, *137*, 49101. [CrossRef]
93. Scalice, R.K.; Becker, D.; Silveira, R.D. Developing a new compatibility table for design for recycling. *Prod. Manag. Dev.* **2009**, *7*, 141–148.
94. Uehara, G.A.; França, M.P.; Canevarolo, S.V. Recycling assessment of multilayer flexible packaging films using design of experiments. *Polímeros* **2015**, *25*, 371–381. [CrossRef]
95. Sustainable Packaging Coalition. Mechanical Recycling Options. Charlottesville, Virginia. Available online: https://sustainablepackaging.org/mechanical-recycling-options/ (accessed on 5 January 2021).
96. Pinzón Moreno, D.D.; Saron, C. Low-density polyethylene/polyamide 6 blends from multilayer films waste. *J. Appl. Polym. Sci.* **2019**, *136*, 47456. [CrossRef]
97. McKinlay, R.; Morrish, L. *Reflex Project: A Summary Report on the Results and Findings from the REFLEX Project*; Innovate UK: Swindon, UK, 2016. Available online: https://ceflex.eu/public_downloads/REFLEX-Summary-report-Final-report-November2016.pdf (accessed on 17 February 2021).
98. Pahl, G.; Beitz, W.; Blessing, L.; Feldhusen, J.; Grote, K.-H.; Wallace, K. *Engineering Design: A Systematic Approach*, 3rd ed.; Springer: London, UK, 2007. [CrossRef]
99. Chen, X.; Kroell, N.; Feil, A.; Pretz, T. Determination of the composition of multilayer plastic packaging with NIR spectroscopy. *Detritus* **2020**, *13*, 62–66. [CrossRef]
100. Ellen MacArthur Foundation. Ellen MacArthur Foundation. *The Initiative*; Ellen MacArthur Foundation: Isle of Wight, UK, 2017. Available online: https://www.newplasticseconomy.org/about/the-initiative (accessed on 17 February 2021).
101. RECOUP. *Plastic Packaging. Recyclability by Design 2020 Update*; RECOUP: Peterborough, UK, 2020. Available online: https://www.recyclingtoday.com/article/recoup-updates-recyclability-by-design-guidelines-film-plastics/ (accessed on 6 October 2020).
102. Packaging SA. *Design for Recycling: For Packaging and Paper in South Africa*; Packaging SA: Bryanston, South Africa, 2017. Available online: https://www.packagingsa.co.za/wp-content/uploads/2019/11/PACSA-Recyclability-by-Design-WEB.pdf (accessed on 17 February 2021).
103. RECOUP. *Plastic Packaging. Recyclability by Design 2021*; RECOUP: Peterborough, UK, 2021. Available online: https://www.recoup.org/p/130/recyclability-by-design (accessed on 27 September 2021).
104. FH Campus Wien. *Circular Packaging Design Guideline: Design Recommendations for Recyclable Packaging*; FH Campus Wien: Vienna, Austria, 2019. Available online: https://www.fh-campuswien.ac.at/fileadmin/redakteure/Studium/01_Applied_Life_Sciences/b_Verpackungstechnologie/Dokumente/FH-Campus-Wien_Circular-Packaging-Design-Guideline_Version-01.pdf (accessed on 17 February 2021).
105. RecyClass. RecyClass Tests Functional Barriers in PP Containers. Available online: https://recyclass.eu/de/recyclass-tests-functional-barriers-in-pp-containers/ (accessed on 5 September 2021).
106. RecyClass. PP Colored Containers Guideline. 2020. Available online: https://recyclass.eu/wp-content/uploads/2020/07/PP-colored-containers_guideline.pdf (accessed on 6 October 2020).
107. RecyClass. PP Natural Containers Guideline. 2020. Available online: https://recyclass.eu/wp-content/uploads/2020/07/PP-natural-containers_guideline-1.pdf (accessed on 6 October 2020).
108. Wohner, B.; Gabriel, V.H.; Krenn, B.; Krauter, V.; Tacker, M. Environmental and economic assessment of food-packaging systems with a focus on food waste. Case study on tomato ketchup. *Sci. Total Environ.* **2020**, *738*, 139846. [CrossRef] [PubMed]
109. Pauer, E.; Wohner, B.; Heinrich, V.; Tacker, M. Assessing the Environmental Sustainability of Food Packaging: An Extended Life Cycle Assessment including Packaging-Related Food Losses and Waste and Circularity Assessment. *Sustainability* **2019**, *11*, 925. [CrossRef]
110. Australian Packaging Covenant Organisation. Available online: https://apco.org.au/ (accessed on 16 February 2021).
111. Ludlow-Palafox, C.; Chase, H.A. Microwave-Induced Pyrolysis of Plastic Wastes. *Ind. Eng. Chem. Res.* **2001**, *40*, 4749–4756. [CrossRef]
112. European Food Safety Authority. *EFSA Journal*; European Food Safety Authority: Parma, Italy, 2020. Available online: https://www.efsa.europa.eu/de/topics/topic/plastics-and-plastic-recycling (accessed on 7 January 2021).
113. Cecon, V.S.; Da Silva, P.F.; Curtzwiler, G.W.; Vorst, K.L. The challenges in recycling post-consumer polyolefins for food contact applications: A review. *Resour. Conserv. Recycl.* **2021**, *167*, 105422. [CrossRef]

 *foods*

MDPI

Article

Reusable Plastic Crates (RPCs) for Fresh Produce (Case Study on Cauliflowers): Sustainable Packaging but Potential *Salmonella* Survival and Risk of Cross-Contamination

Francisco López-Gálvez [1,2], Laura Rasines [1,2], Encarnación Conesa [3], Perla A. Gómez [2], Francisco Artés-Hernández [1,2] and Encarna Aguayo [1,2,*]

1   Postharvest and Refrigeration Group, Escuela Técnica Superior de Ingeniería Agronómica (ETSIA), Universidad Politécnica de Cartagena (UPCT), Paseo Alfonso XIII, 48, 30203 Cartagena, Spain; francisco.lopezgalvez@upct.es (F.L.-G.); laura.rasines@upct.es (L.R.); fr.artes-hdez@upct.es (F.A.-H.)
2   Food Quality and Health Group, Institute of Plant Biotechnology (UPCT), Campus Muralla del Mar, 30202 Cartagena, Spain; perla.gomez@upct.es
3   Plant Production Department, ETSIA, Institute of Plant Biotechnology (UPCT), Paseo Alfonso XIII, 48, 30203 Cartagena, Spain; encarnacion.conesa@upct.es
*   Correspondence: encarna.aguayo@upct.es

Abstract: The handling of fresh fruits and vegetables in reusable plastic crates (RPCs) has the potential to increase the sustainability of packaging in the fresh produce supply chain. However, the utilization of multiple-use containers can have consequences related to the microbial safety of this type of food. The present study assessed the potential cross-contamination of fresh cauliflowers with *Salmonella enterica* via different contact materials (polypropylene from RPCs, corrugated cardboard, and medium-density fiberboard (MDF) from wooden boxes). Additionally, the survival of the pathogenic microorganism was studied in cauliflowers and the contact materials during storage. The life cycle assessment (LCA) approach was used to evaluate the environmental impact of produce handling containers made from the different food-contact materials tested. The results show a higher risk of cross-contamination via polypropylene compared with cardboard and MDF. Another outcome of the study is the potential of *Salmonella* for surviving both in cross-contaminated produce and in contact materials under supply chain conditions. Regarding environmental sustainability, RPCs have a lower environmental impact than single-use containers (cardboard and wooden boxes). To exploit the potential environmental benefits of RPCs while ensuring food safety, it is necessary to guarantee the hygiene of this type of container.

Keywords: pathogenic bacteria; food contact surface; transfer; *Brassica*; life cycle analysis; wooden boxes; environmental impact

Citation: López-Gálvez, F.; Rasines, L.; Conesa, E.; Gómez, P.A.; Artés-Hernández, F.; Aguayo, E. Reusable Plastic Crates (RPCs) for Fresh Produce (Case Study on Cauliflowers): Sustainable Packaging but Potential *Salmonella* Survival and Risk of Cross-Contamination. *Foods* **2021**, *10*, 1254. https://doi.org/10.3390/foods10061254

Academic Editor: Rafael Gavara

Received: 7 May 2021
Accepted: 28 May 2021
Published: 1 June 2021

Publisher's Note: MDPI stays neutral with regard to jurisdictional claims in published maps and institutional affiliations.

1. Introduction

Reusable plastic crates (RPCs) are utilized in different steps of the fruit and vegetable supply chain, including harvest, handling, packaging, and transport operations, as well as in the retail sector [1]. The use of RPCs for the handling of fresh produce has some advantages, such as the potential to improve environmental sustainability [2]. On the other hand, different studies have raised awareness regarding the hygienic status of RPCs and their possible role as a source of microbiological contamination [3–5].

Fruits and vegetables are increasingly being recognized as a source of foodborne outbreaks [6,7]. Pathogenic microorganisms can survive in fresh produce throughout the supply chain, thereby posing a risk to consumers [8]. Cauliflower-containing products have faced recalls due to the potential presence of pathogenic bacteria [9]. Zhang et al. [10] detected *L. monocytogenes* in fresh-cut cauliflower (florets). Quiroz-Santiago et al. [11] detected *Salmonella* in 9% of the cauliflower samples they analyzed (*n* = 100). The contamination of fresh produce can come from several sources, including food-contact surfaces where pathogenic microorganisms can survive and be transferred to food [5]. Bacterial transfer between contact surfaces and food and vice versa is influenced by many factors, including the bacterial species, handling of the inoculum, degree of contamination, type of surface, type of food, temperature, moisture, duration of the contact, and pressure [12,13]. The transfer of microorganisms between fresh produce and equipment surfaces (including harvest bins and packaging boxes or crates) is significant [14,15]. Although the use of RPCs has not been linked directly with foodborne outbreaks, indirect evidence indicates that there is a potential risk when hygiene fails to be properly maintained [1]. Inadequate cleaning can enhance *Salmonella* survival in plastic containers used in harvest operations [16]. The presence of fresh produce residues (e.g., intact tissues, organic matter, decaying plant material) can enable growth and biofilm formation by *Salmonella* in food-contact surfaces [17]. Furthermore, different studies have suggested that there is a higher transfer of microorganisms to fresh produce from plastic containers in comparison with containers made of other materials. Patrignani et al. [18] showed a higher transfer of bacteria from RPCs to peaches compared with cardboard, hypothesizing that such a difference would be caused by the higher entrapment capability of cardboard. Aviat et al. [19] observed a higher transfer of *E. coli* to apples from polypropylene surfaces compared with wood and cardboard surfaces. In their study, apart from the higher entrapping capability of wood and cardboard compared with plastic, the authors also suggested the ability of microorganisms to form biofilms on plastic surfaces as a potential cause for the differences with the other materials. The study by Siroli et al. [20] also indicated that the risk of microbial cross-contamination is higher via plastic surfaces than via cardboard surfaces.

In the present study, events of cross-contamination between inoculated (*Salmonella enterica*) and non-contaminated cauliflowers via different contact materials were simulated and assessed. The materials tested were: polypropylene from RPCs, corrugated cardboard, and medium-density fiberboard (MDF) from wooden boxes. These materials are commonly used in the manufacturing of fresh-produce handling containers. The survival of the pathogenic microorganism in the vegetable and on the contact surfaces under supply chain conditions was also evaluated. Furthermore, a life cycle assessment (LCA) approach was used to evaluate the sustainability of packaging containers made of the different materials studied.

2. Materials and Methods

2.1. Transfer and Survival of Salmonella via Different Fresh-Produce Container Materials

2.1.1. Via Polypropylene

Salmonella Strains and Inoculum Preparation

Three *Salmonella enterica* subsp. *enterica* strains (CECT 443, CECT 4141, and CECT 4372) were used for the preparation of the inoculum. Starting from a refrigerated stock culture, the strains were grown separately in Tryptic Soy Broth (TSB) for 20 h at 37 °C. Subsequently, a cocktail was prepared by mixing 15 mL of each strain, for a total volume of 45 mL. The cocktail was centrifuged at $4500 \times g$ for 20 min, the supernatant was discarded, and the cells were resuspended in saline solution (0.85% NaCl). Finally, the *Salmonella* suspension was used to inoculate 5 L of saline solution at room temperature to reach a level of *Salmonella* of $\approx 10^7$ cfu/mL. The inoculum was used immediately after its preparation.

Plant Material and Inoculation

Cauliflowers (*Brassica oleracea var. botrytis* cv. 'Altair') provided by Jimbofresh International S.L. (La Unión, Murcia, Spain) were used in the experiment. These mini-cauliflowers are harvested when the diameter of the head is in the range of 8–11 cm, so they are smaller than regular cauliflowers (harvested when they reach a diameter of 15–25 cm) [21]. Detailed information on the dimensions of the cauliflowers used in the experiment can be found in the Supplementary Materials (Table S1). After harvesting (the day before the experiment), the cauliflowers were stored under refrigeration (4 °C). On the day of the experiment, they were taken out of the cold room and allowed to reach room temperature before inoculation.

The curds were then immersed (only the apical half) in the inoculated saline solution for 1 min. After draining thoroughly, they were dried for 2 h in a biosafety cabinet until no visible liquid remained on or between the florets. By measuring the weight difference, it was estimated that a mean volume of 4.5 mL of the *Salmonella* suspension was withheld by each cauliflower head after inoculation and draining, and before drying.

Cross-Contamination

Figure 1 shows the steps of the simulated cross-contamination events. The square polypropylene (PP) pieces ($3.5 \times 3.5 = 12.25$ cm^2) utilized in the experiment were obtained by cutting RPCs used for the handling of fruits and vegetables. They were washed using water and dishwasher, rinsed with distilled water, and sterilized by autoclaving before the experiments. The inoculated cauliflowers were placed on top of the sterile PP fragments for 1 h at room temperature to permit the transfer of the inoculated bacteria. The same contact time has been used in other studies assessing microbial cross-contamination of fresh produce via handling container surfaces [19,22]. The cauliflowers were placed upside down, to allow for contact of the inoculated area (apical half) with the PP. Afterwards, the inoculated cauliflowers were removed, and non-inoculated cauliflowers were immediately placed on top of the PP pieces, in the same position (the apical part in contact with the PP). Once again, a contact time of 1 h was used to allow for the transfer of bacteria from the PP surface to the cauliflowers. The cross-contamination of these cauliflowers was studied. Temperature and relative humidity (RH) during these cross-contamination steps were monitored using a thermometer and a psychrometer, respectively. A test was performed to evaluate the real contact area between cauliflowers and the PP pieces [13]. In this test, the apical part of cauliflowers ($n = 24$) was placed in contact with a permanent black ink pad, and then immediately placed on top of PP pieces. Photographs of the stained pieces were taken using a camera, and the blackened area of the PP pieces was measured using the image processing software ImageJ (National Institutes of Health, Bethesda, MD, USA) [23].

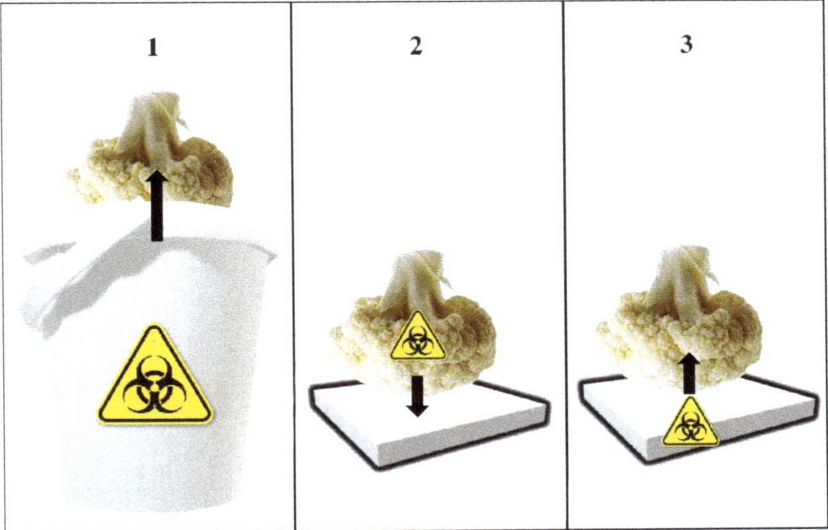

Figure 1. Schematic depiction of the cross-contamination events simulated in the lab. Step 1: Dip inoculation of cauliflowers; Step 2: Transfer of *Salmonella* from inoculated cauliflowers to container pieces; Step 3: Transfer of *Salmonella* from container pieces to non-inoculated cauliflowers. Black arrows represent the direction of the transfer of *Salmonella* cells.

Storage

Inoculated and cross-contaminated cauliflowers were packaged separately in polyethylene terephthalate (PET) trays covered with perforated polyethylene (PE) film as usually performed by the fresh produce industry. Packaged cauliflowers were stored in a cold room at 4 °C for seven days to simulate the storage and transport conditions, followed by six days at 8 °C to simulate supermarket and household conditions. The PP pieces were placed on trays with the inoculated side facing upwards and were stored in the same cold room used for the cauliflowers.

Sampling and Microbiological Analysis

Table 1 shows the types of samples that were analyzed at different moments during the experiment. At each sampling time, three independent samples from each sample type were analyzed. In the case of the cauliflowers, the apical half was cut, and 50 g was taken aseptically for analysis. After adding 200 mL of buffered peptone water (BPW, 20 g/L) to the sample (dilution 1 in 5), it was homogenized using a stomacher for 1 min. The presence of *Salmonella* in the PP pieces and PE films was analyzed using sterile cotton swabs (Aptaca Spa, Canelli, Italy) wetted in BPW (20 g/L). In the case of the PP pieces, the whole area (12.25 cm^2) was swabbed. In the case of the PE films, an area of $\approx$100 cm^2 (10 $\times$ 10 cm) of the zone in contact with the apical area of the inoculated or contaminated cauliflower was swabbed. Swabbing was performed in a standardized way regarding the number and the direction of swab passes. In both cases, swabs were placed in test tubes containing 9 mL of BPW (20 g/L) after use. Serial dilutions in BPW (2 g/L) were prepared as needed, and samples were plated in Xylose Lysine Deoxycholate Agar (XLD; Scharlab, Barcelona, Spain). Apart from the direct plating, an enrichment of the samples was performed by incubation at 37 °C for 24 h. After incubation, the enrichment was also plated in XLD, and the plates were incubated at 37 °C for 24 h before interpretation of results. Red colonies with a black center were considered to be *Salmonella* spp. The detection limit before enrichment was 5 cfu/g in cauliflower, 0.7 cfu/cm^2 in the PP pieces, and 0.09 cfu/cm^2 in the packaging film.

Table 1. Sampling plan of the experiment on transfer and survival of *Salmonella* via polypropylene. Types of samples analyzed at different sampling times.

Sample Type	Sampling Time (Days)						
	0	0.1	1	3	6	9	13
IC [a] before contact with PP pieces	X						
IC after contact with PP pieces	X	X	X	X	X	X	X
PP pieces after contact with IC	X						
PP pieces after contact with IC and non-IC	X	X	X	X	X	X	X
Non-IC before contact with PP pieces	X						
Cross-contaminated cauliflower [b]	X	X	X	X	X	X	X
Polyethylene film from IC		X	X	X	X	X	X
Polyethylene film from cross-contaminated cauliflower		X	X	X	X	X	X

PP: Polypropylene. [a] Inoculated cauliflower. [b] Non-inoculated cauliflower after contact with PP pieces contaminated previously by contact with inoculated cauliflowers. Samples were stored for seven days at 4 °C plus six days at 8 °C.

2.1.2. Effect of the Inoculum Size

The impact of lower inoculum sizes on the transfer from the inoculated product to the PP surface and on the subsequent cross-contamination of uncontaminated cauliflower was studied. The setup of the experiment was similar to that described in Section 2.1.1 albeit with some modifications. In this case, no storage was performed as the goal was to assess if lower inoculum levels could also lead to cross-contamination. In contrast with the inoculated saline solution prepared in the previous experiments ($\approx$10^7 cfu/mL), in this case, two saline solutions containing a level of $\approx$10^6 cfu/mL and $\approx$10^4 cfu/mL of

Salmonella were prepared for the inoculation of the cauliflowers. Five independent samples from each sample type were analyzed at each sampling time.

2.1.3. Via Cardboard and MDF

Transfer and survival of the pathogenic microorganism via other materials were assessed. Cardboard and fiberboard (medium-density fiberboard (MDF)) from wooden boxes were tested as materials commonly used in the manufacturing of vegetable handling containers [24]. The experimental setup was as described in Section 2.1.1 with modifications. In this case, the pieces could not be washed or sterilized by autoclaving but were sanitized by exposure (both sides) to UV light in a biosafety cabinet for 1 h as in Li et al. [25]. In this experiment, the survival of *Salmonella* during storage was assessed in cross-contaminated cauliflower and the container pieces, but not in the inoculated cauliflower or the PE films. In this test, the analysis of cauliflower and pieces during storage was performed at three time points (after 1, 6, and 13 days of storage).

2.1.4. Statistical Analysis

The statistical analyses were executed using IBM SPSS statistics version 26. A level of statistical significance of $p < 0.05$ was used. Data on microbial populations were log-transformed. The Shapiro-Wilk test and Levene's test were used to assess the normality and the homogeneity of variance, respectively. When normality could be assumed, *t*-tests or One-way ANOVA were used to compare treatments, using Tukey's HSD or Dunnett's as post hoc tests depending on the homogeneity of the variances. For data not following a normal distribution, non-parametric tests (Mann–Whitney U and Kruskal–Wallis) were used to search for differences between treatments. Binary logistic regression was used for the analysis of presence/absence data.

2.2. Environmental Impact of Different Types of Fresh Produce Handling Containers

A life cycle assessment (LCA) was performed according to the ISO standards 14040 and 14044 [26,27] using Product Category Rules for Crates for Food [28]. LCA includes four stages: (1) Goal and scope definition, (2) Inventory analysis, (3) Impact assessment, and (4) Interpretation. In LCA studies, the functional unit (FU) is used to normalize all the inputs and outputs. The functional unit in this study was defined as the distribution of 1 kg of cauliflowers in plastic crates, wooden boxes, or cardboard boxes.

2.2.1. Goal and Scope

The goal of this LCA was to compare the environmental impact of reusable plastic crates (RPCs, polypropylene) with that of single-use cardboard (corrugated cardboard) and wooden boxes (poplar wood + pinewood + MDF) using the LCA methodology. Regarding the system boundaries, upstream, core, and downstream processes must be defined. Figure 2 shows the system boundaries of the different types of boxes. In relation to the upstream processes, for wooden and cardboard boxes the life cycle starts in forestry agriculture (production of plants and extraction of resources), while for plastic crates it starts in the extraction of resources and the production of polymer. The next step, in all cases, is the transport of the raw materials to the core process. The core stage covers the manufacture of the final product, including the use of fuel and electricity, emissions generated during manufacturing, machinery maintenance, and treatment of the residues. The final stage (downstream) includes the transport to final disposal and waste treatment, which is different for each type of box.

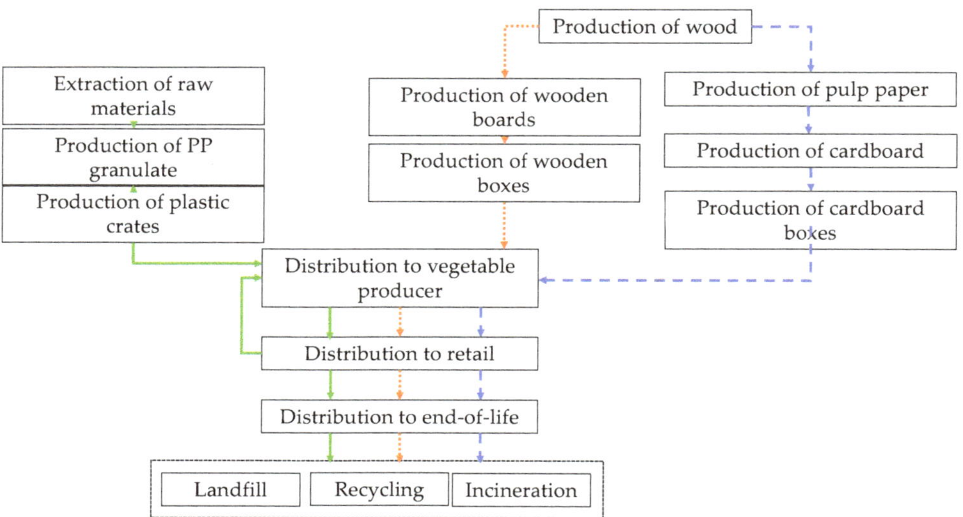

Figure 2. System boundaries of reusable plastic crates (RPCs, polypropylene, green arrows) and single-use cardboard (blue dashed arrows) and wooden (orange dotted arrows) boxes.

2.2.2. Life Cycle Inventory

Cradle-to-grave LCA was performed considering a total of 150 rotations (uses) for plastic crates ([2], and personal communication from a RPC managing company). In other words, it was assumed that each plastic crate is cleaned and reused 150 times, while wood and cardboard boxes are not reused, and it is necessary to produce new boxes for subsequent shipments. Spanish law establishes that wooden and cardboard packaging used for fresh food, regardless of whether they are primary or secondary packaging, can only be used once [29]. In all the cases, the dimensions of the boxes were 40 × 60 × 12 (width × length × height) in cm, and the inner volume was 28.8 L. Each box, regardless of the building material, can be used to carry 6 kg of cauliflower. The list of materials consumed for the manufacturing of each type of box is listed in Table 2. The plastic crates assessed in this LCA are made using primary granulated polypropylene only. The materials in the cardboard boxes evaluated are recycled cardboard (35% in weight) and virgin cardboard (65%). Finally, for the manufacture of the wooden boxes assessed, medium-density fiberboard (MDF) (65.1%), pinewood (20.8%), poplar wood (13.9%), and stainless steel (0.2%) are used. The different transport steps assumed are shown in Table 3. These transport steps included: the shipment of materials to the box manufacturing centers; the transport of crates/boxes to fresh produce packing-houses and retail centers; the return of plastic crates to the cleaning centers; and the transport of crates/boxes to the end-of-life steps. In the case of RPCs, apart from the material for crate production, and the transport steps, the energy and water consumption requirements for the cleaning of the crates during the 150 rotations before disposal were also considered. Based on technical data sheets from RPC washing tunnels [30,31], it was assumed that for the washing of one plastic crate, 0.4 L of water, 0.2% of caustic detergent, and 0.04 kWh of energy are needed. Moreover, scenarios for the waste disposal of the different types of boxes were assessed according to the Spanish annual report on the generation and management of waste [32]. For plastic crates, it was assumed that 79% are recycled, 17% go to landfill, and 4% are incinerated. A total of 65% of cardboard boxes are recycled and 35% go to incineration. Finally, in the case of wooden boxes, 87% of pine and poplar tables are recycled to obtain particle board, 3% finish in the landfill, and 10% are incinerated, whilst in the case of MDF, 77% is incinerated and 23% ends in the landfill.

Table 2. Life cycle inventory of plastic, cardboard, and wooden boxes. MDF: medium density fiberboard.

Type of Box	Materials	Weight (kg)
Plastic	Polypropylene	1.550
Cardboard	Corrugated Cardboard	0.440
	All	1.586
	MDF	1.032
Wooden	Pinewood	0.330
	Poplar wood	0.221
	Stainless Steel	0.003

Table 3. Transport network for single-use cardboard and wooden boxes and reusable plastic crates (RPCs) from manufacture to the end-of-life scenario.

Type of Box	Phase	Mean Distance (km)	Reference
RPCs	Material–manufacturing plant	1000	[2]
RPCs	Manufacturing–packaging center	500	[2]
Cardboard boxes	Material–manufacturing plant	467	[33]
Cardboard boxes	Manufacturing–packaging center	50	[33]
Wooden boxes	Material–manufacturing plant	400	[34]
Wooden boxes	Manufacturing–packaging center	100	[34]
All	Packaging center–logistics platform	400	[2]
All	Logistics platform–retailer	100	[2]
All	Retailer–logistics platform	100	[2]
RPCs	Logistics platform–washing center	100	[2]
RPCs	Washing center–packaging center	200	[2]
RPCs	Washing center–End-of-life	650	[2]
Cardboard boxes	Logistics platform–End-of-life	100	[2]
Wooden boxes	Logistics platform–End-of-life	100	[35]

2.2.3. Impact Assessment

The LCA was performed using SimaPro 9.1 software (PRé Sustainability, Amersfoort, The Netherlands) [36] with the Ecoinvent 3.6 database (Ecoinvent, Zurich, Switzerland) [37]. The CML baseline (Institute of Environmental Sciences, Leiden University, The Netherlands) (Global warming potential, ozone layer depletion, photochemical oxidation, abiotic depletion, acidification eutrophication, freshwater, marine aquatic, and terrestrial ecotoxicity, and human toxicity) and Cumulative Energy Demand (CED) methods were applied.

3. Results

3.1. Transfer and Survival of Salmonella in Cauliflowers via Different Container Materials

3.1.1. Via Polypropylene

Transfer

The level of *Salmonella* detected in the inoculated cauliflowers before contact with the PP fragments was 5.55 ± 0.14 log cfu/g. There were no significant differences between the inoculated cauliflower samples analyzed before and after contact with the PP pieces ($p > 0.05$). This lack of difference is logical, as only a small fraction of the inoculated surface area of each cauliflower was in contact with the PP pieces. Although the total surface of the PP fragments was 12.25 cm^2, the actual contact area between the PP and the cauliflowers was much smaller due to this vegetable's uneven surface. The tests performed to elucidate the actual contact surface between cauliflowers and PP pieces showed that the mean global surface contact was 0.5 ± 0.3 cm^2. Most of the PP pieces analyzed both on the day of the experiment and also during storage showed the presence of *Salmonella* (Table 4). Therefore, transfer of the inoculated microorganism between the inoculated cauliflowers and the PP pieces was detected in most cases. The fact that the pathogenic microorganism was not detected in some of the PP samples could have been due to the limitations of the methods used for the microbiological examination of food contact surfaces [38]. Keeratipibul et al. [39], for example, reported a *Salmonella* recovery efficiency of ≈40% using cotton swabs on dry polyester urethane surfaces. The population of *Salmonella* detected in the PP fragments right after contact with inoculated cauliflower was 0.49 ± 0.71 log cfu/cm^2 (Table 5).

Table 4. Proportion of positive samples (Number of positive samples/Number of samples analyzed) for each type of sample in the experiment of transfer and survival via polypropylene. Storage for seven days at 4 °C plus six days at 8 °C. NA: Not analyzed.

Sample Type	Sampling Time (Days)							Total (*)
	0	0.1	1	3	6	9	13	
Inoculated cauliflower	3/3	3/3	3/3	3/3	3/3	3/3	3/3	21/21 (100%)
PP pieces	3/3	3/3	3/3	3/3	3/3	2/3	1/3	18/21 (33%)
Cross-contaminated cauliflower	3/3	2/3	3/3	3/3	2/3	2/3	2/3	17/21 (35%)
Polyethylene film from inoculated cauliflower	NA	3/3	3/3	3/3	3/3	3/3	2/3	17/18 (71%)
Polyethylene film from cross-contaminated cauliflower	NA	0/3	1/3	0/3	0/3	0/3	0/3	1/18 (0%)

* % of the positive samples detected by direct plating.

Table 5. Prevalence (mean ± standard deviation (proportion of positive samples after enrichment)) of *Salmonella enterica* in container pieces and cross-contaminated cauliflowers during storage (seven days at 4 °C plus six days at 8 °C). Data expressed in log cfu/cm^2 for the container materials, and in log cfu/g for the cauliflowers. PP: polypropylene.

Storage Time (days)	PP	Cardboard	Fiberboard	Cross-Contaminated Cauliflower (PP)	Cross-Contaminated Cauliflower (Cardboard)	Cross-Contaminated Cauliflower (Fiberboard)
0	0.49 ± 0.71 (3/3)	1.14 ± 0.83 (3/3)	0.10 ± 0.40 (3/3)	<0.7 (3/3)	<0.7 (2/3)	<0.7 (2/3)
1	0.40 ± 0.92 (3/3)	0.71 ± 0.15 (3/3)	<−0.14 (3/3)	0.85 ± 0.21 (3/3)	<0.7 (0/3)	<0.7 (0/3)
6	0.20 ± 0.35 (3/3)	<−0.14 (3/3)	<−0.14 (3/3)	<0.7 (2/3)	<0.7 (0/3)	<0.7 (0/3)
13	<−0.14 (1/3)	<−0.14 (3/3)	<−0.14 (3/3)	0.90 ± 0.17 (2/3)	<0.7 (0/3)	<0.7 (0/3)

No *Salmonella* was detected in non-inoculated cauliflowers before contact with the PP pieces (absence after enrichment of samples). After the cross-contamination via the contaminated PP pieces to non-inoculated cauliflowers, in the samples analyzed on the day of the experiment, *Salmonella* was detected only after enrichment (<0.7 log cfu/g). As the pathogenic microorganism was also detected in most cross-contaminated cauliflower samples analyzed during storage (Table 4), we can conclude that there was widespread transfer between the inoculated and non-inoculated cauliflowers via the PP pieces. A larger bacterial transfer from food contact surfaces (plastic, glass, ceramic, stainless steel) to fresh produce than from fresh produce to food contact surfaces had been observed previously [13,40]. More detailed information on the calculations on the transfer of cfu from contact surfaces to non-inoculated cauliflower can be found in the Supplementary Materials (Table S2).

Salmonella Survival

The relative humidity measured in the cold room during storage ranged from 73% to 81% at 4 °C, and from 70% to 77% at 8 °C. Figure 3 shows the slight changes in the populations of *Salmonella* in the inoculated cauliflower during refrigerated storage. The levels remained stable without significant changes throughout that period ($p > 0.05$). Additionally, the change in storage temperature from 4 to 8 °C did not lead to changes in *Salmonella* levels in the inoculated cauliflowers. The stability of the populations of *Salmonella* on vegetables stored in the range of temperatures used in this study (4–8 °C) has been observed in other studies. Kroupitski et al. [41] observed minor changes (<0.5 log cfu/g) in *Salmonella* populations after storage of lettuce leaves at 4 °C for nine days, while the results from Delbeke et al. [42] show stability of *Salmonella* populations in basil leaves stored at 7 °C for one week. Pinton et al. [43] observed survival and even growth of the psychrotrophic pathogen *Listeria monocytogenes* on cauliflower and broccoli stored at 4 °C.

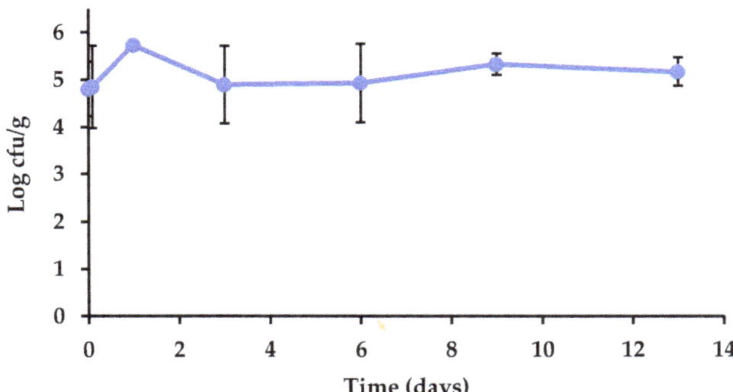

Figure 3. *Salmonella enterica* population (log cfu/g) in inoculated cauliflower during general storage (7 d at 4 °C plus 6 d at 8 °C) in the experiment performed to assess transfer and survival via polypropylene.

Regarding the PP pieces, most of the samples analyzed during the storage were positive for *Salmonella* (Table 4). Throughout the storage, only a few of the positive samples were detected by direct plating (three out of 15 positive samples), whilst the other 12 samples were found positive only after enrichment. Li et al. [24] reported better survival of *Salmonella* in plastic (polyethylene) containers at a refrigeration temperature (3.2 °C; similar to the temperatures used in our tests (4–8 °C)) compared with 22.5 °C.

In the case of the cross-contaminated cauliflowers after contact with contaminated PP, the pathogenic microorganism could not be detected by direct plating during storage, but most of the samples were positive for *Salmonella* after enrichment (Table 4). The proportion of positive samples did not change significantly during storage ($p > 0.05$). These results indicate that, as well as the larger populations present in the inoculated cauliflower ($\approx$5 log cfu/g), the smaller populations present in the cross-contaminated cauliflower (<1 log cfu/g) were able to survive throughout the storage period. Ma et al. [44] reported no effect of the inoculum size (range 0.1–3 log cfu/g) on the survival of *Salmonella* on fresh-cut tropical fruits stored at 4 °C. Strawn and Danyluk [45] also observed stable populations of *Salmonella* on fresh-cut mango inoculated at different initial levels (1, 3, and 5 log cfu/g) and stored at 4 °C.

The packaging film was also analyzed to assess the transfer of *Salmonella* from the inoculated and cross-contaminated cauliflowers to the polyethylene film. In the packaging film from inoculated cauliflowers, *Salmonella* could be detected by direct plating in most of the samples, and only one sample out of 18 was negative both by direct plating and after enrichment. In contrast, in the packaging film from cross-contaminated cauliflowers, no *Salmonella* could be detected by direct plating, and only one sample out of 18 was positive after enrichment (Table 4).

3.1.2. Effect of the Salmonella Inoculum Size

The inoculum size can affect both the number of microorganisms transferred and the transfer rates in the contact between surfaces [46]. Figure 4 shows the results of the test performed to assess the effect of inoculum size on the transfer of *Salmonella* via the PP. The level of *Salmonella* detected in the inoculated cauliflower after drying the inoculum was 4.61 ± 0.11 and 2.58 ± 0.41 log cfu/g for the high and the low inoculum tests, respectively. The PP pieces, after contact with inoculated cauliflower with high inoculum, showed a level of 1.01 ± 0.60 log cfu/cm^2. In the case of the pieces after contact with low-inoculum cauliflower, the pathogen could not be detected by direct plating or by enrichment. In the case of the cross-contaminated cauliflower, both from high and low inoculum tests, *Salmonella* was not detected by direct plating (<0.7 log cfu/g), but it was detected in all the samples after enrichment. The fact that, in the low inoculum test, the pathogen could be detected in cross-contaminated cauliflower but not in the PP pieces suggests a higher efficiency of recovery of the *Salmonella* cells from cross-contaminated cauliflower compared with the PP pieces.

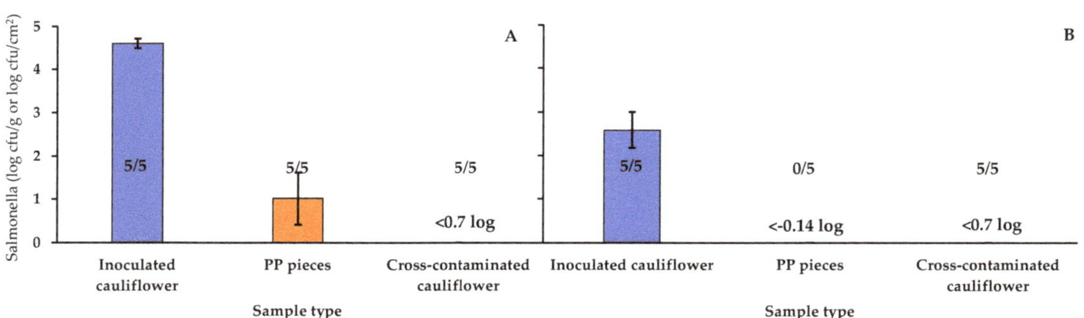

Figure 4. *Salmonella enterica* population (log cfu/g or log cfu/cm^2) and proportion of positive samples (Number of positive samples/Number of samples analyzed) in different types of samples in the experiment performed to assess the effect of inoculum size on the transfer of the pathogen. (**A**): High inoculum; (**B**): Low inoculum. PP: Polypropylene.

3.1.3. Via Cardboard and MDF

The level of *Salmonella* detected in the inoculated cauliflowers before contact with the cardboard and MDF pieces was 5.21 ± 0.06 log cfu/g. Regarding the transfer to the different materials, right after contact with the inoculated cauliflowers the cardboard and the MDF pieces showed a mean level of *Salmonella* of 1.14 ± 0.83 and 0.10 ± 0.40 log cfu/cm^2, respectively (Table 5). In the case of the cross-contaminated cauliflowers analyzed after contact with the cardboard and MDF, no *Salmonella* could be detected by direct plating (<0.7 log cfu/g), but it was detected after enrichment in two out of the three samples, both for cauliflowers cross-contaminated via cardboard and via MDF. The RH in the cold room during storage was between 73% and 81% at 4 °C and between 70% and 77% at 8 °C. In the case of the cardboard pieces, *Salmonella* could be detected by direct plating after one day of storage, but it was detected only after enrichment in the analyses performed after six and 13 days of storage (Table 5). In MDF pieces, the pathogenic microorganism could not be detected by direct plating during storage, but all the samples were positive after enrichment, even after 13 days in the cold room. Regarding the cross-contaminated cauliflowers, *Salmonella* was not detected during storage for both materials; all the samples were negative (by direct plating and also after enrichment) even after only one day of storage (Table 5). The results for the different materials suggest that the transfer of *Salmonella* from the inoculated to non-inoculated cauliflowers was stronger via the PP.

3.2. Environmental Impact of Different Types of Fresh Produce Handling Containers

Tables 6–8 show the results of the different types of boxes in the various impact categories assessed. The wooden boxes showed a higher environmental impact in all the categories assessed. In the global warming category, we obtained values (per FU) of 0.186, 0.059, and 0.006 kgCO$_2$eq for wooden boxes, cardboard boxes, and RPCs, respectively. In all cases, the production step was the stage causing the most greenhouse gas emissions. The materials contributing most to this impact category were the MDF boards used in wooden boxes production, fluting medium and linerboard in cardboard boxes, and obtaining the polypropylene granulate in RPCs. Other activities with a significant contribution to global warming were the end-of-life stage in the case of wooden boxes, and the washing step in the case of the RPCs (electricity and the production of detergent). The contribution of transport in this impact category was negligible in the three cases. The impact of the end-of-life stage of corrugated boxes in the global warming category was negative (this stage reduces the emissions) due to the cardboard production avoided by the recycling process.

Table 6. Life cycle impact per functional unit in reusable plastic crates.

Impact Category	Unit	Total	Plastic Crate Production	Cleaning	Transport	End-of-Life *
Global warming	kg CO_2 eq	6.02×10^{-3}	73%	22%	5%	−25%
Ozone layer depletion	kg CFC-11 eq	1.03×10^{-9}	39%	54%	6%	1%
Photochemical oxidation	kg C_2H_4 eq	1.30×10^{-6}	77%	20%	3%	−23%
Acidification	kg SO_2 eq	2.59×10^{-5}	70%	26%	4%	−16%
Eutrophication	kg PO_4^{3-} eq	1.22×10^{-5}	64%	26%	3%	7%
Abiotic depletion	kg Sb eq	7.82×10^{-8}	65%	26%	9%	−20%
Fresh water aquatic ecotoxicity	kg 1,4-DB eq	3.23×10^{-3}	67%	31%	2%	−3%
Marine aquatic ecotoxicity	kg 1,4-DB eq	7.27×10^{0}	69%	30%	2%	−2%
Terrestrial ecotoxicity	kg 1,4-DB eq	1.55×10^{-5}	48%	49%	3%	0%
Human toxicity	kg 1,4-DB eq	3.46×10^{-3}	69%	27%	4%	−17%
Total energy (non-renewable and renewable)	MJ	1.77×10^{-1}	70%	28%	2%	−35%
Total non-renewable	MJ	1.48×10^{-1}	75%	22%	2%	−39%
Non-renewable, fossil fuels	MJ	1.09×10^{-1}	84%	13%	3%	−45%
Non-renewable, nuclear	MJ	3.92×10^{-2}	35%	64%	0%	−8%
Non-renewable, biomass	MJ	3.68×10^{-6}	64%	32%	4%	4%
Total renewable	MJ	1.80×10^{-2}	39%	61%	0%	−3%
Renewable, biomass	MJ	4.89×10^{-3}	61%	39%	1%	−7%
Renewable, wind, solar, geothermal	MJ	6.13×10^{-3}	20%	80%	0%	−1%
Renewable, water	MJ	7.00×10^{-3}	40%	59%	1%	0%

* The negative value in end-of-life is due to material avoided in the recycling process.

Table 7. Life cycle impact per functional unit in single-use cardboard boxes.

Impact Category	Unit	Total	Cardboard Box Production	Transport	End-of-Life *
Global warming	kg CO_2 eq	5.88×10^{-2}	92%	8%	−26%
Ozone layer depletion	kg CFC−11 eq	7.54×10^{-9}	88%	12%	−24%
Photochemical oxidation	kg C_2H_4 eq	1.22×10^{-5}	95%	5%	−36%
Acidification	kg SO_2 eq	2.59×10^{-4}	92%	8%	−23%
Eutrophication	kg PO_4^{3-} eq	9.62×10^{-5}	97%	3%	−50%
Abiotic depletion	kg Sb eq	2.35×10^{-6}	97%	3%	−48%
Fresh water aquatic ecotoxicity	kg 1,4-DB eq	5.56×10^{-2}	99%	1%	−49%
Marine aquatic ecotoxicity	kg 1,4-DB eq	2.15×10^{2}	99%	1%	−47%
Terrestrial ecotoxicity	kg 1,4-DB eq	3.04×10^{-4}	98%	2%	−44%
Human toxicity	kg 1,4-DB eq	8.29×10^{-1}	97%	3%	−47%
Total energy (non-renewable and renewable)	MJ	5.53×10^{-1}	94%	6%	−71%
Total non-renewable	MJ	9.32×10^{-1}	92%	8%	−26%
Non-renewable, fossil fuels	MJ	8.29×10^{-1}	91%	9%	−26%
Non-renewable, nuclear	MJ	1.02×10^{-1}	99%	1%	−26%
Non-renewable, biomass	MJ	2.40×10^{-4}	100%	0%	−62%
Total renewable	MJ	1.38×10^{-1}	100%	0%	−79%
Renewable, biomass	MJ	1.08×10^{-1}	100%	0%	−82%
Renewable, wind, solar, geothermal	MJ	7.73×10^{-3}	99%	1%	−45%
Renewable, water	MJ	2.15×10^{-2}	98%	2%	−33%

* The negative value in end-of-life is due to material avoided in the recycling process.

Table 8. Life cycle impact per functional unit in single-use wooden boxes.

Impact Category	Unit	Total	Wooden Box Production				Transport	End-of-Life
			MDF	Pine	Poplar	Stainless Steel		
Global warming	kg CO_2 eq	1.86×10^{-1}	51%	9%	3%	1%	14%	23%
Ozone layer depletion	kg $CFC-11$ eq	2.90×10^{-8}	42%	7%	3%	1%	16%	32%
Photochemical oxidation	kg C_2H_4 eq	8.07×10^{-5}	55%	6%	5%	1%	4%	28%
Acidification	kg SO_2 eq	1.16×10^{-3}	42%	7%	2%	1%	9%	40%
Eutrophication	kg PO_4^{3-} eq	2.50×10^{-4}	70%	11%	4%	2%	10%	5%
Abiotic depletion	kg Sb eq	2.47×10^{-5}	89%	2%	1%	0%	2%	5%
Fresh water aquatic ecotoxicity	kg 1,4-DB eq	1.12×10^{-1}	57%	5%	1%	3%	4%	30%
Marine aquatic ecotoxicity	kg 1,4-DB eq	2.53×10^{2}	45%	5%	1%	5%	3%	40%
Terrestrial ecotoxicity	kg 1,4-DB eq	4.80×10^{-4}	67%	11%	7%	2%	8%	6%
Human toxicity	kg 1,4-DB eq	2.47×10^{-1}	41%	4%	1%	1%	5%	48%
Total energy (non-renewable and renewable)	MJ	7.63×10^{0}	41%	28%	17%	0%	5%	8%
Total non-renewable	MJ	4.54×10^{0}	39%	5%	2%	1%	9%	45%
Non-renewable, fossil fuels	MJ	4.05×10^{0}	40%	5%	2%	0%	10%	43%
Non-renewable, nuclear	MJ	4.86×10^{-1}	30%	3%	3%	1%	1%	61%
Non-renewable, biomass	MJ	1.35×10^{-4}	49%	22%	26%	0%	2%	−65%
Total renewable	MJ	3.09×10^{0}	31%	43%	26%	0%	0%	−31%
Renewable, biomass	MJ	2.96×10^{0}	30%	43%	26%	0%	0%	−33%
Renewable, wind, solar, geothermal	MJ	3.79×10^{-2}	53%	5%	2%	1%	2%	38%
Renewable, water	MJ	9.87×10^{-2}	38%	8%	10%	1%	3%	41%

The ozone layer depletion and photochemical ozone oxidation had the same behavior as global warming in cardboard and wooden boxes, i.e., the step that showed the largest impact was production, followed by the end-of-life. In the end-of-life scenario of wooden boxes, the MDF boards contributed to a higher extent to ozone layer depletion while the pine and poplar boards contributed to a higher extent to photochemical ozone oxidation. In RPCs, the major contribution in ozone layer depletion was caused by the detergent used for the cleaning of the crates, while the photochemical oxidation category was affected mainly by the RPC production and cleaning process.

The box production step was the major contributor to the acidification and eutrophication potential categories, except in the case of plastic boxes, in which the cleaning process had a larger impact on eutrophication than the RPC production.

Regarding abiotic depletion, the major impacts were caused by MDF in the case of the wooden boxes and by corrugated cardboard production in the cardboard boxes. In both cases, the contribution of these materials exceeded 90%. In the case of RPCs, the production step was the main contributor in the abiotic depletion category, followed by the cleaning step (mainly because of the Spanish electric mix used in this phase).

Environmental ecotoxicity is divided into freshwater, marine, and terrestrial ecotoxicity. The production of boxes and crates affected the marine ecotoxicity category, whilst the washing of RPCs affected the fresh water and terrestrial ecotoxicity. Independently of the type of packaging, the transport had a greater impact on marine ecotoxicity. In contrast to wooden boxes, the end-of-life step of RPCs and cardboard boxes led to a reduction in the environmental ecotoxicity impact. The end-of-life scenario of wooden boxes mainly affected the marine ecotoxicity due to the incineration of the MDF boards.

During the life cycle of boxes and crates, there are emissions of chemical compounds that are toxic to human beings (human toxicity category). In the case of wooden boxes, the toxic compounds are mainly released in the MDF production step and in the end-of-life stage (in the process of making particle board with pine and poplar wood waste).

For cardboard boxes, these chemicals are mainly released in the corrugated cardboard manufacturing process. In RPCs, the detergent used in the washing step is the main contributor to human toxicity, followed by the production of granulated polypropylene. Most of the energy embedded in plastic and cardboard boxes comes from non-renewable sources, mainly fossil fuels and nuclear power. In wooden boxes, renewable and non-renewable sources are more balanced, although the proportion of non-renewable energy is larger. The wooden and cardboard boxes required more energy per functional unit than the RPCs, 7.63 MJ and 0.55 MJ, respectively (Tables 7 and 8). These energies were needed for the extraction of box building materials (wood and pulp) and manufacturing, as compared with the plastic crates that only demanded 0.18 MJ (Table 6). That value is much lower because plastic boxes have 150 rotations during their life cycle whilst for wooden and cardboard boxes it is necessary to produce more materials in the manufacture of new wooden and cardboard boxes, as they are single-use items. The high value of renewable biomass demand for wooden boxes can be explained by the gross calorific energy embedded in the wood used to manufacture wooden boxes.

Other studies assessing the environmental impact of different types of boxes reached conclusions similar to those presented in our study. Lo-Iacono-Ferreira et al. [47] performed an LCA of different cardboard boxes used to transport fruit and vegetables to different countries and with different end-of-life scenarios. They calculated the global warming potential of each type of box and concluded that the highest impact was linked to the manufacture of cardboard boxes, followed by the transport. On the other hand, in their study, the impact on climate change of the end-of-life stage was found to be negligible, considering that in this scenario nearly 87% of the cardboard boxes are recycled. In our study, by recycling cardboard boxes the carbon footprint showed a 26% decrease. Del Borghi et al. [48] used the LCA approach to compare the impact of RPCs and wooden and cardboard boxes used for food transport. They concluded that the reuse of plastic crates led to a reduction in greenhouse gas emissions compared with single-use plastic crates, thereby reducing the carbon footprint by 96%. In their study, the life cycle of corrugated cardboard contributed the most to the eutrophication potential in comparison with wooden and plastic crates, mainly because of the wastewater from cardboard production. Our results show a larger eutrophication potential in the case of wooden and cardboard boxes compared with RPCs, mainly due to the box production step. In accordance with our results, Abejón et al. [2] also concluded that RPCs have a significantly lower environmental impact than single-use cardboard boxes. In their study, the stages with the highest impact in the case of the cardboard boxes were the manufacturing stage and the recovery of the paper fibers at the end-of-life, while for RPCs the highest environmental impact was linked to sanitation and transport. In our case, the impact of RPCs was caused by crate production and cleaning, whilst the impact due to crate transport was negligible (the contribution in the assessed impact categories was in the range of 2–8%). The consumption of materials avoided by the recycling processes has a beneficial effect on the environment. The recycling of materials during the waste treatment of cardboard boxes and RPCs reduced the impact in all the impact categories assessed, except for ozone layer depletion and eutrophication potential in RPCs. Tua et al. [49] evaluated the environmental performance of RPCs with a different number of rotations (uses) and concluded that a minimum of three rotations is required to improve sustainability, obtaining a 65% carbon footprint reduction. In our scenario, if we change the number of rotations to three we obtain the same reduction in the carbon footprint (65.6%), but a minimum of approximately 15 rotations would be necessary to reduce all the impacts in comparison to single-use cardboard and wooden boxes (Figure 5). Accorsi et al. [35] compared the economic and environmental impact of single-use wooden and corrugated cardboard boxes to that of RPCs from production until the end-of-life in different scenarios. They obtained that the transport stage affected the sustainability of the reusable plastic crates, while for single-use boxes the principal contributor to the environmental impacts was the manufacturing phase. Similar to our results, in their study, the RPC system led to lower emissions in terms of

CO2eq. Albrecht et al. [24] also used the LCA methodology to study the environmental impact of RPCs and single-use wooden and cardboard boxes. Similar to our study, they concluded that the principal contribution to the environmental impacts in single-use boxes (wooden and cardboard) and RPCs is caused by the manufacturing phase. In their study, the activity with the second-greatest impact was the end-of-life in the case of wooden and cardboard, and service life (which involves delivery to the retailer, take-back, inspection, and washing) for RPCs. In our study, the end-of-life was also the second main contributor in the case of wooden boxes, but not for cardboard boxes. The washing step of RPCs also showed a significant impact in our study.

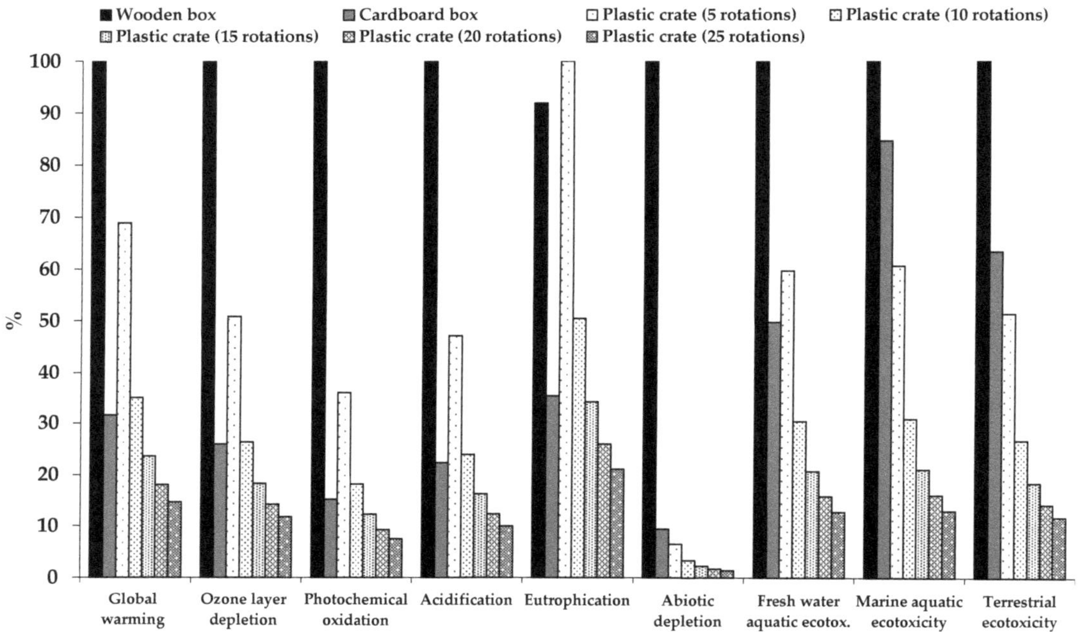

Figure 5. Comparison between single-use wooden and cardboard boxes, and reusable plastic crates with a different number of rotations.

4. Conclusions

The results obtained highlight the risk of fresh produce cross-contamination with pathogenic microorganisms via food-handling containers. Cross-contamination of cauliflowers was more widespread when it occurred via polypropylene than via cardboard or MDF. Furthermore, the survival potential of *Salmonella* under supply chain conditions in the contaminated contact materials and the cross-contaminated cauliflower was demonstrated. The LCA performed showed that RPCs are a better choice to reduce the environmental impacts than single-use cardboard and wooden boxes. The RPCs obtained the lowest impact values for all the categories. Operations used to obtain raw materials for manufacturing wooden and corrugated cardboard boxes have a large impact on marine and terrestrial ecotoxicities and acidification categories. Therefore, the use of RPCs is environmentally beneficial; in fact, in our scenario, a service life of only 15 rotations was sufficient to reduce all the impacts in comparison with single-use cardboard and wooden boxes. However, the hygiene of these reusable containers must be properly maintained to reduce food safety risks.

Supplementary Materials: The following are available online at https://www.mdpi.com/article/10.3390/foods10061254/s1. Table S1: Characteristics of the cauliflowers used in the experiments, Table S2: Detailed assessment of the transfer of *Salmonella* from pieces of different materials to non-inoculated cauliflowers.

Author Contributions: F.L.-G.: Conceptualization; Formal analysis; Methodology; Validation; Investigation; Writing—original draft. L.R.: Methodology; Software; Formal analysis; Investigation. E.C., P.A.G. and F.A.-H.: Conceptualization; Methodology; Writing–review & editing. E.A.: Conceptualization; Methodology; Supervision; Writing–review & editing; Project administration; Funding acquisition. All authors have read and agreed to the published version of the manuscript.

Funding: This research was funded by the Fondo Europeo de Desarrollo Regional/Ministerio de Ciencia e Innovación-Agencia Estatal de Investigación (FEDER/MICINN-AEI), project RTI2018-099139-B-C21. Laura Rasines is grateful for the financial support from the Spanish Ministerio de Ciencia e Innovación through the "Ayudas para contratos predoctorales para la formación de doctores 2019" Program [PRE2019-090573].

Institutional Review Board Statement: Not applicable.

Informed Consent Statement: Not applicable.

Data Availability Statement: The data presented in this study are available upon request from the corresponding author.

Acknowledgments: The authors are grateful to Jimbo Fresh S.L.L. for providing cauliflowers and some specific industrial information, and to Abdenaser Htit for his contribution in the preparation of culture media and other materials for the microbiological analyses.

Conflicts of Interest: The authors declare no conflict of interest. The funders had no role in the design of the study; in the collection, analyses, or interpretation of data; in the writing of the manuscript; or in the decision to publish the results.

References

1. Suslow, T.V. Minimizing Risk in Multiple-Use Containers. Food Safety & Quality Magazine: Suslow UC Davis April 2015 Full Version. Available online: https://ucfoodsafety.ucdavis.edu/sites/g/files/dgvnsk7366/files/inline-files/212397.pdf (accessed on 22 March 2021).
2. Abejón, R.; Bala, A.; Vázquez-Rowe, I.; Aldaco, R.; Fullana-i-Palmer, P. When plastic packaging should be preferred: Life cycle analysis of packages for fruit and vegetable distribution in the Spanish peninsular market. *Resour. Conserv. Recy.* **2020**, *155*, 104666. [CrossRef]
3. Warriner, K. Microbiological Standards for RPCs within Produce Grower Facilities. University of Guelph, Department of Food Science. 2013. Available online: https://26mvtbfbbnv3ruuzp1625r59-wpengine.netdna-ssl.com/wp-content/uploads/PDFs/Package_Cleanliness/RPC_Sanitation_Testing_and_Research_a.pdf (accessed on 22 March 2021).
4. Suslow, T.V. *Assessment of General RPC Cleanliness as Delivered for Use in Packing and Distribution of Fresh Produce*; Assessment of RPC Cleanliness: Final Report; University of California at Davis, Department of Plant Sciences: Davis, CA, USA, 2014; Available online: https://26mvtbfbbnv3ruuzp1625r59-wpengine.netdna-ssl.com/wp-content/uploads/PDFs/Package_Cleanliness/RPC_Sanitation_Testing_and_Research_c.pdf (accessed on 22 March 2021).
5. Zhu, Y.; Wu, F.; Trmcic, A.; Wang, S.; Warriner, K. Microbiological status of RPCs in commercial grower/packer operations and risk of *Salmonella* cross-contamination between containers and cucumbers. *Food Control* **2020**, *110*, 107021. [CrossRef]
6. Truchado, P.; Allende, A. Relevance of fresh fruits and vegetables in foodborne outbreaks and the significance of the physiological state of bacteria | [La implicación de las frutas y hortalizas en las toxiinfecciones alimentarias y la relevancia del estado fisiológico de las bacterias]. *Arbor* **2020**, *196*, 1–9. [CrossRef]
7. Aworh, O.C. Food safety issues in fresh produce supply chain with particular reference to sub-Saharan Africa. *Food Control* **2021**, *123*, 107737. [CrossRef]
8. Jacxsens, L.; Uyttendaele, M.; Luning, P.; Allende, A. Food safety management and risk assessment in the fresh produce supply chain. *IOP Conf. Ser. Mat. Sci.* **2017**, *193*. [CrossRef]
9. U.S. Food and Drugs Administration. Recalls, Market Withdrawals & Safety Alerts. 2021. Available online: https://www.fda.gov/safety/recalls/ (accessed on 22 March 2021).
10. Zhang, H.; Yamamoto, E.; Murphy, J.; Locas, A. Microbiological safety of ready-to-eat fresh-cut fruits and vegetables sold on the Canadian retail market. *Int. J. Food Microbiol.* **2020**, *335*, 108855. [CrossRef]
11. Quiroz-Santiago, C.; Rodas-Suárez, O.R.; Vázquez, C.R.; Fernández, F.J.; Quiñones-Ramírez, E.I.; Vázquez-Salinas, C. Prevalence of *Salmonella* in vegetables from Mexico. *J. Food Prot.* **2020**, *72*, 1279–1282. [CrossRef]

12. Gkana, E.; Chorianopoulos, N.; Grounta, A.; Koutsoumanis, K.; Nychas, G.-J.E. Effect of inoculum size, bacterial species, type of surfaces and contact time to the transfer of foodborne pathogens from inoculated to non-inoculated beef fillets via food processing surfaces. *Food Microbiol.* **2017**, *62*, 51–57. [CrossRef]

13. Villegas, B.M.; Hall, N.O.; Ryser, E.T.; Marks, B.P. Influence of physical variables on the transfer of *Salmonella Typhimurium* LT2 between potato (*Solanum tuberosum*) and stainless steel via static and dynamic contact. *Food Microbiol.* **2020**, *92*, 103607. [CrossRef] [PubMed]

14. Buchholz, A.L.; Davidson, G.R.; Marks, B.P.; Todd, E.C.D.; Ryser, E.T. Transfer of *Escherichia coli* O157:H7 from equipment surfaces to fresh-cut leafy greens during processing in a model pilot-plant production line with sanitizer-free water. *J. Food Prot.* **2012**, *75*, 1920–1929. [CrossRef]

15. Newman, K.L.; Bartz, F.E.; Johnston, L.; Moe, C.L.; Jaykus, L.-A.; Leon, J.S. Microbial load of fresh produce and paired equipment surfaces in packing facilities near the U.S. and Mexico border. *J. Food Prot.* **2017**, *80*, 582–589. [CrossRef] [PubMed]

16. Cotter, J.; Talbert, J.; Goddard, J.; Autio, W.; McLandsborough, L. Influence of soil particles on the survival of *Salmonella* on plastic tomato harvest containers. Technical Abstract. In Proceedings of the IAFP Annual Meeting, Providence, RI, USA, 22–25 July 2012.

17. De Abrew Abeysundara, P.; Dhowlaghar, N.; Nannapaneni, R.; Schilling, M.W.; Mahmoud, B.; Sharma, C.S.; Ma, D.-P. Salmonella enterica growth and biofilm formation in flesh and peel cantaloupe extracts on four food-contact surfaces. *Int. J. Food Microbiol.* **2018**, *280*, 17–26. [CrossRef] [PubMed]

18. Patrignani, F.; Siroli, L.; Gardini, F.; Lanciotti, R. Contribution of two different packaging material to microbial contamination of peaches: Implications in their microbiological quality. *Front. Microbiol.* **2016**, *7*, 938. [CrossRef] [PubMed]

19. Aviat, F.; Le Bayon, I.; Federighi, M.; Montibus, M. Comparative study of microbiological transfer from four materials used in direct contact with apples. *Int. J. Food Microbiol.* **2020**, *333*. [CrossRef] [PubMed]

20. Siroli, L.; Patrignani, F.; Serrazanetti, D.I.; Chiavari, C.; Benevelli, M.; Grazia, L.; Lanciotti, R. Survival of spoilage and pathogenic microorganisms on cardboard and plastic packaging materials. *Front. Microbiol.* **2017**, *8*, 2606. [CrossRef] [PubMed]

21. Bhattacharjee, P.; Singhal, R.S. Broccoli and Cauliflower: Production, Quality, and Processing. In *Handbook of Vegetables and Vegetable Processing*, 2nd ed.; Siddiq, M., Uebersax, M.A., Eds.; John Wiley & Sons Ltd.: Hoboken, NJ, USA, 2018; Volume II, pp. 535–558.

22. Montibus, M.; Ismaïl, R.; Michel, V.; Federighi, M.; Aviat, F.; Le Bayon, I. Assessment of *Penicillium expansum* and *Escherichia coli* transfer from poplar crates to apples. *Food Control* **2016**, *60*, 95–102. [CrossRef]

23. Schneider, C.A.; Rasband, W.S.; Eliceiri, K.W. NIH Image to ImageJ: 25 years of image analysis. *Nat. Methods* **2012**, *9*, 671–675. [CrossRef]

24. Albrecht, S.; Brandstetter, P.; Beck, T.; Fullana, I.; Palmer, P.; Grönman, K.; Baitz, M.; Deimling, S.; Sandilands, J.; Fischer, M. An extended life cycle analysis of packaging systems for fruit and vegetable transport in Europe. *Int. J. Life Cycle Assess.* **2013**, *18*, 1549–1567. [CrossRef]

25. Li, K.; Khouryieh, H.; Jones, L.; Etienne, X.; Shen, C. Assessing farmers market produce vendors' handling of containers and evaluation of the survival of *Salmonella* and *Listeria monocytogenes* on plastic, pressed-card, and wood container surfaces at refrigerated and room temperature. *Food Control* **2018**, *94*, 116–122. [CrossRef]

26. ISO. ISO 14000 Collection 2. In *ISO 14040:2006: Environmental Management. Life Cycle Assessment. Principles and Framework*; International Organization for Standardization: Geneva, Switzerland, 2006.

27. ISO. ISO 14000 Collection 2. In *ISO14044:2006: Life Cycle Assessment. Requirements and Guidelines*; International Organization for Standardization: Geneva, Switzerland, 2006.

28. Environdec. *Produc Category Rules: Crates for Food*; Environtec Limited: Chelmsford, UK, 2020; pp. 1–29.

29. Boletín Oficial del Estado. Real Decreto 888/1988, de 29 de Julio, por el que se Aprueba la Norma General Sobre Recipientes que Contengan Productos Alimenticios Frescos, de Carácter Perecedero, no Envasados o Envueltos. «BOE» núm. 1988, pp. 187, 24293–24294. Available online: https://www.boe.es/diario_boe/txt.php?id=BOE-A-1988-19396 (accessed on 31 May 2021).

30. Betelgeux. Lavadora de Cajas EKW 2500—Betelgeux. 2021. Available online: https://www.betelgeux.es/equipos/maquinas-lavadoras/lavadora-cajas-ekw-2500/ (accessed on 31 May 2021).

31. TEMIC. Technical Sheet of Crates Washing Tunnels TR-42-TEMIC S.L. Available online: http://www.temicsl.com/downloads/tUnel-lavado-de-cajas-tr-424.pdf (accessed on 31 May 2021).

32. MITECO. Memoria Anual de Generación y Gestión de Residuos de Competencia Municipal. 2018. Available online: https://www.miteco.gob.es/es/calidad-y-evaluacion-ambiental/publicaciones/memoriaresiduosmunicipales2018_tcm30-521965.pdf (accessed on 22 April 2021).

33. European Federation of Corrugated Board Manufacturers (FEFCO); CEPI ContainerBoard (CCB). European Database for Corrugated Board Life Cycle Studies. 2018. Available online: https://www.fefco.org/lca (accessed on 6 May 2021).

34. Agence de L'environnement et de la Maîtrise de L'énergie (ADEME). Analyse du Cycle de vie Des Caisses en Bois, Carton Ondulé et Plastique Pour Pommes. 2000. Available online: https://www.ademe.fr/sites/default/files/assets/documents/28246_acvs.pdf (accessed on 6 May 2021).

35. Accorsi, R.; Cascini, A.; Cholette, S.; Manzini, R.; Mora, C. Economic and environmental assessment of reusable plastic containers: A food catering supply chain case study. *Int. J. Prod. Econ.* **2014**, *152*, 88–101. [CrossRef]

36. SimaPro | The World's Leading LCA Software. (n.d.). Available online: https://simapro.com/ (accessed on 18 March 2021).

37. Ecoinvent. (n.d.). Ecoinvent Database 3.7.1. Available online: https://www.ecoinvent.org/ (accessed on 22 March 2021).

38. Jones, S.L.; Ricke, S.C.; Keith Roper, D.; Gibson, K.E. Swabbing the surface: Critical factors in environmental monitoring and a path towards standardization and improvement. *Crit. Rev. Food Sci.* **2020**, *60*, 225–243. [CrossRef] [PubMed]

39. Keeratipibul, S.; Laovittayanurak, T.; Pornruangsarp, O.; Chaturongkasumrit, Y.; Takahashi, H.; Techaruvichit, P. Effect of swabbing techniques on the efficiency of bacterial recovery from food contact surfaces. *Food Control* **2017**, *77*, 139–144. [CrossRef]

40. Jensen, D.A.; Friedrich, L.M.; Harris, L.J.; Danyluk, M.D.; Schaffner, D.W. Quantifying transfer rates of *Salmonella* and *Escherichia coli* O157:H7 between fresh-cut produce and common kitchen surfaces. *J. Food Prot.* **2013**, *76*, 1530–1538. [CrossRef]

41. Kroupitski, Y.; Pinto, R.; Brandl, M.T.; Belausov, E.; Sela, S. Interactions of *Salmonella enterica* with lettuce leaves. *J. Appl. Microbiol.* **2009**, *106*, 1876–1885. [CrossRef] [PubMed]

42. Delbeke, S.; Ceuppens, S.; Jacxsens, L.; Uyttendaele, M. Survival of *Salmonella* and *Escherichia coli* O157:H7 on strawberries, basil, and other leafy greens during storage. *J. Food Prot.* **2015**, *78*, 652–660. [CrossRef]

43. Pinton, S.C.; Bardsley, C.A.; Marik, C.M.; Boyer, R.R.; Strawn, L.K. Fate of *Listeria monocytogenes* on broccoli and cauliflower at different storage temperatures. *J. Food Prot.* **2020**, *83*, 858–864. [CrossRef]

44. Ma, C.; Li, J.; Zhang, Q. Behavior of *Salmonella* spp. on fresh-cut tropical fruits. *Food Microbiol.* **2016**, *54*, 133–141. [CrossRef]

45. Strawn, L.K.; Danyluk, M.D. Fate of *Escherichia coli* O157:H7 and *Salmonella* spp. on fresh and frozen cut mangoes and papayas. *Int. J. Food Microbiol.* **2010**, *138*, 78–84. [CrossRef]

46. Montville, R.; Schaffner, D.W. Inoculum size influences bacterial cross contamination between surfaces. *Appl. Environ. Microb.* **2003**, *69*, 7188–7193. [CrossRef]

47. Lo-Iacono-Ferreira, V.G.; Viñoles-Cebolla, R.; Bastante-Ceca, M.J.; Capuz-Rizo, S.F. Transport of Spanish fruit and vegetables in cardboard boxes: A carbon footprint analysis. *J. Clean. Prod.* **2020**, *244*, 118784. [CrossRef]

48. Del Borghi, A.; Parodi, S.; Moreschi, L.; Gallo, M. Sustainable packaging: An evaluation of crates for food through a life cycle approach. *Int. J. Life Cycle Assess.* **2021**, *26*, 753–766. [CrossRef]

49. Tua, C.; Biganzoli, L.; Grosso, M.; Rigamonti, L. Life cycle assessment of reusable plastic crates (RPCs). *Resources* **2019**, *8*, 110. [CrossRef]

MDPI

St. Alban-Anlage 66

4052 Basel

Switzerland

Tel. +41 61 683 77 34

Fax +41 61 302 89 18

www.mdpi.com

Foods Editorial Office

E-mail: foods@mdpi.com

www.mdpi.com/journal/foods